Plant Cell Monographs

Volume 21

Series Editor:

Peter Nick

Karlsruhe, Germany

For further volumes:
http://www.springer.com/series/7089

Joost T. van Dongen • Francesco Licausi
Editors

Low-Oxygen Stress in Plants

Oxygen Sensing and Adaptive Responses to Hypoxia

Springer

Editors
Joost T. van Dongen
RWTH Aachen University
Aachen, Germany

Francesco Licausi
Institute of Life Sciences
Scuola Superiore Sant´Anna
Pisa, Italy

Series Editor
Peter Nick
Botanisches Institut
Molekulare Zellbiologie
Kaiserstr. 2
76131 Karlsruhe
Germany

ISSN 1861-1370 ISSN 1861-1362 (electronic)
ISBN 978-3-7091-1253-3 ISBN 978-3-7091-1254-0 (eBook)
DOI 10.1007/978-3-7091-1254-0
Springer Wien Heidelberg New York Dordrecht London

Library of Congress Control Number: 2013958222

Cover picture: Image of an *Arabidopsis thaliana* leaf epidermis after modified pseudo-Schiff propidiumiodide (mPS-PI) staining showing basal cells of a removed trichome. Courtesy of Dr. Ruth Eichmann.

Printed on acid-free paper

Springer is part of Springer Science+Business Media (www.springer.com)

Preface

Throughout the history of Earth, a tight relationship was established between the high availability of oxygen in the atmosphere and the biological population. Oxygen enrichment of the Archean Earth's atmosphere was initiated by cyanobacteria and further boosted as a consequence of the colonization of the planet's surface by photosynthetic and multicellular eukaryotes which developed into land plants (Bendall et al. 2008). Nowadays, oxygen accounts for about one-fifth of our atmosphere and represents an essential element, which sustains the life of most multicellular organisms, including fungi, animals, and plants. Not only oxygen is required as terminal electron acceptor to ensure respiratory energy production via oxidative phosphorylation, but it also acts as primary substrate in a majority of metabolic reactions that produce structural and signaling components throughout all kingdoms of life. Consequently, when oxygen availability is reduced below the levels required to sustain these biological processes, a situation of crisis is generated. This is especially serious for sessile organisms, such as plants, which are limited in their possibility to move towards area where oxygen availability is sufficient.

For plants, the most common case of restriction in oxygen availability is caused by submergence, due to the slower diffusion of gases in water than in air and exacerbated by the competition for oxygen consumption by soil microorganisms, whose anaerobic metabolism in turn leads to the accumulation of phytotoxic metabolites (Bailey-Serres and Voesenek 2008). Plants do not need to be completely submerged to suffer from oxygen deficiency: reduced oxygen levels (hypoxia) or complete absence of oxygen (anoxia) in waterlogged soil is sufficient to put plant's survival at stakes and has dramatic effect on crop yield (Ahmed et al. 2013). The consequent reduction in size and functionality of the root apparatus of a flooded or waterlogged plant reduce water and nutrient transport towards the shoot. The high probability that every plant will experience restriction in oxygen availability at sometimes throughout its lifespan suggested the concept that these organisms must have evolved efficient strategies to cope with this situation orchestrated by perception and signaling mechanism that integrated them into developmental and growth programs (Bailey-Serres et al. 2012). Additionally, the notion that oxygen availability is not equal to all plant cell types and certain tissues or organs actually develop at oxygen levels lower than those available in the atmosphere put this element in the perspective of a developmental signal.

Initially, research on plant anaerobiosis developed in the fields of biochemistry, due to its direct effect on primary metabolism, and ecology, as broad differences in tolerance to flooding were traditionally known in wild and cultivated species. Biochemical studies first focused on the metabolic switch from the aerobic respiration to fermentative pathways (reviewed in Davies 1980), but later explored the global adjustment and re-routing of primary metabolic reactions opening the debate over an adaptation of respiratory rates to hypoxia. On the other hand, ecophysiological approaches aimed at the identification and characterization of the strategies adopted by different plant species to restricted oxygen availability throughout evolution (Jackson and Colmer 2005). The characterization of the molecular elements, involved in low oxygen sensing and signaling, initiated at the end of the 1990s and beginning of the current century (Hoeren et al. 1998; Klok et al. 2002) but, in comparison with other abiotic stresses such as heat, cold, and high salinity, knowledge in this field lagged behind. Nevertheless at the descriptive level, very detailed overviews of the transcriptomic adjustment to oxygen deprivation were produced, including time-, oxygen-concentration-, and cell type-specific-resolved analyses (Mustroph et al. 2009; Mustroph et al. 2010). This generated a deep knowledge related to the dynamics of the anaerobic response and greatly contributed to the identification of general and tissue-specific responses. In the last 10 years, two main breakthrough set milestones in the applied and theoretic knowledge with respect to the plant adaptation to low oxygen availability, respectively. First, a joint team of agronomists, physiologists, and molecular biologists revealed the genetic basis for submergence tolerance in wild rice varieties and described for the first time a quiescent metabolic adaptation aimed at saving energy and resources for short-lasting floods (Xu et al. 2006). An opposite strategy was shown a couple of years later to occur in deep-water rice varieties (Hattori et al. 2009). More recently, studies conducted in Arabidopsis converged to the identification of an oxygen-dependent pathway for the degradation of transcription factors that orchestrate the core of the anaerobic response in plants (Gibbs et al. 2011; Licausi et al. 2011). Not surprisingly, these three studies hit on the same class of transcription factors, suggesting that species-specific modes of action to respond to low-oxygen stress evolved from the same basic genetic elements. Concomitantly, the gaseous phytohormone ethylene emerged as a key-regulator of the response to flooding and its interaction with other growth regulators such as gibberellins, auxin, and abscisic acid was shown to shape plant growth (Bailey-Serres and Voesenek 2008).

With this book, we bring together the different fields of research which deal with low oxygen conditions in plants and algae to provide an overview of the deep interconnection between their achievements. The monograph consists of seven sections, starting from the mechanisms adopted by plant cells to *perceive oxygen availability and initiate the signaling cascade* that leads to the activation of conserved and species-specific adaptive responses. In this section, both direct oxygen sensing (Kosmacz and Weits, Chap. 1) and biochemical parameters that are affected as consequence of decreased oxygen availability are discussed, including the level of reactive oxygen species (Blokhina et al., Chap. 2), nitric oxide (Igamberdiev et al., Chap. 3), and pH (Ishizawa, Chap. 4). The *molecular response of plants to hypoxia* is presented in the following section, with a focus at the transcriptional (Giuntoli and Perata, Chap. 5) and the posttranscriptional (Sorenson and Bailey-Serres, Chap. 6) level with an additional chapter dedicated to the hormonal interplay that integrate the adaption to oxygen deficiency into growth and developmental programs (Steffens and Sauter, Chap. 7). The third section of this book is dedicated to the *metabolic adaptations* that take place as consequence of a decrease in the oxygen—and thus energy—availability. This section is not limited to higher plants but takes into consideration also green algae whose anaerobic metabolism is of potential economic interest, such as *Chlamydomonas reinhardtii* (Yang et al., Chap. 8). The role of alternative energy storage units, such as PPi, is discussed by Mustroph et al. (Chap. 9) while the effect of changing oxygen availability on respiratory energy production is described by Paepke et al. (Chap. 10). Oxygen-dependent effect on nitrogen and amino acid metabolism is reviewed by Limami (Chap. 11) and Geigenberger (Chap. 12) describes storage metabolism under oxygen limitations. Most of the molecular and metabolic changes described in the previous sections are ultimately aimed at sustaining prolonged conditions of hypoxia, which is also achieved via *morphological adaptations* that ameliorate oxygen supply and transport within the plant tissues (Armstrong and Armstrong, Chap. 14), namely the formation of aerenchyma (Takahashi et al., Chap. 13) and the production of adventitious roots (Sauter and Steffens, Chap. 15). Species-specific strategies which have been developed by plants to maintain photosynthetic activity under water (Pedersen and Colmer, Chap. 16) and cope with flooding conditions (van Veen et al., Chap. 17) are discussed in a specific section dedicated to the *ecophysiological aspects* of the response to low oxygen. Furthermore, the occurrence and impact of low oxygen responses in agricultural practice are discussed taking into consideration the difficulty of oxygen diffusion into bulky fruits (Nicolai et al., Chap. 18), the oxygen supply in artificial substrates used in horticulture (Wessel et al., Chap. 19), and presenting the effect of herbicides that mimic the hypoxic response in plants (Zabalza and Royuela, Chap. 20). Our book concludes with a review about the state-of-the-art techniques used in the past *to measure oxygen concentrations in vivo* and the novel molecular strategies that are being developed to do so in the least intrusive way (Ast and Draaijer, Chap. 21).

We expect that the detailed survey about the various aspects of low-oxygen stress in plants as it is discussed in this monograph will not just contribute to our understanding of the adaptation of plant to low oxygen stress but also extend its potential to the improvement of crops against the damage caused by flooding. Even more so, we hope it will pave the way towards new discoveries that are expected to further boost our knowledge in this field in the next years.

We would like to express our gratitude to all authors and reviewers that contributed to this book. Furthermore, we acknowledge Christiane Welsch for her excellent help in preparing the final manuscript.

Aachen and Pisa
June 2013

Joost van Dongen
Francesco Licausi

References

Ahmed F, Rafii MY, Ismail MR, Juraimi AS, Rahim HA, Asfaliza R, Latif MA (2013) Waterlogging tolerance of crops: breeding, mechanism of tolerance, molecular approaches, and future prospects. BioMed Res Int 2013:10

Bailey-Serres J, Fukao T, Gibbs DJ, Holdsworth MJ, Lee SC, Licausi F, Perata P, Voesenek LACJ, van Dongen JT (2012) Making sense of low oxygen sensing. Trends Plant Sci 17:129–138

Bailey-Serres J, Voesenek LACJ (2008) Flooding stress: acclimations and genetic diversity. Annu Rev Plant Biol 59:313–339

Bendall DS, Howe CJ, Nisbet EG, Nisbet RER (2008) Introduction. Photosynthetic and atmospheric evolution. Phil Trans Roy Soc Lond B Biol Sci 363(1504):2625–2628

Davies DD (1980) Anaerobic metabolism and the production of organic acids. In: Davies DD (ed) The biochemistry of plants, Vol 2. Academic, New York, p 581–611

Gibbs DJ, Lee SC, Md Isa N, Gramuglia S, Fukao T, Bassel GW, Correia CS, Corbineau F, Theodoulou FL, Bailey-Serres J, Holdsworth MJ (2011) Homeostatic response to hypoxia is regulated by the N-end rule pathway in plants. Nature 479:415–418

Hattori Y, Nagai K, Furukawa S, Song X-J, Kawano R, Sakakibara H, Wu J, Matsumoto T, Yoshimura A, Kitano H, Matsuoka M, Mori H, Ashikari M (2009) The ethylene response factors SNORKEL1 and SNORKEL2 allow rice to adapt to deep water. Nature 460:1026–1030

Hoeren FU, Dolferus R, Wu Y, Peacock WJ, Dennis ES (1998) Evidence for a role for AtMYB2 in the induction of the Arabidopsis alcohol dehydrogenase gene (ADH1) by low oxygen. Genetics 149:479–490

Jackson MB, Colmer TD (2005) Response and adaptation by plants to flooding stress. Ann Bot 96:501–505

Klok EJ, Wilson IW, Wilson D, Chapman SC, Ewing RM, Somerville SC, Peacock WJ, Dolferus R, Dennis ES (2002) Expression profile analysis of the low-oxygen response in Arabidopsis root cultures. Plant Cell 14:2481–2494

Licausi F, Kosmacz M, Weits DA, Giuntoli B, Giorgi FM, Voesenek LACJ, Perata P, Van Dongen JT (2011) Oxygen sensing in plants is mediated by an N-end rule pathway for protein destabilization. Nature 479:419–422

Mustroph A, Lee SC, Oosumi T, Zanetti ME, Yang H, Ma K, Yaghoubi-Masihi A, Fukao T, Bailey-Serres J (2010) Cross-kingdom comparison of transcriptomic adjustments to low-oxygen stress highlights conserved and plant-specific responses. Plant Physiol 152:1484–1500

Mustroph A, Zanetti ME, Jang CJH, Holtan HE, Repetti PP, Galbraith DW, Girke T, Bailey-Serres J (2009) Profiling translatomes of discrete cell populations resolves altered cellular priorities during hypoxia in Arabidopsis. Proc Natl Acad Sci 106:18843–18848

Xu K, Xu X, Fukao T, Canlas P, Maghirang-Rodriguez R, Heuer S, Ismail AM, Bailey-Serres J, Ronald PC, Mackill DJ (2006) Sub1A is an ethylene-response-factor-like gene that confers submergence tolerance to rice. Nature 442:705–708

Contents

Part I
Sensing and Signalling Hypoxic Stress

Oxygen Perception in Plants

Monika Kosmacz and Daan A. Weits

Abstract In aerobic organisms oxygen is a rate-limiting substrate for the efficient production of energy, and therefore they need to adjust their metabolism to the availability of oxygen. For this reason, eukaryotes and prokaryotes independently developed mechanisms to perceive oxygen availability and integrate this into developmental and growth programs. Despite their ability to produce oxygen in the presence of light, plants can experience low oxygen conditions when the oxygen diffusion from the environment cannot satisfy the demand set by metabolic rates. The oxygen-sensing mechanism recently identified in plants shares striking similarities with those previously described in animal cells. While in bacteria the different oxygen-sensing pathways reported involve protein dimerization and phosphorylation cascades, in plants and animals this function is mediated by oxygen-dependent proteolysis. The plant oxygen-sensing pathway is regulated via the oxygen-dependent branch of the N-end rule, which regulates the stability of the group VII of the Ethylene Response Factors, key activators of the anaerobic response. Additionally, constitutively expressed ERF-VII proteins, such as RAP2.12, are bound to the acyl-CoA-binding proteins (ACBPs) at the plasma membrane and protected from aerobic degradation. In hypoxia, RAP2.12 is released from the membrane and relocalizes into the nucleus, where it activates the molecular response to oxygen deficiency. Additional factors, indirectly affected by oxygen availability, have also been suggested to play roles in the fine tuning of oxygen sensing in plants.

M. Kosmacz (✉)
Max Planck Institute of Molecular Plant Physiology, Am Muehlenberg 1, 14476 Potsdam-Golm, Germany
e-mail: Chodasiewicz@mpimp-golm.mpg.de

D.A. Weits
Max Planck Institute of Molecular Plant Physiology, Am Muehlenberg 1, 14476 Potsdam-Golm, Germany

PlantLab, Institute of Life Science, Scuola Superiore Sant'Anna, Piazza Martiri della Liberta 33, 56127 Pisa, Italy

J.T. van Dongen and F. Licausi (eds.), *Low-Oxygen Stress in Plants*, Plant Cell Monographs 21, DOI 10.1007/978-3-7091-1254-0_1, © Springer-Verlag Wien 2014

1 Introduction

Over 2,500 million years ago, the first cyanobacteria, which were able to produce oxygen via photosynthesis, appeared on Earth. The increase in oxygen concentrations into the atmosphere dramatically changed the composition of life forms and led to the emergence of aerobic organisms. These organisms exploit oxygen as final electron acceptor in the oxidative phosphorylation to produce energy in a process called "respiration." Being aerobic organisms, plants also consume oxygen to produce energy and therefore can experience oxygen deficiency when the availability of this element in the environment becomes limiting. This occurs, for example, as a consequence of complete submergence or soil waterlogging, since oxygen diffusion is 10,000 times slower in water than in air. Plant organs also experience low oxygen when oxygen diffusion cannot keep up with its consumption by cells, such as in highly metabolically active tissues. Since, in contrast to animals, plant species lack an efficient oxygen transport system (van Dongen et al. 2009), oxygen-deprived cells have no other option than to adapt energy metabolism to the lower availability of oxygen in order to avoid energy shortage.

The first metabolic response that can be observed upon low oxygen conditions is a reduction of non-essential, energy-consuming processes such as the metabolism of storage products like starch, protein, and lipids (Geigenberger 2003; Bailey-Serres and Voesenek 2008). Moreover, PPi-dependent reactions are favored above those which use ATP as substrate (Greenway and Gibbs 2009). Oxygen consumption via respiration is also reduced, supposedly to avoid the plant to incur anoxic conditions (Zabalza et al. 2009). Indeed, when oxygen is completely depleted, ATP production via oxidative phosphorylation in the mitochondria is impossible and plant cells can only rely on the little ATP that is produced via glycolysis. In this situation, the activity of this pathway is strongly upregulated to maximize energy production, a phenomenon known as the Pasteur Effect (Summers et al. 2000). Since glycolysis consumes NAD^+, continuous NADH oxidation in the absence of oxidative phosphorylation is ensured by the activation of fermentative pathways (Tadege et al. 1999). This regulation of the energy metabolism in plants demands an efficient and tunable sensing mechanism for the oxygen availability, similarly to what has been described for other organism such as bacteria, fungi, and animals.

For long, scientists have searched in plants for homologous systems to the oxygen-sensing pathways described for other organisms. However, no ortholog for sensor proteins was discovered. Oxygen signaling in plants has been suggested to be mediated by metabolic alterations caused by oxygen depletion such as cytosolic pH acidification, production of reactive oxygen species (ROS), NO or Ca^{2+} fluxes (Licausi and Perata 2009). In 2011, a plant-specific asset of proteins was identified as involved in the regulation of gene expression in response to oxygen deficiency (Gibbs et al. 2011; Licausi et al. 2011b). This plant oxygen-sensing pathway shares striking similarities to a branch of the mechanisms that regulates NO-dependent responses in animals (Hu et al. 2005). In this review, we describe the recent discoveries about the molecular mechanisms by which the hypoxic responses in plants are regulated and provide a critical comparison with oxygen-sensing systems in other organisms.

2 Oxygen Sensing in Bacteria Relies on Three Independent Mechanisms

The facultative anaerobes within the Bacteria kingdom have the ability to thrive in both presence and absence of oxygen. In order to do so, these microorganisms need to switch their metabolism between aerobic respiration and fermentation depending on the oxygen availability in the environment.

Three different mechanisms that regulate separate sets of anaerobiosis-related genes were described. The first relies on the Fumarate Nitrate Reductase (FNR) transcriptional regulator, a close homolog of Cyclic AMP Receptor Proteins (CRP) (Guest et al. 1996; Körner et al. 2003). This protein can exist in a monomeric or homodimeric form depending on the oxygen availability (Kiley and Beinert 2003). In its dimeric form, FNR regulates the metabolic switch from aerobic to anaerobic metabolism, inducing Nitrate Reductase and Nitrite Reductase, which are required to utilize nitrate as electron acceptor instead of oxygen. The formation of dimers relies on a cluster of sulfur-containing cysteine residues coordinated to iron atoms ($[4Fe\text{-}4S]^{2+}$ cluster), which is oxidized in presence of oxygen (Khoroshilova et al 1995).

A second, well-studied, mechanism in *E. coli* involved in the control of gene expression under condition of reduced oxygen availability is the two component system consisting of the membrane-bound ArcB kinase and its cognate transcriptional activator ArcA (Iuchi et al. 1988, 1989). When oxygen availability decreases, the increase of reduced ubiquinol promotes ArcB autophosphorylation which, in turn, phosphorylates and activates ArcA (Georgellis et al. 1999; Rodriguez et al. 2004). This results in the up-regulation of operons involved in carbon catabolism and cellular redox status (Unden and Bongaerts 1997; Alexeeva et al. 2003).

A third, wide, class of oxygen sensors in bacteria consists of heme-binding domains (Taylor and Zhulin 1999). The PAS (Per-ARNT-Sim) domain which is contained in the FixL and EcDos (Delgado-Nixon et al. 2000) is an example of heme-dependent perception of oxygen. Usually, these sensory domains are associated with transmitter domains such as histidine kinase, phosphodiesterase, or diguanylate cyclase to direct the signal to transcriptional regulators (Gilles-Gonzalez et al. 2008). A paradigm of this mechanism is provided by FixL, the master regulator of nitrogen fixation in diazotrophic bacteria. In fact, when oxygen concentrations drop below the levels that would inactivate the nitrogen fixing nitrogenases, the deoxygenation of the heme group associated to FixL triggers its autophosphorylation (Da Re et al. 1994). Once active, FixL transfers the phosphate to the transcription factor FixJ causing a change in its conformation, thereby inducing the genes involved in nitrogen fixation (Fischer 1994; Gong et al. 2000). Orthologs of FixL have been identified in plants and were found to contain a functional PAS domain. However, in plants they were shown not to be involved in low oxygen stress, but in transduction of light stress signals (Taylor and Zhulin 1999).

3 The Oxygen-Sensing Mechanism in Animal Cells Relies on the Oxygen-Dependent Stability and Activity of Hypoxia Inducible Factor 1

Opposite to plants and fungi, maintenance of oxygen homeostasis in animals relies on the efficient transport of oxygen through blood vessels. Therefore, the importance of oxygen gradients as cues for specific developmental programs has been widely studied. The key regulator of oxygen homeostasis in animal cells is the Hypoxia Inducible Factor (HIF)1, which is a transcriptional activator, required for the development of new blood vessels (angiogenesis) in response to hypoxia. HIF-1 is a heterodimer of two basic helix loop-helix/PAS proteins containing HIF-1α and the aryl hydrocarbon nuclear translocator (ARNT or HIF1β) (Wang et al. 1995). The molecular regulation of HIF-1 by oxygen is controlled by the alpha subunit. In fact, the HIF1β protein is constitutively expressed and its level is not significantly affected by oxygen (Kallio et al. 1997). In contrast, the HIF-1α protein is stabilized within minutes of exposure to hypoxia. In presence of oxygen, prolines in HIF-1α are hydroxylated and this promotes the polyubiquitination and proteasomal degradation by an E3 ubiquitin ligase complex that contains the von Hippel-Lindau tumor suppressor protein (pVHL), elongin B, elongin C, Cul2, and Rbx (Maxwell et al. 1999 and Ohh et al. 2000). A second oxygen-dependent regulation of HIF-1α relies in hydroxylation of an asparagine residue by the Factor Inhibiting HIF-1 (FIH-1) (Lando et al. 2002). The hydroxylated asparagine residue hinders the interaction of HIF1 with transcriptional co-activators such as CBP and p300 and thereby represses its transcriptional activity (Lando et al. 2002). Upon hypoxia, reduced oxygen availability leads to accumulation of HIF1, and restored interaction with CBP and p300, allowing it to activate its target genes (Lando et al. 2002).

4 Perception of Oxygen Availability in Plants Is Regulated via the Oxygen-Dependent Branch of the N-End Rule Pathway (NERP)

To understand which transcriptional regulators are involved in the anaerobic response to oxygen deficiency in plants, the expression of TF-encoding genes was analyzed using microarrays and qPCR platforms in different plant species, such as

Arabidopsis thaliana (Licausi et al. 2011b; Branco-Price et al. 2005), rice *Oryza sativa* (Lasanthi-Kudahettige et al. 2007), poplar *Populus x canescens* (Kreuzwieser et al. 2009), and cotton *Gossypium hirsutum* L. (Christianson et al. 2010). Among the TF gene families conservatively induced in response to hypoxia, the Ethylene Response Factor (ERF) family attracted the attention of several research groups (Licausi et al. 2010; Hinz et al. 2010). Group VII in particular has been strongly associated with submergence in rice (Xu et al. 2006; Hattori et al. 2009). Indeed, other studies showed that members of this group can activate the expression of the hypoxic genes upon low oxygen (Licausi et al. 2010; Hinz et al. 2010). The unique feature of this group is the highly conserved N-terminal MCGGAI(I/L) motif, which was shown to dictate the stability of proteins belonging to the ERF group VII via the N-end rule in an oxygen-dependent manner (Gibbs et al. 2011; Licausi et al. 2011b). This provided the first demonstration of a mechanism by which plants sense oxygen and regulate the anaerobic response. Among the ERF-VII transcription factors, the constitutively transcribed RAP2.2 and RAP2.12 have been suggested to trigger the initial induction of the anaerobic response (Bailey-Serres et al. 2012).

Upon reoxygenation, ERF VII proteins are quickly degraded via the NERP to repress the anaerobic response. In more detail (Fig. 1), the conserved N-terminal motive MCGGAI/L allows the methionine amino peptidase (MAP) to cleave of the methionine and leave the cysteine exposed (Bradshaw et al. 1998). The sulfur atom on the Cys residue is subsequently oxidized in presence of oxygen yielding cysteine sulfinic or sulfonic acid (Hess et al. 2005). The oxidized cysteine, in turn, is recognized by the arginine-transferases ATE1 and ATE2, which add an arginine to the N-terminus via a peptide bond. Finally, this is recognized by the E3 ligase PRT6 that polyubiquitinates the protein and causes its degradation via the proteasome (Garzón et al. 2007). Indeed, mutants of the enzymes involved in the N-end rule such as *ate1ate2* and *prt6* showed an up-regulation of core hypoxic genes in normoxia and a stronger up-regulation upon hypoxia (Gibbs et al. 2011; Licausi et al. 2011b), providing the first evidence of the involvement of the N-end rule in hypoxia signaling. In addition, Licausi et al. (2011b) showed that transgenic plants expressing a truncated version of RAP2.12 that lacked the conserved N-terminus upregulate the hypoxic genes even in aerobic conditions. In summary, the penultimate cysteine oxidation requires oxygen and this makes the stability of the ERF VII proteins oxygen-dependent: when the oxygen availability decreases, cysteine oxidation is reduced and the ERF VII proteins are stabilized and activate the hypoxic response. To test if ERF group VII members in Arabidopsis are substrates of the NERP, Gibbs et al (2011) tested their stability by performing a heterologous rabbit reticulocyte lysate assay. This in vitro assay showed that all five members are degraded by the proteasome (Gibbs et al. 2011). However, when the cysteine was substituted for an alanine their protein degradation via the NERP was abolished. This experiment strongly indicates that the oxidation of the penultimate cysteine is the key determinant of ERF VII stability (Gibbs et al. 2011; Licausi et al. 2011b).

Interestingly, a member of the ERF VII family in rice, SUB1A, was found to be uncoupled from the N-end rule (Gibbs et al. 2011). This may be explained due to

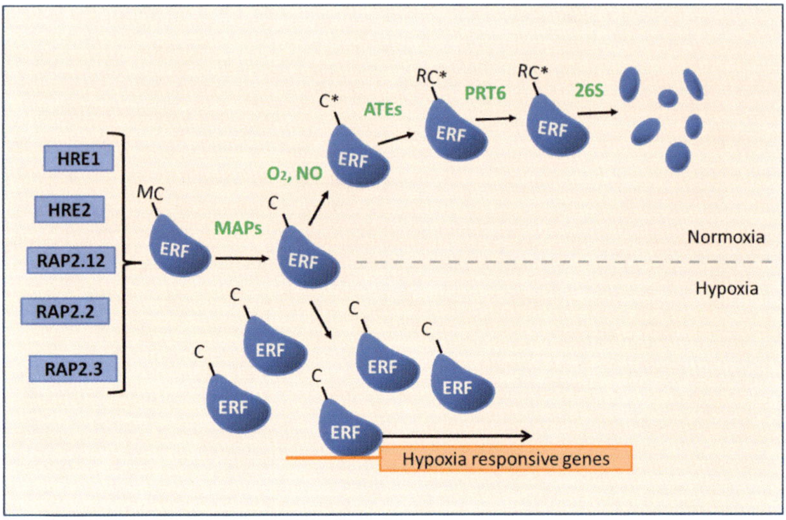

Fig. 1 The induction of the hypoxic response is regulated via the N-end rule pathway for protein degradation. During normoxic conditions ERF group VII TFs are degraded via the N-end rule pathway. The first step is carried out by the methionine aminopeptidase (MAP) which removes the methionine residue, to expose free cysteine for further oxidation. The oxidized form of the cysteine is recognized by the Arginine transferase (ATE) which conjugates an arginine residues to the protein. This in turn triggers the Proteolysis 6 (PRT6) to polyubiquitinate the protein and causes the protein to be degrated by the 26S proteasome. When oxygen becomes limited during hypoxia, cysteine oxidation is reduced and the oxygen-dependent branch of the N-end rule pathway is inhibited. Therefore, ERFs can induce hypoxia response in plants to shift the metabolism from aerobic to fermentation, which results in increased tolerance and survival of plants

the absence of a lysine in the protein sequence downstream of the N-terminus which is, required for its ubiquitination. Another explanation is based on the speculation that the tertiary structure of the SUB1A protein protects it from the N-end rule. Consequently, the higher stability of SUB1A has been associated with submergence tolerance in rice.

5 Role of Acyl-CoA-Binding Proteins (ACBP) in Oxygen Sensing

When Licausi et al. (2011b) studied the subcellular localization of RAP2.12, they found that, in normoxic condition, this protein is localized at the membrane (Fig. 2). This was surprising, since RAP2.12 does not contain any hydrophobic domains. However, former reports in the literature indicated that members of the ERF-VII family can interact with membrane-bound acyl-CoA-binding proteins (ACBPs) (Li and Chye 2004). Using yeast-two-hybrid and biomolecular fluorescence complementation, Licausi et al. (2011b) demonstrated that indeed RAP2.12

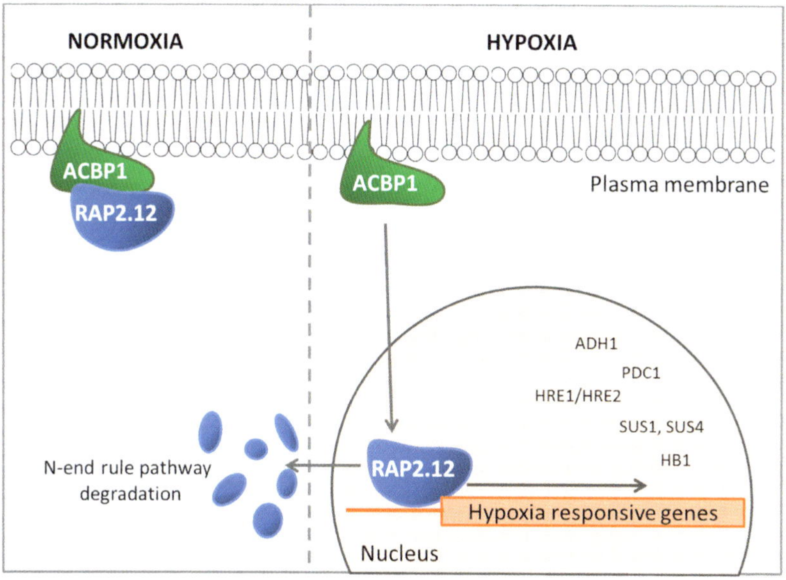

Fig. 2 During hypoxia, RAP2.12 is moving from membrane to the nucleus to trigger the hypoxia response. In normoxic condition RAP2.12 interacts with Acyl-CoA-Binding Proteins (ACBP) and therefore localizes to the membrane. As soon as the oxygen availability in the cell decreases, RAP2.12 moves to the nucleus to induce hypoxia responsive genes such as Hemoglobin1 (HB1), Alcohol Dehydrogenase (ADH1), Pyruvate Decarboxylase (PDC1), Sucrose Synthase 1 and 4 (SUS1 and SUS4), and Hypoxia Responsive Factors 1 and 2 (HRE1 and HRE2). When oxygen level again increases to normoxic condition (reoxygenation), RAP2.12 is degraded via N-end rule pathway

can interact with ACBP1 and ACBP2. Upon hypoxia, RAP2.12 was found to localize to the nucleus suggesting that this translocation from the membrane into the nucleus depends on oxygen as well (Licausi et al. 2011b). It still remains unclear, whether ACBP moves to the nucleus together with RAP2.12, or if, upon low oxygen, RAP2.12 dissociates from the ACBPs and moves to the nucleus alone. The latter hypothesis is supported by the fact that ACBPs contain a membrane-binding domain and therefore is not expected to enter to the nucleus (Li et al. 2003).

Until now, it is still not known whether the interaction of RAP2.12 with the ACBPs is required to prevent its degradation in normoxic conditions or if additional factors are involved. The presence of RAP2.12:GFP at the membrane even under normoxic conditions indicated that a reservoir of RAP2.12 is accumulated in the cell. This reservoir of RAP2.12 units might represent a mechanism which allows plant to respond quickly to a decline in oxygen levels via the relocalization of RAP2.12 from the membrane into the nucleus, instead of relying on *de novo* translation. Indeed the observation that the induction of the hypoxic genes occurs already after 30 min of hypoxia supports this hypothesis (Licausi et al. 2011b).

6 Role of Nitric Oxide in the N-End Rule Pathway

In animals, nitric oxide (NO) is produced by NO synthases and plays a role in a vast range of processes including glycolysis, apoptosis, and cardiovascular homeostasis (Ignarro 2002; Packer et al. 2003; Barouch et al. 2002). Part of NO's effect in animals is mediated by its ability to modify amino acids within proteins (Hess et al. 2005). In particular the role of NO on Cys residues has been extensively studied. It has been shown that NO can convert Cys residues to S-nitrosothiols, via a process called S-nitrosylation (Hu et al. 2005). This reaction alone can directly affect protein function, but in other cases, such as within the N-end rule pathway, additional oxygen-dependent reactions must proceed to yield oxidized Cys (cysteine-sulphinic or sulphonic acid) (Hess et al. 2005)

The Cys-dependent branch of the N-end rule in animals has been characterized as a NO-dependent developmental program. Similarly to the ERF-VII in Arabidopsis, the *Mus musculus* RGS4, RGS5, and RGS16, which are involved in angiogenesis and other tubulogenesis pathways (Kwon et al. 2002) are directed towards proteolysis when their penultimate cysteine residue is oxidized in the concomitant presence of NO and oxygen (Lee et al. 2005; Hu et al. 2005). In fact, in vivo reduction of NO levels through inhibition of NO synthases or direct addition of NO scavengers was shown to increase RGS4 protein levels (Hu et al. 2005), indicating that both NO and O_2 are required for RGS4 degradation via the N-end rule pathway.

In plants, NO has also been implicated to play a role in a wide range of processes, which include pathogen defense, root nodule symbiosis, growth, and development (Wendehenne et al. 2001). Whether NO also affects the arginylation pathway of the N-end rule in plants in not known. Interestingly, low oxygen leads to a strong increase in NO levels (Dordas et al. 2003). As NO was shown to promote arginylation of MC proteins in animals, this could theoretically lead to their destabilization upon low oxygen. At least for the ERF-VII proteins involved in oxygen sensing, this appears to be not the case since they, instead, were shown to be stabilized upon hypoxia (1 % O_2) (Licausi et al. 2011b). Perhaps at these strong hypoxic conditions, it is not NO, but oxygen which becomes the limiting factor. In plants, class 1 hemoglobins were shown to modulate NO levels (Perazzolli et al. 2004) and it may be speculated that through their role in NO homeostasis (Dordas et al. 2003), they may also play a role in Cys oxidation. Interestingly, atHb1 was shown to be a direct target of RAP2.12 and belongs to the core genes which are strongly upregulated in response to hypoxia (Licausi et al. 2011b). Since hypoxia is associated with a strong increase in NO that could potentially promote RAP2.12 degradation, it may be speculated that class 1 hemoglobins boost and support the molecular response to hypoxia by protecting RAP2.12 from degradation (see also Chap. 3, Igamberdiev et al. 2014). According to this hypothesis, it would be expected that AtHb1 overexpressors have a stronger induction of the anaerobic genes in response to hypoxia. This appeared to be not the case: while wild-type Arabidopsis seeds exposed to mild hypoxia (10 % O_2) strongly induce the hypoxic genes, AtHb1 overexpressors show a more moderate induction of these genes (Thiel et al. 2011).

Interestingly though, AtHb1 overexpression increased survival of Arabidopsis seedlings following hypoxia (Hunt et al. 2002). Taken together, these experiments do not support a role for class 1 hemoglobin in control of RAP2.12 stability.

7 Fine Tuning of Oxygen Sensing in Plants is Mediated by Additional Factors

In addition to hemoglobin and NO signaling, other factors can be indirectly involved in the regulation of oxygen homeostasis and signaling in the cell. A very popular example is the phytohormone ethylene. Since gas diffusion is 10,000 times slower in water then in air, under submergence conditions ethylene is trapped and accumulates in the intracellular spaces of plant tissues (Voesenek et al. 1993). This accumulation of ethylene has been shown to promote submergence escape strategies and plays a crucial role in flooding tolerance in rice (Jackson et al. 1985; Voesenek et al. 2004).

Other molecules that change consequentially with a decline in oxygen are also thought to play a role in oxygen sensing. Examples of such molecules include: ROS, pH, and energy status. A paradigmatic example of such an indirect effect is provided by pH on the regulation of fermentation. In fact, initial regeneration of NAD^+ to sustain glycolysis is provided by lactic fermentation (Drew 1997). The accumulation of lactic acid, together with the inhibition of proton extrusion pumps as a consequence of energy crisis, lowers the cytosolic pH to an optimal level for PDC activity, therefore channeling pyruvate towards ethanolic fermentation (Davies 1980). Ethanol production is less harmful in comparison to lactate accumulation as it does not affect pH levels. Pyruvate consumption is also beneficial because this metabolite has been shown to stimulate oxygen consumption via respiration. A decrease in pyruvate levels may therefore aid to reduce respiration and protect the plant tissue from reaching anoxic conditions (Zabalza et al. 2009).

Upon hypoxia, local changes in the Ca^{2+} level occur. In fact, low oxygen levels stimulate the accumulation of cytosolic calcium which is required for induction of ADH1 (Subbaiah et al. 1994), one of the core- hypoxia responsive genes. Interestingly, Ca^{2+} is also involved in the signal transduction that leads to ethylene biosynthesis (He et al. 1996). In addition, changes in calcium level inside the cell can alter gene response, which imply that hypoxia response can be indirectly calcium-dependent.

Microarray studies indicated that also oxidative stress-related genes are induced by oxygen deficiency (Klok et al. 2002; Loreti et al. 2005; Branco- Price et al. 2008; Blokhina et al. 2010). Additionally, the transient accumulation of ROS was shown to be a common response to both anoxia and heat stress (Banti et al. 2008, 2010). Accordingly, Arabidopsis plants overexpressing Heat Shock Factor A2 (HSFA2), a sensor for ROS (Miller et al. 2006), showed increased tolerance to anoxic conditions (Banti et al. 2010). Anoxia induces ROS production via plasma membrane localized NADPH oxidases as well as in mitochondria by affecting the electron transport chain. Baxter-Burrell et al. (2002) proved that the accumulation of hydrogen peroxide (H2O2) under low O2 is required for the expression and activity of alcohol dehydrogenase (ADH) and thus tolerance.

8 Concluding Remarks

Looking at the recently identified plant oxygen-sensing pathway, one cannot avoid noticing its striking functional similarity with the animal mechanism. In fact, both systems rely on the oxygen-dependent regulation of the stability of a master regulator of the anaerobic response. Instead, the pathways downstream activated are extremely divergent between plants and animals, as is expected by the intrinsic properties of each kingdom (autotrophy versus heterotrophy, stillness against motility). Since the MC-branch of the N-end rule pathway exists in both animals and plants, it is tempting to speculate that this mechanism may represent an ancient oxygen perception mechanism, while an additional one, based on HIF-1, was evolved later in animals. However, the functional homology shared by the two mechanisms suggests that a strategy based on oxygen-controlled proteolysis is optimal in both kingdoms.

No orthologs of pVHL, HIF-1, or PHD has been identified in green organisms. Plants do possesses a plethora of prolyl hydroxylase enzymes, some of them also low oxygen-inducible (Mustroph et al. 2010). However, the plant PHDs have been described as involved in cell wall or peptide modifications (Velasquez et al. 2011), rather than in the regulation of the hypoxic response. The fact that molecular oxygen is anyway required by PHD to catalyze the hydroxylation of proline residues has led to the speculation that they may still be involved in oxygen-dependent signaling (Vlad et al. 2007).

Both HIF-1 and RAP2.12 are destabilized in the presence of oxygen. Interestingly, in animals the rate of HIF-1 mRNA translation into protein depends on Target of Rapamycin (mTOR), a protein known to sense cellular energy status through the AMP-activated protein kinase (AMPK) (Wullschleger et al. 2006). An ortholog of the yeast/animal mTOR pathway exists in plants, and has been shown to be directly involved in the control of diverse cellular processes such as autophagy, protein translation, ribosome biogenesis, and actin dynamics (Wullschleger et al. 2006). Until now however, the relation between the mTOR signalling pathway and low oxygen in plants has not been studied and represents an interesting research perspective. Concerning RAP2.12 degradation, one of the most crucial questions deals with the oxidation of its penultimate cysteine: does it occur spontaneously or is it catalyzed enzymatically? Enzymes with cysteine oxidase activity exist in animals and are known to catalyze the oxidation of the free amino acid cysteine to cysteine sulfinic acid. These cysteine dioxygenases are highly induced upon dietary consumption of cysteine and function in maintaining cysteine homeostatis (Stipanuk et al. 2008). *Arabidopsis thaliana* possesses five genes that contain motifs associated with thiol oxidation. The function of these genes has not been studied, but it is known that two of these are strongly induced upon hypoxia (Mustroph et al. 2010). This makes these genes an interesting starting point to look for enzymes with the capacity to oxidize penultimate cysteine. On the other hand, spontaneous oxidation of cysteine in vitro has been reported (Stipanuk et al. 2009). However, the presence of several anti-oxidants and reductants within the cell keep its redox status highly reduced (Kamata and Hirata 1999). Therefore, spontaneous oxidation of cysteine is not likely to occur at high rates within the cell.

Following the finding of oxygen-sensing mechanisms in bacteria, fungi, and animals, also in plants an oxygen-sensing mechanism has now been identified. This mechanism, like the one described in animals, relies on the oxygen-dependent regulation of the protein stability of a master regulator of the hypoxic response. Nevertheless, there are still lots of open questions which provide exciting new research perspectives for the future.

References

Alexeeva S, Hellingwerf KJ, Teixeira de Mattos MJ (2003) Requirement of ArcA for redox regulation in Escherichia coli under microaerobic but not anaerobic or aerobic conditions. J Bacteriol 185(1):204–209

Bailey-Serres J, Voesenek LACJ (2008) Flooding stress: acclimations and genetic diversity. Annu Rev Plant Biol 59:313–339

Bailey-Serres J, Fukao T, Gibbs DJ, Holdsworth MJ, Lee SC, Licausi F, Perata P, Voesenek LA, van Dongen JT (2012) Making sense of low oxygen sensing. Trends Plant Sci 17(3):129–138

Banti V, Loreti E, Novi G, Santaniello A, Alpi A, Perata P (2008) Heat acclimation and cross-tolerance against anoxia in Arabidopsis. Plant Cell Environ 31:1029–1037

Banti V, Mafessoni F, Loreti E, Alpi A, Perata P (2010) The heat-inducible transcription factor HsfA2 enhances anoxia tolerance in Arabidopsis. Plant Physiol 152:1471–1483

Barouch LA, Harrison RW, Skaf MW, Rosas GO, Cappola TP, Kobeissi ZA, Hobai IA, Lemmon CA, Burnett AL, O'Rourke B, Rodriguez ER, Huang PL, Lima JA, Berkowitz DE, Hare JM (2002) Nitric oxide regulates the heart by spatial confinement of nitric oxide synthase isoforms. Nature 416(6878):337–339

Baxter-Burrell A, Yang Z, Springer PS, Bailey-Serres J (2002) RopGAP4-dependent Rop GTPase rheostat control of Arabidopsis oxygen deprivation tolerance. Science 296(5575):2026–2028

Blokhina O, Fagerstedt KV (2010) Oxidative metabolism, ROS and NO under oxygen deprivation. Plant Physiol Biochem 48:359–373

Bradshaw RA, Brickey WW, Walker KW (1998) N-terminal processing: the methionine aminopeptidase and Nα-acetyl transferase families. Trends Biochem Sci 23:263–267

Branco-Price C, Kawaguchi R, Ferreira RB, Bailey-Serres J (2005) Genome-wide analysis of transcript abundance and translation in Arabidopsis seedlings subjected to oxygen deprivation. Ann Bot 96(4):647–660

Branco-Price C, Kaiser KA, Jang CJ, Larive CK, Bailey-Serres J (2008) Selective mRNA translation coordinates energetic and metabolic adjustments to cellular oxygen deprivation and reoxygenation in Arabidopsis thaliana. Plant J 56:743–755

Christianson JA, Llewellyn DJ, Dennis ES, Wilson IW (2010) Global gene expression responses to waterlogging in roots and leaves of cotton (Gossypium hirsutum L.). Plant Cell Physiol 51 (1):21–37

Da Re S, Bertagnoli S, Fourment J, Reyrat JM, Kahn D (1994) Intramolecular signal transduction within the FixJ transcriptional activator: in vitro evidence for the inhibitory effect of the phosphorylatable regulatory domain. Nucleic Acids Res 22:1555–1561

Davies DDD (1980) Anaerobic metabolism and the production of organic acids. In: Davies DDD (ed) The biochemistry of plants, vol 2. Academic, New York, pp 581–611
Delgado-Nixon VM, Gonzalez G, Gilles-Gonzalez MA (2000) Dos, a heme-binding PAS protein from Escherichia coli is a direct oxygen sensor. Biochemistry 39:2685–2691
Dordas C, Hasinoff BB, Igamberdiev AU, Manac'h N, Rivoal J, Hill RD (2003) Expression of a stress-induced hemoglobin affects NO levels produced by alfalfa root cultures under hypoxic stress. Plant J 35(6):763–770
Drew MC (1997) Oxygen deficiency and root metabolism: injury and acclimation under hypoxia and anoxia. Annu Rev Plant Physiol Plant Mol Biol 48:223–250
Fischer HM (1994) Genetic regulation of nitrogen fixation in rhizobia. Microbiol Rev 58 (3):352–386
Garzón M, Eifler K, Faust A, Scheel H, Hofmann K, Koncz C, Yephremov A, Bachmair A (2007) PRT6/At5g02310 encodes an Arabidopsis ubiquitin ligase of the N-end rule pathway with arginine specificity and is not the CER3 locus. FEBS Lett 581(17):3189–3196
Geigenberger P (2003) Response of plant metabolism to too little oxygen. Curr Opin Plant Biol 6:247–256
Georgellis D, Kwon O, Lin EC (1999) Amplification of signaling activity of the Arc two-component system of Escherichia coli by anaerobic metabolites. An in vitro study with different protein modules. J Biol Chem 274:35950–35954
Gibbs DJ, Lee SC, Isa NM, Gramuglia S, Fukao T, Bassel GW, Correia CS, Corbineau F, Theodoulou FL, Bailey-Serres J, Holdsworth MJ (2011) Homeostatic response to hypoxia is regulated by the N-end rule pathway in plants. Nature 479(7373):415–418
Gilles-Gonzalez MA, Gonzalez G, Sousa EHS, Tuckerman J (2008) Oxygen-sensing histidine-protein kinases: assays of ligand binding and turnover of response-regulator substrates. Methods Enzymol 437:173–189
Gong W, Hao B, Chan MK (2000) New mechanistic insights from structural studies of the oxygen-sensing domain of Bradyrhizobium japonicum FixL. Biochemistry 39(14):3955–3962
Greenway H, Gibbs J (2009) Review: mechanisms of anoxia tolerance in plants. II. Energy requirements for maintenance and energy distribution to essential processes. Funct Plant Biol 30(10):999–1036
Guest JR, Green J, Irvine AS, Spiro S (1996) The FNR modulon and FNR-regulated gene expression. In: Lin ECC, Lynch AS (eds) Regulation of gene expression in Escherichia coli. RG Landes and Co, Austin, TX, pp 317–342
Hattori Y, Nagai K, Furukawa S, Song XJ, Kawano R, Sakakibara H, Wu J, Matsumoto T, Yoshimura A, Kitano H, Matsuoka M, Mori H, Ashikari M (2009) The ethylene response factors SNORKEL1 and SNORKEL2 allow rice to adapt to deep water. Nature 460 (7258):1026–1030
He CJ, Morgan PW, Drew MC (1996) Transduction of an ethylene signal is required for cell death and lysis in the root cortex of maize during aerenchyma formation induced by hypoxia. Plant Physiol 112(2):463–472
Hess DT, Matsumoto A, Kim SO, Marshall HE, Stamler JS (2005) Protein S-nitrosylation: purview and parameters. Nat Rev Mol Cell Biol 6(2):150–166
Hinz M, Wilson IW, Yang J, Buerstenbinder K, Llewellyn D, Dennis ES, Sauter M, Dolferus R (2010) Arabidopsis RAP2.2: an ethylene response transcription factor that is important for hypoxia survival. Plant Physiol 153(2):757–772
Hu RG, Sheng J, Qi X, Xu Z, Takahashi TT, Varshavsky A (2005) The N-end rule pathway as a nitric oxide sensor controlling the levels of multiple regulators. Nature 437(7061):981–986
Hunt PW, Klok EJ, Trevaskis B, Watts RA, Ellis MH, Peacock WJ, Dennis ES (2002) Increased level of hemoglobin 1 enhances survival of hypoxic stress and promotes early growth in Arabidopsis thaliana. Proc Natl Acad Sci U S A 99(26):17197–17202
Igamberdiev AU, Stasolla C, Hill RD (2014) Low oxygen stress, non-symbiotic hemoglobins, NO and programmed cell death. In: van Dongen JT, Licausi F (eds) Low oxygen stress in plants. Springer, Heidelberg

Ignarro LJ (2002) Nitric oxide as a unique signalling molecule in the vascular system: a historical overview. J Physiol Pharmacol 53:503–514
Iuchi S, Lin EC (1988) arcA (dye), a global regulatory gene in Escherichia coli mediating repression of enzymes in aerobic pathways. Proc Natl Acad Sci U S A 85:1888–1892
Iuchi S, Cameron DC, Lin EC (1989) A second global regulator gene (arcB) mediating repression of enzymes in aerobic pathways of Escherichia coli. J Bacteriol 171:868–873
Jackson MB (1985) Ethylene and responses of plants to soil waterlogging and submergence. Annu Rev Plant Physiol 36:145–174
Kallio PJ, Pongratz I, Gradin K, McGuire J, Poellinger L (1997) Activation of hypoxia-inducible factor 1alpha: posttranscriptional regulation and conformational change by recruitment of the Arnt transcription factor. Proc Natl Acad Sci U S A 94:5667–5672
Kamata H, Hirata H (1999) Redox regulation of cellular signalling. Cell Signal 11(1):1–14
Khoroshilova N, Beinert H, Kiley PJ (1995) Association of a polynuclear iron-sulfur center with a mutant FNR protein enhances DNA binding. Proc Natl Acad Sci USA 92:2499–2503
Kiley PJ, Beinert H (2003) The role of Fe-S proteins in sensing and regulation in bacteria. Curr Opin Microbiol 6(2):181–185
Klok EJ, Wilson IW, Wilson D, Chapman SC, Ewing RM, Somerville SC, Peacock WJ, Dolferus R, Dennis ES (2002) Expression profile analysis of the low oxygen response in Arabidopsis root cultures. Plant Cell 14(10):2481–2494
Körner H, Sofia HJ, Zumft WG (2003) Phylogeny of the bacterial superfamily of Crp-Fnr transcription regulators: exploiting the metabolic spectrum by controlling alternative gene programs. FEMS Microbiol Rev 27:559–592
Kreuzwieser J, Hauberg J, Howell KA, Carroll A, Rennenberg H, Millar AH, Whelan J (2009) Differential response of gray poplar leaves and roots underpins stress adaptation during hypoxia. Plant Physiol 149(1):461–473. doi:10.1104/pp. 108.125989
Kwon YT, Kashina AS, Davydov IV, Hu RG, An JY, Seo JW, Du F, Varshavsky A (2002) An essential role of N-terminal arginylation in cardiovascular development. Science 297 (5578):96–99
Lando D, Peet DJ, Gorman JJ, Whelan DA, Whitelaw ML, Bruick RK (2002) FIH-1 is an asparaginyl hydroxylase enzyme that regulates the transcriptional activity of hypoxia-inducible factor. Genes Dev 16:1466–1471
Lasanthi-Kudahettige R, Magneschi L, Loreti E, Gonzali S, Licausi F, Novi G, Beretta O, Vitulli F, Alpi A, Perata P (2007) Transcript profiling of the anoxic rice coleoptile. Plant Physiol 144(1):218–231
Lee MJ, Tasaki T, Moroi K, An JY, Kimura S, Davydov IV, Kwon YT (2005) RGS4 and RGS5 are in vivo substrates of the N-end rule pathway. Proc Natl Acad Sci U S A 102(42):15030–15035
Li HY, Chye ML (2003) Membrane localization of Arabidopsis acyl-CoA binding protein ACBP2. Plant Mol Biol 51(4):483–492
Li HY, Chye ML (2004) Arabidopsis Acyl-CoA-binding protein ACBP2 interacts with an ethylene-responsive element-binding protein, AtEBP, via its ankyrin repeats. Plant Mol Biol 54(2):233–243
Licausi F, Perata P (2009) Low oxygen signaling and tolerance in plants. Adv Bot Res 50:189–198
Licausi F, van Dongen JT, Giuntoli B, Novi G, Santaniello A, Geigenberger P, Perata P (2010) HRE1 and HRE2, two hypoxia-inducible ethylene response factors, affect anaerobic responses in Arabidopsis thaliana. Plant J 62(2):302–315
Licausi F, Weits DA, Pant BD, Scheible WR, Geigenberger P, van Dongen JT (2011a) Hypoxia responsive gene expression is mediated by various subsets of transcription factors and miRNAs that are determined by the actual oxygen availability. New Phytol 190(2):442–456
Licausi F, Kosmacz M, Weits DA, Giuntoli B, Giorgi FM, Voesenek LA, Perata P, van Dongen JT (2011b) Oxygen sensing in plants is mediated by an N-end rule pathway for protein destabilization. Nature 479(7373):419–422

Loreti E, Poggi A, Novi G, Alpi A, Perata P (2005) A genome-wide analysis of the effects of sucrose on gene expression in Arabidopsis seedlings under anoxia. Plant Physiol 137:1130–1138

Maxwell PH, Wiesener MS, Chang GW, Clifford SC, Vaux EC, Cockman ME, Wykoff CC, Pugh CW, Maher ER, Ratcliffe PJ (1999) The tumour suppressor protein VHL targets hypoxia-inducible factors for oxygen-dependent proteolysis. Nature 399(6733):271–275

Miller G, Mittler R (2006) Could heat shock transcription factors function as hydrogen peroxide sensors in plants? Ann Bot 98(2):279–288

Mustroph A, Lee SC, Oosumi T, Zanetti ME, Yang H, Ma K, Yaghoubi-Masihi A, Fukao T, Bailey-Serres J (2010) Cross-kingdom comparison of transcriptomic adjustments to low-oxygen stress highlights conserved and plant-specific responses. Plant Physiol 152 (3):1484–1500

Ohh M, Park CW, Ivan M, Hoffman MA, Kim TY, Huang LE, Pavletich N, Chau V, Kaelin WG (2000) Ubiquitination of hypoxia-inducible factor requires direct binding to the beta-domain of the von Hippel-Lindau protein. Nat Cell Biol 2(7):423–427

Packer MA, Stasiv Y, Benraiss A, Chmielnicki E, Grinberg A, Westphal H, Goldman SA, Enikolopov G (2003) Nitric oxide negatively regulates mammalian adult neurogenesis. Proc Natl Acad Sci U S A 100(16):9566–9571

Perazzolli M, Dominici P, Romero-Puertas MC, Zago E, Zeier J, Sonoda M, Lamb C, Delledonne M (2004) Arabidopsis nonsymbiotic hemoglobin AHb1 modulates nitric oxide bioactivity. Plant Cell 16(10):2785–2794

Rodriguez C, Kwon O, Georgellis D (2004) Effect of D-lactate on the physiological activity of the ArcB sensor kinase in Escherichia coli. J Bacteriol 186:2085–2090

Stipanuk MH, Dominy JE Jr, Ueki I, Hirschberger LL (2008) Measurement of cysteine dioxygenase activity and protein abundance. Curr Protoc Toxicol 38:6.15.1–6.15.25

Stipanuk MH, Ueki I, Dominy JE Jr, Simmons CR, Hirschberger LL (2009) Cysteine dioxygenase: a robust system for regulation of cellular cysteine levels. Amino Acids 37(1):55–63

Subbaiah CC, Bush DS, Sachs MM (1994) Elevation of cytosolic calcium precedes anoxic gene expression in maize suspension-cultured cells. Plant Cell 6:1747–1762

Summers JE, Ratcliffe RG, Jackson MB (2000) Anoxia tolerance in the aquatic monocot Potamogeton pectinatus: absence of oxygen stimulates elongation in association with an unusually large pasteur effect. J Exp Bot 51:1413–1422

Tadege M, Dupuis II, Kuhlemeier C (1999) Ethanolic fermentation: new functions for an old pathway. Trends Plant Sci 4:320–325

Taylor BL, Zhulin IB (1999) PAS domains: internal sensors of oxygen, redox potential, and light. Microbiol Mol Biol Rev 63(2):479–506

Thiel J, Rolletschek H, Friedel S, Lunn JE, Nguyen TH, Feil R, Tschiersch H, Müller M, Borisjuk L (2011) Seed-specific elevation of non-symbiotic hemoglobin AtHb1: beneficial effects and underlying molecular networks in Arabidopsis thaliana. BMC Plant Biol 11:48

Unden G, Bongaerts J (1997) Alternative respiratory pathways of Escherichia coli: energetics and transcriptional regulation in response to electron acceptors. Biochim Biophys Acta 1320 (3):217–234

van Dongen JT, Fröhlich A, Ramírez-Aguilar SJ, Schauer N, Fernie AR, Erban A, Kopka J, Clark J, Langer A, Geigenberger P (2009) Transcript and metabolite profiling of the adaptive response to mild decreases in oxygen concentration in the roots of arabidopsis plants. Ann Bot 103(2):269–280

Velasquez SM, Ricardi MM, Dorosz JG, Fernandez PV, Nadra AD, Pol-Fachin L, Egelund J, Gille S, Harholt J, Ciancia M, Verli H, Pauly M, Bacic A, Olsen CE, Ulvskov P, Petersen BL, Somerville C, Iusem ND, Estevez JM (2011) O-glycosylated cell wall proteins are essential in root hair growth. Science 332(6036):1401–1403

Vlad F, Spano T, Vlad D, Daher FB, Ouelhadj A, Fragkostefanakis S, Kalaitzis P (2007) Involvement of Arabidopsis prolyl 4 hydroxylases in hypoxia, anoxia and mechanical wounding. Plant Signal Behav 2(5):368–369

Voesenek L, Banga M, Thier RH, Mudde CM, Harren F, Barendse G, Blom C (1993) Submergence-induced ethylene synthesis, entrapment and growth in two plant species with contrasting flooding resistances. Plant Physiol 103:783–791

Voesenek LACJ, Rijnders JHGM, Peeters AJM, van de Steeg HM, de Kroon H (2004) Plant hormones regulate fast shoot elongation underwater: from genes to communities. Ecology 85:16–27

Wang GL, Jiang B-H, Rue EA, Semenza GL (1995) Hypoxia-inducible factor 1 is a basic-helix-loop-helix-PAS heterodimer regulated by cellular 02 tension. Proc Natl Acad Sci U S A 92:5510–5514

Wendehenne D, Pugin A, Klessig DF, Durner J (2001) Nitric oxide: comparative synthesis and signaling in animal and plant cells. Trends Plant Sci 6(4):177–183

Wullschleger S, Loewith R, Hall MN (2006) TOR signaling in growth and metabolism. Cell 124 (3):471–484

Xu K, Xu X, Fukao T, Canlas P, Maghirang-Rodriguez R, Heuer S, Ismail AM, Bailey-Serres J, Ronald PC, Mackill DJ (2006) Sub1A is an ethylene-response-factor-like gene that confers submergence tolerance to rice. Nature 442(7103):705–708

Zabalza A, van Dongen JT, Froehlich A, Oliver SN, Faix B, Gupta KJ, Schmälzlin E, Igal M, Orcaray L, Royuela M, Geigenberger P (2009) Regulation of respiration and fermentation to control the plant internal oxygen concentration. Plant Physiol 149(2):1087–1098

Oxidative Stress Components Explored in Anoxic and Hypoxic Global Gene Expression Data

Olga B. Blokhina, Petri Törönen, and Kurt V. Fagerstedt

Abstract Global gene expression data were analyzed to search for the genes related to oxidative stress response, to examine the differences between hypoxia and anoxia, and to reveal new components of oxygen deprivation response escaped from the previous analyses. Gene Set Z-score (GSZ) was used to report gene ontology (GO) classes that showed significant regulation and also partial up- and downregulation in Arabidopsis anoxic and hypoxic microarray data sets. Under both anoxia and hypoxia significant upregulation was reported for anaerobic respiration, response to low oxygen levels, and response to hypoxia. Comparable high GSZ scores were shown for several oxidative stress-related GO classes and for functional groups of biological processes known to involve oxygen radical formation such as: cellular respiration, wounding, and response to high light and UV-B. Availability of oxygen in hypoxic experimental sets was marked by upregulation of several oxygenases, including ACC-oxidase responsible for ethylene synthesis. Consistent strong induction of several Fe-dependent ketoglutarate oxygenases (FeKGO) in the majority of hypoxic conditions analyzed suggests an important and yet unidentified function for these enzymes. Based on metabolic and gene expression studies we suggest that FeKGO may function in a bypass route for part of the TCA cycle (citrate-isocitrate) inhibited under hypoxia. This would incorporate 2-ketoglutarate supplied by activated GABA shunt and form succinate, a TCA cycle and mitochondrial electron transport chain substrate. FeKGO turnover is sustained by the putative route coupled to ascorbate–monodehydroascorbate cycling and hemoglobin-dependent NO elimination. The analysis strongly supports earlier findings that formation of activated oxygen and oxidative stress is an integral part of the response to oxygen deprivation. Several novel functional gene groups

O.B. Blokhina (✉) • K.V. Fagerstedt
Department of Biosciences, Division of Plant Biology, University of Helsinki, Helsinki, Finland
e-mail: olga.blokhina@helsinki.fi

P. Törönen
Institute of Biotechnology, University of Helsinki, Helsinki, Finland

J.T. van Dongen and F. Licausi (eds.), *Low-Oxygen Stress in Plants*, Plant Cell Monographs 21, DOI 10.1007/978-3-7091-1254-0_2, © Springer-Verlag Wien 2014

were highlighted by the analysis: upregulation of cysteamine dioxygenase activity and FeKGO and downregulation of circadian rhythm-related genes.

1 Introduction

During the past 2 decades reactive oxygen species (ROS) have developed in our minds from damaging stress molecules to stress-signaling agents important for the development of stress responses in practically all stresses that plants have to endure (Bailey-Serres and Chang 2005; Van Breusegem et al. 2008; Jaspers and Kangasjärvi 2010; Petrov and Van Breusegem 2012). The presence of ROS under oxygen deprivation stress is somewhat a paradox and indeed some oxygen is needed for ROS production under oxygen deprivation stress, and hence it is preferable to talk about severe hypoxia in most cases, and of a reoxygenation period when atmospheric oxygen conditions are again introduced (Branco-Price et al. 2008). The biochemistry behind ROS production and antioxidative defense as well as the damage produced under oxygen deprivation is well documented (Yan et al. 1996; Biemelt et al. 1998, 2000; Blokhina et al. 2000, 2001, 2003; Fukao and Bailey-Serres 2004; Santosa et al. 2007). The evidence for the regulatory role of ROS under oxygen deprivation in the control of gene expression (Pucciariello et al. 2012), the negative feedback regulation of H_2O_2 levels by Rop–RopGAP4 interaction (Baxter-Burrell et al. 2002), and the activation of MAPK kinases in response to mitochondrial ROS resulting in better survival under hypoxia (Chang et al. 2012), all point to a complex relationship between hypoxic metabolic rearrangements, ROS levels, and their cellular localization and affect the physiological outcome of oxidative stress. Naturally, the most important processes from the plant's point of view are the adaptations preserving the adenylate energy charge (Greenway and Gibbs 2003; Bailey-Serres and Voesenek 2008; Lee et al. 2011). During the past few years multiple routes have been elucidated for the production and consumption of ATP such as the pyrophosphate-dependent glycolysis (Huang et al. 2008), nitrate-dependent ATP synthesis under oxygen deprivation (Stoimenova et al. 2007), and ATP hydrolysis in animal mitochondria under anaerobic conditions (St-Pierre et al. 2000). On the other hand, bulky plant organs seem to avoid total anoxia in the tissues by metabolic control of respiration (Borisjuk et al. 2007; Zabalza et al. 2009).

It has been reassuring to note, as we show in this chapter, that bioinformatics analysis of oxygen deprivation arrays picked up multiple classes related to oxidative stress and ROS. We have also noticed that many different stresses such as high light and wounding are leading to the upregulation of oxidative stress-related genes also shared by oxygen deprivation array data.

In addition to ROS, during the recent years a vast amount of data has accumulated in favor of reactive nitrogen species (RNS) in plant tissues and their regulative role in adjusting metabolic events (Qiao and Fan 2008; Igamberdiev et al. 2010; Gupta et al. 2011a, b; Hebelstrup et al. 2012). Plant non-symbiotic hemoglobins are emerging as major players in the regulation of NO levels under oxygen deprivation and therefore controlling the multiple functions exerted by NO, one of the most important being interaction with hormone signaling (Hebelstrup et al. 2007, 2012; Igamberdiev et al. 2010; Hill 2012). In addition, the inhibitory effect of NO on heme- and Fe–S cluster-containing enzymes such as aconitase, cytochrome *c* oxidase, and catalases directly affects cell energy status and defense against ROS (Navarre et al. 2000; Ederli et al. 2006; Blokhina and Fagerstedt 2010b). Furthermore, the accumulation of citrate due to NO-induced aconitase inhibition has been shown to result in activation of the alternative oxidase (Gupta et al. 2012), which controls superoxide formation and NO degradation in the mitochondrial electron transport chain under over-reduced conditions such as under the lack of oxygen (Gupta et al. 2009; Wulff et al. 2009; Blokhina and Fagerstedt 2010b).

When the response to oxidative stress associated with oxygen deprivation is studied on the whole genome level, the complexity of the outcome reflects several strategies used by the organism to overcome the stress without disturbing ROS signaling. We can presume that the same would be applicable also to NO signaling under hypoxic conditions. In plant cells and tissues the most straightforward approach would be a balanced increase in the expression of ROS producers and scavengers. At the onset of the oxidative stress, i.e., when ROS production goes out of control, the upregulation of defense and repair systems and coordinated expression of the corresponding signaling pathways would be beneficial (Luhua et al. 2008; Licausi et al. 2010). And, finally, metabolic alterations caused by oxygen deprivation can indirectly control internal oxygen concentration, preventing the onset of complete anoxia and, therefore control ROS and NO formation (Gupta et al. 2009; Zabalza et al. 2009; Blokhina and Fagerstedt 2010a).

Using novel bioinformatics approaches we have aimed to explore oxidative stress components on global gene expression level in oxygen deprivation arrays by comparing publicly available and in-house array data. We wished to dissect anoxic and hypoxic responses and to elucidate any new components shared by oxidative stress and oxygen deprivation stress responses.

2 Setup for Analyses of Multiple Data Sets on Global Gene Expression Under Oxygen Deprivation: Affymetrix and Agilent Platforms

Recently cross-species analyses of global gene expression and metabolic alterations under oxygen deprivation have been published (Narsai et al. 2011) along with the analysis on ROS-driven transcription under oxygen deprivation (Pucciariello et al. 2012). The latter work is based on the cross-comparison of global gene expression data under oxygen deprivation and ROS stresses (Pucciariello et al. 2012). In the present study we have undertaken a different approach: We used the enrichment analysis to monitor gene ontology (GO) classes that showed significant regulation in anoxic and hypoxic microarray data sets (Table 1). We used two types of gene expression datasets in our analysis: Affymetrix and Agilent platform-based arrays. One of them was our in-house data generated with Agilent microarrays (Blokhina et al. unpublished) and the other was a collection of published gene expression datasets generated with Affymetrix microarrays. Due to the differences in the Affymetrix and Agilent layouts and to experimental conditions (darkness/dim light in Affymetrix and light in Agilent experiments), the data were analyzed separately. Agilent chips were used with three dyes (HyPer5, Cy3, Alexa488) and quantitated using Gene Pix Pro 6.1. After this the data was read into R, where we used background correction (Ritchie et al. 2007) and ComBat normalization (Johnson et al. 2007) to correct the various noise signals. Finally the data was processed using LIMMA package (Smyth 2005) from Bioconductor (Gentleman et al. 2004). From significantly up- and downregulated GO classes in Affymetrix and Agilent-based arrays returned by the analysis, only classes related to oxidative stress and related metabolic pathways were chosen and further discussed. The experiments where the analysis of global gene expression was performed under anoxia or under hypoxia were compared to assess the differences between the two treatments in terms of ROS production/defense. The GO annotations were downloaded from TAIR (http://www.arabidopsis.org) linked to a single locus name. We used an in-house developed enrichment method, Gene Set Z-score (GSZ) (Törönen et al. 2009), for the analysis. The GSZ analysis looks for the GO classes that show strong upregulation or downregulation. The strength of this method is that it can also monitor classes that show partial up- and partial downregulation and detect biological processes which were missed by other analyses (Törönen et al. 2009). GSZ has similarities to other published methods Gene Set Analysis (Efron and Tibshirani 2007) and Allez (Newton et al. 2007), and therefore we tested the enrichment also in a similar manner to these methods. Evaluation of results was done using 120 permutations of GO classes. Permutations were used to generate empirical p-values and also to scale the scores using the estimates for mean and standard deviation obtained from the permutations.

3 GO Classes Upregulated Under Anoxia and Hypoxia

3.1 Anoxia

As expected, in anoxic data represented by five Affymetrix experiments, the analysis returned a set of GO classes directly related to oxygen deprivation (Table 2). In five out of five arrays analyzed the following GO classes BP: 000906: anaerobic respiration; BP: 0036293: response to decreased oxygen levels; BP: 0070482: response to oxygen levels; and BP: 0001666: response to hypoxia, were significantly upregulated. Several ROS-related GO classed with similarly high scores were reported in at least two out of five conditions analyzed: BP: 0006979: response to oxidative stress (632 gene products); BP: 0000302: response to ROS

Table 1 Microarray experiments used for the analyses

Treatment	*Arabidopsis* ecotype	Experimental conditions	Reference	Abbreviation used
Anoxia 6 h	Col-0 seedlings, 4 d	Petri dishes, MS liquid medium, dark	Banti et al. (2010)	BA ax 6h
Anoxia 12 h	Landsberg *erecta*, 7 d	Vertical plates, MS solid medium, dim light. Total RNA[a]	Branco-Price et al. (2005)	BP ax 12h t
Submergence, anoxia 7 h	Col-0, shoot, ten-leaf rosette stage	Pot, soil, dark	Lee et al. (2011)	Lee Sub 7h
Submergence, anoxia 24 h	Col-0, shoot, ten-leaf rosette stage	Pot, soil, dark	Lee et al. (2011)	Lee Sub 24h
Anoxia 6 h	Col glabra, seedlings	Petri dishes, MS liquid medium, dark	Loreti et al. (2005)	LO ax 6h
Hypoxia 4 h	Col-0 seedlings, 7 d	Petri dishes, MS liquid medium, 1 % O_2, dark	Licausi et al. (2010)	LI Hy 4h
Hypoxia 0.5 h, 2 h, 48 h	Col2 roots, 10 d	Vertical plates, MS solid medium, 1 %, 4 % and 8 % O_2, dark	van Dongen et al. (2009)	vD 1% 0.5h, vD 1% 2h, vD 1% 48h, vD 4% 0.5h, vD 4% 2h, vD 4% 48h, vD 8% 0.5h, vD 8% 2h, vD 8% 48h
Hypoxia 9 h	Col-0 seedlings, *35S:His$_6$FLAG-RPL18B*, 7 d	Vertical plates, MS solid medium, dim light. Total RNA[a]	Branco-Price et al. (2008)	BP Hy 9h
Hypoxia 2 h, 24 h	Landsberg *erecta*, shoots, 16 d	Vertical plates, MS solid medium, light. Agilent	Blokhina et al. (unpublished)	BL Hy 2h/0h, BL Hy 24h/0h, BL Hy 2h/c2h, BL 24h/c24h

Data concerning treatments of wild-type plants only were extracted

[a]From these experiments only variants where total RNA was used for hybridisations were analyzed

Table 2 GO classes upregulated under anoxic conditions in Affymetrix arrays

GO class	BA ax 6h	BP ax 12h t	LEE Sub 24h	LEE Sub 7h	LO ax 6h
BP: 0009408 : response to heat	22,92	3,44	0,39	0,43	10,61
BP: 0042542 : response to hydrogen peroxide	**17,21**	**1,66**	**0,50**	**0,71**	**7,19**
BP: 0000302 : response to reactive oxygen species	**16,37**	**1,25**	**-0,06**	**0,02**	**6,41**
BP: 0009644 : response to high light intensity	**16,29**	**1,44**	**-0,10**	**0,20**	**7,52**
BP: 0009061 : anaerobic respiration	15,88	4,52	5,93	9,21	11,85
BP: 0009642 : response to light intensity	14,18	1,59	0,60	1,28	5,92
BP: 0009266 : response to temperature stimulus	14,11	2,75	-1,51	-1,49	6,90
BP: 0006950 : response to stress	12,01	1,25	7,10	5,41	5,98
BP: 0006979 : response to oxidative stress	**11,39**	**2,96**	**2,54**	**1,32**	**3,56**
BP: 0010200 : response to chitin	10,92	2,49	6,08	6,13	1,21
MF: 0001071 : nucleic acid binding transcription factor activity	10,60	3,28	2,86	2,88	4,30
MF: 0003700 : sequence-specific DNA binding transcription factor activity	10,60	3,28	2,86	2,88	4,30
BP: 0009628 : response to abiotic stimulus	9,16	3,28	-0,50	-0,07	6,08
BP: 0045333 : cellular respiration	8,95	1,59	2,34	4,30	7,20
BP: 0010286 : heat acclimation	8,86	1,56	0,65	-0,63	2,93
BP: 0015980 : energy derivation by oxidation of organic compounds	**8,61**	**1,79**	**2,00**	**4,00**	**6,98**
MF: 0009916 : alternative oxidase activity	**8,28**	**1,28**	**-0,36**	**2,50**	**1,50**
BP: 0042221 : response to chemical stimulus	8,23	1,75	2,87	7,04	2,21
BP: 2001141 : regulation of RNA biosynthetic process	7,90	1,91	1,66	1,51	3,34
BP: 0009743 : response to carbohydrate stimulus	7,88	2,15	3,83	5,07	0,15
BP: 0006355 : regulation of transcription, DNA-dependent	7,87	2,00	1,76	1,56	3,39
BP: 0051252 : regulation of RNA metabolic process	7,81	1,89	1,62	1,52	3,31
BP: 0010556 : regulation of macromolecule biosynthetic process	7,43	1,59	1,43	1,50	3,15
BP: 2000112 : regulation of cellular macromolecule biosynthetic process	7,43	1,59	1,43	1,50	3,15
BP: 0019219 : regulation of nucleobase-containing compound metabolic process	7,12	1,54	1,11	1,24	2,92
BP: 0009889 : regulation of biosynthetic process	7,03	1,37	1,14	1,23	2,88
BP: 0036293 : response to decreased oxygen levels	6,96	4,65	9,05	10,89	12,59
BP: 0070482 : response to oxygen levels	6,96	4,65	9,05	10,89	12,59
BP: 0051171 : regulation of nitrogen compound metabolic process	6,93	1,38	1,02	1,13	2,88
BP: 0010035 : response to inorganic substance	6,11	-2,79	-0,69	0,51	3,68
BP: 0001666 : response to hypoxia	5,76	4,50	9,11	10,70	11,02
BP: 0009415 : response to water stimulus	5,53	0,73	-0,94	-1,11	0,20
BP: 0009414 : response to water deprivation	5,49	0,89	-1,18	-1,10	0,16
BP: 0010224 : response to UV-B	**5,42**	**0,64**	**0,84**	**-0,27**	**5,21**
BP: 0009416 : response to light stimulus	5,15	2,20	-2,29	-2,13	3,91
BP: 0006091 : generation of precursor metabolites and energy	**5,11**	**-0,35**	**-0,88**	**0,70**	**3,85**
BP: 0010033 : response to organic substance	5,08	2,48	2,07	5,45	-0,98
BP: 0009411 : response to UV	**4,99**	**1,86**	**0,12**	**-0,97**	**4,93**
BP: 0009611 : response to wounding	**4,96**	**1,99**	**3,07**	**5,16**	**0,36**
BP: 0036294 : cellular response to decreased oxygen levels	4,92	2,84	8,74	10,02	7,75
BP: 0071453 : cellular response to oxygen levels	4,92	2,84	8,74	10,02	7,75
BP: 0071456 : cellular response to hypoxia	4,92	2,84	8,74	10,02	7,75
BP: 0009314 : response to radiation	4,90	2,06	-2,32	-2,29	3,98
BP: 0006952 : defense response	**4,89**	**-0,16**	**7,74**	**6,01**	**2,17**
BP: 0000160 : two-component signal transduction system (phosphorelay)	4,88	1,85	1,37	4,78	1,33
MF: 0016682 : oxidoreductase activity, acting on diphenols and related substances as donors, oxygen as acceptor	4,75	1,47	-0,09	2,74	1,97
BP: 0070301 : cellular response to hydrogen peroxide	4,75	0,30	1,39	-0,15	1,56

The first 50 significantly ($p < 0.01$) upregulated GO classes with the highest score are highlighted. GO classes are arranged according to the score in the first column. Bold font: GO classes related to ROS and oxidative stress. GO class was discussed as regulated if significant changes were observed in at least two anoxic conditions. See Table 1 for the column names

(384 gene products); BP: 0042542: response to hydrogen peroxide (263 gene products); BP: 0006091: generation of precursor metabolites and energy (664 gene products); BP: 0045333: cellular respiration (130 gene products); BP: 0009644: response to high light intensity (221 gene products); BP: 0010224: response to UV-B (104 gene products); and BP: 0009611: response to wounding (336 gene products). The set of significantly upregulated anoxic GO classes revealed not only the induction of genes involved in response to anoxia, but also reflected activation of oxygen and concurrent defense responses, engaging mitochondrial metabolism and respiratory chain in the response to the lack of oxygen and/or ROS.

Cross talk between different stresses such as high light intensity, wounding, and responses to UV light reflected the universal signaling role for ROS under diverse stress situations (Mittler et al. 2004; Van Breusegem et al. 2008; Potters et al. 2009; Petrov and Van Breusegem 2012). Therefore, many of the reported functional gene groups share similar members, heat shock-related genes being one of the examples along with a large group of cytochromes P450. The components of heat shock response have been extensively studied under oxygen deprivation and many are known to be activated under both anoxia and oxidative stress (Li et al. 2005; Loreti et al. 2005; Banti et al. 2010; Inzé et al. 2012).

A large and diverse gene superfamily of cytochromes P450 is a third largest family in Arabidopsis (245 genes) after F-box proteins and receptor-like kinases (Nelson and Werck-Reichhart 2011). The P450 enzymes are heme-containing monooxygenases which produce superoxide anion during their catalytic action and are considered important for the activation of oxygen in signaling events (Lewis 2002). They fulfill multiple functions in plants: biosynthesis of secondary compounds such as flavonoids, isoflavonoids, phytoalexins, and carotenoids, which are known to protect against oxidative stress. The biosynthetic route for the plant hormones brassinosteroids and gibberellins also involve P450, as well as the synthesis of the signaling molecules salicylic and jasmonic acids (Dasgupta et al. 2011). The requirement for molecular oxygen and external electron donor (e.g., NADH) is not absolute for atypical CYP74 class of P450 enzymes, which may be of importance under oxygen deprivation. Members of CYP74 class are responsible for fatty acid hydroperoxide metabolism and participate in oxylipin biosynthesis (Hughes et al. 2009). However, close inspection of gene expression data revealed that two CYP74 genes that are present in oxygen deprivation arrays (hydroperoxidelyase AT4G15440 and allene oxide synthase AT5G42650) are not significantly regulated in most experiments and downregulated only in two conditions out of 23 analyzed. In the study on transcriptomic response of rice (*Oryza sativa*) coleoptiles to anoxia, similar down-regulation of P450 transcripts has been detected and discussed as an energy-saving strategy (Lasanthi-Kudahettige et al. 2007).

While assessing the results of the analysis one should also bear in mind that GO classes are hierarchically organized: e.g., the "response to oxidative stress" includes the response to ROS and the response to hydrogen peroxide. GO class "cellular respiration" contains the members coding for ATP synthesis, TCA cycle enzymes and mitochondrial electron transport chain, and also anaerobic respiration class (GO: 0009061). This class is represented by nine uncharacterized gene products, which have been reported to be induced under anoxia (Gonzali et al. 2005; Mustroph et al. 2010) and almost half of them (AT3G10020, AT2G36220, AT1G05575, AT5G15120, AT5G10040) have been also implicated in response to oxidative stress and hydrogen peroxide (Baxter et al. 2007; Luhua et al. 2008; Inzé et al. 2012). In GO class "response to hydrogen peroxide" the most prominent and universal induction exhibited respiratory burst oxidase RbohD (At5g47910). Plant NADPH oxidases are key components of ROS-mediated signaling under diverse abiotic stresses, in plant–pathogen interactions, and during normal growth and development. It has been shown recently that nitrosylation of a critical cysteine residue negatively regulates RbohD, providing evidence for the cross talk between ROS and NO signaling via this negative feedback loop (Yun et al. 2011). However, other class representatives directly connected to ROS metabolism, such as ascorbate peroxidase 2 (AT3G09640) and monodehydroascorbate reductase (At3g09940), were shown to be moderately upregulated under hypoxic and anoxic conditions.

3.2 Hypoxia

Oxygenases under oxygen deprivation. The hypoxic set of the arrays presented by both Affymetrix and Agilent platforms (analyzed separately) shared with anoxic arrays several significantly upregulated oxygen deprivation-related classes and oxidative stress-related classes: BP: 0015980: energy derivation by oxidation of organic compound, BP: 0045333: cellular respiration, BP: 0009611: response to wounding, BP: 0010224: response to UV-B, BP: 0042542: response to hydrogen peroxide, and BP: 0000302: response to ROS (Tables 3 and 4). The availability of oxygen in hypoxic experiments was signified by a set of GO classes different from anoxia showing the upregulation of oxygenases, the enzymes which incorporate oxygen into organic substrates: GO:0016702:oxidoreductase activity, acting on single donors with incorporation of molecular oxygen, incorporation of two oxygen atoms, GO: 0016701: oxidoreductase activity, acting on single donors with incorporation of molecular oxygen, MF: 0051213: dioxygenase activity, and MF: 0047800: cysteamine dioxygenase activity.

Many microarray studies reveal the intrinsic connection between oxygen deprivation stress and the increased expression of the genes coding for oxygenases, (e.g., Fe-dependent ketoglutarate oxygenases, ACC-oxidase, desaturases, alternative oxidase, etc.). Fe-dependent ketoglutarate oxygenases (FeKGO) showed consistent upregulation over many hypoxic sets analyzed. It is a large gene superfamily which requires Fe^{2+} as a cofactor, and some of the class enzymes utilize ascorbate as an electron donor. Oxidation of a substrate is coupled to decarboxylation of 2-ketoglutarate to yield succinate and CO_2 (Loenarz and Schofield 2008). In mammalian tissues these enzymes are involved in histone and DNA demethylation

Table 3 GO classes upregulated under hypoxic conditions in Affymetrix arrays

GO class	LI Hy 4h	vD 1.% 0.5h	vD 1% 2h	vD1% 48h	vD 4% 0.5h	vD 4% 2h	vD 4% 48h	vD 8% 0.5h	vD 8% 2h	vD 8% 48h	BP Hy 9h
BP: 0009061 : anaerobic respiration	15,96	30,15	31,40	23,09	26,40	42,90	22,38	29,38	29,18	9,24	27,78
BP: 0036293 : response to decreased oxygen levels	14,82	9,27	19,33	11,39	8,17	14,51	9,84	8,69	10,75	5,30	11,61
BP: 0070482 : response to oxygen levels	14,82	9,27	19,33	11,39	8,17	14,51	9,84	8,69	10,75	5,30	11,61
BP: 0036294 : cellular response to decreased oxygen levels	13,75	2,25	13,80	2,52	1,06	4,47	1,13	0,74	2,20	-0,94	5,21
BP: 0071453 : cellular response to oxygen levels	13,75	2,25	13,80	2,52	1,06	4,47	1,13	0,74	2,20	-0,94	5,21
BP: 0071456 : cellular response to hypoxia	13,75	2,25	13,80	2,52	1,06	4,47	1,13	0,74	2,20	-0,94	5,21
BP: 0001666 : response to hypoxia	12,86	7,97	16,90	9,92	6,93	12,79	8,53	8,10	9,78	5,27	10,34
MF: 0043168 : anion binding	8,16	8,22	7,30	8,39	8,25	8,22	10,46	15,73	9,83	5,33	10,90
BP: 0010200 : response to chitin	8,16	4,56	4,44	0,41	-0,39	0,74	-3,38	-1,50	1,59	-2,19	12,70
MF: 0016702 : oxidoreductase activity, acting on single donors with incorporation of molecular oxygen, incorporation of two atoms of oxygen	**6,15**	**5,74**	**8,20**	**7,39**	**6,90**	**8,58**	**9,40**	**6,55**	**9,24**	**7,57**	**5,51**
BP: 0009743 : response to carbohydrate stimulus	5,43	3,71	4,61	0,31	0,52	0,92	-2,52	-0,68	1,83	-2,60	9,94
MF: 0051213 : dioxygenase activity	**5,28**	**3,18**	**5,85**	**3,87**	**4,53**	**5,33**	**4,06**	**4,72**	**6,04**	**3,57**	**3,53**
BP: 0045333 : cellular respiration	**5,18**	**14,74**	**12,36**	**10,69**	**13,48**	**15,14**	**9,98**	**17,67**	**18,59**	**6,21**	**13,73**
MF: 0016701 : oxidoreductase activity, acting on single donors with incorporation of molecular oxygen	**5,17**	**4,44**	**6,24**	**5,64**	**5,17**	**5,63**	**7,36**	**5,26**	**5,87**	**6,85**	**5,51**
MF: 0047800 : cysteamine dioxygenase activity	**5,16**	**14,39**	**12,98**	**12,77**	**15,51**	**21,90**	**16,23**	**19,73**	**22,15**	**10,14**	**8,84**
MF: 0001071 : nucleic acid binding transcription factor activity	5,12	2,57	2,09	-6,21	-1,74	1,13	-6,92	0,63	0,95	-5,67	3,06
MF: 0003700 : sequence-specific DNA binding transcription factor activity	5,12	2,57	2,09	-6,21	-1,74	1,13	-6,92	0,63	0,95	-5,67	3,06
BP: 0015980 : energy derivation by oxidation of organic compounds	**5,08**	**15,55**	**11,84**	**10,41**	**13,91**	**15,12**	**9,78**	**18,51**	**18,22**	**6,21**	**13,53**
BP: 0080167 : response to karrikin	4,88	7,33	3,45	4,10	6,12	3,36	5,20	4,74	6,48	3,00	6,34
MF: 0003950 : NAD+ ADP-ribosyltransferase activity	4,53	11,33	8,17	2,75	10,13	8,43	3,74	7,45	4,91	1,40	5,86
BP: 0009408 : response to heat	4,35	-0,18	-0,46	7,89	0,80	0,50	0,31	1,00	0,08	0,65	18,86
MF: 0016157 : sucrose synthase activity	4,28	7,33	10,18	11,49	10,03	17,56	10,28	13,23	12,65	8,01	3,81
BP: 0006950 : response to stress	4,26	1,80	5,79	4,53	3,64	4,24	1,85	2,96	4,38	-1,21	17,40
BP: 0006355 : regulation of transcription, DNA-dependent	4,26	3,13	1,58	-5,62	-1,23	1,86	-7,53	2,49	0,67	-5,07	2,66
BP: 2001141 : regulation of RNA biosynthetic process	4,21	3,20	1,54	-5,62	-1,18	1,86	-7,42	2,48	0,77	-5,04	2,57
BP: 0051252 : regulation of RNA metabolic process	4,18	3,21	1,55	-5,50	-1,09	1,84	-7,38	2,46	0,74	-5,03	2,62
BP: 0009611 : response to wounding	**4,15**	**3,91**	**3,72**	**2,94**	**4,03**	**4,92**	**0,59**	**3,72**	**5,43**	**0,00**	**5,70**
BP: 0010193 : response to ozone	**4,01**	**-0,23**	**0,25**	**3,40**	**-0,48**	**-0,76**	**-2,40**	**-0,09**	**-1,28**	**-2,00**	**0,96**
MF: 0000287 : magnesium ion binding	3,84	2,46	1,93	2,26	2,10	4,31	2,27	3,54	3,08	1,59	2,42
BP: 0009628 : response to abiotic stimulus	3,79	2,69	5,93	5,85	3,95	2,78	2,93	4,55	2,47	-0,76	11,88
BP: 0019219 : regulation of nucleobase-containing compound metabolic process	3,68	2,91	1,37	-5,46	-1,37	1,76	-7,22	2,49	0,61	-4,96	2,35

The first 50 significantly ($p < 0.01$) upregulated GO classes with the highest score are highlighted. GO classes are arranged according to the score in the first column. Bold font: GO classes related to ROS and oxidative stress. GO class was discussed as regulated when significant score was observed in at least three hypoxic conditions. See Table 1 for the column names

and in other similar hydroxylation and desaturation reactions (Loenarz and Schofield 2008). FeKGO named Prolyl-4-hydroxylases have been implicated in oxygen sensing in humans via hydroxylation of critical prolyl residues in the hypoxia-inducible factor (HIF) transcription factor (Berra et al. 2006). Intriguingly, in plants over-expression of the prolyl-hydroxylase AtP4H1 (AT2G43080) resulted in "hypoxia-in-normoxia" phenotype and concomitant upregulation of growth-, development-, and hypoxia-related genes (Asif et al. 2009). However, in the hypoxic and anoxic arrays analyzed here, the expression pattern for AtP4H1was insignificant.

In the context of oxygen deprivation stress, one of the most important members of the FeKGO family is ACC-oxidase (1-aminocyclopropane-1-carbohylate oxidase, AT2G19590, AT1G03400, AT1G62380, AT2G25450, AT5G43440, AT5G43450), an oxygenase which is involved in the synthesis of ethylene. ACC-oxidase is one of the examples of ascorbate-dependent reaction of FeKGO:

1-aminocyclopropane-1-carboxylate + ascorbate + O_2 → ethylene + cyanide + dehydroascorbate + CO_2 + $2H_2O$

Hence, ascorbate availability and turnover under hypoxic conditions accompanied by oxidative stress can affect not only H_2O_2 detoxification via the ascorbate–glutathione cycle, but also control ethylene biosynthesis. Ethylene is an important

Table 4 GO classes upregulated under hypoxic conditions in Agilent arrays

GO class	BL Hy 2h/0h	BL Hy 24h/0h	BL Hy 2h/c2h	BL Hy 24h/c24h
BP: 0009061 : anaerobic respiration	19,84	18,24	21,37	18,65
BP: 0009266 : response to temperature stimulus	7,81	18,19	13,02	19,92
BP: 0009408 : response to heat	4,40	16,96	14,35	24,10
MF: 0030976 : thiamine pyrophosphate binding	20,48	16,14	18,43	15,97
MF: 0043168 : anion binding	20,69	15,14	18,75	15,06
MF: 0004737 : pyruvate decarboxylase activity	20,72	14,60	20,09	14,62
BP: 0009628 : response to abiotic stimulus	12,28	14,03	18,48	15,17
BP: 0015980 : energy derivation by oxidation of organic compounds	11,02	13,84	10,73	14,29
BP: 0045333 : cellular respiration	**8,91**	**12,21**	**11,37**	**14,09**
BP: 0009642 : response to light intensity	2,97	10,03	9,22	13,98
BP: 0009644 : response to high light intensity	**1,27**	**9,95**	**8,90**	**14,09**
CC: 0043229 : intracellular organelle	10,34	9,88	10,78	10,67
CC: 0043226 : organelle	10,25	9,86	10,74	10,63
BP: 0006091 : generation of precursor metabolites and energy	10,05	9,54	9,85	9,37
BP: 0036293 : response to decreased oxygen levels	10,13	9,48	7,76	7,45
BP: 0070482 : response to oxygen levels	10,13	9,48	7,76	7,45
MF: 0047800 : cysteamine dioxygenase activity	**13,21**	**9,31**	**8,62**	**7,54**
CC: 0043227 : membrane-bounded organelle	11,33	9,14	10,82	8,86
BP: 0042221 : response to chemical stimulus	10,23	9,04	12,70	9,97
BP: 0005983 : starch catabolic process	8,45	8,90	-1,18	2,15
BP: 0044247 : cellular polysaccharide catabolic process	8,45	8,90	-1,18	2,15
BP: 0009251 : glucan catabolic process	8,45	8,90	-1,18	2,15
BP: 0042542 : response to hydrogen peroxide	**2,32**	**8,89**	**7,47**	**11,81**
BP: 0009409 : response to cold	4,58	8,78	3,93	6,19
CC: 0044429 : mitochondrial part	1,87	8,30	3,10	8,65
BP: 0001666 : response to hypoxia	9,36	8,29	6,96	6,21
CC: 0070013 : intracellular organelle lumen	1,50	8,28	7,46	13,67
CC: 0043233 : organelle lumen	1,50	8,28	7,46	13,67
BP: 0000302 : response to reactive oxygen species	**2,37**	**8,28**	**6,74**	**11,21**
CC: 0005739 : mitochondrion	**3,23**	**8,21**	**4,83**	**7,34**
MF: 0005515 : protein binding	6,31	8,09	5,62	7,59
CC: 0005761 : mitochondrial ribosome	0,79	7,82	0,25	8,67
BP: 0010286 : heat acclimation	2,38	7,81	0,81	7,41
BP: 0009631 : cold acclimation	2,88	7,75	-1,38	3,56
CC: 0005634 : nucleus	1,24	7,69	4,45	11,34
CC: 0044428 : nuclear part	1,66	7,64	7,14	13,48
BP: 0044275 : cellular carbohydrate catabolic process	7,36	7,51	-0,66	1,50
BP: 0009416 : response to light stimulus	5,51	7,49	12,69	9,15
CC: 0044446 : intracellular organelle part	5,90	7,32	9,09	9,54
BP: 0010224 : response to UV-B	**5,71**	**7,32**	**13,69**	**12,47**
CC: 0044422 : organelle part	5,86	7,28	9,02	9,46
BP: 0010035 : response to inorganic substance	6,88	7,25	12,64	10,46
BP: 0009314 : response to radiation	5,21	7,23	12,08	8,80
CC: 0030964 : NADH dehydrogenase complex	1,93	7,19	-0,26	6,40
CC: 0045271 : respiratory chain complex I	1,93	7,19	-0,26	6,40
BP: 0006397 : mRNA processing	3,59	7,11	2,58	7,36

The first 50 significantly ($p < 0.01$) upregulated GO classes with the highest score are highlighted. GO classes are arranged according to the score in the first column. Bold font: GO classes related to ROS and oxidative stress. GO class was discussed as regulated when significant score was observed in at least two hypoxic conditions. See Table 1 for the column names

signaling molecule under the lack of oxygen-coordinating morphological adaptations (stem elongation, aerenchyma, and adventitious root formation) and metabolic rearrangements (Fukao and Bailey-Serres 2008; Sairam et al. 2008; Licausi 2011). It has been shown that manipulation of hypoxia-induced Hb1 expression enhanced ethylene levels and ACC-oxidase activity in Hb-downregulated maize lines, possibly via modulation of NO levels by Hb1 (Manac'h-Little et al. 2005). NO can also control ethylene biosynthesis via S-nitrosylation of methionine adenosyl transferase leading to reversible inhibition of the ethylene precursor S-adenosyl methionine synthesis (Lindermayr et al. 2006). Ascorbate and NO levels are not the only cross points between ethylene and oxidative metabolism. An investigation into tocopherol biosynthesis, the main lipid-soluble antioxidant, in *Arabidopsis* mutants defective in ethylene perception and signaling, showed 30 % reduction in tocopherol levels under stress and suggested a regulatory role for ethylene in tocopherol biosynthesis (Cela et al. 2009).

Other examples of Fe2KG oxygenases, equally important under hypoxia as judged by their expression pattern, may be several other members of the oxygenase family: AT4G33910, AT1G20270, and AT3G28480. All of them showed quite strong upregulation in the majority of the conditions analyzed. The specific role for these oxygenases under oxygen deprivation is not clear. Taking into account the transcriptional and metabolic changes occurring under hypoxia, a metabolic association between the TCA cycle, Hbs, and NO under hypoxia can be suggested (Fig. 1). This is supported by strong and ubiquitous upregulation of non-symbiotic Hbs, activation of GABA shunt, moderate induction of ascorbate peroxidase (AT3G09640) and monodehydroascorbate reductase (AT3G09940), the physiological evidence on MDHAR- and ascorbate-sustained NO scavenging by hemoglobin (Igamberdiev et al. 2006), the rearrangements in the TCA cycle and succinate accumulation observed under oxygen deprivation(Rocha et al. 2010), and the dependence of FeKGO on ascorbate in the conversion of 2-ketoglutarate to succinate in Fe2KGO catalyzed reaction (Loenarz and Schofield 2008).

Induction of non-symbiotic class1 hemoglobin under hypoxic conditions is well documented (Dordas et al. 2003; van Dongen et al. 2009; Dordas 2009; Lee et al. 2011), but this upregulation is not specific to the lack of oxygen. Expression patterns of both Hb1 and Hb2 vary in different tissues as well as in response to different types of stress. *Hb1* genes have been reported to be upregulated by abiotic stresses (Trevaskis et al. 1997; Shimoda et al. 2005; Zhao et al. 2008, 2009), treatments with nitrogen compounds including NO, plant hormones, sucrose, and H_2O_2 (Trevaskis et al. 1997; Wang et al. 2000; Sakamoto et al. 2004; Shimoda et al. 2005; Qu et al. 2006). The main physiological function of non-symbiotic hemoglobins under oxygen deprivation relies on NO-binding properties of oxyhemoglobin. The functioning of Hb-NO cycle under hypoxia which controls elevated NO levels and utilizes NADH accumulating under hypoxia has been suggested (Igamberdiev et al. 2005, 2010) and further linked to hypoxically functioning mitochondria (Igamberdiev and Hill 2009). Due to the interference with NO signaling, changes and/or manipulation of Hb expression affects many transcriptional and metabolic events which are under NO control (Lee et al. 2011; Hill

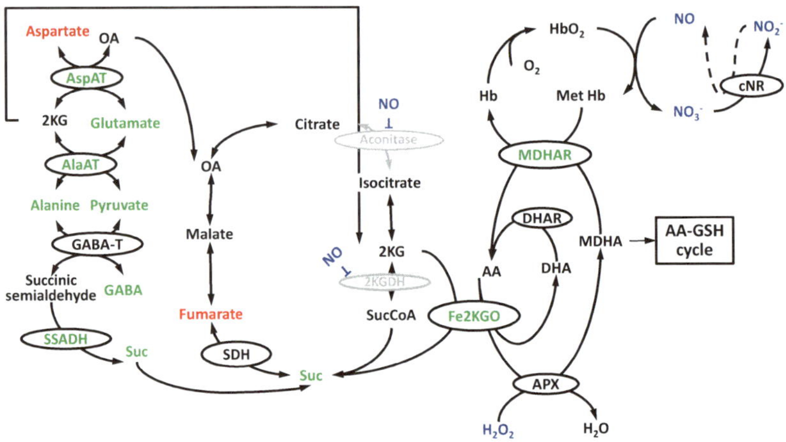

Fig. 1 Hypoxia-induced TCA cycle modification in Arabidopsis shoots: the involvement of Fe^{2+}-dependent ketoglutarate oxidase, GABA shunt, non-symbiotic hemoglobins and NO. NO accumulating under hypoxia is controlled by hypoxically induced non-symbiotic Hbs in an NADH-coupled reaction. To complete the cycle MetHb has to be regenerated by a MetHb reductase. MDHAR supports Hb turnover under hypoxia acting as MetHb reductase (Igamberdiev et al. 2006). In turn, the product of MDHAR reaction, ascorbate, can be oxidized either via the ascorbate–glutathione cycle (operational in the cytosol and mitochondria), or by a novel route suggested by metabolomics and microarray studies under oxygen deprivation. Fe-dependent 2-ketoglutarate oxygenase utilizes 2KG and ascorbate in the presence of oxygen to form a TCA cycle and ETC metabolite, succinate. 2KG needed for the reaction is supplied through the reactions of GABA shunt, which spans the cytosol and mitochondria. These supplementary enzymatic reactions are engaged to feed succinate and oxaloacetate into the TCA cycle bypassing inactivated TCA cycle components, and thus enhance/modify TCA cycle. TCA cycle enzymes are omitted for clarity. *AA* ascorbic acid, *AspAT* aspartate aminotransferase, *AlaAT* alanine aminotransferase, *DHA* dehydroascorbic acid, *GABA-T* GABA transaminase, *OA* oxaloacetate, *Fe2KGO* Fe^{2+}-dependent 2-ketoglutarate oxygenase, *2KG* 2-ketoglutarate, *cNR* cytosolic nitrate reductase, *SSADH* succinic semialdehyde dehydrogenase, *2KGDH* 2-ketoglutarate dehydrogenase, *MetHb* methemoglobin, *MDHAR* monodehydroascorbate reductase. *Green font*—metabolites accumulated and enzymes induced, *red font*—metabolites depleted in Arabidopsis shoots under hypoxia, *gray font*—enzymes inhibited by NO

2012). Hb-sustained NO elimination may have physiological significance also under non-hypoxic conditions via controlling the NO influx into signaling and metabolic events. Such control will affect major stress-related parameters: e.g., the redox state of the cell (NAD(P)H/NAD(P)$^+$, ascorbate/dehydroascorbate, and GSH/GSSG ratios); rate of mitochondrial respiration via direct COX inhibition, and interference with Fe–S TCA cycle enzymes, and may constitute the physiological mechanism for Hb-associated improvement of cellular energy status. Under oxygen deprivation stress Hbs improve the energy status of the cell when MetHb regeneration to Hb is coupled to a cascade of metabolic reactions which result in modification/enhancement of the TCA cycle and incorporates hypoxic metabolites in this adaptive response involving GABA shunt (Fig. 1).

Activation of GABA shunt enzymes and their regulation by hypoxic metabolites is a well-characterized phenomenon during periods with lack of oxygen (Branco-Price et al. 2008; Rocha et al. 2010). In the current study, strong significant upregulation of SSADH (AT1G79440) in hypoxic but not in anoxic arrays was detected. Further evidence for the importance of GABA metabolism controlling oxidative species came from the study of *Arabidopsis* T-DNA knockout mutants of succinic semialdehyde dehydrogenase (SSADH), the mitochondrial enzyme which converts succinic semialdehyde to succinate in the GABA shunt (Bouché et al. 2003). Arabidopsis *ssdh* mutants exhibited a dwarf phenotype, were sensitive to environmental stresses (light, UV-B, heat), and showed enhanced H_2O_2 accumulation (Bouché et al. 2003). The physiological mechanism for GABA shunt-dependent alleviation of oxidative stress can rely on the support of the TCA cycle via succinate and/or NADH and, therefore, decreasing the likelihood of mitochondrial ROS formation (Fait et al. 2005, 2008).

3.3 BP: 0045333: cellular respiration. Relationship between the TCA cycle, mitochondrial electron transport chain, and oxidative stress under hypoxia

The plasticity of the TCA cycle and adjacent metabolic reactions, i.e., the rearrangement of the cycle to a non-cyclic mode in response to stresses including hypoxia, has been recently suggested (Branco-Price et al. 2008; Rocha et al. 2010; Sweetlove et al. 2010). In the latter work hypoxia-induced metabolic changes and inhibition of critical TCA cycle enzymes resulted in adaptive modifications which led to incorporation of hypoxic metabolites into the TCA cycle and production of extra ATP (Rocha et al. 2010). In the study on the rearrangement of mitochondrial protein complexes under oxidative stress, several key metabolites of central carbon metabolism have been reported to associate or dissociate from the protein complexes (Obata et al. 2011). Some of the affected enzymes have been reported also to respond to oxygen deprivation: monodehydroascorbate reductase and alanine aminotransferase (Baxter et al. 2007). Hence, under stress conditions mitochondria as complex dynamic systems respond to the changing environment by metabolic and structural rearrangements. The investigation of expression pattern of TCA cycle and related enzymes in microarray studies revealed either downregulation of the cycle enzymes, or non-significant changes, with significant inhibition of citrate synthase (At2g44350) and isocitrate dehydrogenase (At5g03290). Hence, most of the regulation might occur on post-transcriptional level. Indeed, translational repression of mRNA encoding TCA cycle enzymes has been shown under hypoxia (Branco-Price et al. 2008). Prominent upregulation of the other cellular respiration GO class members, NADH dehydrogenases, was revealed by GSZ analysis and examination of their expression pattern in microarrays. The induction of NADH dehydrogenases occurred irrespective of oxygen levels in the experiment. Induced

NADH dehydrogenases were coded both by nuclear (At2g07711) and mitochondrial genome (Atmg00650, Atmg00060), although the expression of mitochondrial genes cannot be reliably determined with whole genome arrays. Interestingly, the upregulation of At2g07711 and Atmg00060 was strictly attributed to oxygen deprivation, whereas at the onset of reoxygenation these NADH dehydrogenases were downregulated (Branco-Price et al. 2008). The mitochondrial electron transport chain is also an important source of ROS during strongly reduced conditions, such as during oxygen deprivation (Rhoads et al. 2006; Navrot et al. 2007; Hoffman et al. 2007; Blokhina and Fagerstedt 2010b). Several mechanisms can lead to the formation of ROS under unfavorable conditions and, in turn, mitochondrial ROS control NO levels via peroxynitrite (Planchet et al. 2005; Borisjuk et al. 2007; Benamar et al. 2008; Gupta et al. 2011a). The expression of genes and subsequent metabolic changes which affect electron transport, reducing equivalents and substrate supply in mitochondria, will control the rate of ROS/NO formation due to the electron leakage from the electron transport chain, and ultimately the ATP synthesis.

3.4 *Cysteamine dioxygenase activity*

One of the oxygenase classes picked up by analysis in both Affymetrix and Agilent arrays was MF: 0047800: cysteamine dioxygenase activity. The class members catalyze the reaction cysteamine + $O_2 = H^+$ + hypotaurine and are represented by five uncharacterised genes of *Arabidopsis*: AT1G18490, AT2G42670, AT3G58670, AT5G15120, and AT5G39890. The latter two have been associated with the hypoxic response (Mustroph et al. 2010; Branco-Price et al. 2005), regulation of hydrogen peroxide metabolism, salicylic acid signaling, and xylem development. The molecular functions of these proteins in plants are currently unknown. Interestingly, the upregulation of the genes belonging to the GO class "cysteamine dioxygenase activity" was specific for oxygen deprivation. Upon reoxygenation (Branco-Price et al. 2008) this GO class was downregulated and showed the lowest significant score in the analysis (data not shown). In animal tissues hypotaurine is a precursor in taurine biosynthesis (2-aminoethansulfonic acid). Taurine, a sulfur containing amino acid, accumulates to a high level in animal tissues and executes a number of important functions: it can act as antioxidant, as an intracellular osmoregulator, as a neurotransmitter, and can stabilize the membranes and regulate Ca^{2+} entry into the cell (Brosnan and Brosnan 2006). Taurine and hypotaurine metabolism is also closely associated with the enzymatic pathways involving pyruvate and alanine, ketoglutarate and glutamate. However, in plants where taurine content is extremely low, in the range of nmol/gFW (the highest content was found in *Opuntia*, lentil, and red algae) (Kataoka and Ohnishi 1986; Huxtable 1992), it is difficult to predict the physiological role for this metabolite and to ascribe a specific function under oxygen deprivation.

4 Downregulated GO classes under anoxia and hypoxia

The analysis of downregulated GO classes revealed a number of biological processes already known to be inhibited under the lack of oxygen, such as cell wall- and photosynthesis-related classes. None of the GO classes related to formation, processing, and detoxification of ROS were reported as downregulated. In five out of five anoxic conditions in the Affymetrix set the strongest downregulation was detected in the endomembrane system (CC: 0012505), followed by the response to auxin stimulus (BP: 0009733) and the response to hormone stimulus (BP: 0009725). The physiological role for auxin biochemistry depression and particularly auxin binding under anoxia is not completely understood, but the addition of exogenous sucrose has been shown to alleviate anoxia-imposed inhibition on the auxin-related genes (Loreti et al. 2005). It has also been shown that multiple components of auxin-signaling pathway are inhibited by apoplastic ROS (exemplified by O_3 treatment) (Blomster et al. 2011) and that hypoxic stress is accompanied by intensive ROS formation in the apoplastic space (Blokhina et al. 2001). Biological processes downregulated in the hypoxic set (11 conditions in Affymetrix and 4 in Agilent sets) were much more diverse and shared some similarities with anoxic response (CC: 0012505: endomembrane system, BP: 0009733: response to auxin stimulus, BP: 0042430: indole-containing compound metabolic process). The latter class includes genes responsible for, e.g., defense against bacteria and fungi (camalexin biosynthesis), and are represented by a diverse cytochrome P450 group, MYB transcription factors that regulate, e.g., metabolism of aromatic amino acids, and mitogen-activated protein kinases MAPK shown to regulate ethylene signaling pathway downstream the ethylene receptor (Yoo and Sheen 2008). Cytochromes P450 are oxygenases which require molecular oxygen for their functioning and are known to have complicated expression pattern during oxygen deprivation (Lasanthi-Kudahettige et al. 2007). The Affymetrix set of hypoxic experiments also included a number of downregulated GO classes related to biotic stress response and hypoxic metabolism-associated GO classes (also reported as significantly up- or downregulated by GSZ analysis, e.g., BP: 0036294: cellular response to decreased oxygen levels, BP: 0071453: cellular response to oxygen levels, BP: 0071456: cellular response to hypoxia). This fact reflects both the power of statistical approaches to discriminate between induced and repressed genes in the same functional group, and the complexity of response to the lack of oxygen. In the Agilent data set, where hypoxia was performed under light conditions, strong inhibition of photosynthesis was confirmed: the first 20 most downregulated GO classes were related to chloroplast and thylakoid functioning, followed by nucleosome and DNA assembly and organization. Interestingly, there were only eight significantly downregulated GO classes common for Affymetrix and Agilent hypoxic sets: BP: 0009725: response to hormone stimulus; BP: 0009737: response to abscisic acid stimulus; BP: 0009719: response to endogenous stimulus; BP: 0010033: response to organic substance; BP: 0042221:

response to chemical stimulus; BP: 0009628: response to abiotic stimulus; BP: 0007623: circadian rhythm; and BP: 0048511: rhythmic process.

5 Concluding Remarks

In this chapter we have successfully implemented new bioinformatics approaches (Törönen et al. 2009) for the analysis of large sets of global expression data to reveal oxidative stress components in *Arabidopsis* under anoxia and hypoxia on the whole genome level. Response to oxygen deprivation stress-associated ROS and NO formation in terms of gene expression is not straightforward and incorporates the upregulation of some ROS-producing components and antioxidative defense components and relies largely on metabolic rearrangements caused by the lack of oxygen. Judging by the analysis of significantly regulated GO classes, these metabolic alterations include modification of the TCA cycle, incorporation of hypoxic metabolites in efficient energy production, and regulation of mitochondrial electron transport chain as a potent ROS and NO producer.

Interestingly, the analysis returned several significantly regulated GO classes some of which, to our knowledge, have not been discussed previously in connection with oxygen deprivation, for example the downregulated BP: 0007623: circadian rhythm and strongly upregulated MF: 0047800: cysteamine dioxygenase activity. Interestingly, the circadian rhythm classes were reported to be affected in the hypoxic arrays performed under different conditions: most Affymetrix experiments were performed in the dark, whereas Agilent under light conditions. Therefore, it is tempting to speculate that regulation of circadian genes can be part of oxygen deprivation response. The enrichment of a hypoxic set with FeKGO oxygenases suggests a specific role for these enzymes and distinguishes hypoxic and anoxic responses. According to our hypothesis the important role for FeKGO oxygenases lies in the incorporation of hypoxic metabolites into energy production via connection with NO detoxification, MetHb turnover, and ascorbate metabolism. Therefore, this upregulated enzymatic system feeds TCA cycle intermediates succinate and oxaloacetate into the cycle bypassing inactivated TCA cycle components, and thus improves energy production under these harsh conditions.

Future research is needed in the dissection of redundant metabolic adjustments which involve the TCA cycle, GABA shunt, glycolysis, FeKGO oxygenases, and ROS/NO chemistry. These responses seem to vary between plant species and depend on the degree of oxygen deprivation. The plasticity of the TCA cycle can be one of the adaptation mechanisms.

None of the GO classes related to formation, processing, and detoxification of ROS were reported as downregulated, emphasizing the fact that oxidative stress is an integral part of oxygen deprivation response. Accumulation of ROS, especially in the apoplastic space, may underlie the downregulation of auxin signaling under the lack of oxygen and provide further evidence for the signaling roles of ROS. In this chapter we have shown that ROS play an essential role in the formation of hypoxic and anoxic response incorporating gene regulation and oxidative stress as integral parts of this response.

Acknowledgments We wish to thank Julia Bailey-Serres, Joost van Dongen, Pierdomenico Perata, and Chiara Pucciariello for supplying the microarray datasets for our analysis. This work was supported by the Academy of Finland grant no. 123826.

References

Asif M, Trivedi P, Misra P, Nath P (2009) Prolyl-4-hydroxylase (AtP4H1) mediates and mimics low oxygen response in Arabidopsis thaliana. Funct Integr Genomics 9:525–535

Bailey-Serres J, Chang R (2005) Sensing and signalling in response to oxygen deprivation in plants and other organisms. Ann Bot (Lond) 96:507–518

Bailey-Serres J, Voesenek LACJ (2008) Flooding stress: acclimations and genetic diversity. Annu Rev Plant Biol 59:313–339

Banti V, Mafessoni F, Loreti E, Alpi A, Perata P (2010) The heat-inducible transcription factor HsfA2 enhances anoxia tolerance in Arabidopsis. Plant Physiol 152:1471–1483

Baxter CJ, Redestig H, Schauer N, Repsilber D, Patil KR, Nielsen J, Selbig J, Liu J, Fernie AR, Sweetlove LJ (2007) The metabolic response of heterotrophic Arabidopsis cells to oxidative stress. Plant Physiol 143:312–325

Baxter-Burrell A, Yang Z, Springer PS, Bailey-Serres J (2002) RopGAP4-dependent Rop GTPase rheostat control of Arabidopsis oxygen deprivation tolerance. Science 296:2026–2028

Benamar A, Rolletschek H, Borisjuk L, Avelange-Macherel M, Curien G, Mostefai HA, Andriantsitohaina R, Macherel D (2008) Nitrite–nitric oxide control of mitochondrial respiration at the frontier of anoxia. Biochim Biophys Acta—Bioenergetics 1777:1268–1275

Berra E, Ginouves A, Pouyssegur J (2006) The hypoxia-inducible-factor hydroxylases bring fresh air into hypoxia signalling. EMBO Rep 7:41–45

Biemelt S, Keetman U, Albrecht G (1998) Re-aeration following hypoxia or anoxia leads to activation of the antioxidative defense system in roots of wheat seedlings. Plant Physiol 116:651–658

Biemelt S, Keetman U, Mock H, Grimm B (2000) Expression and activity of isoenzymes of superoxide dismutase in wheat roots in response to hypoxia and anoxia. Plant Cell Environ 23:135–144

Blokhina O, Fagerstedt KV (2010a) Oxidative metabolism, ROS and NO under oxygen deprivation. Plant Physiol Biochem 48:359–373

Blokhina O, Fagerstedt KV (2010b) Reactive oxygen species and nitric oxide in plant mitochondria: origin and redundant regulatory systems. Physiol Plant 138:447–462

Blokhina OB, Virolainen E, Fagerstedt KV, Hoikkala A, Wähälä K, Chirkova TV (2000) Antioxidant status of anoxia-tolerant and -intolerant plant species under anoxia and reaeration. Physiol Plant 109:396–403

Blokhina OB, Chirkova TV, Fagerstedt KV (2001) Anoxic stress leads to hydrogen peroxide formation in plant cells. J Exp Bot 52:1179–1190

Blokhina O, Virolainen E, Fagerstedt KV (2003) Antioxidants, oxidative damage and oxygen deprivation stress: a review. Ann Bot (Lond) 91:179–194

Blomster T, Salojärvi J, Sipari N, Brosché M, Ahlfors R, Keinänen M, Overmyer K, Kangasjärvi J (2011) Apoplastic reactive oxygen species transiently decrease auxin signaling and cause stress-induced morphogenic response in Arabidopsis. Plant Physiol 157:1866–1883

Borisjuk L, Macherel D, Benamar A, Wobus U, Rolletschek H (2007) Low oxygen sensing and balancing in plant seeds: a role for nitric oxide. New Phytol 176:813–823

Bouché N, Fait A, Bouchez D, Møller SG, Fromm H (2003) Mitochondrial succinic-semialdehyde dehydrogenase of the γ-aminobutyrate shunt is required to restrict levels of reactive oxygen intermediates in plants. Proc Natl Acad Sci U S A 100:6843–6848

Branco-Price C, Kawaguchi R, Ferreira RB, Bailey-Serres J (2005) Genome-wide analysis of transcript abundance and translation in Arabidopsis seedlings subjected to oxygen deprivation. Ann Bot (Lond) 96:647–660

Branco-Price C, Kaiser KA, Jang CJH, Larive CK, Bailey-Serres J (2008) Selective mRNA translation coordinates energetic and metabolic adjustments to cellular oxygen deprivation and reoxygenation in *Arabidopsis thaliana*. Plant J 56:743–755

Brosnan JT, Brosnan ME (2006) The sulfur-containing amino acids: an overview. J Nutr 136:1636S–1640S

Cela J, Falk J, Munné-Bosch S (2009) Ethylene signaling may be involved in the regulation of tocopherol biosynthesis in Arabidopsis thaliana. FEBS Lett 583:992–996

Chang R, Jang C, Branco-Price C, Nghiem P, Bailey-Serres J (2012) Transient MPK6 activation in response to oxygen deprivation and reoxygenation is mediated by mitochondria and aids seedling survival in *Arabidopsis*. Plant Mol Biol 78:109–122

Dasgupta K, Ganesan S, Manivasagam S, Ayre B (2011) A cytochrome P450 monooxygenase commonly used for negative selection in transgenic plants causes growth anomalies by disrupting brassinosteroid signaling. BMC Plant Biol 11:67

Dordas C (2009) Nonsymbiotic hemoglobins and stress tolerance in plants. Plant Sci 176:433–440

Dordas C, Rivoal J, Hill RD (2003) Plant haemoglobins, nitric oxide and hypoxic stress. Ann Bot (Lond) 91:173–178

Ederli L, Morettini R, Borgogni A, Wasternack C, Miersch O, Reale L, Ferranti F, Tosti N, Pasqualini S (2006) Interaction between nitric oxide and ethylene in the induction of alternative oxidase in ozone-treated tobacco plants. Plant Physiol 142:595–608

Efron B, Tibshirani R (2007) On testing the significance of sets of genes. Ann Appl Stat 1:107–129

Fait A, Yellin A, Fromm H (2005) GABA shunt deficiencies and accumulation of reactive oxygen intermediates: insight from Arabidopsis mutants. FEBS Lett 579:415–420

Fait A, Fromm H, Walter D, Galili G, Fernie AR (2008) Highway or byway: the metabolic role of the GABA shunt in plants. Trends Plant Sci 13:14–19

Fukao T, Bailey-Serres J (2004) Plant responses to hypoxia—is survival a balancing act? Trends Plant Sci 9:449–456

Fukao T, Bailey-Serres J (2008) Ethylene—a key regulator of submergence responses in rice. Plant Sci 175:43–51

Gentleman RC, Carey VJ, Bates DM, Bolstad B, Dettling M, Dudoit S, Ellis B, Gautier L, Ge Y, Gentry J, Hornik K, Hothorn T, Huber W, Iacus S, Irizarry R, Leisch F, Li C, Maechler M, Rossini AJ, Sawitzki G, Smith C, Smyth G, Tierney L, Yang JY, Zhang J (2004) Bioconductor: open software development for computational biology and bioinformatics. Genome Biol 5:R80

Gonzali S, Loreti E, Novi G, Poggi A, Alpi A, Perata P (2005) The use of microarrays to study the anaerobic response in Arabidopsis. Ann Bot (Lond) 96:661–668

Greenway H, Gibbs J (2003) *Review*: mechanisms of anoxia tolerance in plants. II. Energy requirements for maintenance and energy distribution to essential processes. Funct Plant Biol 30:999–1036

Gupta KJ, Zabalza A, van Dongen JT (2009) Regulation of respiration when the oxygen availability changes. Physiol Plant 137:383–391

Gupta KJ, Igamberdiev AU, Manjunatha G, Segu S, Moran JF, Neelawarne B, Bauwe H, Kaiser WM (2011a) The emerging roles of nitric oxide (NO) in plant mitochondria. Plant Sci 181:520–526

Gupta KJ, Hebelstrup KH, Mur LAJ, Igamberdiev AU (2011b) Plant hemoglobins: important players at the crossroads between oxygen and nitric oxide. FEBS Lett 585:3843–3849

Gupta KJ, Shah JK, Brotman Y, Jahnke K, Willmitzer L, Kaiser WM, Bauwe H, Igamberdiev AU (2012) Inhibition of aconitase by nitric oxide leads to induction of the alternative oxidase and to a shift of metabolism towards biosynthesis of amino acids. J Exp Bot 63:1773–1784

Hebelstrup KH, Igamberdiev AU, Hill RD (2007) Metabolic effects of hemoglobin gene expression in plants. Gene 398:86–93

Hebelstrup KH, van Zanten M, Mandon J, Voesenek LACJ, Harren FJM, Cristescu SM, Møller IM, Mur LAJ (2012) Haemoglobin modulates NO emission and hyponasty under hypoxia-related stress in Arabidopsis thaliana. J Exp Bot 63:5581–5591

Hill RD (2012) Non-symbiotic haemoglobins—what's happening beyond nitric oxide scavenging? AoB Plants 2012:pls004

Hoffman DL, Salter JD, Brookes PS (2007) Response of mitochondrial reactive oxygen species generation to steady-state oxygen tension: implications for hypoxic cell signaling. Am J Physiol Heart Circ Physiol 292:H101–H108

Huang S, Colmer TD, Millar AH (2008) Does anoxia tolerance involve altering the energy currency towards PPi? Trends Plant Sci 13:221–227

Hughes RK, De Domenico S, Santino A (2009) Plant cytochrome CYP74 family: biochemical features, endocellular localisation, activation mechanism in plant defence and improvements for industrial applications. ChemBioChem 10:1122–1133

Huxtable RJ (1992) Physiological actions of taurine. Physiol Rev 72:101–163

Igamberdiev AU, Hill RD (2009) Plant mitochondrial function during anaerobiosis. Ann Bot (Lond) 103:259–268

Igamberdiev AU, Baron K, Manac'h-Little N, Stoimenova M, Hill RD (2005) The haemoglobin/nitric oxide cycle: involvement in flooding stress and effects on hormone signalling [Review]. Ann Bot (Lond) 96:557–564

Igamberdiev AU, Bykova NV, Hill RD (2006) Nitric oxide scavenging by barley hemoglobin is facilitated by a monodehydroascorbate reductase-mediated ascorbate reduction of methemoglobin. Planta 223:1033–1040

Igamberdiev AU, Bykova NV, Shah JK, Hill RD (2010) Anoxic nitric oxide cycling in plants: participating reactions and possible mechanisms. Physiol Plant 138:393–404

Inzé A, Vanderauwera S, Hoeberichts FA, vanDorpe M, van Gaever T, van Breusegem F (2012) A subcellular localization compendium of hydrogen peroxide-induced proteins. Plant Cell Environ 35:308–320

Jaspers P, Kangasjärvi J (2010) Reactive oxygen species in abiotic stress signaling. Physiol Plant 138(4):405–413

Johnson WE, Li C, Rabinovic A (2007) Adjusting batch effects in microarray expression data using empirical Bayes methods. Biostatistics 8:118–127

Kataoka H, Ohnishi N (1986) Occurrence of taurine in plants. Agric Biol Chem 50(7):1887–1888

Lasanthi-Kudahettige R, Magneschi L, Loreti E, Gonzali S, Licausi F, Novi G, Beretta O, Vitulli F, Alpi A, Perata P (2007) Transcript profiling of the anoxic rice coleoptile. Plant Physiol 144:218–231

Lee SC, Mustroph A, Sasidharan R, Vashisht D, Pedersen O, Oosumi T, Voesenek LACJ, Bailey-Serres J (2011) Molecular characterization of the submergence response of the Arabidopsis thaliana ecotype Columbia. New Phytol 190:457–471

Lewis DFV (2002) Oxidative stress: the role of cytochromes P450 in oxygen activation. J Chem Technol Biotechnol 77:1095–1100

Li C, Chen Q, Gao X, Qi B, Chen N, Xu S, Chen J, Wang X (2005) AtHsfA2 modulates expression of stress responsive genes and enhances tolerance to heat and oxidative stress in Arabidopsis. Sci China C Life Sci 48:540–550

Licausi F (2011) Regulation of the molecular response to oxygen limitations in plants. New Phytol 190:550–555

Licausi F, Van Dongen JT, Giuntoli B, Novi G, Santaniello A, Geigenberger P, Perata P (2010) HRE1 and HRE2, two hypoxia-inducible ethylene response factors, affect anaerobic responses in Arabidopsis thaliana. Plant J 62:302–315

Lindermayr C, Saalbach G, Bahnweg G, Durner J (2006) Differential inhibition of Arabidopsis methionine adenosyltransferases by protein S-nitrosylation. J Biol Chem 281:4285–4291

Loenarz C, Schofield CJ (2008) Expanding chemical biology of 2-oxoglutarate oxygenases. Nat Chem Biol 4:152–156
Loreti E, Poggi A, Novi G, Alpi A, Perata P (2005) A genome-wide analysis of the effects of sucrose on gene expression in Arabidopsis seedlings under anoxia. Plant Physiol 137:1130–1138
Luhua S, Ciftci-Yilmaz S, Harper J, Cushman J, Mittler R (2008) Enhanced tolerance to oxidative stress in transgenic Arabidopsis plants expressing proteins of unknown function. Plant Physiol 148:280–292
Manac'h-Little N, Igamberdiev AU, Hill RD (2005) Hemoglobin expression affects ethylene production in maize cell cultures. Plant Physiol Biochem 43:485–489
Mittler R, Vanderauwera S, Gollery M, Van Breusegem F (2004) Reactive oxygen gene network of plants. Trends Plant Sci 9:490–498
Mustroph A, Lee SC, Oosumi T, Zanetti ME, Yang H, Ma K, Yaghoubi-Masihi A, Fukao T, Bailey-Serres J (2010) Cross-kingdom comparison of transcriptomic adjustments to low-oxygen stress highlights conserved and plant-specific responses. Plant Physiol 152:1484–1500
Narsai R, Rocha M, Geigenberger P, Whelan J, van Dongen JT (2011) Comparative analysis between plant species of transcriptional and metabolic responses to hypoxia. New Phytol 190 (2):472–487
Navarre DA, Wendehenne D, Durner J, Noad R, Klessig DF (2000) Nitric oxide modulates the activity of tobacco aconitase. Plant Physiol 122:573–582
Navrot N, Rouhier N, Gelhaye E, Jacquot J (2007) Reactive oxygen species generation and antioxidant systems in plant mitochondria. Physiol Plant 129:185–195
Nelson D, Werck-Reichhart D (2011) A P450-centric view of plant evolution. Plant J 66:194–211
Newton MA, Quintana FA, den Boon JA, Sengupta S, Ahlquist P (2007) Random-set methods identify distinct aspects of the enrichment signal in gene-set analysis. Ann Appl Stat 1:85–106
Obata T, Matthes A, Koszior S, Lehmann M, Araújo WL, Bock R, Sweetlove LJ, Fernie AR (2011) Alteration of mitochondrial protein complexes in relation to metabolic regulation under short-term oxidative stress in Arabidopsis seedlings. Phytochemistry 72:1081–1091
Petrov VD, Van Breusegem F (2012) Hydrogen peroxide—a central hub for information flow in plant cells. AoB Plants 2012:pls014
Planchet E, Jagadis Gupta K, Sonoda M, Kaiser WM (2005) Nitric oxide emission from tobacco leaves and cell suspensions: rate limiting factors and evidence for the involvement of mitochondrial electron transport. Plant J 41:732–743
Potters G, Pasternak TP, Guisez Y, Jansen MAK (2009) Different stresses, similar morphogenic responses: integrating a plethora of pathways. Plant Cell Environ 32:158–169
Pucciariello C, Parlanti S, Banti V, Novi G, Perata P (2012) Reactive oxygen species-driven transcription in Arabidopsis under oxygen deprivation. Plant Physiol 159:184–196
Qiao W, Fan L (2008) Nitric oxide signaling in plant responses to abiotic stresses. J Integr Plant Biol 50:1238–1246
Qu Z, Zhong N, Wang H, Chen A, Jian G, Xia G (2006) Ectopic expression of the cotton non-symbiotic hemoglobin gene GhHbd1 triggers defense responses and increases disease tolerance in Arabidopsis. Plant Cell Physiol 47:1058–1068
Rhoads DM, Umbach AL, Subbaiah CC, Siedow JN (2006) Mitochondrial reactive oxygen species. Contribution to oxidative stress and interorganellar signaling. Plant Physiol 141:357–366
Ritchie ME, Silver J, Oshlack A, Holmes M, Diyagama D, Holloway A, Smyth GK (2007) A comparison of background correction methods for two-colour microarrays. Bioinformatics 23:2700–2707
Rocha M, Licausi F, Araujo WL, Nunes-Nesi A, Sodek L, Fernie AR, van Dongen JT (2010) Glycolysis and the tricarboxylic acid cycle are linked by alanine aminotransferase during hypoxia induced by waterlogging of Lotus japonicus. Plant Physiol 152:1501–1513

Sairam R, Kumutha D, Ezhilmathi K, Deshmukh P, Srivastava G (2008) Physiology and biochemistry of waterlogging tolerance in plants. Biol Plant 52:401–412

Sakamoto A, Sakurao S, Fukunaga K, Matsubara T, Ueda-Hashimoto M, Tsukamoto S, Takahashi M, Morikawa H (2004) Three distinct Arabidopsis hemoglobins exhibit peroxidase-like activity and differentially mediate nitrite-dependent protein nitration. FEBS Lett 572:27–32

Santosa I, Ram P, Boamfa E, Laarhoven L, Reuss J, Jackson M, Harren F (2007) Patterns of peroxidative ethane emission from submerged rice seedlings indicate that damage from reactive oxygen species takes place during submergence and is not necessarily a post-anoxic phenomenon. Planta 226:193–202

Shimoda Y, Nagata M, Suzuki A, Abe M, Sato S, Kato T, Tabata S, Higashi S, Uchiumi T (2005) Symbiotic rhizobium and nitric oxide induce gene expression of non-symbiotic hemoglobin in Lotus japonicus. Plant Cell Physiol 46:99–107

Smyth GK (2005) LIMMA: linear models for microarray data. In: Bioinformatics and Computational Biology Solutions Using R and Bioconductor; Gentleman R, Carey V, Huber W, Irizarry R, Dudoit S (eds) Springer, New York, pp 397–420

Stoimenova M, Igamberdiev AU, Gupta K, Hill RD (2007) Nitrite-driven anaerobic ATP synthesis in barley and rice root mitochondria. Planta 226:465–474

St-Pierre J, Brand MD, Boutilier RG (2000) Mitochondria as ATP consumers: cellular treason in anoxia. Proc Natl Acad Sci U S A 97:8670–8674

Sweetlove LJ, Beard KFM, Nunes-Nesi A, Fernie AR, Ratcliffe RG (2010) Not just a circle: flux modes in the plant TCA cycle. Trends Plant Sci 15:462–470

Törönen P, Ojala P, Marttinen P, Holm L (2009) Robust extraction of functional signals from gene set analysis using a generalized threshold free scoring function. BMC Bioinformatics 10:307

Trevaskis B, Watts RA, Andersson CR, Llewellyn DJ, Hargrove MS, Olson JS, Dennis ES, Peacock WJ (1997) Two hemoglobin genes in Arabidopsis thaliana: the evolutionary origins of leghemoglobins. Proc Natl Acad Sci U S A 94:12230–12234

Van Breusegem F, Bailey-Serres J, Mittler R (2008) Unraveling the tapestry of networks involving reactive oxygen species in plants. Plant Physiol 147:978–984

van Dongen JT, Frohlich A, Ramirez-Aguilar SJ, Schauer N, Fernie AR, Erban A, Kopka J, Clark J, Langer A, Geigenberger P (2009) Transcript and metabolite profiling of the adaptive response to mild decreases in oxygen concentration in the roots of Arabidopsis plants. Ann Bot (Lond) 103:269–280

Wang R, Guegler K, LaBrie ST, Crawford NM (2000) Genomic analysis of a nutrient response in Arabidopsis reveals diverse expression patterns and novel metabolic and potential regulatory genes induced by nitrate. Plant Cell Online 12:1491–1509

Wulff A, Oliveira HC, Saviani EE, Salgado I (2009) Nitrite reduction and superoxide-dependent nitric oxide degradation by Arabidopsis mitochondria: influence of external NAD(P)H dehydrogenases and alternative oxidase in the control of nitric oxide levels. Nitric Oxide 21:132–139

Yan B, Dai Q, Liu X, Huang S, Wang Z (1996) Flooding-induced membrane damage, lipid oxidation and activated oxygen generation in corn leaves. Plant Soil 179:261–268

Yoo S, Sheen J (2008) MAPK signaling in plant hormone ethylene signal transduction. Plant Signal Behav 3:848–849

Yun B, Feechan A, Yin M, Saidi NBB, Le Bihan T, Yu M, Moore JW, Kang J, Kwon E, Spoel SH, Pallas JA, Loake GJ (2011) S-nitrosylation of NADPH oxidase regulates cell death in plant immunity. Nature 478:264–268

Zabalza A, van Dongen JT, Froehlich A, Oliver SN, Faix B, Gupta KJ, Schmalzlin E, Igal M, Orcaray L, Royuela M, Geigenberger P (2009) Regulation of respiration and fermentation to control the plant internal oxygen concentration. Plant Physiol 149:1087–1098

Zhao L, Gu R, Gao P, Wang G (2008) A nonsymbiotic hemoglobin gene from maize, ZmHb, is involved in response to submergence, high-salt and osmotic stresses. Plant Cell Tissue Organ Cult 95:227–237

Zhao M, Chen L, Zhang L, Zhang W (2009) Nitric reductase-dependent nitric oxide production is involved in cold acclimation and freezing tolerance in Arabidopsis. Plant Physiol 151:755–767

Low Oxygen Stress, Nonsymbiotic Hemoglobins, NO, and Programmed Cell Death

Abir U. Igamberdiev, Claudio Stasolla, and Robert D. Hill

Abstract Class 1 nonsymbiotic hemoglobins are upregulated during oxygen deprivation in plant cells. Their expression, under conditions of low oxygen stress, has been shown to help maintain energy and redox status, increase the capacity for removing reactive oxygen and nitrogen species, and improve the capability of the plant to withstand the stress. They are efficient scavengers of NO that is produced in significant quantities during oxygen deprivation. NO is a known factor involved in triggering programmed cell death in both plants and animals. It is hypothesized that nonsymbiotic hemoglobins regulate the onset of programmed cell death by scavenging NO. The expression, or non-expression, of a nonsymbiotic hemoglobin within a specific cell would, therefore, determine whether the cell proceeds towards programmed cell death in situations where NO is a factor triggering the process.

Nonsymbiotic plant hemoglobins (nsHbs) have been associated with oxygen deprivation since their discovery in 1994 (Taylor et al. 1994). Physiological research of nsHbs has continued to focus on their role in oxygen deprivation, but interest has also developed in their possible roles in plant disease responses and other plant developmental characteristics (Hill 2012). In this chapter, we will review some of the properties of nsHbs, examine what is known about their physiological function, and propose a mechanism by which they may be involved in programmed cell death (PCD).

A.U. Igamberdiev
Department of Biology, Memorial University of Newfoundland, St. John's, NL, CanadaA1B 3X9,

C. Stasolla • R.D. Hill (✉)
Department of Plant Science, University of Manitoba, Winnipeg, MB, CanadaR3T 2N2,
e-mail: Rob.Hill@umanitoba.ca

J.T. van Dongen and F. Licausi (eds.), *Low-Oxygen Stress in Plants*, Plant Cell Monographs 21, DOI 10.1007/978-3-7091-1254-0_3, © Springer-Verlag Wien 2014

1 Properties of Plant Hemoglobins

Plant hemoglobins have been grouped into three classes: class 1, class 2, and truncated (class 3). In class 2 and, to a lesser extent, in class 1, there is an evolutionary trend from nonsymbiotic hexacoordinated to symbiotic pentacoordinated hemoglobins (Gupta et al. 2011; Vazquez-Limon et al. 2012; Vinogradov et al. 2011).

Nonsymbiotic class 1 hemoglobins (nsHb1) are distinguished from other classes by a low value of the hexacoordination equilibrium constant (K_H). This constant reflects the extent of the interaction of the distal histidine with the heme iron, and describes the equilibrium of pentacoordinated and hexacoordinated species that influence the binding of ligands. The weak hexacoordination of histidine in class 1 hemoglobins causes an extremely high avidity of the hemoglobin for oxygen (Hargrove et al. 2000) resulting in an oxyhemoglobin dissociation constant in the nanomolar range. Rapid oxygen binding results in protein conformational changes that prevent fast oxygen dissociation from the heme site. Upon ligand binding, the distal histidine moves away from the iron atom resulting in an overall more stable conformation (Hoy et al. 2007). This allows a very tight, slowly reversible binding of O_2, which is optimal for oxygen-dependent NO scavenging under near-anaerobic conditions (Igamberdiev and Hill 2004; Igamberdiev et al. 2006b). Steric interactions encourage the formation of nitrate in the reaction of oxyhemoglobin and NO (Nienhaus et al. 2010), along with the accompanying oxidation of hemoglobin (ferrous) to methemoglobin (ferric). This, possibly the main function of nonsymbiotic class 1 hemoglobins, cannot be efficient without an associated reductase that converts the (ferric) methemoglobin back to the ferrous form (Igamberdiev et al. 2006a). One such reductase, a monodehydroascorbate reductase, has been shown to perform this reaction (Igamberdiev et al. 2006a), although there are likely several other plant reductases capable of undertaking the reaction.

Structurally the nonsymbiotic class 1 hemoglobins are protein dimers consisting of two identical subunits, containing one conserved cysteine residue per monomer, with the exception of the Arabidopsis class 1 hemoglobin that has an additional cysteine residue adjacent to the conserved cysteine (Hargrove et al. 2000). The dimer structure may be stabilized through formation of a disulfide bridge (Bykova et al. 2006). In the reduced state, the cysteine may also contribute to maintaining the heme iron in the ferrous form (Igamberdiev et al. 2011). The structural properties of class 1 hemoglobins allow them to serve as soluble electron transport proteins in an enzymatic system scavenging nitric oxide (NO) produced under low oxygen conditions primarily via reduction of nitrite in plants (Dordas et al. 2003). Functionally, in terms of participation in electron transport, class 1 hemoglobins resemble cytochromes more than other hemoglobin classes, the main distinction of class 1 hemoglobins from cytochromes being their solubility, compared to cytochromes anchored to membranes.

As noted earlier, upregulation of nsHb1 barley hemoglobin occurs under hypoxic conditions (Taylor et al. 1994). Increased transcription can be detected as the oxygen atmosphere approaches 5 % and transcription initiates within the first hour of exposure to low oxygen. nsHb transcription, while associated with oxygen deprivation, is not triggered by directly sensing a decline in oxygen. Increased transcription can be observed under normal atmospheric oxygen in the presence of mitochondrial oxidative phosphorylation uncouplers, electron transport inhibitors, or an ATP synthase inhibitor, leading to the possibility that the availability of ATP is a determining factor in nsHb regulation (Nie and Hill 1997). Ca^{2+} also appears to be a factor in nsHb upregulation (Nie et al. 2006) and is released from mitochondria during hypoxia (Subbaiah et al. 1998).

NO is produced when plants are exposed to oxygen deprivation (Dordas et al. 2003, 2004; Perazzolli et al. 2004). NO is barely detectable in aerobic alfalfa root cultures but reaches 120 nmol g^{-1} of fresh weight after 24 h of hypoxia. Down-regulating nsHb expression by anti-sensing increases the levels of detectable NO by at least 1.5-fold, while over-expressing nsHb1 decreases the detectable NO levels by about 50 %. This reaction in which oxyhemoglobin reacts with NO to produce nitrate likely represents the main mechanism by which NO is removed from plants (Igamberdiev et al. 2006a). Because nsHbs1 bind oxygen very tightly (K_d ~ 2 nM) they are capable of scavenging NO at oxygen levels far below oxygen saturation of cytochrome c oxidase (K_m ~ 150 nM), while the reverse reaction of production of NO from nitrite by these hemoglobins can take place only at less than nanomolar oxygen concentrations (Sturms et al. 2011), which are unlikely to be reached even in oxygen-depleted tissues.

Roots are the primary organs experiencing hypoxia and one hypoxia-avoidance mechanism used by plants is the development of aerenchyma, which serves to not only facilitate O_2 diffusion but also to reduce the number of O_2 consuming cells (Drew et al. 2000). A link between aerenchyma formation, NO presence, and nsHb expression has been demonstrated in at least one species (Dordas et al. 2003). NO is known to affect production of ethylene (Mishina et al. 2007; Mur et al. 2008). Transduction of an ethylene signal is a required component for PCD during aerenchyma formation (He et al. 1996) and there is evidence that an ethylene response factor (ERF) may act as an oxygen sensor in plants (Licausi et al. 2011). Ethylene levels in an Hb-down-regulating maize cell line were 5–6.5 times higher than the levels in Hb-over-expressing line and 4–5 times higher than the levels in the wild type (Manac'h-Little et al. 2005). The activity of the ethylene-producing enzyme ACC oxidase in the Hb-down-regulating line increased under hypoxic conditions and upon treatment with NO under normoxic conditions. It was suggested that limiting class-1 hemoglobin synthesis increases ethylene formation in maize suspension cells via the modulation of NO levels.

In contrast to class 1 nsHbs, the class 2 nsHbs are usually not induced under hypoxic conditions but their over-expression can promote survival under hypoxia (Kakar et al. 2010). Class 2 nsHbs are characterized by a tighter hexacoordination than class 1 nsHbs and thus they have lower oxygen affinities (K_d in the order of 100–200 nM). This makes them less efficient in NO scavenging but increases the possibility of functions related to sensing low levels of oxygen and to oxygen storage and diffusion (Kakar et al. 2010). The oxygen binding characteristics of class 2 hemoglobins are similar to those of leghemoglobin, which has been categorized as a symbiotic class 2 hemoglobin. Its oxygen affinity is such that it could facilitate oxygen diffusion for mitochondrial respiration via cytochrome oxidase (Spyrakis et al. 2011; Watts et al. 2001). While class 1 Hbs possess two docking sites for small ligands facilitating the reaction between NO and O_2, class 2 Hbs possess only a single ligand docking site, making the NO dioxygenase reaction for these Hbs less favorable (Vigeolas et al. 2011).

Class 3 plant Hbs are truncated (2-on-2 structure) but the length of polypeptide chain is longer than for class 1 and 2 hemoglobins (Watts et al. 2001). *Arabidopsis thaliana* truncated Hb exhibits unusual concentration-independent binding of O_2 and CO. The protein, which is pentacoordinate in the oxygenated state, forms a transient hexacoordinate structure after reduction and deoxygenation, which slowly converts to a five-coordinate structure. The truncated Hb is expressed throughout the plant but responds to none of the treatments that induce plant 3-on-3 (class 1 and 2) Hbs. It has been suggested that because of a lower O_2 affinity the truncated Hb in Arabidopsis might be an O_2 transport protein with a moderate O_2 affinity (Watts et al. 2001).

Since there are hemoglobins in most biological systems that are capable of efficiently oxygenating NO and NO is a recognized agent in PCD, we suggest that plant nsHbs are factors that regulate the extent of PCD and location at which it occurs within an organ or tissue. In essence, we are proposing that the levels of NO, in relationship to the availability of oxyhemoglobin, within the cell determine whether the cell proceeds towards PCD. While there is no available evidence to support this hypothesis, in the remaining part of this chapter we will attempt to gather the references that suggest this is a viable and testable hypothesis.

2 PCD in Plants

PCD is an important component of plant development and a response to abiotic stress, including oxygen deprivation. Like animal cells which can engage distinct mechanisms of apoptosis (Blaise et al. 2005), the program dismantling plant cells is characterized by elaborate morphological and biochemical events. In plants, PCD can be executed through different mechanisms (van Doorn et al. 2011) although vacuolar cell death appears to be the most prominent event triggered by oxygen deprivation.

2.1 Ultrastructural Characteristics of Vacuolar Cell Death

Plant cells contain two types of vacuoles: storage vacuoles accumulating mainly proteins not possessing enzymatic activity, and lytic vacuoles containing a large array of vacuolar protein enzymes (VPEs) including hydrolytic enzymes, such as nucleases, as well as aspartate and cysteine proteases (van Doorn et al. 2011). Vacuolar cell death involves the participation of the lytic vacuoles, and can manifest itself in either a non-disruptive way, through the fusion of the lytic vacuolar membrane (tonoplast) to the plasma membrane, or a disruptive way, through the collapse of the vacuolar membrane within the cytoplasm (Hara-Nishimura and Hatsugai 2011). While the former mechanism results in the discharge of the VPE in the extracellular matrix, the latter type of vacuolar cell death involves the collapse of the tonoplast and the release of the VPEs into the cytosol. This process is executed through three steps, which have also been described during lysogenous aerenchyma formation in plants grown under hypoxia (reviewed by Evans 2003). The first step is associated with an apparent increase in size of the lytic vacuoles until only a reduced layer of cytoplasm remains. This is followed by small invaginations of the tonoplast which sequester portions of the cytoplasm and degrade them in the vacuolar lumen (van Doorn et al. 2011), through processes analogous to micro- and macro-autophagy of animal cells (Kundu and Thompson 2005). Shrinkage of the plasma membrane from the cell wall, as well as the presence of granular staining within the vacuole and organelles surrounded by the tonoplast, was observed in maize root treated with 3 % oxygen (Gunawardena et al. 2001; Evans 2003). These membrane-bound structures resemble apoptotic bodies in animal cells. The final step involves the permeabilization and rupture of the tonoplast and the release of the VPE, which results in the further degradation of cellular components and organelles, starting with the endoplasmic reticulum, followed lastly by mitochondria and the nucleus (van Doorn et al. 2011). While all these events are often observed during oxygen deprivation-induced PCD, variations in the death program sometime occur. During the early stages of aerenchyma formation in rice, Webb and Jackson (1986) documented the disruption of cell wall components preceding the lysis of the vacuole, whereas Inada et al. (2002) reported that, besides the rupture of the tonoplast, dying aerenchyma cells did not show any distinct morphological feature from the surrounding living cells.

Cytological hallmarks of apoptosis and PCD, including hypoxia-induced death, occur in the nucleus and include the degradation of DNA, chromatin condensation, and nuclear fragmentation. These events facilitate death by preventing DNA replication and transcription (Arends et al. 1990). Systemic fragmentation of DNA is executed by specific nucleases in two distinct steps (Schwartzman and Cidlowski 1993). In the early stages of PCD, fragments of about 50–300 kbp are produced by cleavage of the DNA at the interloop sites of the chromatin at the inter-rosette (Peitsch et al. 1993). Cleavage at the internucleosomal sites during advanced stages of PCD, also mediated by the endonuclease enzyme DNase1 (Peitsch et al. 1993), produces smaller oligonucleosomal DNA fragments (about 200 bp). While both steps are executed in some plant systems (Fojtová and Kovarík 2000; Young and Gallie 2000), only specific fragment sizes are detected in others. This latter case applies to hypoxia-induced PCD in *Pisum sativum* roots where only larger fragments (300 kbp) were observed. In a few instances, varying degrees of DNA fragmentation are reported, and this is consistent with the fact that vacuolar nucleases act non-selectively in the digestion of nucleosomes (reviewed by

Hara-Nishimura and Hatsugai 2011). DNA fragmentation occurs concomitantly to the condensation of chromatin, which requires the depolymerization of nuclear F-actin (Widlak and Garrard 2005). During aerenchyma formation of maize exposed to 3 % oxygen treatments, chromatic condensation is also accompanied by its redistribution along the periphery of the nuclei (Gunawardena et al. 2001). Unlike animal cells, where nuclear fragmentation is typical of apoptosis and uncoupled from DNA degradation, nuclear fragments are rarely visible in plant cells (Eleftheriou 1986; Wojciechowska and Olszewska 2003). While in the majority of the cases collapse of the tonoplast and degradation of cytoplasmic components precede nuclear degeneration, Schussler and Longstreth (2000) documented nuclear fragmentation as the first sign of PCD during aerenchyma formation in oxygen-deprived plants of *Sagittaria lancifolia*. Complete or partial disappearance of the cell wall during PCD also occurs in those cases when cavities must be created within the tissue, such as during tapetum and endosperm development, and aerenchyma initiation, but not in others, i.e., suspensor cells, vessels and tracheids, and in cork cells (van Doorn et al. 2011).

2.2 NO, Hb, and PCD

Reactive oxygen species (ROS) and reactive nitrogen species (RNS) are important factors triggering caspase, a cysteine protease that plays an essential role in apoptosis, and facilitating PCD (Rodríguez-Serrano et al. 2012; Serrano et al. 2012). In particular, NO has emerged as an important factor in stimulating and regulating PCD. Ye et al. (2013) showed that NO promotes MPK6-mediated caspase-3-like activation in cadmium-induced *Arabidopsis thaliana* PCD.

It is possible that NO activates mechanisms of aerenchyma formation only in specific cells. Expression of ACC oxidase, the ethylene forming enzyme, was observed in the root cap, protophloem sieve elements, and companion cells associated with metaphloem sieve elements in maize roots (Geisler-Lee et al. 2010). ACC synthase, which generates the ethylene precursor, was expressed in the root cap and the cortex and its expression was induced in cortical cells following low oxygen treatment. This suggests that maize roots respond to conditions of hypoxia by inducing the spatially restricted expression of the ethylene biosynthetic machinery, resulting in increased ethylene production.

2.3 NO, Gene Regulation, and PCD

Sarkar and Gladish (2012) showed that hypoxic stress triggers a PCD pathway to induce vascular cavity formation in *Pisum sativum* roots. Systematic DNA fragmentation, a feature of many PCD pathways, was detected in the cavity-forming roots after 3 h of flooding. High molecular weight DNA fragments of about

20–30 kb were detected by pulse-field gel electrophoresis, but no low-molecular weight internucleosomal DNA fragments were detected by conventional gel electrophoresis. Release of mitochondrial cytochrome c protein into the cytosol, an integral part of mitochondria-dependent PCD pathways, was detected in the cavity-forming roots within 2 h of flooding by fluorescence microscopy of immunolabeled cytochrome c in situ and in isolated mitochondrial and cytosolic protein fractions by Western blotting. DNA fragmentation and cytochrome c release remained confined to the undifferentiated cells in the center of the root vascular cylinders, even after 24 h of flooding, while outer vascular cylinder cells and cortical cells maintained cellular integrity and normal activity. These findings confirm that hypoxia-induced vascular cavity formation in *P. sativum* roots involves PCD, and provides a chronological model of cytological events involved in this understudied PCD system.

A significant pool of class 1 plant hemoglobins may be localized in the nucleus. This was shown by Seregélyes et al. (2000) for alfalfa nsHb1 and then confirmed for Arabidopsis (Hebelstrup and Jensen 2008). As shown in the latter study, class 1 hemoglobin shares both cytosolic and nuclear localization. Coincidentally, cytoglobin (Geuens et al. 2003) and neuroglobin (Hundahl et al. 2010), mammalian hemoglobins believed to be involved in PCD, also locate in the cytoplasm and nucleus. This would suggest that nsHbs1 have a physiological role in plants beyond scavenging NO produced as a result of using nitrite as an alternative electron acceptor.

An oxygen-sensitive N-end rule pathway of proteolysis has been suggested to regulate events associated with oxygen depletion in plants (Gibbs et al. 2011; Licausi et al. 2011). RAP2.12, identified as one possible plant oxygen sensor, is released from the plasma membrane and migrates to the nucleus under hypoxic conditions where it activates genes associated with hypoxic responses (Licausi et al. 2011). Under aerobic conditions, evidence suggested that the protein rapidly degraded. The N-end rule pathway is also recognized to be involved in NO sensing in mammalian systems under oxygen depletion and other stresses (Hu et al. 2005). Evidence indicates that in the case of N-terminal cysteine proteins, before N-terminal arginylation can occur to initiate the pathway, NO along with oxygen is a required component to oxidize the cysteine. It is, therefore, possible that an avid oxygen and NO scavenger, such as the nsHbs, could reduce the nuclear oxygen concentration and oxygenate the NO produced under hypoxic conditions, preventing the degradation of components required in hypoxia regulation.

No data exists to support the role of a nuclear hemoglobin pool in protection of the plant genome; however, it is known that promoters of several genes are directly regulated by NO and also that NO can nitrosylate chromosomal and other nuclear proteins. NO can activate transcription factors directly or indirectly; in both cases the transcription factors bind to their specific NO-responsive elements. In fact, many of the identified genes modulated by NO were previously reported to be modulated in disease-related experiments (Polverari et al. 2003). Palmieri et al. (2008), by studying the transcriptional changes induced by NO, identified NO-responsive genes that are involved in different functional processes such as signal transduction, defense and cell death, transport, basic metabolism, and ROS production and degradation. They described eight families of common transcription factor-binding sites (TFBS) in promoter regions of NO-regulated genes. Most of these TFBSs have been reported to be involved in particular stress responses such as the promoter regions of genes involved in jasmonic acid biosynthesis that are induced by NO. Also salicylic acid-related signal transduction pathways are regulated by NO at the level of gene expression (Grun et al. 2006) and Terrile et al. (2012) showed that NO influences auxin signaling through S-nitrosylation of the Arabidopsis TRANSPORT INHIBITOR RESPONSE 1 auxin receptor.

Lindermayr et al. (2010) have described an important system of promoter regulation activated by NO connected with NPR1 and TGA1 that are key redox-controlled regulators of systemic acquired resistance in plants. NPR1 monomers interact with the reduced form of TGA1, which target the activation sequence-1 (as-1) element of the promoter region of defense proteins. Both NPR1 and TGA1 are S-nitrosylated after treatment with NO donors and TGA1 is protected by S-nitrosoglutathione from oxygen-mediated modifications enhancing the DNA binding activity of TGA1 to the as-1 element in the presence of NPR1. The translocation of NPR1 into the nucleus is also promoted by NO. These results suggest that NO is a redox regulator of the NPR1/TGA1 system underlining the importance of NO in the plant defense response.

Regulation of several genes by NO may be directly related to its production from nitrate and nitrite and hemoglobin gene induction was shown to be related to this mechanism (Ohwaki et al. 2005). The particular aspects of gene regulation by NO have to be established but it becomes more clear that NO is an important regulator of metabolism not only at the protein but also at genetic levels and triggering PCD involves also its action on gene promoters and chromosomal proteins.

Independent studies have shown that one of the key roles of plant hemoglobins is to scavenge NO during stress conditions, including oxygen deficiency, through an NADH-dependent dioxygenase mechanism (Dordas et al. 2003; Hebelstrup and Jensen 2008; Perazzolli et al. 2004). This notion is well described during Arabidopsis bolting, when applications of NO donors are antagonistic to the effects of hemoglobins (Hebelstrup and Jensen 2008). Since NO is a fundamental signal molecule implicated in apoptosis in animals (Blaise et al. 2005) and PCD in plants (Neill 2005), it is plausible to assume that regulation of the cellular NO level by hemoglobins during oxygen deprivation might influence both processes. In animals the function of NO during apoptosis is controversial since exposure to NO can prevent death in some cell lines, and induce apoptosis in others (reviewed by Wang et al. 2010). This complexity depends on the rate of NO production, as well as its interaction with proteins, ROS, and thiols (Vitecek et al. 2008). The mechanisms of NO-mediated cell death in plants have only begun to emerge in the last few years and they will be reviewed in the next sections.

2.4 Mammalian Hemoglobins and Apoptosis

In mammalian systems, PCD prevention has been suggested to be the main function of neuroglobin (Brittain et al. 2010). It was hypothesized that the central role of neuroglobin in highly metabolically active cells is to reset the trigger level of mitochondrial cytochrome c release necessary to commit the cells to apoptosis (Brittain et al. 2010). Neuroglobin is found in high concentration in some tissues (Schmidt et al. 2003), and has been shown to promote survival of neurons in vitro and thus protect the brain from damage by both stroke and Alzheimer's disease (Greenberg et al. 2008; Khan et al. 2007). Ferrous neuroglobin reacts very rapidly with ferric cytochrome c, which is released from mitochondria during cell death, and it has been proposed that this is a mechanism by which the molecule interferes with the intrinsic pathway of apoptosis (Fago et al. 2006). It has been speculated that neuroglobin, reacting with redox partners and oxygen, provides for the establishment of a redox cycle within cells via certain enzymes serving as methemoglobin reductases capable of employing either glutathione or NADH as re-reductants of ferric neuroglobin (Brittain and Skommer 2012). The authors estimated that the steady state level of antiapoptotic neuroglobin is very sensitive to the cellular oxygen tension and moderately sensitive to the redox status of the cell and hypothesized that this might provide a moderately rapid mechanism for adjusting the antiapoptotic status of a cell, while the reaction of neuroglobin with mitochondrial cytochrome c provides a very rapid, but limited, capacity to intervene in the apoptotic pathway.

This all means that direct reaction of hemoglobin with cytochrome c may play a rapid limited role in prevention of apoptosis while more long-term redox cycle connected with NO and O_2 levels can regulate apoptosis in tissues and organs in longer time durations. Any interaction of plant hemoglobins with cytochrome c is speculative and not documented, while the maintenance of low NO levels under hypoxia and reduction of metHb by corresponding reductases, one of which has been identified (Igamberdiev et al. 2006a), makes it possible to sustain an antiapoptotic mechanism related to the redox-regulating function of Hb, in particular of class 1 Hb. Scavenging of NO and keeping Hb in the ferrous state will be a major factor in plants preventing apoptosis during the hypoxic and other stresses.

2.5 NO-Mediated Mechanisms of Cell Death

While descriptions of NO-mediated PCD in plant cells are accumulating, the precise details of its mode of action remain largely unknown. The pro-apoptotic effect of NO on animal cells can occur through a variety of mechanisms. Some of these include the inhibition of respiration through reduction of membrane potential and release of cytochrome c (Brune 2003), the direct activation of caspases, executors of the death program, possibly through the upregulation of target genes

(CD95 and p53) (Brune et al. 2001; Messmer et al. 1996), and the production of the strong oxidant peroxynitrite ($ONOO^-$) which induces mitochondria swelling and increases the permeability of the mitochondrial membrane, thus favoring the release of cytochrome c which triggers the cell death signaling (reviewed by Blaise et al. 2005). Cellular levels of NO might also induce cell death through the modulation of the antioxidant response. In vascular smooth muscle cells the NO-induced apoptosis is associated with a decline in reduced glutathione, and exogenous applications of this metabolite delay cell death (Zhao et al. 1997). There is evidence that a similar regulation operates in plant cells (Igamberdiev et al. 2006b). In alfalfa cells grown in both normoxic and anoxic conditions over-expression of the NO-scavenger Hb1 increases ascorbate levels and elevates the expression of ascorbate peroxidase and monodehydroascorbate reductase, while a reduction of Hb1 via antisense transformation increases the activity of glutathione reductase, the enzyme forming reduced glutathione. Reduction of the antioxidant response and the concomitant elevation of ROS have been involved in plant PCD (Wilkins et al. 2011).

In animal cells cyclic GMP (cGMP) is an important signal molecule involved in the early signal transduction events leading to apoptosis and NO modulates cellular cGMP by binding to the ferrous heme group of the guanylate cyclase (GC)-coupled receptors resulting in a conformational change of the enzyme which catalyzes the formation of cGMP from GTP (reviewed by Blaise et al. 2005). Binding of NO to GC is a reversible process and GC activity is turned off immediately after removal of NO (Beckman 1996). Since its initial identification by mass spectrometry in plants, cGMP has been recognized as a key down-stream signaling molecule of NO-mediated responses, including pathogen defense responses, where it is required for the activation of plant defense genes leading to PCD (Durner et al. 1998). Transient elevation in cGMP occurs in response to NO treatments (Durner et al. 1998; Pfeiffer et al. 1994). Although several GCs have been identified in higher plants, it was only recently that a GC able to bind to NO was reported in Arabidopsis (Mulaudzi et al. 2011). By using a GC and heme-binding domain-specific search motif, the authors demonstrated that this GC is able to bind to NO at a higher affinity than O_2, and that this binding generates cGMP in an NO-dependent manner. Using a pharmacological approach Clarke et al. (2000) were able to induce PCD in Arabidopsis cells exposed to high levels of NO and this resulted in chromatin condensation and activation of caspase-like activity independent of ROS production. Furthermore, PCD was partially blocked if GC was inhibited, while applications of the cGMP analogue 8Br-cGMP reversed the inhibition (Clarke et al. 2000). The induction of GC activity and production of cGMP by NO is not only restricted to PCD but also observed in other NO-mediated developmental processes including stomatal closure (Neill et al. 2002) and adventitious rooting (Pagnussat et al. 2003). The mechanisms through which cGMP induces PCD are not well understood, although they may rely on an elevation of cytosolic Ca^{++} evoked by the ability of cGMP to open Ca^{++} channels directly, or indirectly, via cADPR which activates the intracellular ryanodine Ca^{++} channel receptors in plants (Besson-Bard et al. 2008). Elevation of cytoplasmic Ca^{++} in tobacco protoplasts treated with the NO donor sodium nitroprusside (SNP) resulted in a reduction in mitochondrial permeability leading to PCD (Wang et al. 2010). NO-induced PCD was prevented by the addition of the Ca^{++} chelator EGTA.

Additional down-stream events of NO-mediated PCD might involve mitogen-activated protein kinases (MAPKs). In human cells NO induction of cell proliferation is associated with the MAPK cascade (Chou et al. 2002), while in plants activation of MAPK activity has been observed during NO-mediated defense responses (Kumar and Klessig 2000) and NO-mediated formation of adventitious roots (Pagnussat et al. 2004). Li et al. (2007) provided evidence that the MAPK signal is involved in initiation of self-incompatibility-induced PCD in *Papaver*.

Emerging information on both animal and plant systems suggests that S-nitrosylation, the addition of NO to a Cys thiol residue to form an S-nitrosothiol (SNO), plays a key role in several NO-mediated responses. In animal cells SNO production has been shown to regulate several steps of the apoptotic program (Hara et al. 2005; Sawa et al. 1997). Several apoptosis-related components, including caspase-3 (Tsang et al. 2009) and the apoptosis signal-regulating kinase (ASK1) (Park et al. 2004), require S-nitrosylation. This is also true in plants where S-nitrosylation regulates the activity of key enzymes participating in hypersensitive response (Romero-Puertas et al. 2007) and NO-induced cell death in rice leaves (Lin et al. 2012). Based on the above evidence, the involvement of SNO during hypoxia-induced PCD cannot be excluded.

Heavy metals, including Cd^{++} and Zn^{++}, have been implicated in animal apoptosis and plant PCD (Lee and Koh 2010; Ma et al. 2010). Elevated levels of Cd^{++} trigger PCD in tobacco BY2 cells (Fojtová and Kovarík 2000), Arabidopsis suspension culture (De et al. 2009). and onion root tips (Behboodi and Samadi 2004). Independent evidence suggests that NO might be involved in these responses. NO is a required factor influencing iron uptake and, through upregulation of genes affecting both iron and Cd transport, also results in increasing Cd^{++} uptake (Besson-Bard and Wendehenne 2009), NO decreases phytochelatin function, through S-nitrosylation, which is required to scavenge Cd^{++} in Arabidopsis cells (De et al. 2009). Reduction of phytochelatin activity favors Cd^{++}-induced PCD. Alternative mechanisms through which NO plays a positive role in Cd^{++}-induced PCD in plant cells are by influencing the uptake of Cd^{++} or by promoting MPK6-mediated caspase-3 like activation (Ma et al. 2010; Ye et al. 2013).

Cellular levels of Zn^{++} modulate apoptosis in animal cells (Lee and Koh 2010). The role of this heavy metal has been investigated especially during brain cell injury and death (Koh et al. 1996). High levels of intracellular free Zn^{++} activate apoptosis through excess autophagy (Lee and Koh 2010), MAPK signaling (He and Aizenman 2010), and NADPH oxidase which induces oxidative stress and cell death (Koh 2001). Nitric oxide increases the intracellular Zn^{++}, released from metallothionine, by destroying zinc-sulfur clusters (Aravindakumar et al. 1999). Unpublished data from our lab further reinforce the link between NO and Zn^{++} during the induction phases of maize somatic embryogenesis. Somatic embryo production is encouraged in maize cells down-regulating Hb2 and repressed in cells down-regulating Hb1. This different behavior in culture is associated with

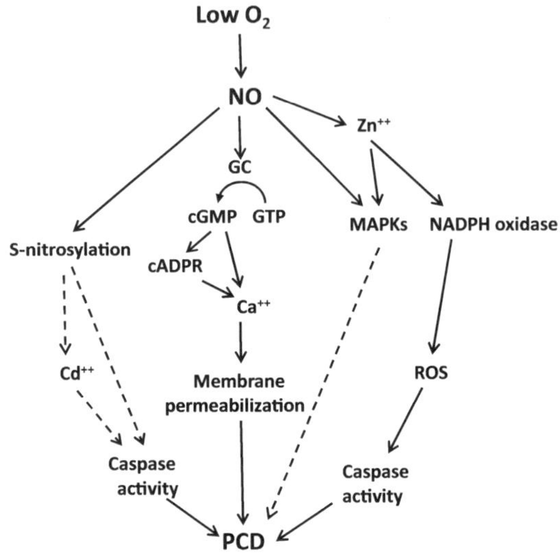

Fig. 1 Possible routes by which NO may influence PCD in plant cells. *Solid arrows* indicate steps described in either animal or plant cells. *Dashed arrows* represent undescribed steps

increasing levels of NO in specific cells in which hemoglobin is suppressed. Accumulation of NO induces free Zn^{++}, metacaspase 2 and 3 expression, caspase 3-like activity, accumulation of ROS, and ultimately PCD. These NO-induced effects are partially abrogated if cellular Zn^{++} is experimentally reduced, and amplified upon supplementation of exogenous Zn^{++}. The difference in somatic embryo yield observed between lines down-regulating Hb1 or 2 is a consequence of the different expression domains of the two genes. Hemoglobin1 is expressed in many cells of the embryo proper and suspensor. The accumulation of NO in these cells, as a result of hemoglobin1 suppression, triggers massive PCD resulting in the premature death of the embryos. This is in contrast to lines down-regulating Hb2 where NO only accumulates in some suspensor cells, where Hb2 is usually expressed. Removal of these cells via PCD "shapes" the embryos and favors the separation of immature somatic embryos which can develop further at higher frequency.

3 Concluding Remarks

Collectively these studies reveal the multifaceted function of NO during apoptosis in animals and PCD in plants and provide experimental models describing mechanisms through which oxygen deprivation can induce cell death (Fig. 1). Deprivation of oxygen leads to increased NO levels resulting in a multitude of cellular responses. Protein nitrosylation potentially may directly increase cell Cd^{++} levels and directly or indirectly upregulate caspase activity leading to PCD. Increased NO is known to lead to increased cell Zn^{++} that upregulates MAPKs and NADPH

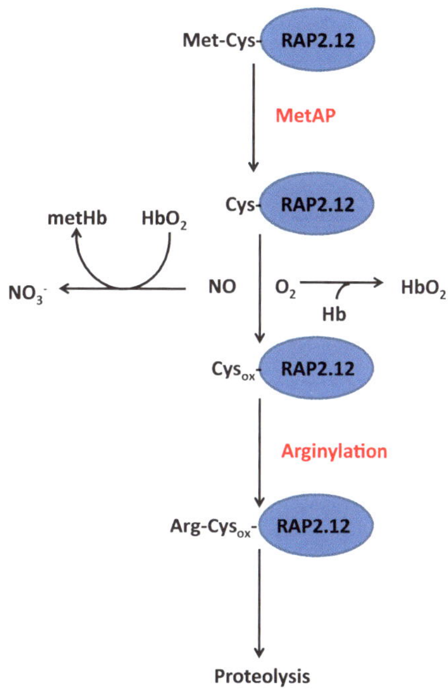

Fig. 2 A hypothetical mechanism by which hemoglobin may disrupt NO-mediated PCD operating via the N-end rule pathway. *MetAP* methionine aminopeptidase, Cys_{ox} cysteinesulfinic or cysteinesulfonic acid

oxidase. There is the possibility that MAPKs, induced due to NO increases, may directly regulate events leading to PCD. NADPH oxidase is known to result in increased ROS, upregulating caspase activity leading to PCD. Increased NO results in upregulation of guanylate cyclase (GC), leading to increased production of cGMP from GTP. cGMP directly or indirectly, via production of cyclic ADP-ribose (cADPR), results in increased cell Ca^{++} levels that influence membrane permeability leading to PCD.

Whether protein nitrosylation by NO is the main or only mechanism by which NO influences cell death is a moot point at this stage. NO does, however, have a central role in PCD in all species that have been studied. The N-end rule pathway has been suggested as a possible route for regulating PCD and Fig. 2 provides a mechanism by which nsHb might influence PCD death in a plant system undergoing oxygen deprivation, using the ERF, RAP2.12, as the protein. The theory is that RAP2.12 undergoes proteolysis possibly via an E3 ubiquitin ligase pathway to trigger PCD. Proteins entering this pathway must be arginylated. RAP2.12 possesses an N-terminal methionine adjacent to a cysteine that can be cleaved by a methionine aminopeptidase (MetAP) that is upregulated during oxygen deprivation, leaving an N-terminal cysteine residue. This cysteine residue must be oxidized before it can be arginylated and this process required both NO and oxygen. Hemoglobins, particularly nsHb1, which is upregulated under oxygen deprivation, because of its strong avidity for oxygen and its efficiency in scavenging NO, may restrict oxidation of the N-terminal Cys of the methionine-cleaved RAP2.12 and its arginylation, preventing the removal of the protein that is required for the initiation of PCD. The concentration of nsHb1 in the nucleus, along with the concentration of oxygen, would be the controlling factors regulating PCD, i.e., nsHbs would be the effective oxygen sensors of the hypoxic cell. By its expression, or non-expression, within a specific cell, an nsHb would, therefore, determine whether the cell proceeds towards PCD in conditions where NO is a factor triggering the process.

References

Aravindakumar CT, Ceulemans J, De LM (1999) Nitric oxide induces Zn2+ release from metallothionein by destroying zinc-sulphur clusters without concomitant formation of S-nitrosothiol. Biochem J 344:253–258

Arends MJ, Morris RG, Wyllie AH (1990) Apoptosis. The role of the endonuclease. Am J Pathol 136:593–608

Beckman JS (1996) Oxidative damage and tyrosine nitration from peroxynitrite. Chem Res Toxicol 9:836–844

Behboodi BS, Samadi L (2004) Detection of apoptotic bodies and oligonucleosomal DNA fragments in cadmium-treated root apical cells of Allium cepa Linnaeus. Plant Sci 167: 411–416

Besson-Bard A, Wendehenne D (2009) NO contributes to cadmium toxicity in Arabidopsis thaliana by mediating an iron deprivation response. Plant Signal Behav 4:252–254

Besson-Bard A, Pugin A, Wendehenne D (2008) New insights into nitric oxide signaling in plants. Annu Rev Plant Biol 59:21–39

Blaise GA, Gauvin D, Gangal M, Authier S (2005) Nitric oxide, cell signaling and cell death. Toxicology 208:177–192

Brittain T, Skommer J (2012) Does a redox cycle provide a mechanism for setting the capacity of neuroglobin to protect cells from apoptosis? IUBMB Life 64:419–422

Brittain T, Skommer J, Henty K, Birch N, Raychaudhuri S (2010) A role for human neuroglobin in apoptosis. IUBMB Life 62:878–885

Brune B (2003) Nitric oxide: NO apoptosis or turning it ON? Cell Death Differ 10:864–869

Brune W, Menard C, Heesemann J, Koszinowski UH (2001) A ribonucleotide reductase homolog of cytomegalovirus and endothelial cell tropism. Science 291:303–305

Bykova NV, Igamberdiev AU, Ens W, Hill RD (2006) Identification of an intermolecular disulfide bond in barley hemoglobin. Biochem Biophys Res Commun 347:301–309

Chou FP, Tseng TH, Chen JH, Wang HC, Wang CJ (2002) Induced proliferation of human MRC-5 cells by nitrogen oxides via direct and indirect activation of MEKK1, JNK, and p38 signals. Toxicol Appl Pharmacol 181:203–208

Clarke A, Desikan R, Hurst RD, Hancock JT, Neill SJ (2000) NO way back: nitric oxide and programmed cell death in *Arabidopsis thaliana* suspension cultures. Plant J 24:667–677

De MR, Vurro E, Rigo C, Costa A, Elviri L, Di VM, Careri M, Zottini M, di Sanita TL, Lo SF (2009) Nitric oxide is involved in cadmium-induced programmed cell death in Arabidopsis suspension cultures. Plant Physiol 150:217–228

Dordas C, Hasinoff BB, Igamberdiev AU, Manac'h N, Rivoal J, Hill RD (2003) Expression of a stress-induced hemoglobin affects NO levels produced by alfalfa root cultures under hypoxic stress. Plant J 35:763–770

Dordas C, Hasinoff BB, Rivoal J, Hill RD (2004) Class 1 hemoglobins, nitrate and NO levels in hypoxic maize cell suspension cultures. Planta 219:66–72

Drew MC, He I, Morgan PW (2000) Programmed cell death and aerenchyma formation in roots. Trends Plant Sci 5:123–127

Durner J, Wendehenne D, Klessig DF (1998) Defense gene induction in tobacco by nitric oxide, cyclic GMP, and cyclic ADP-ribose. Proc Natl Acad Sci U S A 95:10328–10333

Eleftheriou EP (1986) Ultrastructural studies on protopliloem sieve elements in Triticum aestivnm L. nuclear degeneration. J Ultrastruct Mol Struct Res 95:47–60

Evans DE (2003) Aerenchyma formation. New Phytol 161:35–49

Fago A, Mathews AJ, Moens L, Dewilde S, Brittain T (2006) The reaction of neuroglobin with potential redox protein partners cytochrome b5 and cytochrome c. FEBS Lett 580:4884–4888

Fojtová M, Kovarík A (2000) Genotoxic effect of cadmium is associated with apoptotic changes in tobacco cells. Plant Cell Environ 23:531–537

Geisler-Lee J, Caldwell C, Gallie DR (2010) Expression of the ethylene biosynthetic machinery in maize roots is regulated in response to hypoxia. J Exp Bot 61:857–871

Geuens E, Brouns I, Flamez D, Dewilde S, Timmermans JP, Moens L (2003) A globin in the nucleus! J Biol Chem 278:30417–30420

Gibbs DJ, Lee SC, Md Isa N, Gramuglia S, Fukao T, Bassel GW, Correia CS, Corbineau F, Theodoulou FL, Bailey-Serres J, Holdsworth MJ (2011) Homeostatic response to hypoxia is regulated by the N-end rule pathway in plants. Nature 479:415–418

Greenberg DA, Jin K, Khan AA (2008) Neuroglobin: an endogenous neuroprotectant. Curr Opin Pharmacol 8:20–24

Grun S, Lindermayr C, Sell S, Durner J (2006) Nitric oxide and gene regulation in plants. J Exp Bot 57:507–516

Gunawardena AHLAN, Pearce DME, Jackson MB, Hawes CD, Evans DE (2001) Rapid changes in cell wall pectic polysaccharides are closely associated with early stages of aerenchyma formation, a spatially localized form of programmed cell death inroots of maize (Zea mays L.) promoted by ethylene. Plant Cell Environ 24:1369–1375

Gupta KJ, Hebelstrup KH, Mur LA, Igamberdiev AU (2011) Plant hemoglobins: important players at the crossroads between oxygen and nitric oxide. FEBS Lett 585:3843–3849

Hara MR, Agrawal N, Kim SF, Cascio MB, Fujimuro M, Ozeki Y, Takahashi M, Cheah JH, Tankou SK, Hester LD, Ferris CD, Hayward SD, Snyder SH, Sawa A (2005) S-nitrosylated GAPDH initiates apoptotic cell death by nuclear translocation following Siah1 binding. Nat Cell Biol 7:665–674

Hara-Nishimura I, Hatsugai N (2011) The role of vacuole in plant cell death. Cell Death Differ 18:1298–1304

Hargrove MS, Brucker EA, Stec B, Sarath G, Arredondo-Peter R, Klucas RV, Olson JS, Phillips GN (2000) Crystal structure of a nonsymbiotic plant hemoglobin. Structure 8:1005–1014

He K, Aizenman E (2010) ERK signaling leads to mitochondrial dysfunction in extracellular zinc-induced neurotoxicity. J Neurochem 114:452–461

He CJ, Morgan PW, Drew MC (1996) Transduction of an ethylene signal is required for cell death and lysis in the root cortex of maize during aerenchyma formation induced by hypoxia. Plant Physiol 112:463–472

Hebelstrup KH, Jensen EO (2008) Expression of NO scavenging hemoglobin is involved in the timing of bolting in Arabidopsis thaliana. Planta 227:917–927

Hill RD (2012) Non-symbiotic haemoglobins—what's happening beyond nitric oxide scavenging? AoB Plants 2012:1–13. doi:10.1093/aobpla/pls004

Hoy JA, Robinson H, Trent JT III, Kakar S, Smagghe BJ, Hargrove MS (2007) Plant hemoglobins: a molecular fossil record for the evolution of oxygen transport. J Mol Biol 371:168–179

Hu RG, Sheng J, Qi X, Xu Z, Takahashi TT, Varshavsky A (2005) The N-end rule pathway as a nitric oxide sensor controlling the levels of multiple regulators. Nature 437:981–986

Hundahl CA, Allen GC, Hannibal J, Kjaer K, Rehfeld JF, Dewilde S, Nyengaard JR, Kelsen J, Hay-Schmidt A (2010) Anatomical characterization of cytoglobin and neuroglobin mRNA and protein expression in the mouse brain. Brain Res 1331:58–73

Igamberdiev AU, Hill RD (2004) Nitrate, NO and haemoglobin in plant adaptation to hypoxia: an alternative to classic fermentation pathways. J Exp Bot 55:2473–2483

Igamberdiev AU, Bykova NV, Hill RD (2006a) Nitric oxide scavenging by barley hemoglobin is facilitated by a monodehydroascorbate reductase-mediated ascorbate reduction of methemoglobin. Planta 223:1033–1040
Igamberdiev AU, Stoimenova M, Seregélyes C, Hill RD (2006b) Class-1 hemoglobin and antioxidant metabolism in alfalfa roots. Planta 223:1041–1046
Igamberdiev AU, Bykova NV, Hill RD (2011) Structural and functional properties of class 1 plant hemoglobins. IUBMB Life 63:146–152
Inada N, Sakai A, Kuroiwa H, Kuroiwa T (2002) Three-dimensional progression of programmed death in the rice coleoptile. Int Rev Cytol 218:221–258
Kakar S, Hoffman FG, Storz JF, Fabian M, Hargrove MS (2010) Structure and reactivity of hexacoordinate hemoglobins. Biophys Chem 152:1–14
Khan AA, Mao XO, Banwait S, Jin K, Greenberg DA (2007) Neuroglobin attenuates beta-amyloid neurotoxicity in vitro and transgenic Alzheimer phenotype in vivo. Proc Natl Acad Sci U S A 104:19114–19119
Koh JY (2001) Zinc and disease of the brain. Mol Neurobiol 24:99–106
Koh JY, Suh SW, Gwag BJ, He YY, Hsu CY, Choi DW (1996) The role of zinc in selective neuronal death after transient global cerebral ischemia. Science 272:1013–1016
Kumar D, Klessig DF (2000) Differential induction of tobacco MAP kinases by the defense signals nitric oxide, salicylic acid, ethylene, and jasmonic acid. Mol Plant Microbe Interact 13:347–351
Kundu M, Thompson CB (2005) Macroautophagy versus mitochondrial autophagy: a question of fate? Cell Death Differ 12(suppl 2):1484–1489
Lee SJ, Koh JY (2010) Roles of zinc and metallothionein-3 in oxidative stress-induced lysosomal dysfunction, cell death, and autophagy in neurons and astrocytes. Mol Brain 3:30
Li S, Samaj J, Franklin-Tong VE (2007) A mitogen-activated protein kinase signals to programmed cell death induced by self-incompatibility in Papaver pollen. Plant Physiol 145:236–245
Licausi F, Kosmacz M, Weits DA, Giuntoli B, Giorgi FM, Voesenek LACJ, Perata P, van Dongen JT (2011) Oxygen sensing in plants is mediated by an N-end rule pathway for protein destabilization. Nature 479:419–422
Lin A, Wang Y, Tang J, Xue P, Li C, Liu L, Hu B, Yang F, Loake GJ, Chu C (2012) Nitric oxide and protein S-nitrosylation are integral to hydrogen peroxide-induced leaf cell death in rice. Plant Physiol 158:451–464
Lindermayr C, Sell S, Muller B, Leister D, Durner J (2010) Redox regulation of the NPR1-TGA1 system of Arabidopsis thaliana by nitric oxide. Plant Cell 22:2894–2907
Ma W, Xu W, Xu H, Chen Y, He Z, Ma M (2010) Nitric oxide modulates cadmium influx during cadmium-induced programmed cell death in tobacco BY-2 cells. Planta 232:325–335
Manac'h-Little N, Igamberdiev AU, Hill RD (2005) Hemoglobin expression affects ethylene production in maize cell cultures. Plant Physiol Biochem 43:485–489
Messmer UK, Reed UK, Brune B (1996) Bcl-2 protects macrophages from nitric oxide-induced apoptosis. J Biol Chem 271:20192–20197
Mishina TE, Lamb C, Zeier J (2007) Expression of a nitric oxide degrading enzyme induces a senescence programme in Arabidopsis. Plant Cell Environ 30:39–52
Mulaudzi T, Ludidi N, Ruzvidzo O, Morse M, Hendricks N, Iwuoha E, Gehring C (2011) Identification of a novel Arabidopsis thaliana nitric oxide-binding molecule with guanylate cyclase activity in vitro. FEBS Lett 585:2693–2697
Mur LA, Laarhoven LJ, Harren FJ, Hall MA, Smith AR (2008) Nitric oxide interacts with salicylate to regulate biphasic ethylene production during the hypersensitive response. Plant Physiol 148:1537–1546
Neill S (2005) NO way to die–nitric oxide, programmed cell death and xylogenesis. New Phytol 165:5–7
Neill SJ, Desikan R, Clarke A, Hancock JT (2002) Nitric oxide is a novel component of abscisic acid signaling in stomatal guard cells. Plant Physiol 128:13–16
Nie XZ, Hill RD (1997) Mitochondrial respiration and hemoglobin gene expression in barley aleurone tissue. Plant Physiol 114:835–840

Nie X, Durnin DC, Igamberdiev AU, Hill RD (2006) Cytosolic calcium is involved in the regulation of barley hemoglobin gene expression. Planta 223:542–549

Nienhaus K, Dominici P, Astegno A, Abbruzzetti S, Viappiani C, Nienhaus GU (2010) Ligand migration and binding in nonsymbiotic hemoglobins of Arabidopsis thaliana. Biochemistry 49:7448–7458

Ohwaki Y, Kawagishi-Kobayashi M, Wakasa K, Fujihara S, Yoneyama T (2005) Induction of class-1 non-symbiotic hemoglobin genes by nitrate, nitrite and nitric oxide in cultured rice cells. Plant Cell Physiol 46:324–331

Pagnussat GC, Lanteri ML, Lamattina L (2003) Nitric oxide and cyclic GMP are messengers in the indole acetic acid-induced adventitious rooting process. Plant Physiol 132:1241–1248

Pagnussat GC, Lanteri ML, Lombardo MC, Lamattina L (2004) Nitric oxide mediates the indole acetic acid induction activation of a mitogen-activated protein kinase cascade involved in adventitious root development. Plant Physiol 135:279–286

Palmieri MC, Sell S, Huang X, Scherf M, Werner T, Durner J, Lindermayr C (2008) Nitric oxide-responsive genes and promoters in Arabidopsis thaliana: a bioinformatics approach. J Exp Bot 59:177–186

Park HS, Yu JW, Cho JH, Kim MS, Huh SH, Ryoo K, Choi EJ (2004) Inhibition of apoptosis signal-regulating kinase 1 by nitric oxide through a thiol redox mechanism. J Biol Chem 279: 7584–7590

Peitsch MC, Polzar B, Stephan H, Crompton T, MacDonald HR, Mannherz HG, Tschopp J (1993) Characterization of the endogenous deoxyribonuclease involved in nuclear DNA degradation during apoptosis (programmed cell death). EMBO J 12:371–377

Perazzolli M, Dominici P, Romero-Puertas MC, Zago E, Zeier J, Sonoda M, Lamb C, Delledonne M (2004) Arabidopsis nonsymbiotic hemoglobin AHb1 modulates nitric oxide bioactivity. Plant Cell 16:2785–2794

Pfeiffer S, Janistyn B, Jessner G, Pichorner H, Ebermann R (1994) Gaseous nitric oxide stimulates guanosine-3′,5′-cyclic monophosphate (cGMP) formation in spruce needles. Phytochemistry 36:259–262

Polverari A, Molesini B, Pezzotti M, Buonaurio R, Marte M, Delledonne M (2003) Nitric oxide-mediated transcriptional changes in Arabidopsis thaliana. Mol Plant Microbe Interact 16:1094–1105

Rodríguez-Serrano M, Barany I, Prem D et al (2012) NO, ROS, and cell death associated with caspase-like activity increase in stress-induced microspore embryogenesis of barley. J Exp Bot 63:2007–2014

Romero-Puertas MC, Laxa M, Matte A, Zaninotto F, Finkemeier I, Jones AM, Perazzolli M, Vandelle E, Dietz KJ, Delledonne M (2007) S-nitrosylation of peroxiredoxin II E promotes peroxynitrite-mediated tyrosine nitration. Plant Cell 19:4120–4130

Sarkar P, Gladish DK (2012) Hypoxic stress triggers a programmed cell death pathway to induce vascular cavity formation in Pisum sativum roots. Physiol Plant 146:413–426

Sawa A, Khan AA, Hester LD, Snyder SH (1997) Glyceraldehyde-3-phosphate dehydrogenase: nuclear translocation participates in neuronal and nonneuronal cell death. Proc Natl Acad Sci U S A 94:11669–11674

Schmidt M, Giessl A, Laufs T, Hankeln T, Wolfrum U, Burmester T (2003) How does the eye breathe? Evidence for neuroglobin-mediated oxygen supply in the mammalian retina. J Biol Chem 278:1932–1935

Schussler EE, Longstreth DJ (2000) Changes in cell structure during the formation of root aerenchyma in Sagittaria lancifolia (Alismataceae). Am J Bot 87(1):12–19

Schwartzman RA, Cidlowski JA (1993) Apoptosis: the biochemistry and molecular biology of programmed cell death. Endocr Rev 14:133–151

Seregélyes C, Mustárdy L, Ayaydin F, Sass L, Kovács L, Endre G, Lukács N, Kovács I, Vass I, Kiss GB, Horváth GV, Dudits D (2000) Nuclear localization of a hypoxia-inducible novel non-symbiotic hemoglobin in cultured alfalfa cells. FEBS Lett 482:125–130

Serrano I, Romero MC, Rodriguez-Serrano M, Pelliccione S, Sandalio L, Olmedilla A (2012) Peroxynitrite mediates programmed cell death both in papillar cells and in self-incompatible pollen in the olive (*Olea enropaea* L.). J Exp Bot 63:1479–1493

Spyrakis F, Bruno S, Bidon-Chanal A, Luque FJ, Abbruzzetti S, Viappiani C, Dominici P, Mozzarelli A (2011) Oxygen binding to Arabidopsis thaliana AHb2 nonsymbiotic hemoglobin: evidence for a role in oxygen transport. IUBMB Life 63:355–362

Sturms R, DiSpirito AA, Hargrove MS (2011) Plant and cyanobacterial hemoglobins reduce nitrite to nitric oxide under anoxic conditions. Biochemistry 50:3873–3878

Subbaiah CC, Bush DS, Sachs MM (1998) Mitochondrial contribution to the anoxic Ca^{2+} signal in maize suspension-cultured cells. Plant Physiol 118:759–771

Taylor ER, Nie XZ, MacGregor AW, Hill RD (1994) A cereal haemoglobin gene is expressed in seed and root tissues under anaerobic conditions. Plant Mol Biol 24:853–862

Terrile MC, París R, Calderón-Villalobos LIA, Iglesias MJ, Lamattina L, Estelle M, Casalongué CA (2012) Nitric oxide influences auxin signaling through S-nitrosylation of the Arabidopsis Transport Inhibitor Response 1 auxin receptor. Plant J 70:492–500

Tsang AH, Lee YI, Ko HS, Savitt JM, Pletnikova O, Troncoso JC, Dawson VL, Dawson TM, Chung KK (2009) S-nitrosylation of XIAP compromises neuronal survival in Parkinson's disease. Proc Natl Acad Sci U S A 106:4900–4905

van Doorn WG, Beers EP, Dangl JL, Franklin-Tong VE, Gallois P, Hara-Nishimura I, Jones AM, Kawai-Yamada M, Lam E, Mundy J, Mur LAJ, Petersen M, Smertenko A, Taliansky M, Van Breusegem F, Wolpert T, Woltering E, Zhivotovsky B, Bozhkov PV (2011) Morphological classification of plant cell deaths. Cell Death Differ 18:1241–1246

Vazquez-Limon C, Hoogewijs D, Vinogradov SN, Arredondo-Peter R (2012) The evolution of land plant hemoglobins. Plant Sci 191–192:71–81

Vigeolas H, Huhn D, Geigenberger P (2011) Non-symbiotic hemoglobin-2 leads to an elevated energy state and to a combined increase in polyunsaturated fatty acids and total oil content when over-expressed in developing seeds of transgenic Arabidopsis plants. Plant Physiol 155:1435–1444

Vinogradov SN, Hoogewijs D, Arredondo-Peter R (2011) What are the origins and phylogeny of plant hemoglobins? Commun Integr Biol 4:443–445

Vitecek J, Reinohl V, Jones RL (2008) Measuring NO production by plant tissues and suspension cultured cells. Mol Plant 1:270–284

Wang Y, Chen C, Loake GJ, Chu C (2010) Nitric oxide: promoter or suppressor of programmed cell death? Protein Cell 1:133–142

Watts RA, Hunt PW, Hvitved AN, Hargrove MS, Peacock WJ, Dennis ES (2001) A hemoglobin from plants homologous to truncated hemoglobins of microorganisms. Proc Natl Acad Sci U S A 98:10119–10124

Webb J, Jackson MB (1986) A transmission and cryo-scanning electron microscopy study of the formation of aerenchyma (cortical gas-filled space) in adventitious roots of rice (Oryza sativa). J Exp Bot 37:832–841

Widlak P, Garrard WT (2005) Discovery, regulation, and action of the major apoptotic nucleases DFF40/CAD and endonuclease G. J Cell Biochem 94:1078–1087

Wilkins KA, Bancroft J, Bosch M, Ings J, Smirnoff N, Franklin-Tong VE (2011) Reactive oxygen species and nitric oxide mediate actin reorganization and programmed cell death in the self-incompatibility response of papaver. Plant Physiol 156:404–416

Wojciechowska M, Olszewska MJ (2003) Endosperm degradation during seed development of Echinocystis lobata (Cucurbitaceae) as a manifestation of programmed cell death (PCD) in plants. Folia Histochem Cytobiol 41:41–50

Ye Y, Li Z, Xing D (2013) Nitric oxide promotes MPK6-mediated caspase-3-like activation in cadmium-induced Arabidopsis thaliana programmed cell death. Plant Cell Environ 36:1–15. doi:10.1111/j.1365-3040.2012.02543.x

Young TE, Gallie DR (2000) Programmed cell death during endosperm development. Plant Mol Biol 44:283–301

Zhao Z, Francis CE, Welch G, Loscalzo J, Ravid K (1997) Reduced glutathione prevents nitric oxide-induced apoptosis in vascular smooth muscle cells. Biochim Biophys Acta 1359:143–152

Intracellular pH Regulation of Plant Cells Under Anaerobic Conditions

Kimiharu Ishizawa

Abstract The intracellular pH of living cells is strictly controlled in each compartment. Under normal conditions, the cytoplasmic pH (pHc) and the vacuolar pH (pHv) of typical plant cells are maintained at slightly alkaline (typically 7.5) and acidic (typically 5.5) values, respectively. A failure to maintain the pH homeostasis of cells leads to cell death. In general, anaerobic conditions induce acidosis in the cytoplasm of plant cells and thereby prolonged anoxia causes cell death. As a result, the regulation of intracellular pH has been an important topic for research in studies of the anoxia tolerance of plant cells (Plant Physiol 100:1–6, 1992; Annu Rev Plant Physiol Plant Mol Biol 48:223–250, 1997; Funct Plant Biol 30:1–47, 2003; Funct Plant Biol 30:999–1036, 2003; Plant Stress 2:1–19, 2008; Annu Rev Plant Biol 59:313–339, 2008). To date many researchers have published review articles to discuss acidosis and pH regulation of plant cells exposed to anaerobic conditions (Encyclopedia of plant physiology, Springer, Berlin, pp. 317–346, 1976; Annu Rev Plant Physiol 30:289–311, 1979; Int Rev Cytol 127:111–173, 1991; Ann Bot 79:39–48, 1997; Regulation of tissue pH in plants and animals, Cambridge University Press, Cambridge, pp. 193–213, 1999; Int Rev Cytol 206:1–44, 2001; Ann Bot 96:519–532, 2005; Plant roots: the hidden half, CRC Press, Boca Raton, Chapter 23, pp. 1–18, 2013). In this review, I will summarize the proposed mechanisms to control intracellular pH and include a brief discussion about anoxia tolerance on the basis of the limited information available for plant cells possessing extremely strong tolerance to anoxia.

K. Ishizawa (✉)
Miyagi University of Education, 149 Aramaki-Aza-Aoba, Aoba-Ku, Sendai 980-0845, Japan
e-mail: kimiharu@staff.miyakyo-u.ac.jp

J.T. van Dongen and F. Licausi (eds.), *Low-Oxygen Stress in Plants*, Plant Cell Monographs 21, DOI 10.1007/978-3-7091-1254-0_4, © Springer-Verlag Wien 2014

1 Mechanisms of pH Maintenance in Cells

The regulation of intracellular pH is governed by ion-exchange at the boundary membrane(s) of the compartment and by the ionic composition of the compartment (Raven 1985; Kurkdjian and Guern 1989). Mechanisms to maintain intracellular pH homeostasis can be divided into biochemical and biophysical regulation. The biophysical pH-stat refers not only to the removal of acids produced in cells from a compartment but also to the exchange of various ions across the membrane(s) that bound the compartment. The biochemical pH-stat refers to the metabolic control of the synthesis and consumption of acidic and alkaline compounds in the compartment, and firstly was proposed as a pH-stat theory by Davies (1980). Both mechanisms are mutually compatible. Roos and Boron (1981) argued that the rapidly responding buffering mechanisms in animal cells included physicochemical buffering, consumption of nonvolatile acids, and the transfer of acid or alkali between the cytosol and the organelles.

Stewart (1978) proposed that the hydrogen ion concentration in biological solutions at equilibrium is determined by the three independent factors, the carbon dioxide partial pressure (pCO_2), the strong ion difference (SID), and the total weak acid present (A_{tot}). The strong ions are defined as ions that are more than 99 % dissociated across the physiological range of acidity. SID means the difference between the strong cations and anions (i.e., cations minus anions). Weak acids are much less than 99 % dissociated but the degree of dissociation is in dynamic equilibrium with the prevailing acidity of the solution. Weak electrolytes are either volatile (carbonic acid and bicarbonate ion—i.e., in equilibrium with volatile CO_2) or nonvolatile (protein and phosphate). Lloyd (2004) introduced the application of the strong ion calculator to determine the pH of physiological solutions as a practical tool for clinicians. The application of the SID concept to pH regulation in plants has been discussed by some researchers (Gerendás and Schurr 1999; Greenway and Gibbs 2003; Felle 2005).

Methods to measure intracellular pH are important for investigating the mechanism of intracellular pH regulation. Three different methods have been mainly used to measure intracellular pH of plant cells. Microelectrodes are a useful tool to monitor real-time changes in the internal and external pH of a cell and simultaneously the membrane potential (Felle 1993). Fluorescence probes to measure intracellular pH have been improved by the development of recombinant fluorescent pH indicators (Schulte et al. 2006) and ester-type fluorescence probes (Kosegarten et al. 1997), which are easily introduced into cells, and by the popularization of confocal laser scanning microscopy for direct, noninvasive, serial optical sectioning of intact, thick, living specimens. The most widely used method to have been used for measuring the intracellular pH of anoxic plant cells is nuclear magnetic resonance (NMR) spectroscopy (Ratcliffe 1995).

2 Changes in Intracellular pH in Response to Anaerobic Conditions

Plant cells exposed to anaerobic conditions manifest various kinds of symptoms. Most of these symptoms are related to the deterioration of energy metabolism under anoxia and the resulting collapse of intracellular homeostasis, which leads to cell death. Many studies have suggested that cell death in plants under hypoxic and anoxic conditions is closely associated with acidification of the cytoplasm. The assessment of cell death itself is a difficult problem. Usually the extent to which growth recovers after the transfer from hypoxic or anoxic conditions to normoxic conditions is used as a measure of survival. In the case of *Arabidopsis*, death of roots is induced after 36-h hypoxia or 24-h anoxia, whereas death of shoots needs a longer treatment (Ellis et al. 1999). Relationships between death and acidosis of the cytoplasm were examined in maize root tips treated with nitrogen-saturated medium (Roberts et al. 1984a). Root tips lacking the activity of alcohol dehydrogenase (ADH) died after 12-h hypoxia, whereas tips expressing ADH at the normal level survived even after 24-h hypoxia. The pHc of the root tips lacking ADH fell from 7.5 to around 6.2, whereas pHc remained closed to 6.8 after 12-h hypoxia in the normal root tips. This is convincing evidence to show that cytoplasmic acidosis is a determinant of cell death (Roberts et al. 1984a, b).

Seminal roots of presoaked rice seedlings form lysigenous aerenchyma in the cortex. The acidification of cortical cells precedes cell death and induces cell collapse to form aerenchyma (Kawai et al. 1998). This is a good example that shows the link between cell acidification and cell death. On the other hand, some aquatic plants, including rice seedlings (Tsuji 1972), *Potamogeton pectinatus* shoots (Summers and Jackson 1994), and *P. distinctus* turions (Ishizawa et al. 1999), which are specialized overwintering organs, show strong anoxia tolerance, growing under anoxia. In the case of *P. distinctus*, the turions, an overwintering organ, survive in a state of anoxia for more than 1 month (Ishizawa unpublished data). As describing below, the pHc of these plants is maintained at higher values under anoxia than that of anoxia-intolerant plants. However, there are no data to show how long the pHc of these anoxia-tolerant plants remains alkaline.

Figure 1 shows schematic profiles for the changes in the intracellular pH of plant cells exposed to different concentrations of oxygen. When a plant cell is exposed to anoxic conditions, the pHc (*a*) drops rapidly and reaches the minimum pH (*b*) after the time (α). Subsequently the pH value increases slightly and reaches a transient stationary value (*c*). The strength of anoxia tolerance can be expressed as the duration (β) of the stationary phase. For the determination of this duration (β), however, it is necessary to define a threshold value for pHc that leads to cell death. In anoxic sycamore cells, death was reported to occur when the pHc decreased below 6.5 (Gout et al. 2001). Whether this pH threshold can be universally adopted remains unclear. When anoxic cells are transferred to normoxic conditions before death, pHc recovers to a more alkaline value. In contrast the pHv of cells under

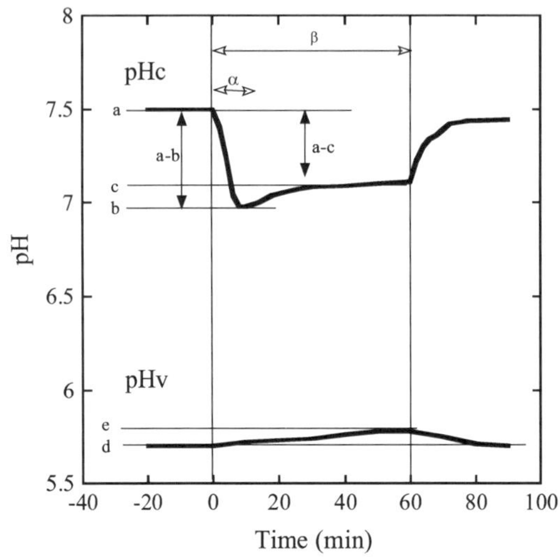

Fig. 1 Schematic profiles of changes in the intracellular pH (pHc and pHv) of a plant cell exposed to anoxia. *Left* and *right vertical lines* in the figure show the start and end, respectively, of anoxia. Horizontal lines show the pH levels of the plant cell (*a*: pHc under normoxia, *b*: pHc at initial fall after anoxic conditions, *c*: pHc in a stationary phase under anoxia, *d*: pHv under normoxia, and *e*: pHv under anoxia). *Double-pointed*, *vertical*, and *close arrows* show the degree of cytoplasmic acidification ($a - b$: at the initial fall, $a - c$: in the quasi-equilibrium). *Double-pointed*, *horizontal*, and *open arrows* show the duration of cytoplasmic acidosis (α: the time to reach the initial fall, β: the duration in quasi-equilibrium before cell death). The figure was drawn with reference to data reported by Roberts et al. (1984a, b, 1985, 1992), Mathieu et al. (1986), Menegus et al. (1991), Saint-Ges et al. (1991), Xia and Saglio (1992), Xia and Roberts (1994, 1996), Xia et al. (1995), Felle (1996, 2006), Gout et al. (2001), Dixon et al. (2006), Kulichikhin et al. (2007, 2009), Couldwell et al. (2009), and Koizumi et al. (2011)

anoxia is usually fairly stable, but sometimes it drifts from its initial value (*d*) to a less acidic value (*e*).

Values for the parameters *a*, *b*, *c*, *d*, *e*, and α are summarized in Table 1 from data reported for a range of plant tissues. The data allow the tissues to be divided into two groups: weak anoxia-tolerant plants which cannot grow under anoxia, and strong anoxia-tolerant plants which can grow under anoxia, such as rice and two *Potamogeton* species. Irrespective of the degree of anoxia tolerance, the pHc under normoxia (*a*) ranges from 7.3 to 7.7 with an average pHc of 7.5; whereas pHv (*d*) varies from 5.0 to 6.0 with an average pHv of 5.5.

An overshoot (shown as *b* in Fig. 1) of the cytoplasmic acidification within the first several minutes after the onset of anoxia, followed by a partial recovery to reach a relative stable pH value (*c*), was recorded in maize root tips (Roberts et al. 1985, 1992; Saint-Ges et al. 1991; Fox et al. 1995), sycamore culture cells (Gout et al. 2001), and rice and wheat shoots (Menegus et al. 1991; Kulichikhin et al. 2007). However, the initial minimum value (*b*) is not always observed and seems to be dependent on the plant tissue and the experimental conditions.

Table 1 Cytoplasmic (pHc) and vacuolar (pHv) pH values of various plant tissues under normoxia or anoxia

	pHc					pHv		
Plant species and tissue	$a^{a,b}$	$b^{a,b}$	$c^{a,b}$	$a-c^{b}$	α (min)[a]	$d^{a,b}$	$e^{a,b}$	References
Weak tolerance								
Corn root tip	7.5	–	7.0	0.5	–	–	–	Roberts et al. (1982)
Corn root tip	7.3	–	6.8	0.5	20	–	–	Roberts et al. (1984a)
Pea root tip	7.4	7.2	7.3	0.2	<30	6.0	6.2	Roberts et al. (1984b)
Corn root tip	7.4	–	6.8	0.6	<1	5.6	6.0	Roberts et al. (1984b)
Black-eyed pea root tip	7.4	–	7.1	0.4	<240	–	–	Roberts et al. (1985)
Navy beans root tip	7.4	–	6.9	0.5	<400	–	–	Roberts et al. (1985)
Soybean root tip	7.4	–	7.0	0.4	<240	–	–	Roberts et al. (1985)
Sycamore cell	7.5	–	7.3	0.2	<110	–	–	Mathieu et al. (1986)
Wheat shoot	7.4	–	6.6	0.8	<120	5.0	5.0	Menegus et al. (1991)
Corn root tip	7.6	7.0	7.1	0.5	<10	–	–	Saint-Ges et al. (1991)
Sycamore cell	7.5	6.8	7.1	0.4	<10	5.7	5.7	Gout et al. (1992)
Corn root tip	7.5	–	6.9	0.6	–	–	–	Xia and Saglio (1992)
Corn root tip	7.6	–	6.9	0.7	30	–	–	Xia and Roberts (1994)
Corn root tip (intact)	7.5	–	6.6	0.9	120	–	–	Xia and Roberts (1994)
Corn root tip	7.6	–	6.6	1.0	120	–	–	Xia et al. (1995)
Corn root tip (pH 6.0)	7.6	7.0	7.2	0.4	45	–	–	Fox et al. (1995)
Corn root tip	7.5	6.8	6.9	0.7	30	–	–	Xia and Roberts (1996)
Meducago root hair	7.3	–	6.8	0.5	<2	–	–	Felle (1996)
Corn root tip	7.5	–	7.0	0.5	<30	–	–	Chang et al. (2000)
Pea internode	7.5	–	6.8	0.7	–	–	–	Summers et al. (2000)
Tobacco root	7.5	–	6.7	0.8	<45	–	–	Stoimenova et al. (2003)
Arabidopsis root	7.7	–	7.2	0.5	<10	–	–	Tournaire-Roux et al. (2003)
Corn root	7.5	–	6.7	0.8	<30	5.3	5.2	Libourel et al. (2006)
Barley leaf	7.3	–	6.6	0.7	<20	–	–	Felle (2006)
Wheat coleoptiles	7.7	7.0	7.1	0.6	<18	–	–	Kulichikhin et al. (2007)
Tobacco leaf	7.4	–	6.3	1.1	<50	–	–	Couldwell et al. (2009)
Strong tolerance								
Rice shoot	7.5	6.9	7.0	0.5	<30	5.0	6.1	Menegus et al. (1991)
Rice root tip	7.5	–	7.3	0.2	–	–	–	Fan et al. (1992)
Potamogeton pectinatus	7.5	–	7.3	0.2	<60	–	–	Summers et al. (2000)
Potamogeton pectinatus	7.4	–	7.1	0.3	<60	5.7	5.3	Dixon et al. (2006)
Rice coleoptiles	7.6	7.0	7.2	0.4	<10	–	–	Kulichikhin et al. (2007)
Rice coleoptiles	7.7	–	7.3	0.4	–	5.3	5.9	Kulichikhin et al. (2009)
Potamogeton distinctus	7.6	–	7.4	0.2	<50	5.9	6.0	Koizumi et al. (2011)

[a]pH values and the times to reach a quasi-equilibrium were estimated by reading directly from the figures in cited references

[b]The values are rounded off to the first decimal place

Such an initial response of pHc implies the operation of fine control for a disturbance of pHc homeostasis by anoxic conditions. However, varying the external pH has a marked effect on the behavior of the changes in pHc of maize root tips immediately after the onset of anoxia (Fox et al. 1995). At pH 4.0, the partial recovery ($b - c$: Fig. 1) of pHc under anoxic conditions disappeared, whereas at pH 10, it was exaggerated. Thus, the response of pHc to anoxia is related to the proton gradient between the external medium and the cytoplasm, suggesting that H^+ extrusion from the cytoplasm occurs during the recovery period.

The difference between (*a*) and (*c*) is different for anoxia-intolerant and -tolerant plant species/organs. The average difference for the former is 0.6 pH unit and that for the latter is 0.3 pH unit, implying that anoxia-tolerant plants possess a stronger system to compensate cytoplasmic acidosis under anoxia. There is no published data to show how long the stationary pH value (*c*) persists in anoxia-tolerant plants, although pHc of *P. pectinatus* stem tissues recorded anaerobically after a 7–10 day period of anoxic growth was 7.4 (Dixon et al. 2006). Therefore, it is confidently expected that anoxia-tolerant plants will resist cytoplasmic acidosis for much longer than anoxia-intolerant plants, implying that regulatory mechanisms are able to function for a longer period in anoxia-tolerant plants.

3 Biochemical Regulation

At an initial stage of hypoxia or anoxia, energy metabolism is switched from aerobic respiration to anaerobic fermentation to supply ATP. Consequently, produced and consumed metabolites during this process have been proposed to act as factors influencing intracellular pH.

3.1 Lactate

The drop induced in pHc that occurs immediately after transfer to anoxic or hypoxic conditions is thought to be mostly due to the net production of protons associated with fermentation to lactate. The initial rapid decrease in pHc and the transient stabilization can be explained by the Davies biochemical pH-stat (Davies 1980). This model states that under anaerobiosis, a transient lactate fermentation occurs until this acidification of the cytoplasm activates pyruvate decarboxylase (PDC), thus bringing on a steady state of mainly ethanol production. This idea has been supported by several experimental results. Roots of mutant maize, lacking a functional *ADH1* gene, had a reduced ability to carry out glycolysis and fermentation and showed a more rapid cytoplasmic pH drop (Roberts et al. 1982, 1984a, b). Other evidence comes from the effects of hypoxic pretreatment of excised (Xia and Saglio 1992) and intact maize root tips (Xia and Roberts 1994) on the efflux of lactic acid or lactate. The larger efflux of lactic acid could account for the higher pHc values that were found in pretreated tips in comparison with non-pretreated tips. The mechanism of lactate efflux had been unclear, but recently the gene (*AtNIP2;1*) encoding a lactic acid transporter was cloned from *Arabidopsis* (Choi and Roberts 2007). AtNIP2;1 supports the transport of lactic acid, with a preference for the protonated acidic form, and the transcript level increased by 300 times 2 h after the start of anoxia. This transporter may be involved in the regulation of pHc under anoxia. Furthermore, clear evidence for the role of the pHc in triggering the switch to ethanol production under anoxia was presented by Fox et al. (1995). In maize root tip cells manipulated with acid and base in the external medium, the cytoplasmic acidification was able to trigger ethanol production, and the amount and duration of ethanol production could be controlled by manipulation of the pHc value.

3.2 Nucleotide Triphosphate

Some reports argue against a dominant role for lactate fermentation in the acidification caused by anoxia. Saint-Ges et al. (1991) reported that the pHc of maize root tips followed an oscillatory time course that was almost identical to the time course of the nucleotide triphosphate (NTP) content, but not identical to that of the lactate content. In rice seedlings subject to anoxia, the pHc dropped suddenly without the activation of lactate dehydrogenase (LDH) to produce lactate, suggesting that lactate is not involved in the acidosis of the cytoplasm (Rivoal et al. 1991). Similarly the acidification of the pHc in rice and wheat shoots subject to hypoxia and anoxia did not correspond to changes in the content of lactate (Menegus et al. 1991). Moreover, when the effects of altered levels of PDC and LDH on the pH response to anoxia were examined in transgenic potato tubers (reduced levels of LDH) and tobacco leaves (over-expressed PDC), the results suggested that the biochemical pH-stat model alone could not provide a full explanation of anoxic pH regulation (Couldwell et al. 2009). Time-courses of pHc, pHv, the pH of the medium and the NTP content of suspension cultured cells of sycamore showed marked correlation between the variations of pHc and NTP during anoxia. Therefore, Gout et al. (2001) concluded that the release of H^+ accompanying the Pi-liberating hydrolysis of NTP was the principal cause of the initial pHc drop and that this cytoplasmic acidosis was not overcome by H^+ extrusion. However, using calculations based on the SID concept, Greenway and Gibbs (2003) estimated that NTP hydrolysis only explained one third of the initial acidification.

3.3 Other Organic Acids Produced Under Anaerobiosis

Plant cells activate some metabolic pathways under anoxic conditions and produce various end products such as malate, succinate, gamma-aminobutyric acid

(GABA), and Ala, in addition to ethanol and lactate, which are possibly involved in the fine control of intracellular pH (Davies 1980; Kennedy et al. 1992). Menegus et al. (1989) examined the relationship between cell sap pH and succinate to lactate ratios in plants having different anoxia tolerances. They found that rice and *Echinochloa* showing high resistance to anoxia accumulated more succinate than lactate and cell saps became alkaline, and conversely, wheat, rye, and barley, showing low resistance to anoxia, accumulated more lactate and cell saps acidified. Roberts et al. (1992) suggested several proton-consuming metabolic pathways in hypoxic maize root tips: (1) conversion of malate to lactate or ethanol, or malate/Gln to Ala/Glu; (2) GABA from Glu; and (3) conversion of carbon and nitrogen of Asp to Ala. Ala is known to be a predominant fermentation end product of maize root tips acclimated with hypoxia (Xia and Roberts 1994; Miyashita and Good 2008). Moreover, the accumulation of GABA in anoxic plant tissues supports the involvement of GABA synthesis in pHc regulation (Fan et al. 1992; Carroll et al. 1994; Shelp et al. 1999).

3.4 Nitrate and Nitrite

The potential impact of nitrate on pH regulation during anoxia has been examined in flooding-tolerant rice roots (Fan et al. 1988) and flooding-intolerant maize roots (Roberts et al. 1985). However, the relationship between nitrate and pHc regulation during anoxia is unclear. Exposure to nitrate was found to reduce the acidification of the cytoplasm in anoxic maize root tips (Roberts et al. 1985). Stoimenova et al. (2003) showed that while the acidification of pHc in wild-type tobacco roots was smaller than that in a recombinant lacking root nitrate reductase (NR), this result was caused by the greater metabolic rate of the recombinant roots rather than by the inability to reduce nitrate. Libourel et al. (2006) found that the nitrate-induced improvement in the pHc regulation under anoxia was caused by a metabolite derived from nitrate, rather than by nitrate itself. Nitrite is released during anoxia from maize roots treated with nitrate and exposure to low concentrations of nitrite reduced the acidification to the cytoplasm under anoxia. It was suggested that the availability of nitrite under anoxia gives rise to an NO-mediated reduction in metabolic rate, with a consequent decrease in acidification. In support of this explanation, nitrite is readily converted to NO under anoxia by the action of cytosolic NR (Stoimenova et al. 2003).

4 Biophysical Regulation

Proton pumps create the proton-motive force that drives the transport of many solutes necessary for living plant cells (Sze 1985; Michelet and Boutry 1995). Proton pumps on the plasma membrane and tonoplast have been proposed to play an important role in intracellular pH regulation in plants (Kurkdjian and Guern 1989; Gout et al. 1992; Rea and Poole 1993; Gaxiola et al. 2007; Duby and Boutry 2009). Three distinct proton pumps are involved in intracellular pH regulation. The plasma membrane H^+-ATPase (PM H^+-ATPase) works to generate a proton gradient across the plasma membrane. The vacuolar-type H^+-ATPase (V-ATPase) and the vacuolar-H^+-pumping pyrophosphatase (H^+-PPase) maintain a proton electrochemical gradient across endomembrane compartments (Sze and Palmtren 1999). These proton pumps are thought to play a crucial role in the biophysical pH-stat (Greenway and Gibbs 2003; Felle 2005). Nevertheless, definite evidence of the involvement of proton pumps in some of these roles is rather limited.

4.1 PM H^+-ATPase and Secondary Active Transport of Ions

The PM H^+-ATPase plays a major role in counteracting cytoplasmic acidification in well-oxygenated cells (Gout et al. 1992). However, under conditions of low oxygen concentrations proton pumps are generally thought to be severely inhibited through the shortage of intracellular ATP. Hence, the direct involvement of proton pump has been questioned in the regulation of intracellular pH under anoxia.

Sycamore cells can recover from cytoplasmic acidification induced by acid-load. This recovery occurs even under anoxia, where the PM H^+-ATPase could be restricted, suggesting that other systems besides the PM H^+-ATPase are involved in the excretion of proton-equivalents, for example, Na^+ and HCO_3^- exchanges in the regulation of pHc of acid-loaded cells (Mathieu et al. 1986). Maize root tips acclimated to a low-oxygen environment show stronger tolerance to anoxia (Saglio et al. 1988; Xia and Roberts 1994). The effects of the acclimation to hypoxia were not dependent on the cellular ATP content and the acclimated roots could survive and regulate pHc even after the ATP content and energy charge value had been significantly reduced by metabolic inhibitors (Xia et al. 1995). In root hairs of *Medicago sativa*, the switch from pHc in the "aerobic" state to that in the "anaerobic" state was not correlated with proton pump activity, showing that changes in pump activity did not affect pHc (Felle 1996). Gerendás and Schurr (1999) theoretically assessed a role of the proton pump in intracellular pH regulation. They concluded that the electrogenic pumps cannot directly control the pH of adjacent compartments, but that the pumps energize secondary active ion transport that results in pH changes. Glucose uptake into sycamore cells under normoxia was driven by the proton gradient formed by the proton pump, whereas under anoxic conditions no glucose uptake occurred because the PM H^+-ATPase was not functioning (Gout et al. 2001). However, glucose uptake recovered soon after the transfer to normoxic conditions. It has also been proposed that the diminution of the ATP pool is not the cause of the inhibition of proton pump under anoxia and that Ca^{2+} released into the cytoplasm (Subbaiah et al. 1994, 1998) may act as a messenger to inhibit the proton pump.

On the other hand, there is indirect evidence to support the idea that the PM H^+-ATPase does play a role in regulating intracellular pH under anoxia. Saint-Ges et al. (1991) proposed that the initial cytoplasmic acidosis induced by anoxia is attributed to the hydrolysis of NTP and that proton extrusion by the PM H^+-ATPase and V-ATPase could be involved in the subsequent fluctuations of pHc. The effects of vanadate, a potent inhibitor of PM H^+-ATPase, on apoplastic alkalinization of potato leaves were similar to those of anoxic conditions, suggesting that PM H^+-ATPase is involved in the intracellular pH regulation (Savchenko et al. 2000). In barley leaves exposed to anoxic conditions, apoplastic alkalinization caused by the inhibition of the PM H^+-ATPase continued for as long as the supply of energy remained (Felle 2006). He concluded that anoxia tolerance depends on maintaining the proton-motive force and H^+ turnover at a level that guarantees sufficient energy is harvested to overcome the crisis. Thus, the residual activity of the PM H^+-ATPase is important under anoxic conditions not only for H^+ removal from the cytoplasm, but also to provide further secondary transport.

Some researchers have been interested in the strong anoxia tolerance of aquatic plants. Changes of pHc in root tips of rice, an anoxia-tolerant plant, under hypoxic and anoxic conditions were compared with those of wheat, an anoxia-intolerant plant (Kulichikhin et al. 2007). The initial acidification was similar in two plants following the onset of anoxia. Subsequently the rice pHc partially recovered and stabilized, but the wheat pHc continued to fall, showing that rice has a stronger system to prevent acidosis of the cytoplasm. These results suggested that both the PM H^+-ATPase and the action of the biochemical pH-stat were important in stabilizing pHc in rice roots. It has also been observed that rice coleoptile cells continued to take up K^+ under anoxia and that the K^+ flux and pHc regulation were maintained at a pH as low as 3.5 (Kulichikhin et al. 2009). It was proposed that a possible mechanism contributing to pHc regulation under anoxia is the action of proton pumping on the plasma membrane and tonoplast.

Overwintering asexual organs of two *Potamogeton* species are known to survive and grow under anoxia for long periods, stabilizing pHc above pH 7.0 (Summers et al. 2000; Dixon et al. 2006; Koizumi et al. 2011). The elongation of *Potamogeton* stems is stimulated under anoxia. The anoxic elongation of *P. distinctus* turions is concomitant with the active extrusion of H^+ equivalent from the tissues, maintaining high enzymatic activity of the PM H^+-ATPase (Koizumi et al. 2011). Stimulators of the PM H^+-ATPase, IAA and fusicoccin (FC), promoted both anoxic elongation and H^+ extrusion, and potent inhibitors of the PM H^+-ATPase, vanadate and erythrosine B, inhibited them. Unlike *P. distinctus*, pea is a typical plant showing weak anoxia tolerance. Anoxia stops the elongation of epicotyls completely, severely inhibits the activity of the PM H^+-ATPase, and leads to cell death. Interestingly, pretreatment with FC greatly improves the survival rate of anoxic pea epicotyls (Hara and Ishizawa, unpublished data). These results provided strong evidence that the activity of the PM H^+-ATPase is involved in the survival of plant tissues under anoxia.

4.2 Vacuolar Proton Pump

The vacuole can also act as a source or sink for H^+ and H^+ equivalents. One of the causes for cell acidification under anoxia is thought to be the leakage of acid from the vacuole, because pHv is lower than pHc (Roberts et al. 1984b). The electrogenic H^+ pumps, V-ATPase and V-PPase, are major components of the tonoplast (Maeshima 2000; Martinoia et al. 2007). The proton gradient between the cytosol and the vacuole is a crucial parameter as it affects not only most biochemical reactions but also secondary active transport across the tonoplast. V-ATPase and V-PPase represent up to 30 % of total tonoplast protein and there have been many studies on the functions of these pumps (Davies 1997; Beyenbach and Wieczorek 2006; Schumacher 2006; Gaxiola et al. 2007; Silva and Gerós 2009; Schumacher and Krebs 2010). Usual pHv of plant cells is kept around 5.5. A critical role of the V-ATPase in vacuolar acidification has been demonstrated in two yeast mutants, in which the genes encoding the subunits of the V-ATPase were disrupted. The two mutants were restricted to growth within a very narrow pH range around 5.5 (Nelson and Nelson 1990). However, information on the role of the V-ATPase in the regulation of intracellular pH under anoxia is limited in plant cells. Bafilomycin A_1 (BAF), a specific inhibitor of the V-ATPase, caused alkalinization of the pHv of maize root hair cells under normoxia. However, under anoxia BAF was ineffective in disrupting vacuolar acidity, suggesting that the V-ATPase does not contribute to maintenance of the pHv regulation in the absence of oxygen (Brauer et al. 1997).

Recently, a role of the V-PPase in regulation of intracellular pH under anoxia has received special attention because plants can use PPi as an energy source when the availability of ATP is limited (Huang et al. 2008; Igamberdiev and Kleczkowski 2011). In rice seedlings, the transcript and protein level of the V-PPase increased greatly under anoxia and the enzymatic activity in a tonoplast fraction increased progressively with the duration of anoxia, suggesting that the V-PPase, rather than the V-ATPase that acts under normoxia, plays an important role in the survival of rice plants under anoxia (Carystinos et al. 1995; Huang et al. 2005). Liu et al. (2010) confirmed that only *OVP3*, one of six V-PPase genes, is consistently upregulated by anoxia. Although direct evidence to show that proton pumps located on the tonoplast are involved in intracellular pH regulation has not yet been obtained, it is possible that the V-PPase alleviates acidification of the cytosol by translocating protons across the tonoplast into vacuoles.

5 Concluding Remarks

Both biophysical and biochemical pH-stat mechanisms are undoubtedly involved in pHc and pHv regulation under anaerobic conditions. These mechanisms respond immediately after anoxia, and soon the values of pHc and pHv reach a stationary level, depending on plant species. Supply of energy to maintain intracellular homeostasis is essential for plant cells to survive under prolonged anoxia. Plant species that have a strong tolerance to anoxia, for example rice and *Potamogeton*, are likely to be able to keep on supplying energy for a longer period. Generally ethanol fermentation is the main source to supply energy in anoxic plant cells. Starch degradation and sucrose metabolism to supply the substrate for ethanol fermentation are very important for *P. distinctus* turions to survive under anoxia (Sato et al. 2002; Harada et al. 2005), and also for rice coleoptiles (Huang et al. 2003). Simultaneously it is essential to maintain the proton gradients across the plasma membrane and the tonoplast. These proton gradients are believed to be produced by the action of the PM H^+-ATPase, V-ATPase, and V-PPase. However, direct evidence to show the involvement of these proton pumps in intracellular pH regulation under anoxia is still limited. Further physiological and molecular biological experiments are needed to examine the actions of proton pumps under anoxia and these should lead to a better understanding of their role in intracellular pH regulation and anoxia tolerance of plant cells.

Acknowledgments The author thanks Professor R. George Ratcliffe, Oxford University, Professor Timothy D. Colmer, and Dr. Hank Greenway, The University of Western Australia, for a critical reading of the manuscript.

References

Bailey-Serres J, Voesenek LACJ (2008) Flooding stress: acclimations and genetic diversity. Annu Rev Plant Biol 59:313–339

Beyenbach KW, Wieczorek H (2006) The V-type H^+ ATPase: molecular structure and function, physiological roles and regulation. J Exp Biol 209:577–589

Brauer D, Uknalis J, Triana R, Shachar-Hill Y, Tu S-I (1997) Effects of bafilomycin A_1 and metabolic inhibitors on the maintenance of vacuolar acidity in maize root hair cells. Plant Physiol 113:809–816

Carroll AD, Fox GG, Laurie S, Phillips R, Ratcliffe RG, Stewart GR (1994) Ammonium assimilation and the role of γ-aminobutyric acid in pH homeostasis in carrot cell suspensions. Plant Physiol 106:513–520

Carystinos GD, MacDonald HR, Monroy AF, Dhindsa RS, Poole RJ (1995) Vacuolar H^+-translocating pyrophosphatase is induced by anoxia or chilling in seedlings of rice. Plant Physiol 108:641–649

Chang WWP, Huang L, Shen M, Webster C, Burlingame AL, Roberts JKM (2000) Patterns of protein synthesis and tolerance of anoxia in root tips of maize seedlings acclimated to a low-oxygen environment, and identification of proteins by mass spectrometry. Plant Physiol 122:295–317

Choi W-G, Roberts DM (2007) *Arabidopsis* NIP2;1, a major intrinsic protein transporter of lactic acid induced by anoxic stress. J Biol Chem 282:24209–24218

Couldwell DL, Dunford R, Kruger NJ, Lloyd DC, Ratcliffe RG, Smith AMO (2009) Response of cytoplasmic pH to anoxia in plant tissues with altered activities of fermentation enzyme: application of methyl phosphonate as an NMR pH probe. Ann Bot 103:249–258

Davies DD (1980) Anaerobic metabolism and the production of organic acids. In: Davies DD (ed) The biochemistry of plants. A comprehensive treatise, vol 2, Metabolism and respiration. Academic, New York, pp 581–611

Davies JM (1997) Vacuolar energization: pumps, shunts and stress. J Exp Bot 48:633–641

Dixon MH, Hill SA, Jackson MB, Ratcliffe RG, Sweetlove LJ (2006) Physiological and metabolic adaptations of *Potamogeton pectinatus* L. tubers support rapid elongation of stem tissue in the absence of oxygen. Plant Cell Physiol 47:128–140

Drew MC (1997) Oxygen deficiency and root metabolism: injury and acclimation under hypoxia and anoxia. Annu Rev Plant Physiol Plant Mol Biol 48:223–250

Duby G, Boutry M (2009) The plant plasma membrane proton pump ATPase: a highly regulated P-type ATPase with multiple physiological roles. Pflugers Arch 457:645–655

Ellis MH, Dennis ES, Peacock WJ (1999) Arabidopsis roots and shoots have different mechanisms for hypoxic stress tolerance. Plant Physiol 119:57–64

Fan TW-M, Higashi RM, Lane AN (1988) An *in vivo* ^{1}H and ^{31}P NMR investigation of the effect of nitrate on hypoxic metabolism in maize roots. Arch Biochem Biophys 266:592–606

Fan TW-M, Lane AN, Higashi RM (1992) Hypoxia does not affect rate of ATP synthesis and energy metabolism in rice shoot tips as measured by ^{31}P NMR *in vivo*. Arch Biochem Biophys 294:314–318

Felle HH (1993) Ion-selective microelectrode: their use and important in modern plant cell biology. Bot Acta 106:5–12

Felle HH (1996) Control of cytoplasmic pH under anoxic conditions and its implication for plasma membrane proton transport in *Medicago sativa* root hair. J Exp Bot 47:967–973

Felle HH (2005) pH regulation in anoxic plants. Ann Bot 96:519–532

Felle HH (2006) Apoplastic pH during low-oxygen stress in barley. Ann Bot 98:1085–1093

Fox GG, McCallan NR, Ratcliffe RG (1995) Manipulating cytoplasmic pH under anoxia: a critical test of the role of pH in the switch from aerobic to anaerobic metabolism. Planta 195:324–330

Gaxiola RA, Palmgren MG, Schumacher K (2007) Plant proton pumps. FEBS Lett 581:2204–2214

Gerendas J, Ratcliffe RF (2013) Root pH regulation. In: Eshel A, Beeckman T (eds) Plant roots: the hidden half, 4th edn. CRC Press, Boca Raton, Chapter 23, pp 1–18

Gerendás J, Schurr U (1999) Physicochemical aspects of ion relations and pH regulation in plants—a quantitative approach. J Exp Bot 50:1101–1114

Gibbs J, Greenway H (2003) Mechanisms of anoxia tolerance in plants. I. Growth, survival and anaerobic catabolism. Funct Plant Biol 30:1–47

Gout E, Bligny R, Douce R (1992) Regulation of intracellular pH values in higher plant cells. Carbon-13 and phosphorus-31 nuclear magnetic resonance studies. J Biol Chem 267:13903–13909

Gout E, Boisson A-M, Aubert S, Douce R, Bligny R (2001) Origin of the cytoplasmic pH changes during anaerobic stress in higher plant cells. Carobon-13 and phosphorous-31 nuclear magnetic resonance studies. Plant Physiol 125:912–925

Greenway H, Gibbs J (2003) Mechanisms of anoxia tolerance in plants. II. Energy requirements for maintenance and energy distribution to essential processes. Funct Plant Biol 30:999–1036

Guern J, Felle H, Mathieu Y, Kurkdjian A (1991) Regulation of intracellular pH in plant cells. Int Rev Cytol 127:111–173

Harada T, Satoh S, Yoshioka T, Ishizawa K (2005) Expression of sucrose synthase genes involved in enhanced elongation of pondweed (*Potamogeton distinctus*) turions under anoxia. Ann Bot 96:683–692

Huang S, Greenway H, Colmer TD (2003) Anoxia tolerance in rice seedlings: exogenous glucose improves growth of an anoxia-'intolerant' but not of a 'tolerant', genotype. J Exp Bot 54:2363–2373

Huang S, Greenway H, Colmer TD, Millar AH (2005) Protein synthesis by rice coleoptiles during prolonged anoxia: implications for glycolysis, growth and energy utilization. Ann Bot 96:703–715

Huang S, Colmer TD, Millar AH (2008) Does anoxia tolerance involve altering the energy currency towards PPi? Trends Plant Sci 13:221–227

Igamberdiev AU, Kleczkowski LA (2011) Magnesium and cell energetics in plants under anoxia. Biochem J 437:373–379

Ishizawa K, Murakami S, Kawakami Y, Kuramochi H (1999) Growth and energy status of arrowhead tubers, pondweed turions and rice seedlings under anoxic conditions. Plant Cell Environ 22:505–514

Kawai M, Samarajeewa PK, Barrero RA, Nishiguchi M, Uchimiya H (1998) Cellular dissection of the degradation pattern of cortical cell death during aerenchyma formation of rice roots. Planta 204:277–287

Kennedy RA, Rumpho ME, Fox TC (1992) Anaerobic metabolism in plants. Plant Physiol 100:1–6

Koizumi Y, Hara Y, Yazaki Y, Sakano K, Ishizawa K (2011) Involvement of plasma membrane H^+-ATPase in anoxic elongation of stems in pondweed (*Potamogeton distinctus*) turions. New Phytol 190:421–430

Kosegarten H, Grolig F, Wieneke J, Wilson G, Hoffmann B (1997) Differential ammonia-elicited changes of cytosolic pH in root hair cells of rice and maize as monitored by 2′,7′-bis-(2-carboxyethyl)-5 (and -6)-carboxyfluorescein-fluorescence ratio. Plant Physiol 113:451–461

Kulichikhin KY, Aitio O, Chirkova TV, Fagerstedt KV (2007) Effect of oxygen concentration on intracellular pH, glucose-6-phosphate and NTP content in rice (*Oryza sativa*) and wheat (*Triticum aestivum*) root tips: in vivo ^{31}P-NMR study. Physiol Plant 129:507–518

Kulichikhin KY, Greenway H, Byrne L, Colmer TD (2009) Regulation of intracellular pH during anoxia in rice coleoptiles in acidic and near neutral conditions. J Exp Bot 60:2119–2128

Kurkdjian A, Guern J (1989) Intracellular pH: measurement and importance in cell activity. Annu Rev Plant Physiol Plant Mol Biol 40:271–303

Libourel IGL, van Bodegom PM, Fricker MD, Ratcliffe RG (2006) Nitrite reduces cytoplasmic acidosis under anoxia. Plant Physiol 142:1710–1717

Liu Q, Zhang Q, Burton RA, Shirley NJ, Atwell BJ (2010) Expression of vacuolar H^+-pyrophosphatase (*OVP3*) is under control of an anoxia-inducible promoter in rice. Plant Mol Biol 72:47–60

Lloyd P (2004) Strong ion calculator—a practical bedside application of modern quantitative acid-base physiology. Crit Care Resusc 6:285–294

Maeshima M (2000) Vacuolar H^+-pyrophosphatase. Biochim Biophys Acta 1465:37–51

Martinoia E, Maeshima M, Neuthaus HE (2007) Vacuolar transporters and their essential role in plant metabolism. J Exp Bot 58:83–102

Mathieu Y, Guern J, Pean M, Pasquier C, Beloeil J-C, Lallemand J-Y (1986) Cytoplasmic pH regulation in *Acer pseudoplatanus* cells. II. Possible mechanisms involved in pH regulation during acid-load. Plant Physiol 82:846–852

Menegus F, Cattaruzza L, Chersi A, Fronza G (1989) Differences in the anaerobic lactate-succinate production and in the changes of cell sap pH for plants with high and low resistance to anoxia. Plant Physiol 90:29–32

Menegus F, Cattaruzza L, Mattana M, Beffagna N, Ragg E (1991) Response to anoxia in rice and wheat seedlings. Changes in the pH of intracellular compartments, glucose-6-phosphate level, and metabolic rate. Plant Physiol 95:760–767

Michelet B, Boutry M (1995) The plasma membrane H^+-ATPase. A highly regulated enzyme with multiple physiological functions. Plant Physiol 108:1–6

Miyashita Y, Good AG (2008) Contribution of the GABA shunt to hypoxia-induced alanine accumulation in roots of *Arabidopsis thaliana*. Plant Cell Physiol 49:92–102

Nelson H, Nelson N (1990) Disruption of genes encoding subunits of yeast vacuolar H^+-ATPase causes conditional lethality. Proc Natl Acad Sci U S A 87:3503–3507

Ratcliffe RG (1995) Metabolic aspects of the anoxic response in plant tissue. In: Smirnoff N (ed) Environment and plant metabolism: flexibility and acclimation environment and plant biology series. BIOS Scientific Publishers, Oxford, pp 111–127

Ratcliffe RG (1997) *In vivo* NMR studies of the metabolic response of plant tissues of anoxia. Ann Bot 79(suppl A):39–48

Ratcliffe RG (1999) Intracellular pH regulation in plants under anoxia. In: Egginton S, Taylor EW, Raven JA (eds) Regulation of tissue pH in plants and animals, A reappraisal of current techniques. Cambridge University Press, Cambridge, pp 193–213

Raven JA (1985) Regulation of pH and generation of osmolarity in vascular plants: a cost-benefit analysis in relation to efficiency of use of energy, nitrogen and water. New Phytol 101:25–77

Rea PA, Poole RJ (1993) Vacuolar H^+-translocating pyrophosphatase. Annu Rev Plant Physiol Plant Mol Biol 44:157–180

Rivoal J, Ricard B, Pradet A (1991) Lactate dehydrogenase in *Oryza sativa* L. seedlings and roots. Identification and partial characterization. Plant Physiol 95:682–686

Roberts JKM, Wemmer D, Ray PM, Jardetzky O (1982) Regulation of cytoplasmic and vacuolar pH in maize root tips under different experimental conditions. Plant Physiol 69:1344–1347

Roberts JKM, Callis J, Wemmer D, Walbot V, Jardetzky O (1984a) Mechanism of cytoplasmic pH regulation in hypoxic maize root tips and its role in survival under hypoxia. Proc Natl Acad Sci U S A 81:3379–3383

Roberts JKM, Callis J, Jardetzky O, Walbot V, Freeling M (1984b) Cytoplasmic acidosis as a determinant of flooding intolerance in plants. Proc Natl Acad Sci U S A 81:6029–6033

Roberts JKM, Abdrade FH, Anderson IC (1985) Further evidence that cytoplasmic acidosis is a determinant of flooding intolerance in plants. Plant Physiol 77:492–494

Roberts JKM, Hooks MA, Miaullis AP, Edwards S, Webster C (1992) Contribution of malate and amino acid metabolism to cytoplasmic pH regulation in hypoxic maize root tips studied using nuclear magnetic resonance spectroscopy. Plant Physiol 98:480–487

Roos A, Boron WF (1981) Intracellular pH. Physiol Rev 61:296–434

Saglio PH, Drew MC, Pradet A (1988) Metabolic acclimation to anoxia induced by low (2–4 kPa partial pressure) oxygen pretreatment (hypoxia) in root tips of *Zea mays*. Plant Physiol 86:61–66

Saint-Ges V, Roby C, Bligny R, Pradet A, Douce R (1991) Kinetic studies of the variations of cytoplasmic pH, nucleotide triphosphates (^{31}P-NMR) and lactate during normoxic and anoxic transitions in maize root tips. Eur J Biochem 200:477–482

Sakano K (2001) Metabolic regulation of pH in plant cells: role of cytoplasmic pH in defense reaction and secondary metabolism. Int Rev Cytol 206:1–44

Sato T, Harada T, Ishizawa K (2002) Stimulation of glycolysis in anaerobic elongation of pondweed (*Potamogeton distinctus*) turions. J Exp Bot 53:1847–1856

Savchenko G, Wiese C, Neimanis S, Hedrich R, Heber U (2000) pH regulation in apoplastic and cytoplasmic cell compartments of leaves. Planta 211:246–255

Schulte A, Lorenzen I, Böttcher M, Plieth C (2006) A novel fluorescent pH probe for expression in plants. Plant Methods 2:7–19

Schumacher K (2006) Endomembrane proton pumps: connecting membrane and vesicle transport. Curr Opin Plant Biol 9:595–600

Schumacher K, Krebs M (2010) The V-ATPase: small cargo, large effects. Curr Opin Plant Biol 13:724–730

Shelp BJ, Bown AW, McLean MD (1999) Metabolism and functions of gamma-aminobutyric acid. Trends Plant Sci 4:446–452

Silva P, Gerós H (2009) Regulation by salt of vacuolar H^+-ATPase and H^+- pyrophosphatase activities and Na^+/H^+ exchange. Plant Signal Behav 4:718–726

Smith FA, Raven JA (1976) H^+ transport and regulation of cell pH. In: Lüttge U, Pitman MG (eds) Encyclopedia of plant physiology. New series, vol 2 part A Cells. Springer, Berlin, pp 317–346

Smith FA, Raven JA (1979) Intracellular pH and its regulation. Annu Rev Plant Physiol 30:289–311

Stewart PA (1978) Independent and dependent variables of acid-base control. Respir Physiol 33:9–26

Stoimenova M, Libourel IGL, Ratcliffe RG, Kaiser WM (2003) The role of nitrate reduction in the anoxic metabolism of roots. II. Anoxic metabolism of tobacco roots with or without nitrate reductase activity. Plant Soil 253:155–167

Subbaiah CC, Zhang JK, Sachs MM (1994) Involvement of intracellular calcium in anaerobic gene expression and survival in maize seedlings. Plant Physiol 105:369–376

Subbaiah CC, Buch DS, Sachs MM (1998) Mitochondrial contribution to anoxic Ca^{2+} signal in maize suspension-cultured cells. Plant Physiol 118:759–771

Summers JE, Jackson MB (1994) Anaerobic conditions strongly promote extension by stems of overwintering tubers of *Potamogeton pectinatus* L. J Exp Bot 45:1309–1318

Summers JE, Raticliffe RG, Jackson MB (2000) Anoxia tolerance in the aquatic monocot *Potamogeton pectinatus*: absence of oxygen stimulates elongation in association with an unusually large Pasteur effect. J Exp Bot 51:1413–1422

Sze H (1985) H^+-translocating ATPases: advances using membrane vesicles. Annu Rev Plant Physiol 36:175–208

Sze H, Li X, Palmtren MG (1999) Energization of plant cell membranes by H^+-pumping ATPases: regulation and biosynthesis. Plant Cell 11:677–689

Tournaire-Roux C, Sutka M, Javot H, Gout E, Gerbeau P, Luu D-T, Bligny R, Maurel C (2003) Cytosolic pH regulates root water transport during anoxic stress through gating of aquaporins. Nature 425:393–397

Tsuji H (1972) Respiratory activity in rice seedlings germinated under strictly anaerobic conditions. Bot Mag Tokyo 85:207–218

Vartapetian BB, Sachs MM, Fagerstedt KV (2008) Plant anerobic stress. II. Strategy of avoidance of anaerobiosis and other aspects of plant life under hypoxia and anoxia. Plant Stress 2:1–19

Xia J-H, Roberts JKM (1994) Improved cytoplasmic pH regulation, increased lactate efflux, and reduced cytoplasmic lactate levels are biochemical traits expressed in root tips of whole maize seedlings acclimated to a low-oxygen environment. Plant Physiol 105:651–657

Xia J-H, Roberts JKM (1996) Regulation of H^+ extrusion and cytoplasmic pH in maize root tips acclimated to a low-oxygen environment. Plant Physiol 111:227–233

Xia J-H, Saglio PH (1992) Lactic acid efflux as a mechanism of hypoxic acclimation of maize root tips to anoxia. Plant Physiol 100:40–46

Xia J-H, Saglio P, Roberts JKM (1995) Nucleotide levels do not critically determine survival of maize root tips acclimated to a low-oxygen environment. Plant Physiol 108:589–595

Part II
Molecular Responses

Transcriptional Regulation Under Low Oxygen Stress in Plants

Beatrice Giuntoli and Pierdomenico Perata

Abstract Adaptations to low oxygen envision common strategies aimed at either optimizing energy production and consumption, or limiting the extent of oxidative stress and reducing the exposure to low oxygen. Coordination of gene expression is essential to achieve the broad transcriptional reconfiguration that takes place in all organism subjected to oxygen limitations. Recent discoveries have highlighted that, in Arabidopsis, transcriptional adjustments to oxygen depletion are triggered by the ethylene response factor (ERF) RAP2.12 that, being subjected to an oxygen-dependent proteolytic mechanism, can act like a bridge directly linking oxygen sensing and regulation of low oxygen-responsive genes. This low oxygen sensing system bears some resemblance with the well-known HIF pathway operating in metazoans, which similarly involves the post-translational stabilization of a master activator of hypoxic gene expression. RAP2.12 is constitutively accumulated in the cell and quickly relocated into the nucleus upon hypoxia to promote early transcriptional responses. Several additional hypoxia-inducible transcription factors have been identified in plants, which are supposed to regulate downstream responses: although their functions remain largely undetermined, some of them have been associated to plant adaptations and survival to submergence, flooding, hypoxic or anoxic stresses. Information about the interplay among different components of the transcriptional regulatory network is now required to gain deeper understanding of the way in which plants are able to shape transcription in response to changing oxygen availability.

B. Giuntoli (✉) • P. Perata
Plantlab, Institute of Life Sciences, Scuola Superiore Sant'Anna, 56124 Pisa, Italy
e-mail: b.giuntoli@sssup.it

J.T. van Dongen and F. Licausi (eds.), *Low-Oxygen Stress in Plants*, Plant Cell Monographs 21, DOI 10.1007/978-3-7091-1254-0_5,

1 Evolution of Coordinated Regulatory Systems of Transcription in Response to Changing Oxygen Levels

Life on Earth originated under strictly anaerobic conditions: the primordial earth atmosphere was composed by what we would nowadays consider a very toxic mixture of carbon dioxide, hydrogen, methane and reduced nitrogen gases (Nisbet and Fowler 2011). For about two billion years, during the Archean and Paleoproterozoic eras (between around 3.7 and 2.4 billion years ago), prokaryotic organisms evolved in a reducing environment (Liu et al. 2012). Therefore, the fundamental features of cell biology, including primary metabolic pathways (glycolysis), nucleic acids synthesis and regulation of transcription and translation, were developed under such reducing conditions.

The acquisition of photosynthetic capacities by cyanobacteria was the turning point triggering oxygen replenishment in the atmosphere, although oxygen levels during the Precambrian period were limited to 2–4 % pO_2, due to oxygen incorporation into earth minerals and other buffering factors (Stamati et al. 2011). After the evolution of the first eukaryotic green cells (around 1.5 billion years ago) and in the subsequent billion years, oxygen levels raised up to 10–20 % pO_2, so that multicellular eukaryotes had to face the presence of increasingly high oxygen levels in the atmosphere. Oxygen availability as the final electron acceptor for bioenergetic processes, while allowing the cell to acquire more efficient energy-producing pathways to couple with glycolysis, also poses serious challenges for the protection of sensitive molecules from oxidation. Indeed, mass extinctions in the Precambrian era have been attributed to the toxicity of reactive oxygen by-products of the oxidative metabolism (Webster 2003). Ambient oxygen has thus been hypothesized to be a major selective force driving both the symbiotic integration of respiratory and photosynthetic bacteria and the formation of separate compartments inside the eukaryotic cells (Flück et al. 2007).

During evolution, gene regulatory pathways were adapted to changing oxygen levels. The transition to a facultative anaerobic or even fully aerobic lifestyle is reflected by the adoption of various strategies to convey information regarding oxygen availability towards ultimate transcriptional responses. Signal transduction pathways converge onto early transcription factors (TFs) that promote transcriptional responses and, in some cases, transcriptional cascades: this general mechanism, for example, is what enables channelling the fate of cells during cell differentiation (e.g., the conversion of fibroblast-like precursor cells into mature adipocytes in mammals; Siersbæk et al. 2012) and after developmental decisions (e.g., the transition to the mitotic phase in yeast cells; Gurevich and Kassir 2010), or mediates hormonal responses (see the coordinated transcriptional network that lies downstream of ethylene signalling in plants; Schaller 2012). In the hypoxic response, gene expression can be adjusted onto the cellular oxygen status directly through oxygen-sensitive early transcription factors (e.g., the iron-sulfur cluster-containing protein FNR; Kiley and Beinert 2003), or rather upon oxygen-dependent regulation of early TFs mediated by non-TF oxygen sensors (e.g., heme-containing two component bacterial systems such as FixL, DosT and globins; the ArcB kinase; prolyl hydroxylases in mammals; the Ofd1 dioxygenase in fungi; Georgellis et al. 2001; Aragonés et al. 2009; Green et al. 2009; Lee et al. 2011).

It should be appreciated that diverse regulatory strategies converge onto the early TFs to meet the cell needs. In the hypoxic responses, for instance, the functions of genes related to anaerobic metabolism can be down-regulated in the presence of oxygen, while those needed for the aerobic metabolism are stimulated; or, on the other hand, anaerobic responses can be induced or derepressed as oxygen levels drop. One example of the first mechanism is the oxygen-dependent degradation or functional inactivation of constitutive TFs, which are able to promote the hypoxic metabolism: this can be viewed as a convergent evolutionary mechanism acting upon the FNR, ArcA/B and FixJ/L proteins in eubacteria, the Sre1 (SREBP) factor in fission yeast, HIF in metazoans and RAP2.12 in plants (see below). As a different strategy, in a facultative anaerobe like baker's yeast, normoxia can activate the heme-binding TF Hap1 and cause the downstream accumulation of Rox1, which in turn represses anaerobic gene transcription by binding a conserved CCATTGTTCTC *cis*-sequence (Klinkenberg et al. 2005).

Early oxygen-regulated TFs have the task of orchestrating the broad spectrum of processes that forms the anaerobic response of the organism. In bacteria, coordinated gene regulation is achieved by means of operon systems, whereby clusters of functionally related genes can be transcribed as a unit and controlled by a single operator promoter element. In *Escherichia coli*, the FNR and ArcA/B systems (which, along with the FixJ/L system, constitute main switches for aerobic/anaerobic metabolism) control over 100 bacterial operons each (Gunsalus and Park 1994; Constantinidou et al. 2006). FNR is able to bind typical helix-turn-helix factor-specific promoter motifs (NTTGATNNNNATCAAN; Webster 2003). Targeted operons are involved in respiratory metabolism, TCA cycle, glyoxylate shunt, and fatty acid degradation. No glycolytic operon is, however, controlled by these systems: notably, evidence is missing of direct low oxygen regulation of glycolytic genes, which constitute the cornerstone of the anaerobic metabolism, in either bacteria or fungi. This seems to represent, then, a later acquisition, possibly related to multicellularity and genome reorganization and expansion in higher eukaryotes (Webster 2003).

Orchestrating transcription of functionally related genes is a more challenging issue in the case of eukaryotic genomes, where linkage of genes belonging to a single pathway is seldom observed, as a consequence of multiple rearrangement events occurred during genome expansion (Ben-Shahar et al. 2007; Ling et al. 2009; Mozes-Koch et al. 2012). Coordinated transcription of scattered genes relies on the action of master regulators and on the presence of their cognate DNA *cis*-elements in the promoters of target genes, which lead to the formation of eukaryotic regulons (Keene 2007). In all metazoans, a central pathway coordinates the anaerobic response and involves a transcriptional regulator known as HIF, hypoxia-inducible factor (Semenza 2007a, b). HIF works as a heterodimer of a constitutive β subunit and an O_2-regulated α subunit, which is controlled by two distinct oxygen-dependent mechanisms affecting both its abundance and transcriptional activity

(Aragonés et al. 2009). On one side, in fact, the presence of oxygen enables HIF-α hydroxylation by prolyl hydroxylase domain proteins (PHDs), targeting an HIF-α O_2-dependent degradation domain (ODDD), which initiates a VHL-dependent polyubiquitylation and proteasomal degradation process. On the other side, the C-terminal transcriptional activation domain (CAD) of HIF-α can be made inactive through hydroxylation by the FIH-1 (asparaginyl hydroxylase factor-inhibiting HIF) protein, which prevents HIF-α interaction with its co-activators p300/CBP. Upon low oxygen, then, HIF-α becomes stabilized and functional. Parallel post-translational and multiple regulation determines the activity of the key hypoxic SREBP (sterol regulatory element-binding proteins) bHLH TFs in fission yeast and filamentous fungi (Porter et al. 2012); SREBP are able to bind mammalian SRE *cis*-elements (TCACNCCAC) and regulate the nonrespiratory oxygen-requiring pathways of ergosterol, heme, sphingolipid, and ubiquinone biosynthesis (Todd et al. 2006).

Transcriptional activation of target genes is achieved through recognition by HIF of a conserved hypoxia response promoter element (HRE, consensus [G/A] CGTG; Wenger et al. 2005), originally mapped onto the erythropoietin gene promoter (Wang et al. 1995). Computational surveys of the distribution of HREs and chromatin immunoprecipitation studies identified hundreds of HIF-binding sites in the human genome (Schoedel et al. 2011). Confirmed direct targets of HIF include genes involved in the central energy metabolism (glycolysis, glucose uptake, mitochondrial respiration), antioxidant systems, oxygen transport, angiogenesis, cell proliferation and apoptosis (Bruick 2003), globally highlighting the central role of HIF in the maintenance of oxygen homeostasis in animal cell and its importance in mediating the consequences of many disease states (Semenza 2007a).

A general feature of biological systems is the capacity to exploit a combination of multiple regulatory factors in order to facilitate tight and dynamic modulation of the responses to fluctuating stimuli. Coherently, a picture is emerging for the transcriptional regulation of the hypoxic response, whereby a network of growing complexity is established around HIF. Selectivity and specificity to HIF action can, thus, be provided by cooperative interaction with additional transcription factors, targeting dedicated *cis*-elements in the vicinity of the HRE, that concur to shaping the hypoxic response in animals. Integration of promoter-specific studies, genome-wide analyses of HIF-binding sites and computational predictions of enriched sequence motifs underpinned cooperative promoter regulation by AP-1 family members, Sp1/Sp3 zinc finger factors, Smads, CREBP (cAMP response element-binding proteins) and USFs (Sánchez-Elsner et al. 2004; Liao et al. 2007; Archer 2011; Villar et al. 2012).

The HIF pathway constitutes the best available description of how living organism can tune gene transcription and adapt oxygen homeostasis to variable oxygen conditions. The observation that the HIF pathway appeared early during the evolution of metazoans, (dating back to the Precambrian period, according to some estimates; Loenarz et al. 2010; Taylor and McElwain 2010), but is absent in fungi and plants points out a connection between establishment of the HIF pathway and evolution of highly mobile sea and land species (Webster 2003). A large comparison of transcriptional adjustments across kingdoms confirmed that low-oxygen signalling components and TFs are much less conserved than the downstream elements of the response (Mustroph et al. 2010). Plant genomes encode hypoxia-inducible prolyl hydroxylases (Vlad et al. 2007), but to date they are not known to catalyze the modification of any transcription factor. Moreover, plants did not evolved homologues of the HIF-α bHLH-PAS (Per/ARNT/Sim) family protein. PAS domains, which are responsible for HIF dimerization, are widely-distributed sensing and interaction modules of ancient bacterial origin (Möglich et al. 2009) that can be found, among the others, in bacterial heme-containing proteins. The observation that, in plants, PAS domains have only been associated to light perception (Möglich et al. 2010), while they are never coupled with bHLH factors, raised the hypothesis that the association of PAS and bHLH domains might be a metazoan feature that played a crucial role for the building of the HIF pathway (Rytkönen et al. 2011). Therefore, evidence of the evolutionary divergence of low oxygen sensing and transduction mechanisms in plant organisms has, reasonably, channeled intensive research into the reconstruction of the transcriptional networks that mould the response in the plant kingdom and, in particular, in higher plants.

2 Anaerobic *Cis*-Elements and *Trans*-Acting Factors in Plant Low Oxygen Responses

Research on the regulatory mechanisms driving plant transcription under oxygen deprivation dates back to 30 years ago. The root system of maize was initially selected as a model to describe the molecular events triggered by oxygen deprivation in plant tissues and organs, since deep penetration into the soil by plant root systems entails the presence of broad adaptations to minimal oxygen tensions in their cells. Exposure of maize root tips to hypoxic conditions caused quick repression of aerobic protein synthesis and the concurrent onset of two 33 kDa transition polypeptides, which was followed by the synthesis of about 20 anaerobic proteins (ANPs) in the subsequent 90 min (Sachs et al. 1980; Bailey-Serres and Freeling 1990). Most ANPs were related to glycolytic or fermentative carbohydrate metabolism and included two alcohol dehydrogenases, glucose phosphate isomerase, aldolase and lactate dehydrogenase. The observation that the absence of such an anaerobic response in maize leaves correlated with their extreme sensitivity to hypoxia boosted the quest for promoter sequences accounting for this organ-specific stress response.

Investigations on the promoter of maize *Adh1* gene, combining in vivo footprinting with dimethyl sulfate, gel retardation assays and mutational analyses of proximal *Adh1* promoter fragments, lead to the first identification of hypoxia-responsive sequences in plant promoters (Paul and Ferl 1997). A 40 bp-long region, lying between positions −140 and −99 from the transcription start site (TSS), was found to be responsible for the anaerobic activation of the gene and designated anaerobic response element (ARE) (Walker et al. 1987). Its further dissection revealed the involvement of two different types of *cis*-elements, namely GC- and GT-elements, in the response. Afterward, AREs with similar organization were recovered in the promoters of other maize hypoxia-responsive genes, such as *aldolase* (Dennis et al. 1988), *lactate dehydrogenase* (Good and Paetkau 1992) and *GapC4* (Geffers et al. 2000), and in the Arabidopsis genes *ADH*, *PDC1* and *GAPDH* (Dolferus et al. 1994; Sachs 1991).

In maize, two GC-elements (GC[G/C]CC) were protected from dimethyl sulfate degradation upon binding by a hypothetical factor, called GCBP-1 (GC-binding protein), present in nuclear extracts of either hypoxic or aerobic maize cells (Olive et al. 1990, 1991). The GCBP-1 factor was neither purified, nor appeared sufficient for the anaerobic activation of the downstream sequence, suggesting the requirement for additional ARE-binding factors (Olive et al. 1991). In fact, two in vivo footprinted GT-elements ([T/C]GGTTT), or half G-boxes, were related to *Adh1* response to hypoxia (de Vetten and Ferl 1995). A maize protein was isolated that was able to recognize *Adh1* G-boxes, namely the GBF1 hypoxia-inducible leucine zipper protein; bZIP factors are, indeed, known to bind plant G-box elements (Izawa et al. 1993). GBF1 is able to form heterodimers with another maize bZIP factor, GIP1, that may enhance GBF DNA-binding activity (Sehnke et al. 2005), as well as with GF14ω, a 14-3-3 protein equipped with a Ca^{2+}-binding domain (EF hand). It has been suggested that, as a part of the trans-acting complex that binds the G-box *cis*-element, GF14ω might act as a nexus interrelating both calcium- and kinase-dependent pathways during hypoxic responses (de Vetten et al. 1992). Also in this case it was concluded that GBF1 alone was not sufficient to control the induction of the gene by hypoxia.

The Arabidopsis ARE was, furthermore, shown to mediate *ADH* induction through binding of the MYB2 transcription factor to a specific MYB-binding site (MBS-1) (Hoeren et al. 1998). During anoxia, MYB2 needs to interact with its partner MYC2, a bHLH G-box-binding factor, in a transcriptional complex that is also responsible for ABA-dependent gene up-regulation in response to drought and cold stress (Abe et al. 2003). However, anoxic *ADH* induction is not affected in *myb2* knock-out plants (Licausi et al. 2010) suggesting that functional redundancy with other MYB or MYB-like factors could account for these contrasting results. Additional MYB factors have been connected to hypoxia in plants, namely TaMyb in wheat (Lee et al. 2007) and Mybleu in rice (Magaraggia et al. 1997; Locatelli et al. 2000; Mattana et al. 2007). Finally, in rice MYBS1 regulates the starvation-dependent hydrolysis of starch during hypoxia through the *Ramy3D* α-amylase gene (Lee et al. 2009).

The search for hypoxia-related TFs was enhanced by the possibility to study genome-wide changes in transcription, as many microarray studies indicated the differential expression of numerous TF-coding genes as a hallmark of the hypoxic response in Arabidopsis (Klok et al. 2002; Liu et al. 2005; Loreti et al. 2005; van Dongen et al. 2009; Mustroph et al. 2009; Hsu et al. 2011), rice (Lasanthi-Kudahettige et al. 2007; Narsai et al. 2009), cotton (Christianson et al. 2010) and poplar (Kreuzwieser et al. 2009). Scanning of the upstream sequences of hypoxia-responsive genes, moreover, lead to the identification of enriched DNA motifs (Mohanty et al. 2005; Licausi et al. 2011a) and consequent in silico prediction of transcription factors involved in the hypoxic signalling cascade. These efforts confirmed the role of well-known TF families in the low-oxygen signalling cascade (ERF, MYB and MYC TFs), uncovered the involvement of new factors, such as ANAC102 (Christianson et al. 2009), LBD and MYB-like trihelix proteins (Licausi 2011), and shed light on the participation of possible regulators belonging to the DOF, homeobox, ARF, AUX/IAA and zinc finger families (Licausi et al. 2011a).

Among the differentially expressed genes found in the available microarray datasets, Arabidopsis members of the bHLH family are now known to be widely represented in the set of both up- and down-regulated genes upon hypoxia (Licausi et al. 2011a). Nonetheless, while in metazoans PAS-domain-conjugated bHLH HIF proteins play a unique role in metazoan hypoxia sensing, in plants neither MYC2 nor any other bHLH factor has been related to low oxygen tolerance. GC-box elements, recognized by bHLH TFs, are found in the AREs, but in plants are rather related to *alcohol dehydrogenase* gene induction by cold, dehydration and UV light rather than hypoxia (Dolferus et al. 1994). The hypothesis has been put forward that bHLH may take part to plant hypoxic responses through their ability to heterodimerize with other hypoxia-inducible TFs, such as LBD proteins (Licausi et al. 2011a).

G-box-binding bZIP factors are largely affected in low oxygen microarray datasets, although only a handful of them has been examined individually, namely GBF1 (see above) and mLIP15 (Kusano et al. 1995) in maize and ABZ1 in tomato (Sell and Hehl 2004). Intriguingly, some degree of functional complementarity has been demonstrated between maize GBF1 and the human zinc finger factor Sp1 in their ability to interact with maize *Adh1* G-boxes, suggesting a hypothetic conservation of the ancient, HIF-independent pathway of hypoxic genes modulation in plants (Discher et al. 1998). Recent evidence indicates that energy depletion signals arising from low oxygen, darkness, starvation and related stresses, after being integrated by SnRK/AMPK/Snf sensor kinases, might be transmitted to gene-specific transcriptional regulation through physical interaction between the non-DNA-binding kinases and GBF DNA-binding proteins (Baena-González et al. 2007). In particular, GBF5 and other partially redundant bZIP factors (such as ATB2, AtbZIP53 and AtbZIP44) are believed to mediate SnRK1 interaction with the promoter of its target gene *PDC1* in Arabidopsis (Cho et al. 2012).

Down-regulation of anaerobic gene expression under normal conditions is conceivable; however, no constitutively expressed repressor has been found hindering the interaction of the basal transcriptional machinery with anaerobic promoters. Some labile transcriptional repressor (similar to fungal Rox1) might carry out this task; hypoxia lability, and consequent derepression of the responsive promoters, might arise from protein turnover in the absence of ongoing transcription, or rather from some active mechanisms triggered by hypoxia. Alternatively, chromatin at the anaerobic promoters might be kept in a nonpermissive state during normoxia. Chromatin structure plays a crucial role during transcription

(Li et al. 2007a); therefore, the mechanisms of chromatin modification upon hypoxia in plants are gaining increasing attention. Reversible Lys4 methylation of histone H3, associated with the permissive state of chromatin (di-methylation) or ongoing transcription (tri-methylation), occurs in *ADH1* and *PDC1* promoters in submerged rice plants (Tsuji et al. 2006). VIN3, a histone-modifying protein that contributes to the epigenetic repression of the *FLC* locus during vernalization, has been indeed associated to Arabidopsis survival under hypoxia. However, since its loss does not affect the expression of any known component of low-oxygen response pathways, VIN3 might take part to novel mechanisms of hypoxic response (Bond et al. 2009).

Low-oxygen signalling in plants involves the production of reactive oxygen species (ROS), which interplay with ethylene during programmed cell death and aerenchyma formation in submerged tissues (Pucciariello et al. 2012; Parlanti et al. 2011; Steffens et al. 2011; Mühlenbock et al. 2007). ROS-related transcription factors, such as the zinc finger proteins ZAT10 and ZAT12 and HsfA2 (Banti et al. 2010), are known to be induced by anoxia in Arabidopsis. HsfA2, in particular, has been proposed as the only factor promoting the expression of *HSPs* genes, whose products contribute to the protection of cellular components from the oxidative stress arising from anoxia and heat. Extensive analysis of microarray datasets identifies the HsfA2 pathway as an N-end rule-independent (see below) response in plants, triggered by hydrogen peroxide sensing, that only acts under complete anoxia (Pucciariello et al. 2012). These observations reinforced the view that plants are able to discriminate between partial depletion and complete absence of oxygen.

The impact of individual signalling pathways and components on the shaping of gene expression during low oxygen stresses has been intensively investigated in the past 2 decades. From these efforts, a picture of the involved transcription factors has started delineating. Recently, our knowledge of low-oxygen signalling was considerably improved by the finding that the core transcriptional changes occurring upon oxygen deprivation rely on the group VII ERF RAP2.12 protein, whose activity as a positive transcription regulator is controlled by an oxygen-dependent mechanism of protein destabilization.

3 Group VII ERFs Are Master Regulators of Plant Transcription in Hypoxia

AP2-ERFs are a large family of plant-specific transcription factors (Sakuma et al. 2002), possibly evolved from bacterial or virus endonucleases (Magnani et al. 2004), whose implication in plant development and in the responses to abiotic and biotic stresses has been widely investigated (Xu et al. 2011; Licausi et al. 2013; Mizoi et al. 2012). Among them, members of the group VII ERF subfamily (Nakano et al. 2006) have been linked to plant adaptive responses to flooding and oxygen deprivation conditions.

The main connection between hypoxia signalling and ERFs ensues from ethylene entrapment in the tissues being a distinctive feature of plant submergence (Voesenek et al. 1993). Remarkable adaptations taking place in wetland species, such as rice and *Rumex* spp., are now known to be driven by ERF proteins (Bailey-Serres and Voesenek 2008). In rice species, in particular, different ERF members of the same group VII can mediate opposite adaptive strategies to flooding, by regulating GA-dependent shoot elongation processes: briefly, on the one hand submergence-tolerant rice varieties are able to limit their underwater shoot elongation by means of the Sub1-A protein, which promotes a quiescence strategy entailing conservative ways of carbohydrate consumption (Fukao et al. 2006; Xu et al. 2006), while on the other hand fast elongating deepwater rice varieties can trigger enhanced GA responses upon ethylene-induced accumulation of the SNORKEL1/2 ERF factors (Hattori et al. 2009). Manipulation of the abundance of group VII ERF proteins lead to altered expression of the hypoxic transcriptome also in Arabidopsis. The first candidates to be studied as possible regulators of the hypoxic response in Arabidopsis were two low oxygen-inducible ERFs, deemed HRE1 and HRE2 (Hypoxia-Responsive ERFs; Licausi et al. 2010), the closest homologs to rice Sub1-A. Absence of both *HRE1* and *2* expression, however, only caused an accelerated decline of anaerobic mRNAs during the progression of the stress, but proved to have no impact on their levels in normoxic conditions, which points out the requirement of low oxygen-dependent mechanisms for HRE functional activation.

A crucial step forward was represented, thus, by the finding that group VII ERF proteins can be directly regulated by oxygen in a post-translational fashion, so that, in the presence of normal oxygen levels, their accumulation in the nucleus can be prevented by the cell (Gibbs et al. 2011; Licausi et al. 2011b). As was demonstrated by two independent studies, Arabidopsis ERFs VII are subjected to an N-end rule mechanism for selective protein degradation through the 26S proteasome (Graciet et al. 2010): essentially, the presence of oxygen results in the oxidation of an N-terminal cysteine moiety (Cys2) of the ERFs, whose destabilization turns it into a substrate for sequential further modifications, catalyzed by the N-end rule pathway (NERP) enzymes ATE1/2 and PRT6; ERF proteins are subsequently polyubiquitinylated and eventually subjected to constitutive proteolysis (Sasidharan and Mustroph 2011; Bailey-Serres et al. 2012; Licausi 2013). Therefore, even if the gene are over-expressed, ERF proteins fail to accumulate in aerobic conditions, unless their N-terminus is removed or at least protected from redox modifications (i.e., by conjugation of an N-terminal peptide such as the HA tag, or by substitution of the redox-sensitive Cys2 residue). Evidence for this was obtained in regard to the HRE factors, thereby explaining the molecular phenotypes found by Licausi et al. (2010), and also to the remaining members of the Arabidopsis group VII subfamily, which includes three RAP2 (Related to AP2) factors called RAP2.2, RAP2.12 and RAP2.3 (Gibbs et al. 2011).

Being encoded by constitutively expressed genes (Licausi et al. 2011b), the latter factors could be considered as better candidates for early oxygen sensing than the hypoxia-inducible HREs. This hypothesis was reinforced by the discovery of a protein sequestration mechanism, specifically acting on RAP2.12, that preserves this constitutively synthesized protein from constitutive proteasomal degradation (Licausi et al. 2011b). In normoxia, RAP2.12 is targeted to the plasma membrane, where it can be stored, thanks to its interaction with the ACBP1/2 integral proteins and from which it can be quickly released upon hypoxia, to rapidly relocate into the nucleus and activate transcription. The same mechanism might extend to RAP2.12 homolog, RAP2.2, which is believed to act redundantly with RAP2.12 in the regulation of the hypoxic genes (Papdi et al. 2008).

Once RAP2.12 moves to the nucleus, the transcriptional response it drives is largely overlapping with that promoted by hypoxia (Licausi et al. 2011b). ERF subfamily members in general display some variability in their DNA-binding preferences (Licausi et al. 2013), a behavior that might be linked to their ability to coordinate disparate biological responses (Mizoi et al. 2012). The most typical GCC-box motif is not generally found in the anaerobic promoters, however, enrichment of hypoxia target datasets with a new ATCTA motif, known to be bound by RAP2.2 (Welsch et al. 2007) has been revealed by in silico analyses (Licausi et al. 2011a).

Altogether, these findings have thus made it possible to build up the first integrative model of anaerobic gene regulation in Arabidopsis (Sasidharan and Mustroph 2011; Bailey-Serres et al. 2012), where the primary oxygen signal is interpreted through a post-translational system of redox regulation that limits the function of a class of master transcriptional regulators to periods of low oxygen stress. A short conserved N-terminal motif (consensus MCGGAI/L) mediates Cys2 proteins processing through the NERP (Licausi et al. 2011b). Remarkably, rice ERF VII factors display peculiar features: SNORKEL proteins possess variant N-terminal sequences that should make them unsusceptible to oxygen-dependent proteolysis, whereas Sub1-A, despite bearing a complete consensus sequence, does not behave as an NERP substrate (Gibbs et al. 2011), likely because of other determinants located outside the consensus motif, such as crucial lysine residues. It is, therefore, tempting to link the divergence of group VII ERFs to the adaptive strategies implemented by plant species with different levels of tolerance to oxygen depletion stresses (Sasidharan and Mustroph 2011; Licausi 2013). In species that adapted to submergence, such as rice and *Rumex palustris*, ERF responses to ethylene seem to prevail over redox regulation of ERF activity by oxygen (Bailey-Serres et al. 2012); in Arabidopsis, instead, such an hormonal regulation on ERF VII proteins is retained only by the Sub1A homologs HRE1 and 2 (Yang et al. 2011). Transcriptional regulation of Arabidopsis ERFs by ethylene appears to be important in mediating systemic responses to root flooding in the unsubmerged shoots (Hsu et al. 2011).

Additional regulatory functions might be taken on by RAP2.12 as a consequence of cysteine sensitivity to the redox status of the cell (Miki and Funato 2012). However, RAP2.12 is unlikely to convey nitric oxide (NO) signals onto gene transcription, differently from what is done by HIF in metazoans, where NO acts as a fundamental signalling molecule in hypoxia and has been proven to lead to HIF stabilization (Li et al. 2007b). In plants, instead, the anaerobic accumulation of NO might have a negative impact on RAP2.12 stability, since S-nitrosylation of Cys2 residues by NO can prime NERP substrates for proteolysis (Hu et al. 2005). It has been speculated that RAP-dependent anaerobic induction of a gene encoding non-symbiotic hemoglobin 1 (*HB1*), an NO-scavenging enzyme, might represent a mechanism used by the plant to antagonize Cys2 oxidation in ERF VII factors (Licausi et al. 2013).

4 Conclusions

Our knowledge of how transcriptional networks coordinate gene expression has greatly expanded in the last few years and has culminated with the identification of the TF RAP2.12 as the oxygen sensor of plants. This discovery represents a new starting point for the research in this field, as it raises many-fold questions regarding the modulation of RAP2.12 activity and its interplay with additional regulators.

Hypoxic gene expression is initiated by RAP2.12; however, as *RAP2.12* does not respond to hypoxia, it is not likely to play a role during prolonged stresses: rather, transcription might be sustained by downstream hypoxia-inducible TFs. A possible transcriptional cascade involves the HRE factors, which have been proven to be necessary for the ongoing expression of hypoxia targets. Additional downstream activators might be looked for in the large set of TF-encoding genes identified by genome-scale transcriptional analyses.

Moreover, it will be important to evaluate the possibility of synergistic interactions among hypoxic TFs in plants. For instance, the interaction with repressor TFs could be used as a strategy to achieve quick and reversible down-regulation of the hypoxic activator(s), in light of a dynamic response of plant cells to fluctuations of oxygen levels. As a growing body of evidence demonstrates in the case of HIF, selective DNA binding by RAP2.12 and flexibility in the modulation of targets expression is likely to be facilitated either by co-binding of other transcription factors to DNA or by RAP2.12 protein–protein interactions. Therefore, the combination of high-throughput chromatin immunoprecipitation data and available hypoxic gene expression data could be exploited to identify regions in the Arabidopsis genome recognized by RAP2.12 under selected experimental conditions, which would subsequently enable evaluating the presence of enriched TF-binding sites (TFDBSs) in the surrounding genomic environment. Besides this, reconstruction of RAP2.12 protein–protein interactions could be needed to pinpoint important co-factors that participate in the regulation of hypoxic transcription without binding to DNA.

References

Abe H, Urao T, Ito T, Seki M, Shinozaki K, Yamaguchi-Shinozaki K (2003) Arabidopsis AtMYC2 (bHLH) and AtMYB2 (MYB) function as transcriptional activators in abscisic acid signalling. Plant Cell 15:63–78

Aragonés J, Fraisl J, Baes M, Carmeliet P (2009) Oxygen sensors at the crossroad of metabolism. Cell Metab 9:11–22

Archer MC (2011) Role of Sp transcription factors in the regulation of cancer cell metabolism. Genes Cancer 2:712–719

Baena-González E, Rolland F, Thevelein JM, Sheen J (2007) A central integrator of transcription networks in plant stress and energy signalling. Nature 448:938–942

Bailey-Serres J, Freeling M (1990) Hypoxic stress-induced changes in ribosomes of maize seedling roots. Plant Physiol 94:1237–1243

Bailey-Serres J, Voesenek LA (2008) Flooding stress: acclimations and genetic diversity. Annu Rev Plant Biol 59:313–339

Bailey-Serres J, Fukao T, Gibbs DJ, Holdsworth MJ, Lee SC, Licausi F, Perata P, Voesenek LA, van Dongen JT (2012) Making sense of low oxygen sensing. Trends Plant Sci 17:129–138

Banti V, Mafessoni F, Loreti E, Alpi A, Perata P (2010) The heat-inducible transcription factor HSFA2 enhances anoxia tolerance in *Arabidopsis*. Plant Physiol 152:1471–1483

Ben-Shahar Y, Nannapaneni K, Cavasant TL, Scheetz TE, Welsh MJ (2007) Eukaryotic operon-like transcription of functionally related genes in *Drosophila*. Proc Natl Acad Sci U S A 104: 222–227

Bond DM, Wilson IW, Dennis ES, Pogson BJ, Finnegan EJ (2009) VERNALIZATION INSENSITIVE 3 (VIN3) is required for the response of *Arabidopsis thaliana* seedlings exposed to low oxygen conditions. Plant J 59:576–587

Bruick RK (2003) Oxygen sensing in the hypoxic response pathway: regulation of the hypoxia-inducible transcription factor. Genes Dev 17:2614–2623

Cho YH, Hong JW, Kim EC, Yoo SD (2012) Regulatory functions of SnRK1 in stress-responsive gene expression and in plant growth and development. Plant Physiol 158:1955–1964

Christianson JA, Wilson IW, Llewellyn DJ, Dennis ES (2009) The low-oxygen-induced NAC domain transcription factor *ANAC102* affects viability of Arabidopsis seeds following low-oxygen treatment. Plant Physiol 149:1724–1738

Christianson JA, Llewellyn DJ, Dennis ES, Wilson IW (2010) Comparisons of early transcriptome responses to low-oxygen environments in three dicotyledonous plant species. Plant Signal Behav 5:1006–1009

Constantinidou C, Hobman JL, Griffiths L, Patel MD, Penn CW, Cole JA, Overton TW (2006) A reassessment of the FNR regulon and transcriptomic analysis of the effects of nitrate, nitrite, NarXL, and NarQP as *Escherichia coli* K12 adapts from aerobic to anaerobic growth. J Biol Chem 281:4802–4815

de Vetten NC, Ferl RJ (1995) Characterization of a maize G-box binding factor that is induced by hypoxia. Plant J 7:589–601

de Vetten NC, Lu G, Ferl RJ (1992) A maize protein associated with the G-box binding complex has homology to brain regulatory proteins. Plant Cell 4:1295–1307

Dennis E, Gerlach WL, Walker JC, Lavin M, Peacock WJ (1988) Anaerobically regulated aldolase gene of maize. A chimaeric origin? J Mol Biol 202:759–767

Discher DJ, Bishopric NH, Wu X, Peterson CA, Webster KA (1998) Hypoxia regulates β-enolase and pyruvate kinase-M promoters by modulating Sp1/Sp3 binding to a conserved GC element. J Biol Chem 273:26087–26093

Dolferus R, Jacobs M, Peacock WJ, Dennis ES (1994) Differential interactions of promoter elements in stress responses of the Arabidopsis *Adh* gene. Plant Physiol 105:1075–1087

Flück M, Webster KA, Graham J, Giomi F, Gerlach F, Schmitz A (2007) Coping with cyclic oxygen availability: evolutionary aspects. Integr Comp Biol 47:524–531

Fukao T, Xu K, Ronald PC, Bailey-Serres J (2006) A variable cluster of ethylene response factor-like genes regulates metabolic and developmental acclimation responses to submergence in rice. Plant Cell 18:2021–2034

Geffers R, Cerff R, Hehl R (2000) Anaerobiosis-specific interaction of tobacco nuclear factors with *cis*-regulatory sequences in the maize *Gap4* promoter. Plant Mol Biol 43:11–21

Georgellis D, Kwon O, Lin EC (2001) Quinones as the redox signal for the Arc two-component system of bacteria. Science 292:2314–2316

Gibbs DJ, Lee SC, Md Isa N, Gramuglia S, Fukao T, Bassel GW, Correia CS, Corbineau F, Theodoulou FL, Bailey-Serres J, Holdsworth MJ (2011) Homeostatic response to hypoxia is regulated by the N-end rule pathway in plants. Nature 479:415–418

Good AG, Paetkau DH (1992) Identification and characterization of a hypoxically induced maize *lactate dehydrogenase* gene. Plant Mol Biol 19:693–697

Graciet E, Mesiti F, Wellmer F (2010) Structure and evolutionary conservation of the plant N-end rule pathway. Plant J 61:741–751

Green J, Crack JC, Thomson AJ, LeBrun NE (2009) Bacterial sensors of oxygen. Curr Opin Microbiol 12:145–151

Gunsalus RP, Park SJ (1994) Aerobic-anaerobic gene regulation in *Escherichia coli*: control by the ArcAB and Fnr regulons. Res Microbiol 145:437–450

Gurevich V, Kassir Y (2010) A switch from a gradient to a threshold mode in the regulation of a transcriptional cascade promotes robust execution of meiosis in budding yeast. PLoS One 5: e11005

Hattori Y, Nagai K, Furukawa S, Song X-J, Kawano R, Sakakibara H, Wu J, Matsumoto T, Yoshimura A, Kitano H, Matsuoka M, Mori H, Ashikari M (2009) The ethylene response factors SNORKEL1 and SNORKEL2 allow rice to adapt to deep water. Nature 460:1026–1030

Hoeren FU, Dolferus R, Wu Y, Peacock WJ, Dennis ES (1998) Evidence for a role for AtMYB2 in the induction of Arabidopsis alcohol dehydrogenase gene (*ADH1*) by low oxygen. Genetics 149:479–490

Hsu F-C, Chou M-Y, Peng H-P, Chou S-J, Shih M-C (2011) Insights into hypoxic systemic responses based on analyses of transcriptional regulation in *Arabidopsis*. PLoS One 6:e28888

Hu R-G, Sheng J, Qi X, Xu Z, Takahashi TT, Varshavsky A (2005) The N-end rule pathway as a nitric oxide sensor controlling the levels of multiple regulators. Nature 437:981–986

Izawa T, Foster R, Chua NH (1993) Plant bZIP protein DNA binding specificity. J Mol Biol 230: 1131–1144

Keene JD (2007) RNA regulons: coordination of post-transcriptional events. Nat Rev Genet 8: 533–543

Kiley PJ, Beinert H (2003) The role of Fe–S proteins in sensing and regulation in bacteria. Curr Opin Microbiol 6:181–185

Klinkenberg LG, Mennella TA, Luetkenhaus K, Zitomer RS (2005) Combinatorial repression of the hypoxic genes of Saccharomyces cerevisiae by DNA binding proteins Rox1 and Mot3. Eukaryot Cell 4:649–660

Klok EJ, Wilson IW, Wilson D, Chapman SC, Ewing RM et al (2002) Expression profile analysis of the low-oxygen response in *Arabidopsis* root cultures. Plant Cell 14:2481–2494

Kreuzwieser J, Hauberg J, Howell KA, Carroll A, Rennenberg H, Millar AH, Whelan J (2009) Differential response of gray poplar leaves and roots underpins stress adaptation during hypoxia. Plant Physiol 149:461–473

Kusano T, Berberich T, Harada M, Suzuki N, Sugawara K (1995) A maize DNA binding factor with bZIP motif is induced by low temperature. Mol Gen Genet 248:1–8

Lasanthi-Kudahettige R, Magneschi L, Loreti E, Gonzali S, Licausi F, Novi G, Beretta O, Vitulli F, Alpi A, Perata P (2007) Transcript profiling of the anoxic rice coleoptile. Plant Physiol 144:218–231

Lee TG, Jang CS, Kim JY, Kim DS, Park JH, Kim DY, Seo YW (2007) A Myb transcription factor (TaMyb1) from wheat roots is expressed during hypoxia: roles in response to the oxygen concentration in root environment and abiotic stresses. Physiol Plant 129:375–385

Lee K-W, Chen P-W, Lu C-A, Chen S, Ho T-H D, Yu S-M (2009) Coordinated responses to oxygen and sugar deficiency allow rice seedlings to tolerate flooding. Sci Signal 2:ra61

Lee C-YS, Yeh T-L, Hughes BT, Espenshade PJ (2011) Regulation of the Sre1 hypoxic transcription factor by oxygen-dependent control of DNA binding. Mol Cell 44:225–234

Li B, Carey M, Workman JL (2007a) The role of chromatin during transcription. Cell 128: 707–719

Li F, Sonveaux P, Rabbani ZN, Liu S, Yan B, Huang Q, Vujaskovic Z, Dewhirst MW, Li CY (2007b) Regulation of HIF-1 alpha stability through S-nitrosylation. Mol Cell 26:63–74
Liao H, Hyman MC, Lawrence DA, Pinsky DJ (2007) Molecular regulation of the PAI-1 gene by hypoxia: contributions of Egr-1, HIF-1alpha, and C/EBPalpha. FASEB J 21:935–949
Licausi F (2011) Regulation of the molecular response to oxygen limitations in plants. New Phytol 190:550–555
Licausi F (2013) Molecular elements of low-oxygen signalling in plants. Physiol Plant 148:1–8
Licausi F, van Dongen JT, Giuntoli B, Novi G, Santaniello A, Geigenberger P, Perata P (2010) HRE1 and HRE2, two hypoxia-inducible ethylene response factors, affect anaerobic responses in *Arabidopsis thaliana*. Plant J 62:302–315
Licausi F, Weits DA, Pant BD, Scheible W-R, Geigenberger P, van Dongen JT (2011a) Hypoxia responsive gene expression is mediated by various subsets of transcription factors and miRNAs that are determined by the actual oxygen availability. New Phytol 190:442–456
Licausi F, Kosmacz M, Weits DA, Giuntoli B, Giorgi FM, Voesenek LACJ, Perata P, Van Dongen JT (2011b) Oxygen sensing in plants is mediated by an N-end rule pathway for protein destabilization. Nature 479:419–422
Licausi F, Ohme-Takagi M, Perata P (2013) APETALA2/Ethylene Responsive Factor (AP2/ERF) transcription factors: mediators of stress responses and developmental programs. New Phytol 199(3):639–649. doi:10.1111/nph.12291
Ling X, He X, Xin D (2009) Detecting gene clusters under evolutionary constraint in a large number of genomes. Bioinformatics 25:571–577
Liu F, Van Toai T, Moy LP, Bock B, Linford LD, Quackenbush J (2005) Global transcription profiling reveals comprehensive insights into hypoxic response in Arabidopsis. Plant Physiol 137:1115–11129
Liu Y, Beer LL, Whitman WB (2012) Methanogens: a window into ancient sulfur metabolism. Trends Microbiol 20:251–258
Locatelli F, Bracale M, Magaraggia F, Faoro F, Manzocchi LA, Coraggio I (2000) The product of rice myb7 unspliced mRNA dimerizes with the maize leucine zipper Opaque2 and stimulate its activity in a transient expression assay. J Biol Chem 275:17619–17625
Loenarz C, Coleman ML, Boleininger A, Schierwater B, Holland PWH, Ratcliffe PJ, Schofield CJ (2010) The hypoxia-inducible transcription factor pathway regulates oxygen sensing in the simplest animal, *Trichoplax adhaerens*. EMBO Rep 12:63–70
Loreti E, Poggi A, Novi G, Alpi A, Perata P (2005) A genome-wide analysis of the effects of sucrose on gene expression in Arabidopsis seedlings under anoxia. Plant Physiol 137: 1130–1138
Magaraggia F, Solinas G, Valle G, Giovinazzo G, Coraggio I (1997) Maturation and translation mechanisms involved in the expression of a myb gene of rice. Plant Mol Biol 35:1003–1008
Magnani E, Sjoelander K, Hakea S (2004) From endonucleases to transcription factors: evolution of the AP2 DNA binding domain in plants. Plant Cell 16:2265–2277
Mattana M, Vannini C, Espen L, Bracale M, Genga A, Marsoni M, Iriti M, Bonazza V, Romagnoli F, Baldoni E, Coraggio I, Locatelli F (2007) The rice Mybleu transcription factor increases tolerance to oxygen deprivation in Arabidopsis plants. Physiol Plant 131:106–121
Miki H, Funato Y (2012) Regulation of intracellular signalling through cysteine oxidation by reactive oxygen species. J Biochem 151:255–261
Mizoi J, Shinozaki K, Yamaguchi-Shinozaki K (2012) AP2/ERF family transcription factors in plant abiotic stress responses. Biochim Biophys Acta 1819:86–96
Möglich A, Ayers RA, Moffat K (2009) Structure and signalling mechanism of Per-ARNT-Sim domains. Structure 17:1282–1294
Möglich A, Yang X, Ayers RA, Moffat K (2010) Structure and function of plant photoreceptors. Annu Rev Plant Biol 61:21–47
Mohanty B, Krishnan SPT, Swarup S, Bajic VB (2005) Detection and preliminary analysis of motifs in promoters of anaerobically induced genes of different plant species. Ann Bot 96: 669–681

Mozes-Koch R, Gover O, Tanne E, Peretz Y, Maori E, Chernin L, Sela I (2012) Expression of an entire bacterial operon in plants. Plant Physiol 158:1883–1892

Mühlenbock P, Plaszczyca M, Plaszczyca M, Mellerowicz E, Karpinski S (2007) Lysigenous aerenchyma formation in Arabidopsis is controlled by LESION SIMULATING DISEASE1. Plant Cell 19:3819–3830

Mustroph A, Zanetti ME, Jang CJH, Holtan HE, Repetti PP, Galbraith DW, Girke T, Bailey-Serres J (2009) Profiling translatomes of discrete cell populations resolves altered cellular priorities during hypoxia in *Arabidopsis*. Proc Natl Acad Sci U S A 106:18843–18848

Mustroph A, Lee SC, Oosumi T, Zanetti ME, Yang H, Ma K, Yaghoubi-Masihi A, Fukao T, Bailey-Serres J (2010) Cross-kingdom comparison of transcriptomic adjustments to low-oxygen stress highlights conserved and plant-specific responses. Plant Physiol 152: 1484–1500

Nakano T, Suzuki K, Fujimura T, Shinshi H (2006) Genome-wide analysis of the ERF gene family in Arabidopsis and rice. Plant Physiol 140:411–432

Narsai R, Howell KA, Carrol A, Ivanova A, Millar AH, Whelan J (2009) Defining core metabolic and transcriptomic responses to oxygen availability in rice embryos and young seedlings. Plant Physiol 151:306–322

Nisbet E, Fowler CMR (2011) The evolution of the atmosphere in the archaean and early Proterozoic. Chinese Sci Bull 56:4–13

Olive MR, Walker JC, Singh K, Dennis ES, Peacock WJ (1990) Functional properties of the anaerobic responsive element of the maize *Adh1* gene. Plant Mol Biol 15:593–604

Olive MR, Peacock WJ, Dennis ES (1991) The anaerobic responsive element contains two GC-rich sequences essential for binding a nuclear protein and hypoxic activation of the maize *Adh1* promoter. Nucleic Acids Res 19:7053–7060

Papdi C, Ábrahám E, Joseph MP, Popescu C, Koncz C, Szabados L (2008) Functional identification of Arabidopsis stress regulatory genes using the controlled cDNA overexpression system. Plant Physiol 147:528–542

Parlanti S, Kudahettige NP, Lombardi L, Mensuali-Sodi A, Alpi A, Perata P, Pucciariello C (2011) Distinct mechanisms for aerenchyma formation in leaf sheaths of rice genotypes displaying a quiescence or escape strategy for flooding tolerance. Ann Bot 107(8):1335–1343

Paul A-L, Ferl R (1997) The hypoxic response of three *Alcohol Dehydrogenase* genes: in vivo and in vitro footprinting of DNA/protein interactions describes multiple signalling connections. Ann Bot 79:33–37

Porter JR, Lee CY, Espenshade PJ, Iglesias PA (2012) Regulation of SREBP during hypoxia requires Ofd1-mediated control of both DNA binding and degradation. Mol Biol Cell 23: 3764–3774

Pucciariello C, Parlanti S, Banti V, Novi G, Perata P (2012) Reactive oxygen species-driven transcription in Arabidopsis under oxygen deprivation. Plant Physiol 159:184–196

Rytkönen KT, Williams TA, Renshaw GM, Primmer CR, Nikinmaa M (2011) Molecular evolution of the metazoan PHD-HIF oxygen-sensing system. Mol Biol Evol 28:1913–1926

Sachs MM (1991) Molecular response to anoxic stress. In: Jackson MB, Davis DD, Lambers H (eds) Plant life under oxygen deprivation. SPB Academic Publishing BV, The Hague, pp 129–139

Sachs MM, Freeling M, Okimoto R (1980) The anaerobic proteins of maize. Cell 20:761–767

Sakuma Y, Liu Q, Dubouzet JG, Abe H, Shinozaki K, Yamaguchi-Shinozaki K (2002) DNA-binding specificity of the ERF/AP2 domain of Arabidopsis DREBs, transcription factors involved in dehydration- and cold-inducible gene expression. Biochem Biophys Res Commun 290:998–1009

Sánchez-Elsner T, Ramírez JR, Sanz-Rodriguez F, Varela E, Bernabéu C, Botella LM (2004) A cross-talk between hypoxia and TGF-beta orchestrates erythropoietin gene regulation through SP1 and Smads. J Mol Biol 36:9–24

Sasidharan R, Mustroph A (2011) Plant oxygen sensing is mediated by the N-end rule pathway: a milestone in plant anaerobiosis. Plant Cell 23:4173–4183

Schaller GE (2012) Ethylene and the regulation of plant development. BMC Biol 10:9

Schoedel J, Oikonomopoulos S, Ragoussis J, Pugh CW, Ratcliffe PJ, Mole DR (2011) High-resolution genome-wide mapping of HIF-binding sites by ChIP-seq. Blood 117:e207–e217

Sehnke PC, Laughner BJ, Lyerly Linebarger CL, Gurley WB, Ferl RJ (2005) Identification and characterization of GIP1, an *Arabidopsis thaliana* protein that enhances the DNA binding affinity and reduces the oligomeric state of G-box binding factors. Cell Res 15:567–575

Sell S, Hehl R (2004) Functional dissection of a small anaerobically induced bZIP transcription factor from tomato. Eur J Biochem 271:4534–4544

Semenza GL (2007a) Oxygen-dependent regulation of mitochondrial respiration by hypoxia-inducible factor 1. Biochem J 405:1–9

Semenza GL (2007b) Life with oxygen. Science 318:62–64

Siersbæk R, Nielsen R, Mandrup S (2012) Transcriptional networks and chromatin remodeling controlling adipogenesis. Trends Endocrinol Metab 23:56–64

Stamati K, Mudera V, Cheema U (2011) Evolution of oxygen utilization in multicellular organisms and implications for cell signalling in tissue engineering. J Tissue Eng 2:2041731411432365

Steffens B, Geske T, Sauter M (2011) Aerenchyma formation in the rice stem and its promotion by H_2O_2. New Phytol 190:369–378

Taylor CT, McElwain JC (2010) Ancient atmospheres and the evolution of oxygen sensing via the hypoxia-inducible factor in metazoans. Physiology 25:272–279

Todd BL, Stewart EV, Burg JS, Hughes AL, Espenshade PJ (2006) Sterol regulatory element binding protein is a principal regulator of anaerobic gene expression in fission yeast. Mol Cell Biol 26:2817–2831

Tsuji H, Saika H, Tsutsumi N, Hirai A, Nakazono M (2006) Dynamic and reversible changes in histone H3-lys4 methylation and H3 acetylation occurring at submergence-inducible genes in rice. Plant Cell Physiol 47:995–1003

van Dongen JT, Froehlich A, Ramírez-Aguilar SJ, Schauer N, Fernie AR, Erban A, Kopka J, Clark J, Langer A, Geigenberger P (2009) Transcript and metabolite profiling of the adaptive response to mild decreases in oxygen concentration in the roots of Arabidopsis plants. Ann Bot 103:269–280

Villar D, Ortiz-Barahona A, Gómez-Maldonado L, Pescador N, Sánchez-Cabo F, Hackl H, Rodriguez BAT, Trajanoski Z, Dopazo A, Huang THM, Yan PS, del Peso L (2012) Cooperativity of stress-responsive transcription factors in core hypoxia-inducible factor binding regions. PLoS One 7:e45708

Vlad F, Spano T, Vlad D, Daher FB, Ouelhadj A, Fragkostefanakis S, Kalaitzis P (2007) Involvement of prolyl 4 hydroxylases in hypoxia, anoxia and mechanical wounding. Plant Signal Behav 2:368–369

Voesenek LACJ, Banga M, Thier RH, Mudde CM, Harren FJM, Barendse GWM, Blom CWPM (1993) Submergence-induced ethylene synthesis, entrapment, and growth in two plant species with contrasting flooding resistances. Plant Physiol 103:783–791

Walker JC, Howard EA, Dennis ES, Peacock WJ (1987) DNA sequences required for anaerobic expression of the maize *ADH1* gene. Proc Natl Acad Sci U S A 84:6624–6629

Wang GL, Jiang BH, Rue EA, Semenza GL (1995) Hypoxia-inducible factor 1 is a basic-helix-loop-helix-PAS heterodimer regulated by cellular O_2 tension. Proc Natl Acad Sci U S A 92: 5510–5514

Webster KA (2003) Evolution and coordinate regulation of glycolytic enzyme genes by hypoxia. J Exp Biol 206:2911–2922

Welsch R, Maass D, Voegel T, Dellapenna D, Beyer P (2007) Transcription factor RAP2.2 and its interacting partner SINAT2: stable elements in the carotenogenesis of Arabidopsis leaves. Plant Physiol 145:1073–1085

Wenger RH, Stiehl DP, Camenisch G (2005) Integration of oxygen signalling at the consensus HRE. Sci STKE 306:re12

Xu K, Xu X, Fukao T, Canlas P, Maghirang-Rodriguez R, Heuer S, Ismail AM, Bailey-Serres J, Ronald PC, Mackill DJ (2006) Sub1A is an ethylene-response-factor-like gene that confers submergence tolerance to rice. Nature 442:705–708

Xu Z-S, Chen M, Li L-C, Ma Y-Z (2011) Functions and application of the AP2/ERF transcription factor family in crop improvement. J Integr Plant Biol 53:570–585

Yang CY, Hsu FC, Li JP, Wang NN, Shih MC (2011) The AP2/ERF transcription factor AtERF73/HRE1 modulates ethylene responses during hypoxia in Arabidopsis. Plant Physiol 156:202–212

Selective mRNA Translation Tailors Low Oxygen Energetics

Reed Sorenson and Julia Bailey-Serres

Abstract In plant cells, severe oxygen deprivation (hypoxia/anoxia) dramatically curtails the efficient production of adenosine triphosphate (ATP) via mitochondrial oxidative phosphorylation. Survival is mediated in part by limiting energy consumption and maximizing the substrate level production of ATP. This strategy entails regulation of gene expression at transcriptional, posttranscriptional, and translational levels. Based on studies on *Arabidopsis thaliana*, low oxygen stress promotes an energy conservation program that tailors cytosolic protein synthesis to the proteins required for anaerobic metabolism to maintain cell viability. For example, mRNAs that encode enzymes associated with energy-efficient sucrose catabolism and anaerobic metabolism are efficiently translated. Despite maintained abundance during low oxygen stress, mRNAs encoding proteins associated with growth, such as components of the ribosome, translational machinery, and cell walls, are poorly translated. This chapter summarizes the robust evidence of selective translation during hypoxia and proposes that an important component of this process is the sequestration of mRNAs from the translational machinery until reoxygenation. Cytosolic domains known as stress granules and processing bodies, which in animals contain translationally repressed RNA and RNA-binding proteins, are found in plants. Their possible role in mRNA sequestration and regulation of translational activity is discussed.

Abbreviations

ADH	Alcohol dehydrogenase
AlaAT	Alanine aminotransferase
ANP	Anaerobic polypeptide

R. Sorenson • J. Bailey-Serres (✉)
Department of Botany and Plant Science, Center for Plant Cell Biology, University of California, Riverside, CA 92521-0124, USA
e-mail: serres@ucr.edu

J.T. van Dongen and F. Licausi (eds.), *Low-Oxygen Stress in Plants*, Plant Cell Monographs 21, DOI 10.1007/978-3-7091-1254-0_6, © Springer-Verlag Wien 2014

ANT	Adenine nucleotide translocator
ATP	Adenosine triphosphate
CDS	Coding sequence
DCP	Decapping protein
DHH1p	DExD/H-box helicase 1
DNP	Dinitrophenol
eEF	Eukaryotic elongation factor
eIF	Eukaryotic initiation factor
GUS	β-Glucuronidase
HUP	Hypoxia-responsive unknown protein
LDH	Lactate dehydrogenase
LUC	Luciferase
mRNP	Messenger RNA ribonucleoprotein complex
P	Peptidyl
PB	Processing body
PDC	Pyruvate decarboxylase
PRD	Prion-related domain
RBP	RNA-binding protein
RH12	RNA helicase 12
RPS6	Ribosomal protein S6
SG	Stress granule
SUS	Sucrose synthase
TIA-1	T-cell intracellular antigen 1
TIAR	T-cell intracellular antigen 1-related
UBP1	Oligouridylate-binding protein 1
UTR	Untranslated region
VCS	Varicose

1 Tailoring Protein Synthesis During a Low Oxygen Energy Crisis

In plants, insufficient oxygen limits efficient production of adenosine triphosphate (ATP) by mitochondrial oxidative phosphorylation. As the terminal electron acceptor in the mitochondrial electron transport chain, oxygen completes the circuit of energetic electrons, which originate in carbon–carbon bonds of organic molecular fuel, the bulk of which is lipid or carbohydrate. Energy of the electron is harnessed and transferred to phosphoanhydride bonds of ATP, the cellular energy currency, requiring an electrochemical gradient across the inner mitochondrial membrane driving ATP synthesis by the mitochondrial F_1-ATPase. Lack of oxygen for cytochrome *c* oxidase ($K_m = 0.14$–$1.7\ \mu M$; Drew 1997), the final electron acceptor, breaks this circuit. Consequentially, levels of NADH reductant and semiubiquinone build up, and oxygenic phosphorylation slows. ATP generation from carbohydrate becomes limited to the substrate level (i.e., ATP yielding reactions of glycolysis and the tricarboxylic acid cycle). It has been proposed that cell survival of severe oxygen deprivation (i.e., <1 % oxygen) is largely controlled by the management of the cellular energy budget (Branco-Price et al. 2008; Edwards et al. 2012). Low oxygen tolerance mechanisms associated with flooding tolerance enhance the ability to maintain the cellular energy supply and reduce consumption (Bailey-Serres and Voesenek 2008, 2010).

Protein synthesis consumes a large portion of the plant cell energy currency. It involves three distinct phases: initiation, elongation, and termination (Kawaguchi and Bailey-Serres 2002). Initiation entails the selection of the mRNA by initiation factors and the scanning of its 5′-untranslated region (UTR) in the 3′ direction by the 43S preinitiation complex. This is comprised of the small ribosomal subunit, a methionine-charged tRNA, and eukaryotic initiation factor (eIF)2. The elongation phase follows recognition of the initiator methionine and the joining of the 60S ribosomal subunit through the action of several eIFs. Elongation entails the peptidyl transferase reaction catalyzed by the ribosome and translocation facilitated by eukaryotic elongation factor (eEF)2. Termination occurs when the first stop codon in the mRNA enters the peptidyl (P) site of the ribosome and involves a Release Factor. The process of translation requires ATP or GTP in steps of initiation and elongation. For example, successful scanning of the 43S preinitiation complex requires GTP to form the preinitiation complex and ATP in the scanning process, dependent upon the amount of secondary structure of the transcript 5′-UTR. In the elongation phase, at least four energetic equivalents of ATP are consumed in the addition of each amino acid to the nascent peptide. The average coding sequence (CDS) length of genes of the model plant *A. thaliana* is 1,350 bp (450 codons) (Alexandrov et al. 2006) and requires >1,800 ATP equivalents per transcript per round of translation. This roughly requires catabolism of 50 glucose molecules, under oxygen-replete conditions (30–36 mol ATP mol^{-1} glc aerobic yield) or as many as 900 glucose molecules when ATP is generated in the absence of oxygen (estimated 2–4 mol ATP mol^{-1} glc anaerobic yield).

During periods of growth, there are high demands for synthesis of core components of the cell: ribosomes, other organelles, the plasma membrane, and cell wall. After maturation of the cell, energy is invested in the production of secondary metabolites and to maintain homeostasis in a dynamic environment. For over 40 years, it has been known that oxygen deprivation causes a strong global decrease in translating polyribosomes (polysomes: multiple-ribosome loaded mRNAs). This was observed for roots of soybean (*Glycine max* L.) (Lin and Key 1967), coleoptiles and roots of maize (*Zea mays* L.) (Bailey-Serres and Freeling 1990), and whole seedlings of *A. thaliana* (Branco-Price et al. 2005). Mustroph et al. (2009) demonstrated that the reduction in protein synthesis was twofold greater in the root than the shoot of *Arabidopsis* seedlings grown on medium containing sucrose and deprived of oxygen and carbon dioxide in dim light. Thus, plant cells appear to selectively modulate protein synthesis as a means to budget energy.

Regulation of the amount of energy expended on protein synthesis benefits the stressed plant in two ways: (1) decreased energy consumption for biosynthetic activities allows reallocation to processes required for cell homeostasis, such as maintenance of the proton gradient across the plasma and vacuolar membranes and (2) restriction of protein synthesis to the newly synthesized transcripts encoding proteins and enzymes important for efficient anaerobic metabolism tailors energy investment to survival processes. This strategy prioritizes energy investment in proteins that optimize cellular metabolism for substrate level phosphorylation. Evidence of an mRNA discriminatory mechanism of translation initiation during low oxygen was evident in the seminal molecular description of the anaerobic polypeptides (ANPs) of maize (Sachs et al. 1980) and has remained an outstanding question. Here, we review the dynamic regulation of protein synthesis by oxygen availability within the context of the fundamental shift in basal metabolism and discuss the nature of selective translation in terms of ribosomal protein modification, mRNA-polysome association, and mRNA sequestration.

2 Low Oxygen Acclimation: Time Is of the Essence

Coordination of gene regulation in response to changes in oxygen availability is complex. The velocity and severity of oxygen deprivation can limit acclimation at the level of gene regulation or metabolism. The temporal dynamics of low oxygen stress imposed on an organism or organ may depend upon a variety of environmental factors, such as the rate or depth of a flood, temperature of floodwaters, amount of light and carbon dioxide available for photosynthesis the size and density of the affected organ, or the presence of other aerobic organisms. Selection pressure for flooding survival differs in riparian environments where rains slowly raise the water table over a season as compared to flood plains where plants experience rapid but transient partial to complete submergence. Indeed antithetical genetic controls of submergence responses in rice and wild species have been elucidated (Voesenek et al. 2006; Bailey-Serres and Voesenek 2008, 2010). These include the low-oxygen escape strategy that enables growth out of a slow flood and the low-oxygen quiescence strategy that enables energy conservation during a short term but deep flood. Particularly in the case of the quiescence strategy, cells of aerial and root tissue balance efficient usage of diminishing ATP to prolong survival.

Plants require energy to transition from aerobic to anaerobic metabolism because the key enzymes of fermentation are predominantly produced during the stress. Because of the requirement for energy expenditure, it can be hypothesized that reductions in oxygen levels may initiate the changes in gene expression that enable anaerobic metabolism prior to the onset of full anoxia. Although the notion that plants respond proactively at the level of respiration to declining levels of oxygen through changes in metabolism is controversial (Armstrong and Beckett 2011), there is evidence that genes associated with anaerobic metabolism are upregulated under conditions of modest oxygen deprivation (8 % O_2) prior to reaching oxygen values below the K_m of cytochrome *c* oxidase (van Dongen et al. 2009; Licausi et al. 2011). However, the recent finding that the stabilization of key transcription factors, the group VII Ethylene Response Factor proteins, suggests that as oxygen levels decline genes associated with low oxygen survival may be activated prior to the occurrence of significant inhibition of mitochondrial electron transport. Resolution of this debate might be achieved by measurement of cellular concentrations of oxygen by noninvasive technologies at the same time as monitoring gene expression. Nonetheless, evidence of pronounced activation of genes encoding proteins relevant to metabolic adaptation at intermediate external oxygen conditions was demonstrated in older literature. For example, the activation of expression of the two *alcohol dehydrogenase* (*Adh*) genes of maize was observed in cells cultured under 15–0 % oxygen. Cells incubated under 15 % oxygen showed a fourfold induction in *Adh1* and *Adh2* transcripts, with higher levels of accumulation at lower oxygen concentrations (Paul and Ferl 1991). The pretreatment of maize seedling roots or root tips with low oxygen (2–4 kPa) prior to anoxia dramatically enhanced glucose consumption, maintenance of ATP levels, ethanol biosynthesis, and cellular survival (Saglio et al. 1988; Johnson et al. 1989, 1994; Hole et al. 1992). The enhanced metabolic acclimation was partially explained by higher accumulation of transcripts encoding ADH and other key enzymes of anaerobic metabolism (Andrews et al. 1994). Of the two maize genes encoding this enzyme, *Adh1* was constitutively expressed at low levels in the root tip and strongly inducible by low oxygen stress, whereas *Adh2* transcript accumulation was only induced by low oxygen (Andrews et al. 1994).

The importance of ADH for anaerobic acclimation was demonstrated in root tips using an *adh1* loss-of-function mutant (*adh1*$^-$) (Andrews et al. 1994). In the *adh1*$^-$ genotype, the benefit of hypoxia pretreatment was magnified. Survival of rapid anoxia, in which ADH activity was not induced in the mutant or wild type, was reduced to less than 12 h as compared to 24 h of survival by wild type (*Adh1*$^+$) root tips (Roberts et al. 1984). In the mutant, pretreatment induced *Adh2*, resulting in an increase in ADH-specific activity to about half the level in untreated *Adh*$^+$ root tips. This was sufficient for *Adh1*$^-$ mutant viability, demonstrating that preinduction of modest levels of ADH is necessary and sufficient for low oxygen survival of maize seedling roots (Johnson et al. 1994). Of note, hypoxia pretreatment also enhanced viability of *Adh1*$^+$ seedlings, suggesting that *Adh1* and other genes induced or proteins synthesized during hypoxia pretreatment facilitate acclimation and survival of anoxia (Andrews et al. 1994).

3 Targeted Translation of Anaerobic Polypeptides Facilitates Metabolic Adjustment

By monitoring in vivo protein synthesis and polysome levels in maize roots deprived of oxygen it was recognized that altered protein synthesis contributed to a specific reallocation of cellular resources that benefit stress acclimation. Sachs et al. (1980) identified a group of ~20 soluble proteins that were synthesized in maize root tips during oxygen deprivation. These were designated as ANPs of maize. In this study, roots of seedlings incubated in buffer sparged with N_2(g) or Ar (g) were pulse labeled with ^{3}H-leucine or ^{35}S-methionine. To resolve the newly synthesized proteins, a cell extract of soluble proteins was prepared and separated by two-dimensional gel electrophoresis using native and denaturing dimensions, and radiolabeled proteins were detected by autoradiography. Only minor changes in total protein accumulation were observed. However, the aerobic roots synthesized de novo a more complex pattern of proteins than the oxygen-deprived roots. The 20 ANPs were observed after 6 h and for at least 70 h of the stress.

The ANPs produced in the oxygen-starved seedling roots were estimated to represent 70 % of all protein synthesis (Sachs et al. 1980). The discovery of the ANPs preceded the development of high-throughput technologies for analysis of gene expression and protein identification. Consequentially, a laborious combination of genetics and biochemistry was required to identify these proteins. The major ANPs were recognized as the products of the two *Adh* genes, encoding an enzyme required for regeneration of NAD^+ from NADH to sustain anaerobic metabolism (Sachs et al. 1980). Other ANPs that were identified included enzymes involved in carbohydrate catabolism, glycolysis, and fermentation: sucrose synthase 1 and 2 (SUS1/2) (Bailey-Serres et al. 1988; Manjunath et al. 1998), phosphoglucomutase (Manjunath et al. 1998), glyceraldehyde-3-phosphate dehydrogenase (Russell and Sachs 1991), fructose-1,6-diphosphate aldolase (Kelley and Tolan 1986), glucose-6-phosphate isomerase (Kelley and Freeling 1984), enolase (Lal et al. 1998), and pyruvate decarboxylase (PDC) (Laszlo and St. Lawrence 1983). In many cases, the increase in the synthesis of the ANPs was concomitant with elevation of their transcripts (Sachs and Ho 1986).

Despite the very clear distinctions in de novo protein synthesis from nonstressed and anoxia-treated roots, Sachs et al. (1980) found a generally similar spectrum of mRNAs present under the two conditions based on in vitro translation of the poly $(A)^+$ mRNAs in the presence of ^{35}S-methionine and the evaluation of the proteins produced by one-dimensional gel electrophoresis. This provided the first hint of a mechanism of selective synthesis of proteins during low oxygen stress. Based on the characterization of the maize ANPs, the translated mRNAs were primarily proteins that contribute to anaerobic metabolism, and hence low oxygen survival. The evaluation of the low oxygen responses in soybean roots, in a system similar to that used to study maize, also showed translational regulation based on selective synthesis in planta relative to the total cellular mRNAs translatable in vitro. In this species, only four major proteins were synthesized under anoxia, one of which was ADH (Sachs et al. 1980; Russell et al. 1990). Later studies on rice coleoptiles (*Oryza sativa* L.) provided evidence that the metabolic adjustments to low oxygen stress tend to maximize the output of substrate level ATP production, or minimize its consumption through use of pyrophosphate or UTP (Bailey-Serres and Voesenek 2008; Huang et al. 2008; see also Mustroph et al. 2013).

Key to anaerobic metabolism is the consumption of stored or soluble carbohydrates. In some species, starch catabolism to hydrolysable sugars occurs through the activity of α-amylase and is promoted by oxygen deprivation. As for the enzymes responsible for anaerobic fermentation, increases in α-amylase activity in oxygen-starved organs require protein synthesis. Intriguingly, the regulation of changes in metabolism in response to low oxygen stress was shown to be distinct between flooding-sensitive tubers of *Solanum tuberosum* and rhizomes of the flooding-tolerant species *Acorus calamus*. Although each maintained α-*AMYLASE* mRNA within their respective storage organs, only *Acorus* displayed a high level of α-*AMYLASE* mRNA translation and enzymatic activity under anoxia (Arpagaus and Braendle 2000). By contrast, *S. tuberosum* tubers had little α-*AMYLASE* mRNA translation and activity. This suggested that selective translation of mRNAs plays a role in mobilization of stored carbohydrates and metabolic activity in submerged organs of these species.

Sucrose released from storage can be consumed locally or translocated via the phloem to sink tissue where it is imported and metabolized. The hydrolysis of sucrose into two hexose phosphates can be accomplished by either of two enzymes, invertase or sucrose synthase (SUS). Two SUS genes of maize (*Sh1* and *Sus1*) were recognized as ANPs and shown to be upregulated at the mRNA (Springer et al. 1986) and protein level (Bailey-Serres et al. 1988) in seedling roots. Upon oxygen deprivation, multiple species increase levels of SUS and repress the activity of invertase. Invertase requires two equivalents of ATP to phosphorylate the two hexoses produced from sucrose. Alternatively, the SUS pathway consumes UDP and pyrophosphate and yields UTP, providing a high-energy molecule for fructose phosphorylation that can substitute for ATP (Bailey-Serres et al. 2012). Thus, the net requirement for sucrose catabolism by SUS is pyrophosphate and a dinucleotide. Pyrophosphate is the byproduct of many biochemical reactions such as the charging of tRNA by aminoacyl tRNA synthetases and, therefore, the shift to use of SUS under low oxygen stress is believed to be a mechanism of energy conservation (Zeng et al. 1999). This enzymatic shift occurs in many plants including potato in which the K_m for invertase is 7–15 mM and the K_m for SUS is 40–200 mM (Bologa et al. 2003) indicating that higher sucrose as well as increased synthesis of SUS may help to shift sucrose catabolism from INV to SUS.

Some low oxygen scenarios promote an increase in carbohydrate flux through glycolysis (the Pasteur Effect). This may occur in order to meet energy demands because the energetic yield from metabolized sugar is reduced by ~18-fold following a shift to fermentation, involving pyruvate catabolism to lactate by lactate dehydrogenase (LDH) or to alcohol by PDC production of acetaldehyde which is converted to ethanol by ADH. Both of these fermentative reactions regenerate NAD^+ needed to sustain glycolysis, although LDH activity is transient and was proposed earlier to promote a change in cytosolic pH that favors PDC activity in maize roots (Davies et al. 1974). As mentioned, PDC and ADH are ANPs of maize seedling roots.

Pyruvate can also be metabolized to other end products during low oxygen stress, including Alanine, GABA, and succinate (Limami et al. 2008; Rocha et al. 2010; Bailey-Serres et al. 2012). The production of these end-products suggests that additional ATP can be generated and NAD^+ regenerated by conversion of succinyl CoA to succinate from 2-oxoglutarate via reactions of the tricarboxylic acid cycle. This pathway requires glutamine and aspartate inputs. The finding of hypoxic accumulation of *ALANINE AMINOTRANSFERASE* (*AlaAT*) in several species, including barley (*Hordeum vulgare*), maize (*Zea mays*), soybean (*Glycine max*), and *Medicago truncatula* (Ricoult et al. 2006; see Miyashita et al. 2007), led to the discovery that AlaAT1 activity in *Arabidopsis* contributes to the rapid reconversion of accumulated Alanine to pyruvate upon reoxygenation (Miyashita et al. 2007). Hence, the production of Alanine may result in less carbon loss than production of ethanol, which escapes by diffusion out of cells.

4 Early Discovery That Low Oxygen Stress Limits Protein Synthesis

In the early 1960s, there was a basic understanding that protein synthesis involved the ribosome that catalyzed the process, mRNA that determined the sequence of the polypeptide, and tRNA that functioned as an interpreter between the ribosome and mRNA. A major breakthrough in the understanding of this process came when the interaction between the mRNA and ribosomes was deciphered. Polysomes were discovered as the dominant structure of protein synthesis in a rabbit reticulocyte lysate by Warner et al. (1962, 1963). These stable complexes could be detected when cells were osmotically lysed and cell contents sedimented through a 15–30 % sucrose density gradient. The differential centrifugation separated peaks of an optical density typical of nucleic acids (A_{260}), representing one to many ribosomes connected by a single RNA molecule. Electron micrographs of the predominant complexes showed ribosome aggregates with the number of ribosomes corresponding to the sequential order of peak density. Incorporation of ^{3}H-leucine into protein occurred in the more dense complexes, which became the size of the less active and less dense complex after brief digestion with ribonuclease. Shortly following these studies, conditions regulating polysome abundance were identified, including inhibition of mitochondrial oxidative phosphorylation and anoxia.

The first characterization of plant polysomes utilized soybean seedling roots and contributed to the general understanding of polysome function (Lin et al. 1966). These experiments included use of low oxygen stress to study polysome disassociation and association. mRNA, known as DNA-like RNA at the time, was shown to associate with polysomes. It was hypothesized that protein synthesis was energy dependent; therefore, anaerobiosis was used as a treatment to study mRNA-ribosome association. Corroborating these data, experiments with chemical treatments were used to dissect the dependencies and dynamics of translation. Lin and Key (1967) used dinitrophenol (DNP) to deplete proton gradients and thereby ATP synthesis. DNP treatment caused a dramatic depletion in polysome levels within minutes of application to soybean roots. Oxygen deprivation was also known to limit aerobic ATP production. These studies of Lin and Key (1967) revealed a number of key aspects of polysomal protein synthesis (1) both DNP and anaerobiosis caused polysome dissociation; (2) polysome depletion was not caused by a decrease in mRNA abundance as depleted polysomes recovered even when 90 % of RNA synthesis was blocked by actinomycin D, an inhibitor of RNA polymerase II; (3) the translational elongation inhibitor cycloheximide prevented polysome-mRNA dissociation upon completion of polypeptide synthesis, indicating that initiation (the recruitment of mRNA to the ribosome) was likely the control point in protein synthesis; and (4) 30 min of reaeration of oxygen-deprived roots was sufficient for recovery of large polysomes. Of note, the oxygen deprivation was applied by bubbling N_2(g) through media in a stationary flask and aeration was achieved by shaking the submerged seedlings in the flask without bubbling N_2(g).

The next steps toward understanding the nature of translational control were made using maize seedling roots. The roots of intact seedlings were treated as described by Sachs et al. (1980) and pulse labeled with either ^{35}S-methionine or ^{32}P-inorganic phosphate (Bailey-Serres and Freeling 1990). Ribosomes were isolated by differential centrifugation and fractionated on one- or two-dimensional gels to compare newly synthesized ribosome-associated proteins produced under control or oxygen deprivation conditions. The stress dramatically reduced the synthesis of most of the ribosomal proteins, with only four proteins of the complex labeled during the stress. In addition, the phosphorylation of a 31 kDa ribosomal protein was dramatically reduced by the stress. This protein was identified later as ribosomal protein S6 (RPS6) (Williams et al. 2003). In addition, Bailey-Serres and Freeling (1990) included a quantitative comparison of polysome distribution between free 40S and 60S subunits, free ribosomes (80S), small polysomes (2–5 ribosomes), and large polysomes ($\geq$5 ribosomes). The levels of large polysomes decreased following hypoxia, whereas small polysome levels changed little and free subunits dramatically increased. After 2 h reaeration of the seedlings, polysome levels partially recovered. Polysome levels can recovery rapidly upon reaeration in *A. thaliana* seedlings, as only 5 min of reaeration was sufficient to recover >50 % of the polysome decrease resulting from 2 h of hypoxia (Fig. 1a).

The finding that few proteins were synthesized and polysome levels declined during oxygen deprivation led to experiments that investigated dynamics of translation of individual mRNAs in roots of maize seedlings. The first systematic study evaluated the expression of two proteins, the low oxygen-induced ADH1 and the essential mitochondrial adenine nucleotide translocator (ANT) (Fennoy and Bailey-Serres 1995). *Adh1* was chosen as a representative ANP, and *Ant* was selected as a representative gene expressed during normal growth conditions. A comparison was made of the regulation of *Adh1* and *Ant* expression at multiple levels including evaluation of run-on nuclear transcription, steady-state transcript abundance, sedimentation density of ribosome complexes associated with the transcripts, and de novo synthesis of the two proteins. The results confirmed that hypoxia activates *Adh1* transcription, producing a stable transcript that is highly translated. By contrast, *Ant* was transcribed and its transcript was initially stable, but slowly degraded over 10–24 h of hypoxia. Under control conditions, *ANT* mRNA sedimented with large polysomes, indicating it was actively translated. By contrast, after 12 h of oxygen deprivation, *Adh1* and *Ant* gene transcripts showed a different distribution across polysome-gradient samples, with *ADH1* mRNA predominantly

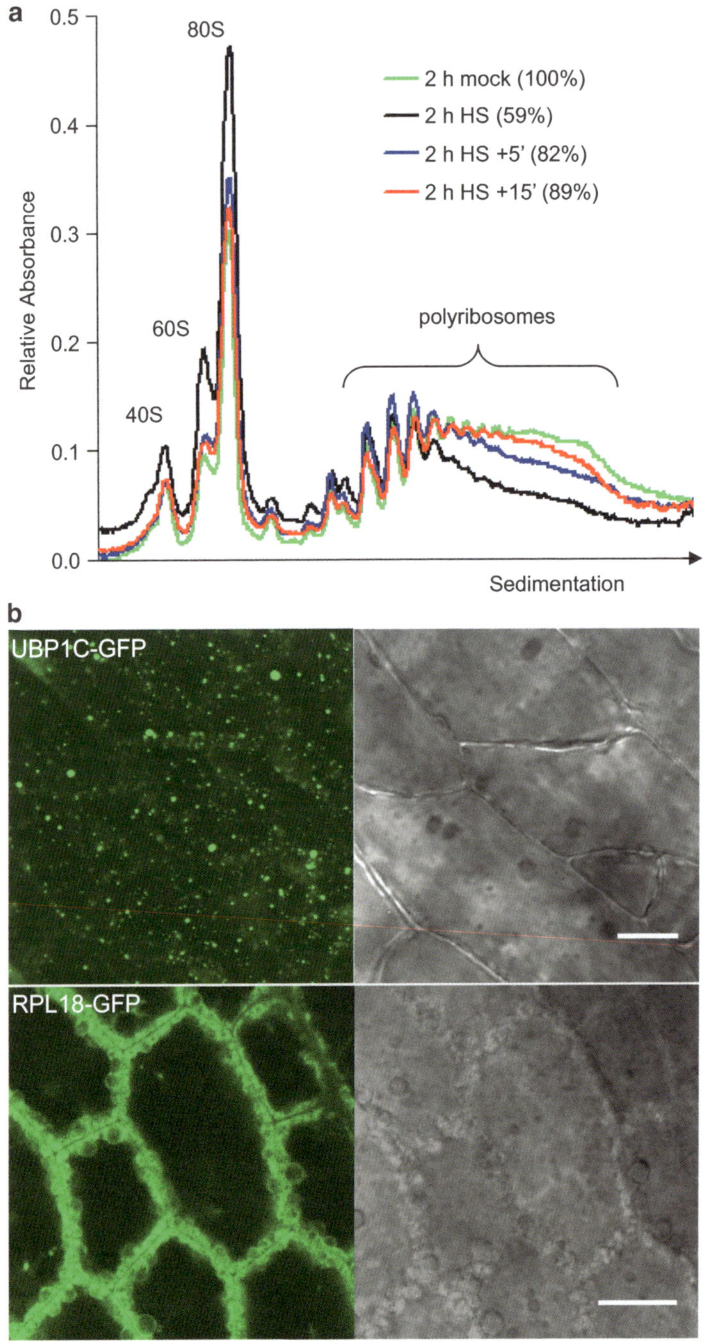

Fig. 1 Translational regulation during oxygen deprivation. (**a**) Dynamic polyribosome (polysome) regulation by oxygen availability. Polysomes of 7-day-old Arabidopsis seedlings treated

associated with large polysomes and *ANT* mRNA associated with small polysomes. The reduced association of *ANT* mRNA with polysomes corresponded to reduced de novo synthesis of ANT, as measured by pulse labeling followed by immunoprecipitation with a specific antiserum. The roots synthesized high levels of ADH1 during the stress but not under well-aerated conditions. A subsequent study evaluated the run-on transcription, steady-state mRNA levels, and polysome association of gene transcripts encoded by *Adh1*, *Adh2*, *Aldolase1*, *Enolase1*, *Sh1* and *Sus1*, and an unidentified ANP (all ANP genes) along with *Ant*, *Actin*, *Catalase2*, *Eukaryotic initiation factor4A*, *Phosphogluconate dehydrogenase1*, *Ribosomal protein P2a*, and *Ubiquitin polyprotein1* (all housekeeping genes) (Fennoy et al. 1998). This report provided additional evidence of induced transcription and selective translation of ANP mRNAs during low oxygen stress in seedling roots of maize. These directed gene studies furthered the suggestion from the results of Sachs et al. (1980) that ANP mRNAs are selectively translated during low oxygen stress.

5 Clues of mRNA Translational Selection: Ribosomal Protein Modifications

How *ADH1* and other hypoxia responsive mRNAs are selected for translation when thousands of other mRNAs are not is still largely unknown. This discrimination most likely includes both distinguishing characteristics of the mRNAs selected for translation as well as changes in the ribosome and translation factor machinery, which direct mRNAs in initiation and other posttranslational processes. Bailey-Serres and Dawe (1996) used luciferase (LUC) and β-glucuronidase (GUS) reporter systems to identify regions of the maize *ADH1* mRNA that enable its preferential translation under low oxygen stress. This was accomplished by coupling a series of truncated 5′-UTR and N-terminal coding regions of the maize *Adh1* gene to the GUS reporter. The contribution of the *ADH1* mRNA's 3′-UTR was also evaluated. mRNAs were synthesized in vitro that included the 5′-m^{7}Gppp cap and 3′-poly (A) tail required for efficient translation (Kawaguchi and Bailey-Serres 2002). The mRNAs were electroporated into protoplasts generated from cultured maize cells, along with a LUC gene with a 5′-UTR containing the viral 5′-UTR of tobacco mosaic virus that is routinely used as a translational enhancer. For the control

Fig. 1 (continued) with 2 h hypoxia stress (HS), or with an additional 5 or 10 min reaeration (R) were sedimented through a 15–60 % sucrose gradient. A_{254} was measured along the sedimentation axis. As a control, seedlings were treated under the same light, temperature, and humidity conditions in air (mock). Percentages represent the RNA in polysomes relative to the mock sample. (**b**) Granular subcellular localization of putative stress granule protein UBP1C. Seedlings overexpressing UBP1C-GFP or RPL18-GFP were submerged for 60 min under a coverslip and fluorescent protein localization was evaluated by scanning confocal microscopy. Bar = 10 μm

condition, protoplasts were gently perfused with 40 % oxygen, ensuring far lower ADH enzyme activity than in 20 % oxygen. The presence of the 5′-UTR and first 18 codons and 3′-UTR of *ADH1* mRNA enhanced reporter activity by 57-fold in transiently transformed protoplasts cultured under 5 % oxygen. Removal of only the 3′-UTR reduced the accumulation by one-third, and sequential truncation from the 3′-end of the 5′-UTR reduced the enhancement in a step-wise manner. By contrast, the LUC control mRNA was translated at tenfold lower levels under the stress condition. Importantly, the different culture conditions had no significant effect on the stability of the mRNAs tested. These data demonstrate that *ADH1* mRNA sequences are sufficient to enable a chimeric RNA with a reporter gene CDS to circumvent repression of translation.

Alterations in abundance and posttranslational modification of ribosomal proteins have also been observed in response to hypoxia and may reflect changes in ribosome activity. A reduction in a group of small, acidic ribosomal proteins was detected among the ribosomal proteins in polysome complexes (Bailey-Serres and Freeling 1990). Later these were confirmed to be the ribosomal P proteins, including one that is novel to plants (Szick et al. 1998). The P proteins form a stalk that protrudes from the large ribosomal subunit and assemble onto the ribosome in the cytoplasm and not the nucleus, as do other ribosomal proteins. In maize, four proteins make up this acidic stock, P0, a 36 kDa protein, as well as P1 (a and b), P2 (a and b), and P3, all of which are 12–15 kDa (Szick et al. 1998). The complex formed by these proteins can be salt-washed from polysomes (Bailey-Serres et al. 1997). Altered electrophoretic mobility of P1 and P3 was shown in response to hypoxia promoting the hypothesis that their posttranslational modification is involved in polysome regulation during hypoxia (Szick-Miranda and Bailey-Serres 2001). However, the ramifications of heterogeneity in P protein phosphorylation or composition that is promoted by hypoxia was not investigated in greater depth because of lack of tractable mutant resources in maize.

In addition to the ribosomal P proteins, another ribosomal protein and several initiation factors are posttranslationally modified in response to hypoxia in seedling roots or root tips of maize. Five phosphorylation sites of the carboxyl terminus of RPS6 were characterized, along with the demonstration that the protein is dephosphorylated in response to anoxia and heat (Williams et al. 2003). The reduction in phosphorylation was dependent on okadaic acid-inhibited phosphatase activity and the increase in phosphorylation was blocked by LY-294002, a compound that inhibits phosphatidylinositol-3 kinases (Williams et al. 2003). Other studies on maize seedling roots showed that phosphorylation of eIF4A and eEF2 rapidly increased and that of eIF4B decreased in response to low oxygen stress (Webster et al. 1991; Szick-Miranda et al. 2003). Also modified is the m^7-Gppp cap-binding protein, eIF4E. Two phosphorylated isoforms of eIF4E accumulated following 1 or 6 h hypoxia treatment in seedling roots (Manjunath et al. 1999). While the direct effects of altered translation factor phosphorylation in oxygen-starved seedlings are not known, these changes largely coincide with the global decrease in protein synthesis and enhanced selective mRNA translation. An increasing number of examples of ribosome heterogeneity and activity of distinct or modified ribosomes associated with specific aspects of development or environmental responses (Kim et al. 2004, 2007; Rosado et al. 2010; Szakonyi and Byrne 2011; Chen et al. 2012; Hummel et al. 2012) elevates the question of whether these modifications are functionally significant, individually and/or collectively, during hypoxia.

6 Clues of mRNA Translational Selection: Limited mRNA-Polysome Association

Hypoxia acclimation entails complex changes in gene expression in plants. These have been cataloged in numerous studies globally profiling total cellular mRNA transcripts using various microarray hybridization platforms (e.g., Klok et al. 2002; Agarwal and Grover 2005; Branco-Price et al. 2005; Liu et al. 2005; Loreti et al. 2005; Mohanty et al. 2005; Lasanthi-Kudahettige et al. 2007). Selective translation of stress-relevant gene transcripts was effectively demonstrated on a global scale in *A. thaliana* employing a technology pioneered for this purpose. Polysome immunoprecipitation was developed as a means to compare total cellular mRNA steady-state abundance with that of polysome-associated mRNAs (Zanetti et al. 2005). This method utilizes transgenic *Arabidopsis* overexpressing a His_6-FLAG-epitope-tagged ribosomal protein L18 (FLAG-RPL18), which is incorporated into fully functional ribosomes. Total cellular mRNAs and immunopurified polysomal mRNAs were compared by microarray hybridization in nonstressed and hypoxia-stressed tissue following 2 or 9 h, as well as 9 h hypoxia plus 1 h recovery (Branco-Price et al. 2008). This study further illuminated the gene expression landscape of hypoxic whole seedlings. The results validated trends in mRNA abundance and polysome association that were predicted based on polysome sedimentation experiments and work on ANPs. Polysome association decreased for a majority of cytosolic transcripts. Within this group, the total abundance of most of the individual mRNAs was maintained, demonstrating transcript stability for at least 9 h of oxygen deprivation for a large pool (65 %) of the total cellular content of mRNAs. On the other hand, a smaller pool (21 %) of translationally repressed mRNA was unstable and decreased in total abundance during the period of translational repression (Fig. 2). Selective translation was further validated by the significant induction and polysome association (presumably mRNAs undergoing active translation into protein) of a relatively large number of mRNAs, including numerous orthologs of ANPs (106 genes compared to the 20 ANPs of maize roots or four ANPs of soybean roots).

The polysome immunopurification method was further employed to probe mRNAs of specific cell types of *Arabidopsis* by limiting expression of FLAG-RPL18 by promoter fusion (Mustroph et al. 2009). This analysis revealed a group of 49 core hypoxia-responsive genes induced by all cell types. Of these, more that 50 % encode proteins of no known biological function, termed HYPOXIA RESPONSIVE UNKNOWN PROTEINS (HUPs). The investigation of the low

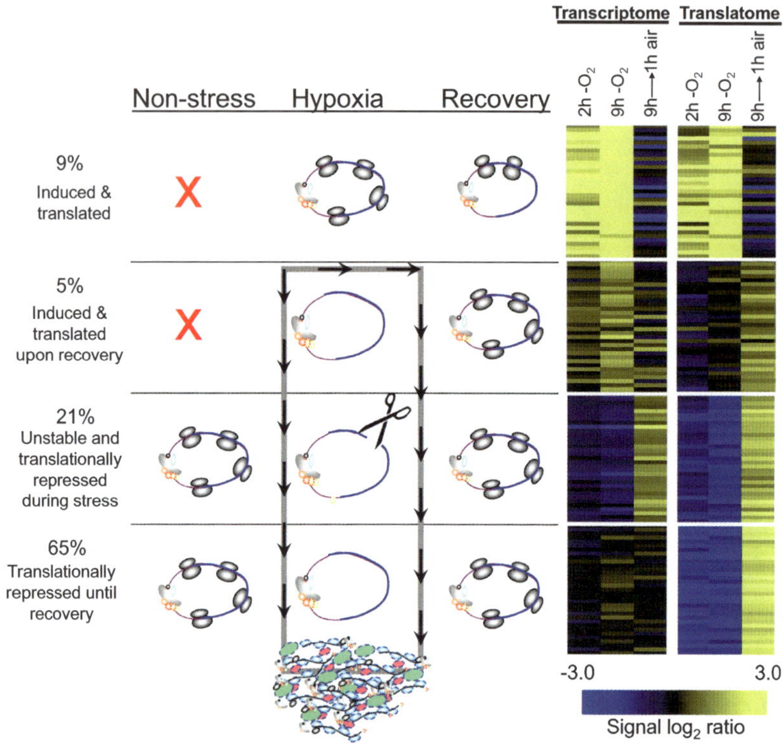

Fig. 2 Hypoxia regulation of the transcriptome and translatome. Three of four major groups of mRNAs are translationally repressed during hypoxia and predicted to be sequestered (*black arrows*) into RNA granules for storage and/or degradation. Major patterns of translational regulation were determined by Branco-Price et al. (2008) by global evaluation using Affymetrix ATH1 DNA microarrays to evaluate dynamics in total and polysome-associated RNA after 2 or 9 h of mock or oxygen deprivation ($-O_2$), or 9 h $-O_2$ plus 1 h aeration (air). The results are displayed in a heat map

oxygen and submergence survival of lines overexpressing a number of *HUP*s demonstrated that many have a functional role in low oxygen responses (Mustroph et al. 2010; Lee et al. 2011). Of the 49 core hypoxia responsive genes, nine corresponded to genes associated with anaerobic metabolism, including *ADH1*, *AlaAT1/2*, *PDC1*, *PDC2*, and *SUS4* (Mustroph et al. 2009, 2010). This finding supports the conclusion that cells are energetically autonomous and independently respond to generate their own energy currency (i.e., ATP) when oxygen deprived.

7 Clues of mRNA Translational Selection: mRNA Sequestration

Outside of regulation of the initiation, elongation, and termination phases of translation by initiation and elongation factors as well as the ribosome, the cellular domains referred to stress granules (SGs) and processing bodies (PBs) have been identified as sites of stress-triggered, translationally repressed mRNA accumulation in both animals and plants. It has been proposed that at these sites, particularly SGs, mRNAs are actively sequestered during stress manifesting their translational repression (Kedersha et al. 2005; Parker and Sheth 2007; Anderson and Kedersha 2008). These mRNA–protein complexes (messenger RNA ribonucleoprotein complexes, mRNPs) appear as granules when visualized by use of fluorescent tagging of individual protein components, and range in size from 1 to 2 μm in diameter and are up to 500,000 times more voluminous than a ribosome. Dense inclusions appear in transmission electron microscopy and can have ribosome-like structures adjacent to them (Gilks et al. 2004). The internal organization of SG/PBs remains a "black box," but the packaging of mRNAs into these granules would physically decrease their accessibility to ribosomes, driving the hypothesis that mRNA sequestration is a means of controlling protein synthesis. Evidence suggests these mRNPs are active sites of mRNA sorting, storage, turnover, and possibly silencing (Kedersha et al. 2005; Anderson and Kedersha 2008). Condition-specific translational repression, whether at the genome or gene level, can result in a failure of reinitiation of translation and subsequent sequestration into an SG, which may then facilitate storage or transfer to a PB where mRNA degradation occurs (Decker et al. 2007; Anderson and Kedersha 2008). SG and PB mRNPs may be heterogeneous in both protein and mRNA content; therefore, it may be most informative to consider specific proteins and their colocalization and dynamics.

PBs contain the enzymes that facilitate removal of the 5′-m^{7}Gppp cap (decapping proteins, DCPs) and several 5′→3′ exonucleases that digest decapped mRNA, scaffold proteins, and RNA helicases. Mutational analyses indicate that the components of plant PBs are required for proper development and response to dehydration stress. For example, the protein VARICOSE (VSC) of *Arabidopsis*, a partner of DCP1 is required for postembryonic development (Xu et al. 2006) as are DCP1 and DCP2 (Xu et al. 2006; Goeres et al. 2007; Iwasaki et al. 2007). Homozygous null mutations in genes encoding any of these three proteins lead to an arrest of postembryonic development (Xu et al. 2006), confirming that they function in indispensable processes. Several confocal laser microscopy studies that utilize fluorescent proteins for visualization have demonstrated colocalization of other proteins in cytoplasmic foci that contain DCP1. Other evidence suggests that DCP1, DCP2, and VCS interact in *bona fide* PBs. *Arabidopsis* DCP2, DCP1, and VCS coimmunoprecipitate, colocalize in cytoplasmic foci, and interact in the yeast two-hybrid system. Also, DCP2 colocalizes in protoplasts with yeast PB protein DExD/H-box helicase 1 (DHH1p) and its plant equivalent RNA helicase 12 (RH12) in transient expression assays (Xu et al. 2006). An interaction between *Arabidopsis* Exoribonuclease (XRN)4, the cytosolic ortholog of human XRN1, and both DCP2 and DCP1 was confirmed by bimolecular fluorescence complementation in tobacco (*Nicotiana plumbaginifolia*) protoplasts (Xu et al. 2006; Goeres et al. 2007; Iwasaki et al. 2007; Xu and Chua 2011). A compensatory mechanism for loss of decapping was found in the Landsberg *erecta* ecotype, giving it a phenotype of suppression in *vcs* mutants (Zhang et al. 2010). It was subsequently found that the mRNAs that were not degraded in decapping mutants were targeted for degradation by the

Suppressor of Varicose (SOV) protein. This ribonuclease was localized to cytoplasmic granules, but association with other PB proteins was not reported (Zhang et al. 2010). It will be interesting to determine if SOV is also a component of PBs.

There is evidence that PBs form in response to heat shock and hypoxia, since these stresses promoted fluorescently tagged DCP1 to accumulate in granular cytoplasmic foci in leaf protoplasts of *Arabidopsis* (Weber et al. 2008). This change in cytoplasmic distribution of DCP1 was not observed if the translational inhibitor cycloheximide was applied along with the stress, indicating that their formation requires completion of protein synthesis. The visual apparentness of DCP1-containing granules was exacerbated in a mutant with reduced cytosolic $5' \rightarrow 3'$ exonuclease activity (*xrn4-5*), consistent with the hypothesis that these foci are the site of mRNA turnover.

SGs function in mRNA sequestration, sorting and storage in animals and possibly in plants (Anderson and Kedersha 2008). In animals, these dynamic granular structures are reportedly composed of mRNAs and proteins including T-cell intracellular antigen 1 (TIA1) and T-cell intracellular antigen 1-related (TIAR), eIF4E, eIF4G, eIF4A, eIF3, PABP, G3BP1, and 40S ribosomal subunits (Anderson and Kedersha 2008). TIA1 contains a prion-related domain (PRD) that is critical for SG aggregation following stress-triggered eIF2α phosphorylation, which blocks initiation of translation under conditions of nutrient starvation (Gilks et al. 2004). Besides its role in SGs, TIA1 has also been characterized in inhibition of translation, regulation of alternative splicing, mRNA decay (Del Gatto-Konczak et al. 2000; see Yamasaki et al. 2007), and in nucleocytoplasmic transport (Zhang et al. 2005; Eisinger-Mathason et al. 2008).

In plants, the gene families of oligouridylate-binding protein 1 (UBP1), RNA-binding protein (RBP)45, and RBP47 of *A. thaliana* share amino acid sequence conservation with animal TIA-1. These proteins have demonstrated influence in pre-mRNA processing and stability. UBP1 was localized to the nucleus and cytoplasm, shown to enhance splicing of low-efficiency U-rich introns and maintain abundance of transcripts driven by specific promoters (Lambermon et al. 2000, 2002; Lorković et al. 2000). UBP1 and RBP47 colocalized with one another, eIF4E, and mRNAs in cytoplasmic granules in transfected protoplasts following heat shock. This recruitment of UBP1 to granules upon heat was dependent on conservation of RRM domain sequences and was decreased by truncation of the glutamine-rich region reminiscent of the TIA-1 PRD (Weber et al. 2008). The similarity of these characteristics with animal SG points to their near conservation across eukaryotes (Anderson and Kedersha 2008). This suggests that mRNA management is facilitated by a ubiquitous, complex, and organized system of regulated RNA-binding proteins that, complementary to regulation of polysome activity, might regulate mRNA accessibility to translational machinery and recruitment to degradation machinery.

It is hypothesized that plants cells enduring hypoxia likely employ these structures to manage translationally inactive mRNAs, as observed in mammals (Gottschald et al. 2010) (Fig. 2). Hinting at this, *Arabidopsis* UBP1C-GFP localizes in granules following 60 min coverslip submergence in stable transgenic seedlings, whereas the 60S ribosomal subunit labeled with RPL18-GFP does not (Fig. 1b). The sequestration of mRNAs from the initiation machinery could add an additional layer of translational control. Plant orthologs of mammalian RBPs that localize to PB or SG should be targeted for study, especially in the context of translational regulation in response to low oxygen stress. The studies could include monitoring the dynamics in cytosolic location of RBPs and their colocation with other proteins and mRNA. By tagging specific RBPs or production of specific antisera, the mRNPs could be isolated for evaluation of associated proteins and mRNAs. Finally, mutant analysis might reveal if SG/PB contribute to survival during oxygen deprivation.

In conclusion, energy management is a key strategy for survival of transient low oxygen stress. Activation of transcription of genes encoding proteins that facilitated anaerobic metabolism is observed in diverse species. Accompanying this transcriptional regulation is a less well-known inhibition of translation of the majority of cellular mRNAs. The mechanisms of selective translation of ANP mRNAs and *en mass* repression of most normal cellular mRNAs remain an interesting and important topic for future research.

References

Agarwal S, Grover A (2005) Isolation and transcription profiling of low-O_2 stress-associated cDNA clones from the flooding-stress-tolerant FR13A rice genotype. Ann Bot 96:831–844

Alexandrov N, Troukhan M, Brover V, Tatarinova T, Flavell R, Feldmann K (2006) Features of Arabidopsis genes and genome discovered using full-length cDNAs. Plant Mol Biol 60:69–85

Anderson P, Kedersha N (2008) Stress granules: the Tao of RNA triage. Trends Biochem Sci 33:141–150

Andrews DL, MacAlpine DM, Johnson JR, Kelley PM, Cobb BG, Drew MC (1994) Differential induction of mRNAs for the glycolytic and ethanolic fermentative pathways by hypoxia and anoxia in maize seedlings. Plant Physiol 106:1575–1582

Armstrong W, Beckett PM (2011) The respiratory down-regulation debate. New Phytol 190: 276–278

Arpagaus S, Braendle R (2000) The significance of {alpha}-amylase under anoxia stress in tolerant rhizomes (Acorus calamus L.) and non-tolerant tubers (Solanum tuberosum L., var. Desiree). J Exp Bot 51:1475–1477

Bailey-Serres J, Dawe RK (1996) Both 5′ and 3′ sequences of maize adh1 mRNA are required for enhanced translation under low-oxygen conditions. Plant Physiol 112:685–695

Bailey-Serres J, Freeling M (1990) Hypoxic stress-induced changes in ribosomes of maize seedling roots. Plant Physiol 94:1237–1243

Bailey-Serres J, Voesenek LAC (2008) Flooding stress: acclimations and genetic diversity. Annu Rev Plant Biol 59:313–339

Bailey-Serres J, Voesenek LA (2010) Life in the balance: a signaling network controlling survival of flooding. Curr Opin Plant Biol 13:489–494

Bailey-Serres J, Kloeckener-Gruissem B, Freeling M (1988) Genetic and molecular approaches to the study of the anaerobic response and tissue specific gene expression in maize. Plant Cell Environ 11:351–357

Bailey-Serres J, Vangala S, Szick K, Lee C (1997) Acidic phosphoprotein complex of the 60S ribosomal subunit of maize seedling roots (components and changes in response to flooding). Plant Physiol 114:1293–1305

Bailey-Serres J, Fukao T, Gibbs DJ, Holdsworth MJ, Lee SC, Licausi F, Perata P, Voesenek LACJ, van Dongen JT (2012) Making sense of low oxygen sensing. Trends Plant Sci 17:129–138

Bologa KL, Fernie AR, Leisse A, Ehlers Loureiro M, Geigenberger P (2003) A bypass of sucrose synthase leads to low internal oxygen and impaired metabolic performance in growing potato tubers. Plant Physiol 132:2058–2072

Branco-Price C, Kawaguchi R, Ferreira R, Bailey-Serres J (2005) Genome-wide analysis of transcript abundance and translation in Arabidopsis seedlings subjected to oxygen deprivation. Ann Bot (Lond) 96:647–660

Branco-Price C, Kaiser KA, Jang CJH, Larive CK, Bailey-Serres J (2008) Selective mRNA translation coordinates energetic and metabolic adjustments to cellular oxygen deprivation and reoxygenation in Arabidopsis thaliana. Plant J 56:743–755

Chen M-Q, Zhang A-H, Zhang Q, Zhang B-C, Nan J, Li X, Liu N, Qu H, Lu C-M, Sudmorgen et al (2012) Arabidopsis NMD3 is required for nuclear export of 60S ribosomal subunits and affects secondary cell wall thickening. PLoS One 7:e35904

Davies DD, Grego S, Kenworthy P (1974) The control of the production of lactate and ethanol by higher plants. Planta 118:297–310

Decker CJ, Teixeira D, Parker R (2007) Edc3p and a glutamine/asparagine-rich domain of Lsm4p function in processing body assembly in Saccharomyces cerevisiae. J Cell Biol 179:437–449

Del Gatto-Konczak F, Bourgeois CF, Le Guiner C, Kister L, Gesnel M-C, Stévenin J, Breathnach R (2000) The RNA-binding protein TIA-1 is a novel mammalian splicing regulator acting through intron sequences adjacent to a 5′ splice site. Mol Cell Biol 20:6287–6299

Drew MC (1997) Oxygen deficiency and root metabolism: injury and acclimation under hypoxia and anoxia. Annu Rev Plant Physiol Plant Mol Biol 48:223–250

Edwards JM, Roberts TH, Atwell BJ (2012) Quantifying ATP turnover in anoxic coleoptiles of rice (Oryza sativa) demonstrates preferential allocation of energy to protein synthesis. J Exp Bot 63:4389–4402

Eisinger-Mathason TSK, Andrade J, Groehler AL, Clark DE, Muratore-Schroeder TL, Pasic L, Smith JA, Shabanowitz J, Hunt DF, Macara IG et al (2008) Codependent functions of RSK2 and the apoptosis-promoting factor TIA-1 in stress granule assembly and cell survival. Mol Cell 31:722–736

Fennoy SL, Bailey-Serres J (1995) Post-transcriptional regulation of gene expression in oxygen-deprived roots of maize. Plant J 7:287–295

Fennoy SL, Nong T, Bailey-Serres J (1998) Transcriptional and post-transcriptional processes regulate gene expression in oxygen-deprived roots of maize. Plant J 15:727–735

Gilks N, Kedersha N, Ayodele M, Shen L, Stoecklin G, Dember LM, Anderson P (2004) Stress granule assembly is mediated by prion-like aggregation of TIA-1. Mol Biol Cell 15:5383–5398

Goeres DC, Van Norman JM, Zhang W, Fauver NA, Spencer ML, Sieburth LE (2007) Components of the Arabidopsis mRNA decapping complex are required for early seedling development. Plant Cell 19:1549–1564

Gottschald OR, Malec V, Krasteva G, Hasan D, Kamlah F, Herold S, Rose F, Seeger W, Hänze J (2010) TIAR and TIA-1 mRNA-binding proteins co-aggregate under conditions of rapid oxygen decline and extreme hypoxia and suppress the HIF-1α pathway. J Mol Cell Biol 2:345–356

Hole DJ, Cobb BG, Hole PS, Drew MC (1992) Enhancement of anaerobic respiration in root tips of Zea mays following low-oxygen (hypoxic) acclimation. Plant Physiol 99:213–218

Huang S, Colmer TD, Millar AH (2008) Does anoxia tolerance involve altering the energy currency towards PPi? Trends Plant Sci 13:221–227

Hummel M, Cordewener JHG, de Groot JCM, Smeekens S, America AHP, Hanson J (2012) Dynamic protein composition of *Arabidopsis thaliana* cytosolic ribosomes in response to sucrose feeding as revealed by label free MSE proteomics. Proteomics 12:1024–1038

Iwasaki S, Takeda A, Motose H, Watanabe Y (2007) Characterization of Arabidopsis decapping proteins AtDCP1 and AtDCP2, which are essential for post-embryonic development. FEBS Lett 581:2455–2459

Johnson J, Cobb BG, Drew MC (1989) Hypoxic induction of anoxia tolerance in root tips of Zea mays. Plant Physiol 91:837–841

Johnson JR, Cobb BG, Drew MC (1994) Hypoxic induction of anoxia tolerance in roots of Adh1 null Zea mays L. Plant Physiol 105:61–67

Kawaguchi R, Bailey-Serres J (2002) Regulation of translational initiation in plants. Curr Opin Plant Biol 5:460–465

Kedersha N, Stoecklin G, Ayodele M, Yacono P, Lykke-Andersen J, Fritzler MJ, Scheuner D, Kaufman RJ, Golan DE, Anderson P (2005) Stress granules and processing bodies are dynamically linked sites of mRNP remodeling. J Cell Biol 169:871–884

Kelley P, Freeling M (1984) Anaerobic expression of maize glucose phosphate isomerase I. J Biol Chem 259:673–677

Kelley PM, Tolan DR (1986) The complete amino acid sequence for the anaerobically induced aldolase from maize derived from cDNA clones. Plant Physiol 82:1076–1080

Kim T-H, Kim B-H, Yahalom A, Chamovitz DA, von Arnim AG (2004) Translational regulation via 5′ mRNA leader sequences revealed by mutational analysis of the Arabidopsis translation initiation factor subunit eIF3h. Plant Cell 16:3341–3356

Kim B-H, Cai X, Vaughn JN, von Arnim AG (2007) On the functions of the h subunit of eukaryotic initiation factor 3 in late stages of translation initiation. Genome Biol 8:R60

Klok EJ, Wilson IW, Wilson D, Chapman SC, Ewing RM, Somerville SC, Peacock WJ, Dolferus R, Dennis ES (2002) Expression profile analysis of the low-oxygen response in Arabidopsis root cultures. Plant Cell 14:2481–2494

Lal SK, Lee C, Sachs MM (1998) Differential regulation of enolase during anaerobiosis in maize. Plant Physiol 118:1285–1293

Lambermon MHL, Simpson GG, Kirk DAW, Hemmings-Mieszczak M, Klahre U, Filipowicz W (2000) UBP1, a novel hnRNP-like protein that functions at multiple steps of higher plant nuclear pre-mRNA maturation. EMBO J 19:1638–1649

Lambermon MHL, Fu Y, Wieczorek Kirk DA, Dupasquier M, Filipowicz W, Lorković ZJ (2002) UBA1 and UBA2, two proteins that interact with UBP1, a multifunctional effector of pre-mRNA maturation in plants. Mol Cell Biol 22:4346–4357

Lasanthi-Kudahettige R, Magneschi L, Loreti E, Gonzali S, Licausi F, Novi G, Beretta O, Vitulli F, Alpi A, Perata P (2007) Transcript profiling of the anoxic rice coleoptile. Plant Physiol 144:218–231

Laszlo A, St. Lawrence P (1983) Parallel induction and synthesis of PDC and ADH in anoxic maize roots. Mol Gen Genet 192:110–117

Lee SC, Mustroph A, Sasidharan R, Vashisht D, Pedersen O, Oosumi T, Voesenek LAC, Bailey-Serres J (2011) Molecular characterization of the submergence response of the Arabidopsis thaliana ecotype Columbia. New Phytol 190:457–471

Licausi F, Weits DA, Pant BD, Scheible W-R, Geigenberger P, van Dongen JT (2011) Hypoxia responsive gene expression is mediated by various subsets of transcription factors and miRNAs that are determined by the actual oxygen availability. New Phytol 190:442–456

Limami AM, Glévarec G, Ricoult C, Cliquet J-B, Planchet E (2008) Concerted modulation of alanine and glutamate metabolism in young Medicago truncatula seedlings under hypoxic stress. J Exp Bot 59:2325–2335

Lin CY, Key JL (1967) Dissociation and reassembly of polyribosomes in relation to protein synthesis in the soybean root. J Mol Biol 26:237–247

Lin CY, Key JL, Bracker CE (1966) Association of D-RNA with polyribosomes in the soybean root. Plant Physiol 41:976–982

Liu F, VanToai T, Moy LP, Bock G, Linford LD, Quackenbush J (2005) Global transcription profiling reveals comprehensive insights into hypoxic response in Arabidopsis. Plant Physiol 137:1115–1129

Loreti E, Poggi A, Novi G, Alpi A, Perata P (2005) A genome-wide analysis of the effects of sucrose on gene expression in Arabidopsis seedlings under anoxia. Plant Physiol 137:1130–1138

Lorković ZJ, Wieczorek Kirk DA, Klahre U, Hemmings-Mieszczak M, Filipowicz W (2000) RBP45 and RBP47, two oligouridylate-specific hnRNP-like proteins interacting with poly(A)+ RNA in nuclei of plant cells. RNA 6:1610–1624

Manjunath S, Lee C-HK, Winkle PV, Bailey-Serres J (1998) Molecular and biochemical characterization of cytosolic phosphoglucomutase in maize expression during development and in response to oxygen deprivation. Plant Physiol 117:997–1006

Manjunath S, Williams AJ, Bailey-Serres J (1999) Oxygen deprivation stimulates Ca2+-mediated phosphorylation of mRNA cap-binding protein eIF4E in maize roots. Plant J 19:21–30

Miyashita Y, Dolferus R, Ismond KP, Good AG (2007) Alanine aminotransferase catalyses the breakdown of alanine after hypoxia in Arabidopsis thaliana. Plant J 49:1108–1121

Mohanty B, Krishnan SPT, Swarup S, Bajic VB (2005) Detection and preliminary analysis of motifs in promoters of anaerobically induced genes of different plant species. Ann Bot 96:669–681

Mustroph A, Zanetti ME, Jang CJH, Holtan HE, Repetti PP, Galbraith DW, Girke T, Bailey-Serres J (2009) Profiling translatomes of discrete cell populations resolves altered cellular priorities during hypoxia in Arabidopsis. Proc Natl Acad Sci U S A 106:18843–18848

Mustroph A, Lee SC, Oosumi T, Zanetti ME, Yang H, Ma K, Yaghoubi-Masihi A, Fukao T, Bailey-Serres J (2010) Cross-kingdom comparison of transcriptomic adjustments to low oxygen stress highlights conserved and plant-specific responses. Plant Physiol 152:1484–1500

Mustroph A, Hess N, Sasidharan R (2013) Hypoxic energy metabolism and PPi as an alternative energy currency. In: Dongen JT, Licausi F (eds) Low oxygen stress in plants. Springer, New York

Parker R, Sheth U (2007) P bodies and the control of mRNA translation and degradation. Mol Cell 25:635–646

Paul A-L, Ferl RJ (1991) Adh1 and Adh2 regulation. Maydica 36:129–134

Ricoult C, Echeverria LO, Cliquet J-B, Limami AM (2006) Characterization of alanine aminotransferase (AlaAT) multigene family and hypoxic response in young seedlings of the model legume Medicago truncatula. J Exp Bot 57:3079–3089

Roberts JKM, Callis J, Jardetzky O, Walbot V, Freeling M (1984) Cytoplasmic acidosis as a determinant of flooding intolerance in plants. Proc Natl Acad Sci U S A 81:6029–6033

Rocha M, Licausi F, Araújo WL, Nunes-Nesi A, Sodek L, Fernie AR, van Dongen JT (2010) Glycolysis and the tricarboxylic acid cycle are linked by alanine aminotransferase during hypoxia induced by waterlogging of Lotus japonicus. Plant Physiol 152:1501–1513

Rosado A, Sohn EJ, Drakakaki G, Pan S, Swidergal A, Xiong Y, Kang B-H, Bressan RA, Raikhel NV (2010) Auxin-mediated ribosomal biogenesis regulates vacuolar trafficking in Arabidopsis. Plant Cell 22:143–158

Russell DA, Sachs MM (1991) The maize cytosolic glyceraldehyde-3-phosphate dehydrogenase gene family: organ-specific expression and genetic analysis. Mol Gen Genet 229:219–228

Russell DA, Wong DM-L, Sachs MM (1990) The anaerobic response of soybean. Plant Physiol 92:401–407

Sachs MM, Ho THD (1986) Alteration of gene expression during environmental stress in plants. Annu Rev Plant Physiol 37:363–376

Sachs MM, Freeling M, Okimoto R (1980) The anaerobic proteins of maize. Cell 20:761–767

Saglio PH, Drew MC, Pradet A (1988) Metabolic acclimation to anoxia induced by low (2–4 kPa partial pressure) oxygen pretreatment (hypoxia) in root tips of Zea mays. Plant Physiol 86:61–66

Springer B, Werr W, Starlinger P, Bennett DC, Zokolica M, Freeling M (1986) The Shrunken gene on chromosome 9 of Zea mays L is expressed in various plant tissues and encodes an anaerobic protein. Mol Gen Genet 205:461–468

Szakonyi D, Byrne ME (2011) Ribosomal protein L27a is required for growth and patterning in Arabidopsis thaliana. Plant J 65:269–281

Szick K, Springer M, Bailey-Serres J (1998) Evolutionary analyses of the 12-kDa acidic ribosomal P-proteins reveal a distinct protein of higher plant ribosomes. Proc Natl Acad Sci U S A 95:2378–2383

Szick-Miranda K, Bailey-Serres J (2001) Regulated heterogeneity in 12-kDa P-protein phosphorylation and composition of ribosomes in maize (Zea mays L.). J Biol Chem 276:10921–10928

Szick-Miranda K, Jayachandran S, Tam A, Werner-Fraczek J, Williams AJ, Bailey-Serres J (2003) Evaluation of translational control mechanisms in response to oxygen deprivation in maize. Russ J Plant Physiol 50:774–786

van Dongen JT, Fröhlich A, Ramírez-Aguilar SJ, Schauer N, Fernie AR, Erban A, Kopka J, Clark J, Langer A, Geigenberger P (2009) Transcript and metabolite profiling of the adaptive response to mild decreases in oxygen concentration in the roots of Arabidopsis plants. Ann Bot 103:269–280

Voesenek LACJ, Colmer TD, Pierik R, Millenaar FF, Peeters AJM (2006) How plants cope with complete submergence. New Phytol 170:213–226

Warner JR, Rich A, Hall CE (1962) Electron microscope studies of ribosomal clusters synthesizing hemoglobin. Science 138:1399–1403

Warner JR, Knopf PM, Rich A (1963) A multiple ribosomal structure in protein synthesis. Proc Natl Acad Sci U S A 49:122–129

Weber C, Nover L, Fauth M (2008) Plant stress granules and mRNA processing bodies are distinct from heat stress granules. Plant J 56:517–530

Webster C, Gaut RL, Browning KS, Ravel JM, Roberts JK (1991) Hypoxia enhances phosphorylation of eukaryotic initiation factor 4A in maize root tips. J Biol Chem 266:23341–23346

Williams AJ, Werner-Fraczek J, Chang I-F, Bailey-Serres J (2003) Regulated phosphorylation of 40S ribosomal protein S6 in root tips of maize. Plant Physiol 132:2086–2097

Xu J, Chua NH (2011) Processing bodies and plant development. Curr Opin Plant Biol 14:88–93

Xu J, Yang J-Y, Niu Q-W, Chua N-H (2006) Arabidopsis DCP2, DCP1, and VARICOSE form a decapping complex required for postembryonic development. Plant Cell 18:3386–3398

Yamasaki S, Stoecklin G, Kedersha N, Simarro M, Anderson P (2007) T-cell intracellular antigen-1 (TIA-1)-induced translational silencing promotes the decay of selected mRNAs. J Biol Chem 282:30070–30077

Zanetti ME, Chang I-F, Gong F, Galbraith DW, Bailey-Serres J (2005) Immunopurification of polyribosomal complexes of Arabidopsis for global analysis of gene expression. Plant Physiol 138:624–635

Zeng Y, Wu Y, Avigne WT, Koch KE (1999) Rapid repression of maize invertases by low oxygen. Invertase/sucrose synthase balance, sugar signaling potential, and seedling survival. Plant Physiol 121:599–608

Zhang T, Delestienne N, Huez G, Kruys V, Gueydan C (2005) Identification of the sequence determinants mediating the nucleo-cytoplasmic shuttling of TIAR and TIA-1 RNA-binding proteins. J Cell Sci 118:5453–5463

Zhang W, Murphy C, Sieburth LE (2010) Conserved RNaseII domain protein functions in cytoplasmic mRNA decay and suppresses Arabidopsis decapping mutant phenotypes. Proc Natl Acad Sci U S A 107:15981–15985

Role of Ethylene and Other Plant Hormones in Orchestrating the Responses to Low Oxygen Conditions

Bianka Steffens and Margret Sauter

Abstract Flooding results in an altered physical environment with a drastically reduced diffusion rate for gases into and from the submerged plant parts. The gaseous hormone ethylene inevitably accumulates in submerged tissue even at low rates of synthesis. Not surprisingly, ethylene is used by plants to regulate many of the adaptations plants have evolved to cope with submergence stress. Even more so, the ethylene precursor 1-aminocyclopropane-1-carboxylic acid acts as a mobile signal to promote a systemic response in root waterlogged plants. Regulation of ethylene synthesis and signaling contribute to timely and coordinated adaptive responses. In many instances ethylene acts in concert with gibberellic acid (GA) and abscisic acid (ABA) signaling either by interacting with the biosynthetic or catabolic pathways or with signaling pathways. Changes in ethylene, GA and ABA levels, and consequently in hormone signaling bring about many of the adaptive responses to flooding including acceleration or deceleration of shoot growth, adventitious root growth, and aerenchyma formation.

1 Introduction

Ethylene is a gaseous hormone that moves by diffusion. The rate of gas diffusion is dependent on the surrounding medium and is 10.000-fold reduced in water over that in a gas atmosphere (Armstrong 1979). Lowered diffusion drastically reduces the escape of ethylene gas from flooded plant tissues such that ethylene accumulates in submerged plants even at low rates of synthesis due to its gaseous nature and the property of the physical environment. Since ethylene is not biodegradable by plants its concentration will increase by default in submerged plant tissues. It is therefore

B. Steffens • M. Sauter (✉)
Plant Developmental Biology and Plant Physiology, Institute of Botany, University of Kiel, 24118 Kiel, Germany
e-mail: bsteffens@bot.uni-kiel.de; msauter@bot.uni-kiel.de

J.T. van Dongen and F. Licausi (eds.), *Low-Oxygen Stress in Plants*, Plant Cell Monographs 21, DOI 10.1007/978-3-7091-1254-0_7, © Springer-Verlag Wien 2014

not surprising that ethylene has evolved as a key hormone in submergence adaptation (Yasumura et al. 2012).

Plants adapt to flooding at multiple levels which range from metabolic reorganization to anatomical adaptations and morphological reprogramming. Among the adaptations to flooding which are controlled by ethylene are the formation of aerenchyma, enhanced or repressed shoot growth, and growth of adventitious roots. These adaptations alter plant structure or morphology to improve uptake and endogenous distribution of oxygen. Other adaptations such as the formation of a rhizodermal diffusion barrier in roots aim to reduce loss of oxygen. Still, another line of response aims to improve the plants' ability to cope with reduced oxygen tension. This is most obviously seen in metabolic changes, although aerenchyma formation not only improves gas exchange but also reduces the number of cells that consume O_2. Hence, aerenchyma also contributes to improve the energy balance. Metabolic adjustment is a general feature of hypoxic plant tissues. It can go along with the quiescent strategy, which is manifest in shoot growth arrest accompanied by reduced energy consumption or with rapid shoot elongation typical of an escape response. All modes of adjustment rely on transcriptional reprogramming and many are controlled by ethylene which in turn interacts with gibberellic acid (GA) and abscisic acid (ABA) synthesis and signaling in various ways as will be discussed in this chapter.

2 Ethylene as a Key Hormone in Submergence Adaptation

2.1 Transcriptional Regulation of ACS and ACO Genes

The volatile plant hormone ethylene is synthesized from *S*-adenosylmethionine, an activated form of methionine (Fig. 1). The ethylene-specific synthesis pathway is initiated by the conversion of *S*-adenosylmethionine to 1-aminocyclopropane-1-carboxylic acid (ACC) by ACC synthase (ACS). Synthesis of ACC is considered the rate-limiting step in ethylene biosynthesis in most tissues and is thus critical for the regulation of ethylene abundance (Yang and Hoffman 1984). ACS proteins are encoded in higher plants by small gene families with eight members in Arabidopsis, six members in rice, and nine members in tomato (Rzewuski and Sauter 2008; Harpaz-Saad et al. 2012). ACS enzymes are divided into three groups based on their divergent C-terminal domains. Type-1 ACS proteins have a relatively long C-terminus with designated phosphorylation sites for modification by mitogen-activated protein kinase (MAPK) and calcium-dependent protein kinase (CDPK). Type-2 ACS proteins have a single predicted CDPK phosphorylation site and type-3 ACS proteins share a very short C-terminus with no predicted modification site (Harpaz-Saad et al. 2012). The second enzymatic step committed to ethylene synthesis is catalyzed by ACC oxidase (ACO), a dioxygenase which converts ACC to ethylene, CO_2, HCN, and H_2O in the presence of molecular oxygen and

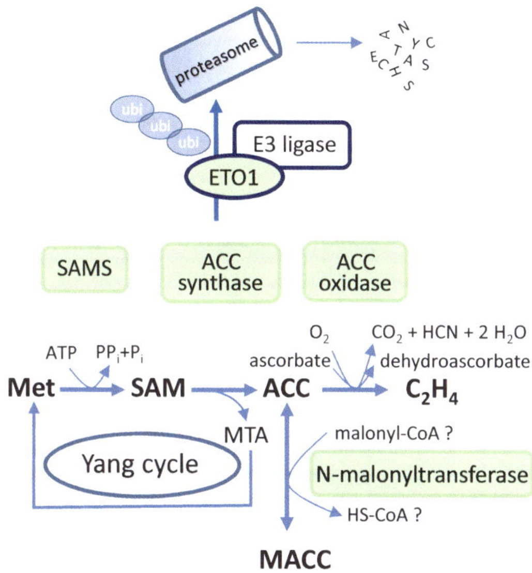

Fig. 1 Ethylene metabolic pathway. Ethylene is synthesized from methionine (Met) which is activated to *S*-adenosylmethionine (SAM) by SAM synthetases (SAMS). In the ethylene-specific biosynthesis steps, 1-aminocyclopropane-1-carboxylic acid (ACC) is synthesized by ACC synthetase (ACS) releasing the by-product 5′-methyl-thio-adenosine (MTA) which is recycled to Met via the Yang cycle. ACC oxidase (ACO) converts ACC and dioxygen to ethylene, CO_2, HCN, and H_2O in the presence of ascorbate. Alternatively, ACC can be conjugated by N-malonyltransferase to form malonyl-ACC (MACC) in a reversible reaction. Type-2 ACS proteins are targeted for ubiquitination (ubi) and degradation in the proteasome pathway by the E3 ligase subunit ETHYLENE OVERPRODUCER1 (ETO1)

ascorbate (Dorling and McManus 2012). ACO proteins are likewise encoded by small gene families with five members in Arabidopsis and tomato and seven members in rice.

Both, ACS and ACO genes were shown to be subject to differential expression during development and in response to external cues including flooding or hypoxia (Tsuchisaka and Theologis 2004). Regulation of ACS gene expression is generally considered as a major mechanism to regulate ACS activity. In partially submerged rice, *OsACS5* is induced within 1 h and *OsACS1* transcript levels increase within 6 h (Zarembinski and Theologis 1997; Van Der Straeten et al. 2001). An experimental setup in which Arabidopsis roots were immersed in media flushed with 3 % oxygen, while shoots remained in air revealed that hypoxia imposed on roots affects ACS gene expression not only in roots but also in the shoot (Peng et al. 2005). Transcript levels of *AtACS9* steadily increased over a 12-h period in hypoxic roots and within 6 h in the fully aerated shoot. More so, expression of *AtACS2* was elevated in the normoxic shoot, but not in hypoxic roots indicating that ethylene may serve to coordinate shoot adaptation to hypoxic stress of roots.

A detailed qPCR and in situ hybridization study on hypoxic (4 % O_2) maize roots revealed that ACS and ACO genes are regulated differentially not only with respect to degree and kinetics of transcript abundance, but also with high spatial resolution in a cell-type and tissue-specific manner (Geisler-Lee et al. 2010). Transcript levels increased in response to low oxygen treatment for instance in cortical cells indicating that a specific and localized ethylene response such as cortical cell death and subsequent aerenchyma formation may be promoted by local induction of ethylene synthesis. Regulation of ethylene synthesis at the transcriptional level was supported in maize by the finding that ethylene production increased upon hypoxic (4 % O_2) treatment in parallel with ACS as well as ACO activities (He et al. 1996). Enhanced expression of *OsACO1* and down-regulation of the ETO1-like *OsEOL1* (*ETHYLENE OVERPRODUCER-LIKE1*) gene were observed in defined epidermal cells at the rice stem node which are situated above adventitious root primordia upon ethylene treatment suggesting local autocatalytic enhancement of ethylene synthesis in these cells which are known to undergo cell death upon submergence (Steffens and Sauter 2009).

The flooding-tolerant plant *Rumex palustris* responds to flooding with rapid petiole elongation of young leaves. The stimulation of petiole growth is regulated by ethylene. Within the first 24 h of flooding, ACC levels but not ethylene production increased indicative of limiting ACO activity (Banga et al. 1996). Within 4–5 h of submergence *Rp-ACO1* transcript levels increased followed by an increase in ACO activity between 12 and 24 h (Vriezen et al. 1999).

2.2 Posttranslational Mechanisms Controlling Ethylene Evolution and Activity

Aside from transcriptional regulation, ACS activity was shown to be regulated posttranslationally. Distinct phosphorylation and dephosphorylation at the C-terminus alter type-1 and type-2 ACS protein stability by either promoting or preventing ubiquitination and subsequent degradation through the proteasomal pathway (Kamiyoshihara et al. 2010; Skottke et al. 2011). Analysis of ethylene overproducing mutants in Arabidopsis revealed this regulatory mechanism at the protein level. The *eto1* (*ethylene overproducer1*) mutant of Arabidopsis is defective in a BTB (Bric a brac, Tramtrack and Broad Complex) protein which confers substrate specificity as an adapter protein subunit of E3 ubiquitin ligase. ETO1 recognizes type-2 ACS proteins which are targeted for degradation through the proteasome pathway resulting in low ACS activity. Knockout of ETO1 function in the *eto1* mutant results in ACS protein stabilization, in elevated ACS activity, and as a consequence in ethylene overproduction (Guzman and Ecker 1990; Yoshida et al. 2005). As described above, transcript levels of the ETO1-like *OsEOL1* gene are reduced in an ethylene-dependent manner in rice epidermal cells which undergo cell death upon submergence suggestive of enhanced ethylene production through stabilization of ACS protein (Mergemann and Sauter 2000; Steffens and Sauter 2009). *eto2* has a disrupted C-terminus in *AtACS5* which causes stabilization of the respective ACS protein, and *eto3* has a point mutation in *AtACS9* causing an amino acid exchange in the C-terminus again resulting in protein stabilization (Chae and Kieber 2005).

ACS proteins form homo- and heterodimers. The homodimeric ACS complexes from Arabidopsis were shown to possess distinct biochemical properties and most heterodimers between ACS types were inactive (Tsuchisaka et al. 2009). Dimerization thus provides an additional posttranslational level to regulate ACS activity. Synthesis of ACC can hence be controlled at the level of ACS gene expression, ACS protein stability, and complex formation.

ACC synthesis is considered as rate-limiting for ethylene production in most tissues where ACO activity is present in excess. However, at conditions of low oxygen the ACO activity can become rate-limiting due to limitation of its co-substrate O_2. The K_m value of ACO for O_2 was found to be dependent on the concentration of ACC. Supply of exogenous ACC between zero and 10 mM to apple tissue resulted in an increase of the estimated K_m value for O_2 from 0.3 to 6.2 %, indicating that a high ACC concentration favors ethylene formation particularly at low oxygen conditions (Yip et al. 1988). Consequently, ethylene production from apple tissue not supplied with exogenous ACC was only half-maximal at 5 % oxygen as compared to that from aerated tissue. *Rumex palustris* plants held at 5 % oxygen during a 1-h sampling period released only 35 % of the ethylene released by plants kept at normoxic conditions indicative of limited ACO activity at hypoxic conditions (Vriezen et al. 1999). Submergence of *Rumex* plants causes elevated *ACO1* transcription and ACO activity. Incubation of previously submerged plants elevated ethylene release even at 5 % oxygen, further supporting importance of ACO activity for ethylene synthesis at oxygen-deficient conditions *in planta*.

Transcription factors of subgroup VII of the APETALA2/ETHYLENE RESPONSE FACTOR (AP2/ERF) family regulate hypoxia-responsive gene expression and anoxic or submergence stress tolerance in Arabidopsis and rice. Arabidopsis subgroup VII ERFs were recently shown to be regulated at the level of protein stability by a mechanism known as the N-end rule pathway (Sasidharan and Mustroph 2011; Bailey-Serres et al. 2012). At normoxic conditions RAP2.12 and ERF73/HRE1 proteins are modified N-terminally, recognized by the E3 ligase PROTEOLYSIS6 (PRT6) to be continuously degraded. At hypoxic conditions, proteins are no longer modified and hence are no longer targeted for degradation. The Arabidopsis ERF gene *ERF73/HRE1* is induced by hypoxia and by ethylene (Hess et al. 2011), suggesting that regulation of protein stability not only controls ethylene synthesis but also ethylene-controlled hypoxic stress adaptation. Root-waterlogged Arabidopsis plants displayed altered expression of genes encoding RING finger/U-box and F-box proteins in the shoot depending on ethylene signaling through the signaling protein ETHYLENE INSENSITIVE2 (EIN2; Hsu et al. 2011), revealing involvement of ubiquitin-dependent protein degradation also in the systemic response to hypoxia.

2.3 *Storage of the Ethylene Precursor ACC*

No mechanisms for the degradation or inactivation of ethylene are known such that regulation of ethylene levels can occur only at the level of synthesis. ACC is the only intermediary metabolite in the ethylene-specific biosynthetic pathway (Fig. 1). Hence, it may not come as a surprise that ACC levels are subject to regulation at various levels. Aside from synthesis by ACS and oxidation to ethylene by ACO, ACC can be conjugated to an inactive storage form. 1-Malonyl-ACC (MACC) is the major ACC conjugate reported (Hoffman et al. 1982; Liu et al. 1983; Peiser and Yang 1998), while minor formation of γ-glutamyl-ACC (GACC) was reported in green tomato fruit (Peiser and Yang 1998). The conjugating enzyme ACC *N*-malonyltransferase was identified and characterized from tomato fruit and was shown to be induced by ethylene treatment (Martin and Saftner 1995). MACC is stable and therefore considered an inactive form with respect to ethylene synthesis, but can be enzymatically reconverted to ACC as reported for senescing peanut plants (Jiao et al. 1986). A role for remobilized ACC for ethylene synthesis in flooded plants has not been explored to our knowledge. However, a comparison between flooding-tolerant *Rumex palustris* and flooding-sensitive *Rumex acetosella* revealed that the tolerant species produced ACC in roots and shoots upon partial root submergence, whereas the sensitive species accumulated ACC in the shoot only. Conjugated ACC accumulated in roots and shoots of both species indicating that ACC formed in roots of the sensitive species was either inactivated through conjugation or exported to the shoot (Banga et al. 1996). Transport of ACC from roots to shoots was reported to occur for instance in flooded tomato as described in the next paragraph.

In summary, the concentration of ethylene in a particular plant tissue or organ is determined by *de novo* synthesis of ACC from *S*-adenosylmethionine via ACS, formation of and release from MACC, efficient conversion of ACC to the active hormone by ACO, and by the physical nature of the plant environment. Particularly in submerged plants with a low ethylene diffusion rate, it is conceivable that local ethylene synthesis controls rapid adaptations to flooding stress.

3 Systemic Responses to Root Hypoxia Require Root-to-Shoot Signaling

Soil water logging and flooding most often affect the plant root system. Uptake of mineral nutrients and of water relies on functional roots. Consequently, roots which are impaired in oxygen supply and hence in energy production also affect the shoot. It is therefore not surprising that shoots display hypoxic stress adaptation in response to root flooding. Leaf epinasty and stoma closure are among these responses. Some efforts were made to identify the signals which are exchanged between root and shoot to coordinate plant adaptation.

Ethylene is not actively transported in plants but moves by diffusion. Local separation of ethylene synthesis has been recognized as one way of communicating hypoxic stress of the root to the shoot to optimize adaptation of the whole plant to root flooding. ACC synthesized in the root is exported to the shoot where it is converted to ethylene (Bradford and Yang 1980). ACC transport from roots to the shoot was reported in sunflower (Finlayson et al. 1991) and tomato (English et al. 1995) and was observed as a response to diverse stresses such as rehydration after drought treatment as well as flooding (Tudela and Primo-Millo 1992; English et al. 1995). In tomato, ACC produced in the root as a result of water logging was shown to be transported through the xylem to the shoot (Bradford and Yang 1980). In the shoot, ACC is converted to ethylene where it causes epinastic leaf movement. Elevated ethylene production was accompanied by increased ACO activity in petioles of epinastic leaves in tomato raising the question how ACO activity was promoted in the shoot upon root flooding (English et al. 1995). Leaf epinasty is typical of hypoxically stressed tomato plants and is thought to reduce transpirational water loss of stressed plants (Bradford et al. 1982). The *LE-ACS7* gene of tomato was upregulated within 1 h of root flooding which was early enough to account for the observed export of ACC from the root to the shoot (Shiu et al. 1998). In situ analysis of maize roots revealed that *ZmACS2/7* transcript levels increased within 3 h of hypoxic (4 % O_2) treatment in the inner cortex surrounding the vascular cylinder of the root and remained elevated in cortex cells surrounding the endodermis (Geisler-Lee et al. 2010). Local elevation of ACC synthesis in the root can contribute to locally enhanced ethylene synthesis and to the adaptation of roots to hypoxic conditions. Elevated ACC production near the root stele may likewise support ACC export to the shoot via xylem transport. Reversed transport of ACC from tomato leaves to roots was also observed suggesting that ACC can be transported through the phloem as well (Amrhein et al. 1982).

Root flooding causes a reduction in hydraulic conductance and hence limits water supply to the shoot. A reduction in water loss from leaves of root-flooded plants is achieved through stomatal closure. Stomatal closure is known to be controlled by ABA (Joshi-Saha et al. 2012) and therefore stoma closure induced in leaves by soil flooding was long suspected to be under control of root-derived ABA. However, no conclusive evidence was provided so far that supports transport of ABA from roots to the shoot prior to flooding-induced stoma closure (Else et al. 2001; Jackson 2002; Else et al. 2006; Rodriguez-Gamir et al. 2011). Thus, the nature of the mobile signal that mediates stoma closure remains elusive (Else et al. 2006).

Light can be shed on signaling of systemic responses of root-flooded plants by molecular-genetic studies. A transcriptome study carried out in Arabidopsis identified genes which are regulated in the shoot upon root flooding. The study revealed three classes of genes (Hsu et al. 2011). Those which were regulated in roots only, those which were regulated in both roots and shoot, and genes which showed changes in transcript abundance in the shoot upon hypoxic treatment of roots. The microarray study was carried out on 14-day-old Columbia wild-type plants and on *ethylene insensitive 2-5* (*ein2-5*) plants which are deficient in ethylene signaling (Rzewuski and Sauter 2008; Vandenbussche et al. 2012). Microarray and qPCR expression analysis revealed up-regulation of ABA biosynthesis genes in leaves and down-regulation in roots. Concomitant with the expression of biosynthetic genes ABA levels in the shoot rose with root flooding, while ABA levels in the root dropped. However, ABA levels also increased in shoots of control plants. Interestingly, the *HIGHLY ABA-INDUCED PP2C GENE 1* (*HAI1*) gene and a gene involved in ABA biosynthesis, *NINE-CIS-EPOXYCAROTENOID DIOXYGENASE 3* (*NCED3*), were less induced in *ein2-5* indicating that ethylene alters ABA signaling and ABA biosynthesis in systemic responses. These data support the idea that ABA contributes to systemic flooding stress adaptation. Detailed analysis of mutants impaired in ABA synthesis or signaling may clarify the contribution of this hormone to systemic flooding stress adaptation.

Overrepresented genes which were upregulated in the shoot of root-flooded wild-type Arabidopsis plants encoded for proteins involved in major carbohydrate metabolism pointing to starch and sugar degradation, nucleotide metabolism, gluconeogenesis/glyoxylate cycle involved in carbohydrate synthesis from fatty acids, lipid metabolism, and secondary metabolism (Hsu et al. 2011). Downregulated genes were assigned to tetrapyrrole synthesis, pentose phosphate pathway, *N*-metabolism, S-assimilation, minor carbohydrate metabolism, and photosynthesis. As also more carbohydrate transporter genes were upregulated in the shoot than in roots, authors suggested that starch and lipid reserves are mobilized in leaves and transported to the roots (Hsu et al. 2011). A decrease in leaf sucrose levels and an elevated sucrose concentration in the phloem were observed in waterlogged 3-month-old poplar trees. Neither starch nor glucose or fructose content was altered in the shoot and it was hypothesized that transport of sucrose from leaves to flooded roots occurs in waterlogged poplar plants (Kreuzwieser et al. 2009).

In short, plants respond to root flooding with a systemic response. The ethylene precursor ACC and possibly ABA act as mobile signals for root-to-shoot communication to control leaf epinasty, stoma closure, and possibly metabolite fluxes from the shoot to the root.

4 Molecular Interaction of Ethylene with Gibberellin Signaling

Rapid shoot elongation is an escape mechanism observed in a number of flooding-adapted plants (Bailey-Serres and Voesenek 2008). In deepwater rice varieties, the youngest internode elongates rapidly in response to partial submergence as a result of ethylene signaling. Rapid internodal elongation allows rice plants to keep part of their foliage in air even when flooded by several meters and thus survive longlasting flooding periods over several months. The growth response is mediated by gibberellic acid (GA) (Kende et al. 1998; Hattori et al. 2009). In the growth-responsive internode of deepwater rice plants, the level of GA_1, which is the most abundant active GA in rice, increased fourfold after 3 or 12 h of partial submergence, depending on the genotype analyzed (Hoffmann-Benning and Kende 1992; Hattori et al. 2009). By contrast, GA_1 levels did not increase in the non-deepwater rice variety T65, despite of a comparable accumulation of ethylene in both genotypes. The ERF PROTEIN ASSOCIATED WITH TILLERING AND BRANCHING (EATB) from rice belongs to the APETALA2/ETHYLENE RESPONSE FACTOR (AP2/ERF) family of transcription factors (Nakano et al. 2006), but lacks defined features that would allow assignment to a specific subgroup (Cao et al. 2006). *EATB* over-expression resulted in down-regulation (~11-fold) of the *ent-kaurene synthase A* (*CPS*) gene (Qi et al. 2011). CPS catalyzes the cyclization of geranylgeranyl diphosphate (GGDP) to *ent*-copalyl diphosphate (CDP), which is then converted to *ent*-kaurene by *ent*-kaurene synthase B in the early GA biosynthesis pathway (Sakamoto et al. 2004). *EATB* over-expressing plants contained reduced levels of GA_1 with no changes in ABA or auxin content as compared to wild-type. Treatment of rice plants with ethephon resulted in down-regulation of *EATB* expression and concomitant up-regulation of *CPD*, supporting the idea that ethylene promotes GA_1 synthesis through the AP2/ERF transcription factor EATB. Over-expression of EATB promotes GA synthesis, but does not alter the dose-dependent growth response of GA.

In deepwater rice, GA levels increase and the sensitivity of the internode toward GA is enhanced in the presence of ethylene such that GA promotes rice stem elongation to the same degree at about threefold lower concentrations in the presence of ethylene than in its absence (Raskin and Kende 1984). Enhanced responsiveness to GA is a result of altered ABA metabolism leading to an increase in the GA to ABA ratio (see below). Ethylene signaling of growth in submerged deepwater rice plants is mediated by two members of subgroup VII of the AP2/ERF family, SNORKEL1 (SK1) and SNORKEL2 (SK2) (Nakano et al. 2006; Hattori et al. 2009). The youngest rice internode possesses an active intercalary meristem at its base. Internodal elongation is accelerated around tenfold when plants are partially submerged or when stems are treated with GA (Sauter and Kende 1992; Lorbiecke and Sauter 1999). The increase in growth rate is achieved through induction of the cell division cycle. GA upregulates CYCLIN DEPENDENT KINASE (CDK) and CDK activating (CAK) kinase, key regulators of cell cycle progression resulting in about threefold higher cell production rate from the meristem (Sauter et al. 1995; Sauter 1997). In addition, each new cell that is produced in submerged or GA-treated internodes grows threefold longer than in uninduced rice internodes (Sauter and Kende 1992). Rapid growth is hence achieved by GA through activation of the cell division cycle and through enhanced elongation of newly produced cells.

Interestingly, members of subgroup VII of the AP2/ERF family can also suppress shoot elongation. This was described for the *Sub1A-1* gene which is present in a lowland rice variety that survives fairly short (up to 2 weeks) and shallow (up to 50 cm in depth) floods by saving resources through a quiescence response (Xu et al. 2006; Bailey-Serres and Voesenek 2008). Both strategies, quiescence and escape, are successful to survive floods in the respective habitats. Ectopic expression of *Sub1A-1* reduced plant height and shoot elongation in response to submergence and in response to exogenous GA application (Fukao and Bailey-Serres 2008). DELLA proteins are repressors of GA signaling and belong to the GRAS protein family (Gao et al. 2011). They act in the nucleus and are degraded in response to GA binding to its receptor GIBBERELLIN INSENSITIVE DWARF1 (GID1). GID1 is an F-box subunit of an SCF E3 ubiquitin ligase which targets the DELLA proteins for degradation in the proteasome pathway (Gao et al. 2011; Sun 2011). Thus, GA promotes growth through degradation of growth repressing proteins. The GRAS proteins SLENDER RICE-1 (SLR1) and SLR1 like-1 (SLRL1) of rice are both negative regulators of GA signaling even though only SLR1 possesses a DELLA domain (Itoh et al. 2005). Knockout of *SLR1* or *SLRL1* promotes shoot growth through constitutive GA activity resulting in a slender phenotype, whereas plants over-expressing *SLR1* or *SLRL1* show constitutive GA insensitivity (Ikeda et al. 2001; Itoh et al. 2002, 2005). Ethylene-inducible *Sub1A-1* promotes *SLR1* and *SLRL1* gene expression and protein accumulation, indicating that growth suppression of the quiescence response is achieved through inhibition of GA signaling via SLR1 and SLRL1 (Fukao and Bailey-Serres 2008). In rice plants over-expressing *EATB*, expression of *SLR1* was downregulated indicating that EATB supports stem growth not only through enhanced GA biosynthesis but possibly also through de-repression of GA signaling (Qi et al. 2011).

Flooding-induced stem growth in rice is altered by ethylene via AP2/ERF transcription factors. Subgroup VII ERFs SK1 and SK2 present in deepwater rice promote rapid growth, whereas Sub1A-1 present in submergence-tolerant lowland rice inhibits growth in either case by altering GA activity. Growth inhibition by Sub1A-1 is achieved through the GA signaling repressors SLR1 and SLRL1. Enhanced synthesis of GA is a feature of plants which show an escape response and is controlled in rice by the ERF protein EATB which also enhances GA activity through down-regulation of *SLR1*.

5 Molecular Interaction of GA with ABA Signaling

ABA acts as a growth inhibitory plant hormone. Lowered ABA levels promote shoot elongation and adventitious root growth in response to flooding (Bailey-Serres and Voesenek 2008; Steffens et al. 2006). ABA levels can be reduced through reduced biosynthesis and/or enhanced catabolism. *CYP707A5* encodes an ABA 8′-hydroxylase (ABA8ox) that catalyzes oxidation of ABA to 8′-hydroxy ABA which is spontaneously converted to phaseic acid (PA), a largely inactive ABA catabolite. In deepwater rice stems, *CYP707A5* was upregulated within 1.5 h by ethylene, while ABA levels dropped and the stem growth rate increased (Yang and Choi 2006).

The AP2/ERF transcription factor AP2-39 from rice mediates interactions between ABA and GA in shoot growth regulation (Yaish et al. 2010). Overexpression of *AP2-39* results in reduced bioactive GA and increased ABA levels and in a pleiotropic phenotype including shorter internodes. AP2-39 binds to the promoter of the *NCED-1* gene predicted to encode the ABA biosynthetic enzyme 9-*cis*-epoxycarotenoid dioxygenase (Schwartz et al. 1997). NCED catalyzes the first committed step in ABA synthesis. Expression of *NCED-1* by AP2-39 is enhanced by GA and inhibited by ABA, indicating feedback inhibition of ABA biosynthesis by ABA and up-regulation of ABA synthesis by GA. The *slr1-1* mutant of rice that shows a constitutive GA response of shoot elongation has a tenfold higher ABA content, suggesting that up-regulation of ABA synthesis by GA occurs via degradation of this DELLA protein (Ikeda et al. 2002). AP2-39 also binds to the promoter of the *ELONGATED UPPERMOST INTERNODE* (*EUI*) gene and promotes *EUI* expression in the absence of a hormone. EUI is a cytochrome P450 monooxygenase, CYP714D1, which deactivates gibberellins (Zhu et al. 2006). Transcripts of *NCED-1* and *EUI* are expressed at higher levels in *AP2-39ox* plants, indicating that AP2-39 reduces the GA to ABA ratio to promote growth arrest. Independent from AP2-39, *EUI* expression is induced by ABA in wild-type plants, indicating that growth inhibition by ABA is achieved through elevated ABA levels and a reduction in GA levels (Yaish et al. 2010).

Submergence-intolerant rice plants which are rendered submergence-tolerant through ectopic expression of *Sub1A-1* are growth-inhibited upon submergence. Despite the growth arrest observed in *Sub1A-1* expressing plants, submergence stimulated a decrease in ABA content which was however also observed in the submergence-intolerant NIL M202 lacking *Sub1A-1,* indicating that regulation of ABA levels was not controlled by Sub1A-1 (Fukao and Bailey-Serres 2008). Growth inhibition was brought about in these plants through regulation of GA synthesis and signaling rather than through elevated ABA levels.

The submergence-tolerant dicot *Rumex palustris* responds to flooding with hyponasty and enhanced petiole elongation to ensure leaf contact to the atmosphere (Voesenek et al. 2006). Petiole growth is induced by ethylene via reduction of ABA levels. Ethylene inhibits for one expression of biosynthetic *NCED* genes. At the same time enhanced degradation of ABA to phaseic acid was observed (Benschop et al. 2005). Applied ABA in turn prevented accumulation of bioactive GA_1 which promotes petiole growth rate (Cox et al. 2004).

The interplay between ethylene, GA, and ABA is a general feature of and plays a prominent role in flooding stress adaptation. Changes in hormone levels are central to the regulation of defined flooding responses such as acceleration or arrest of shoot growth. AP2/ERF transcription factors surfaced as key players in submergence adaptation. Yet, still much is to be learned about the molecular mechanisms that govern changes in hormone homeostasis and activity.

References

Amrhein N, Breuing F, Eberle J, Skorupka H, Tophof S (1982) The metabolism of 1-aminocyclopropane-1-carboxylic acid. In: Waering PF (ed) Plant growth substances. Academic, New York, pp 249–258

Armstrong W (1979) Aeration in higher plants. Adv Bot Res 7:225–232

Bailey-Serres J, Voesenek LACJ (2008) Flooding stress: acclimations and genetic diversity. Annu Rev Plant Biol 59:313–339

Bailey-Serres J, Fukao T, Gibbs DJ, Holdsworth MJ, Lee SC, Licausi F, Perata P, Voesenek LA, van Dongen JT (2012) Making sense of low oxygen sensing. Trends Plant Sci 17:129–138

Banga M, Slaa EJ, Blom C, Voesenek L (1996) Ethylene biosynthesis and accumulation under drained and submerged conditions (A comparative study of two Rumex species). Plant Physiol 112:229–237

Benschop JJ, Jackson MB, Gühl K, Vreeburg RA, Croker SJ, Peeters AJ, Voesenek LA (2005) Contrasting interactions between ethylene and abscisic acid in Rumex species differing in submergence tolerance. Plant J 44:756–768

Bradford KJ, Yang SF (1980) Xylem transport of 1-aminocyclopropane-1-carboxylic acid, an ethylene precursor, in waterlogged tomato plants. Plant Physiol 65:322–326

Bradford KJ, Hsiao T, Yang SF (1982) Inhibition of ethylene synthesis in tomato plants subjected to anaerobic root stress. Plant Physiol 70:1503–1507

Cao Y, Song F, Goodman RM, Zheng Z (2006) Molecular characterization of four rice genes encoding ethylene-responsive transcriptional factors and their expressions in response to biotic and abiotic stress. J Plant Physiol 163:1167–1178

Chae HS, Kieber JJ (2005) Eto Brute? The role of ACS turnover in regulating ethylene biosynthesis. Trends Plant Sci 10:291–296

Cox MC, Benschop JJ, Vreeburg RA, Wagemaker CA, Moritz T, Peeters AJ, Voesenek LA (2004) The roles of ethylene, auxin, abscisic acid, and gibberellin in the hyponastic growth of submerged Rumex palustris petioles. Plant Physiol 136:2948–2960

Dorling SJ, McManus MT (2012) The fate of ACC in higher plants. In: McManus M (ed) The plant hormone ethylene. Annual plant reviews, vol 44. Wiley-Blackwell Press, Massey, New Zealand, pp 83–115

Else MA, Coupland D, Dutton L, Jackson MB (2001) Decreased root hydraulic conductivity reduces leaf water potential, initiates stomatal closure and slows leaf expansion in flooded plants of castor oil (Ricinus communis) despite diminished delivery of ABA from the roots to shoots in xylem sap. Physiol Plant 111:46–54

Else MA, Taylor JM, Atkinson CJ (2006) Anti-transpirant activity in xylem sap from flooded tomato (Lycopersicon esculentum Mill.) plants is not due to pH-mediated redistributions of root- or shoot-sourced ABA. J Exp Bot 57:3349–3357

English PJ, Lycett GW, Roberts JA, Jackson MB (1995) Increased 1-aminocyclopropane-1-carboxylic acid oxidase activity in shoots of flooded tomato plants raises ethylene production to physiologically active levels. Plant Physiol 109:1435–1440

Finlayson SA, Foster KR, Reid DM (1991) Transport and metabolism of 1-aminocyclopropane-1-carboxylic acid in sunflower (Helianthus annuus L.) seedlings. Plant Physiol 96:1360–1367

Fukao T, Bailey-Serres J (2008) Submergence tolerance conferred by Sub1A is mediated by SLR1 and SLRL1 restriction of gibberellin responses in rice. Proc Natl Acad Sci U S A 105:16814–16819

Gao XH, Xiao SL, Yao QF, Wang YJ, Fu XD (2011) An updated GA signaling 'relief of repression' regulatory model. Mol Plant 4:601–606

Geisler-Lee J, Caldwell J, Gallie DR (2010) Expression of the ethylene biosynthetic machinery in maize roots is regulated in response to hypoxia. J Exp Bot 61:857–871

Guzman P, Ecker JR (1990) Exploiting the triple response of Arabidopsis to identify ethylene related mutants. Plant Cell 2:513–523

Harpaz-Saad S, Yoon GM, Mattoo AK, Kieber JJ (2012) The formation of ACC and competition between polyamines and ethylene for SAM. In: McManus M (ed) The plant hormone ethylene. Annual plant reviews, vol 44. Wiley-Blackwell Press, Massey, New Zealand, pp 53–81

Hattori Y, Nagai K, Furukawa S, Song XJ, Kawano R, Sakakibara H, Wu J, Matsumoto T, Yoshimura A, Kitano H, Matsuoka M, Mori H, Ashikari M (2009) The ethylene response factors SNORKEL1 and SNORKEL2 allow rice to adapt to deep water. Nature 460:1026–1030

He CJ, Finlayson SA, Drew MC, Jordan WR, Morgan PW (1996) Ethylene biosynthesis during aerenchyma formation in roots of maize subjected to mechanical impedance and hypoxia. Plant Physiol 112:1679–1685

Hess N, Klode M, Anders M, Sauter M (2011) The hypoxia responsive transcription factor genes ERF71/HRE2 and ERF73/HRE1 of Arabidopsis are differentially regulated by ethylene. Physiol Plant 143:41–49

Hoffman NE, Yang SF, McKeon T (1982) Identification of 1-(malonylamino)cyclopropane-1-carboxylic acid as a major conjugate of 1-aminocyclopropane-1-carboxylic acid, an ethylene precursor in higher plants. Biochem Biophys Res Commun 104:765–770

Hoffmann-Benning S, Kende H (1992) On the role of abscisic acid and gibberellin in the regulation of growth in rice. Plant Physiol 99:1156–1161

Hsu FC, Chou MY, Peng HP, Chou SJ, Shih MC (2011) Insights into hypoxic systemic responses based on analyses of transcriptional regulation in Arabidopsis. PLoS One 6(12):e28888

Ikeda A, Ueguchi-Tanaka M, Sonoda Y, Kitano H, Koshioka M, Futsuhara Y, Matsuoka M, Yamaguchi J (2001) Slender rice, a constitutive gibberellin response mutant, is caused by a null mutation of the SLR1 gene, an ortholog of the height-regulating gene GAI/RGA/RHT/D8. Plant Cell 13:999–1010

Ikeda A, Sonoda Y, Vernieri P, Perata P, Hirochika H, Yamaguchi J (2002) The slender rice mutant, with constitutively activated gibberellin signal transduction, has enhanced capacity for abscisic acid level. Plant Cell Physiol 43:974–979

Itoh H, Ueguchi-Tanaka M, Sato Y, Ashikari M, Matsuoka M (2002) The gibberellin signaling pathway is regulated by the appearance and disappearance of SLENDER RICE1 in nuclei. Plant Cell 14:57–70

Itoh H, Asako Shimada A, Ueguchi-Tanaka M, Kamiya N, Hasegawa Y, Ashikari M, Matsuoka M (2005) Overexpression of a GRAS protein lacking the DELLA domain confers altered gibberellin responses in rice. Plant J 44:669–679

Jackson MB (2002) Long-distance signalling from roots to shoots assessed: the flooding story. J Exp Bot 53:175–181

Jiao XZ, Philosophhadas S, Su LY, Yang SF (1986) The conversion of 1-(malonylamino)cyclopropane-1-carboxylic acid to 1-aminocyclopropane-1-carboxylic acid in plant tissues. Plant Physiol 81:637–641

Joshi-Saha A, Valon C, Leung J (2012) A brand new START: abscisic acid perception and transduction in the guard cell. Sci Signal 4(201):re4

Kamiyoshihara Y, Iwata M, Fukaya T, Tatsuki M, Mori H (2010) Turnover of LeACS2, a wound-inducible 1-aminocyclopropane-1-carboxylic acid synthase in tomato, is regulated by phosphyrylation/dephosphorylation. Plant J 64:140–150
Kende H, van der Knaap E, Cho HT (1998) Deepwater rice: a model plant to study stem elongation. Plant Physiol 118:1105–1110
Kreuzwieser J, Hauberg J, Howell KA, Carroll A, Rennenberg H, Millar AH, Whelan J (2009) Differential response of gray poplar leaves and roots underpins stress adaptation during hypoxia. Plant Physiol 149:461–473
Liu Y, Hoffman NE, Yang SF (1983) Relationship between the malonylation of 1-aminocyclopropane-1-carboxylic acid and D-amino acids in mung-bean hypocotyls. Planta 158:437–441
Lorbiecke R, Sauter M (1999) Adventitious root growth and cell cycle induction in deepwater rice (Oryza sativa L.). Plant Physiol 119:21–29
Martin MN, Saftner RA (1995) Purification and characterization of 1-aminocyclopropane-1-carboxylic acid N-malonyltransferase from tomato fruit. Plant Physiol 108:1241–1249
Mergemann H, Sauter M (2000) Programmed death of epidermal cells facilitates emergence of adventitious roots in deepwater rice. Plant Physiol 124:609–614
Nakano T, Suzuki K, Fujimura T, Shinshi H (2006) Genome-wide analysis of the ERF gene family in Arabidopsis and rice. Plant Physiol 140:411–432
Peiser G, Yang SF (1998) Evidence for 1-(malonylamino)cyclopropane-1-carboxylic acid being the major conjugate of aminocyclopropane-1-carboxylic acid in tomato fruit. Plant Physiol 116:1527–1532
Peng HP, Lin T-Y, Wang N-N, Shih MC (2005) Differential expression of genes encoding 1-aminocyclopropane-1-carboxylate synthase in Arabidopsis during hypoxia. Plant Mol Biol 58:15–25
Qi W, Sun F, Wang Q, Chen M, Huang Y, Feng YQ, Luo X, Yang J (2011) Rice ethylene-response AP2/ERF factor OsEATB restricts internode elongation by down-regulating a gibberellin biosynthetic gene. Plant Physiol 157:216–228
Raskin I, Kende H (1984) Role of gibberellin in the growth response of submerged deep water rice. Plant Physiol 76:947–950
Rodriguez-Gamir J, Ancillo G, Gonzalez-Mas MC, Millo E, Iglesias DJ, Forner-Giner MA (2011) Root signalling and modulation of stomatal closure in flooded citrus seedlings. Plant Physiol Biochem 49:636–645
Rzewuski G, Sauter M (2008) Ethylene biosynthesis and signaling in rice. Plant Sci 175:32–42
Sakamoto T, Miura K, Itoh H, Tatsumi T, Ueguchi-Tanaka M, Ishiyama K, Kobayashi M, Agrawal GK, Takeda S, Abe K et al (2004) An overview of gibberellin metabolism enzyme genes and their related mutants in rice. Plant Physiol 134:1642–1653
Sasidharan R, Mustroph A (2011) Plant oxygen sensing is mediated by the N-end rule pathway: a milestone in plant anaerobiosis. Plant Cell 23:4173–4183
Sauter M (1997) Differential expression of a CAK (cdc2-activating kinase)-like protein kinase, cyclins and cdc2 genes from rice during the cell cycle and in response to gibberellin. Plant J 11:181–190
Sauter M, Kende H (1992) Gibberellin-induced growth and regulation of the cell division cycle in deepwater rice. Planta 188:362–368
Sauter M, Mekhedov SL, Kende H (1995) Gibberellin promotes histone H1 kinase activity and the expression of cdc2 and cyclin genes during the induction of rapid growth in deepwater rice internodes. Plant J 7:623–632
Schwartz SH, Tan BC, Gage DA, Zeevaart JA, McCarty DR (1997) Specific oxidative cleavage of carotenoids by VP14 of maize. Science 276:1872–1874
Shiu OY, Oetiker JH, Yip WK, Yang SF (1998) The promoter of LE-ACS7, an early flooding-induced 1-aminocyclopropane-1-carboxylate synthase gene of the tomato, is tagged by a Sol3 transposon. Proc Natl Acad Sci U S A 95:10334–10339

Skottke KR, Yoon GM, Kieber JJ (2011) Protein phosphatase 2A controls ethylene biosynthesis by differentially regulating the turnover of ACC synthase isoforms. PLoS Genet 7:e1001370

Steffens B, Sauter M (2009) Epidermal cell death in rice is confined to cells with a distinct molecular identity and is mediated by ethylene and H2O2 through an autoamplified signal pathway. Plant Cell 21:184–196

Steffens B, Wang J, Sauter M (2006) Interactions between ethylene, gibberellin and abscisic acid regulate emergence and growth rate of adventitious roots in deepwater rice. Planta 223:604–612

Sun TP (2011) The molecular mechanism and evolution of the GA-GID1-DELLA signaling module in plants. Curr Biol 21:R338–R3345

Tsuchisaka A, Theologis A (2004) Unique and overlapping expression patterns among the Arabidopsis 1-aminocyclopropane-1-carboxylate synthase gene family members. Plant Physiol 136:2982–3000

Tsuchisaka A, Yu G, Jin H, Theologis A (2009) A combinatorial interplay among the 1-aminocyclopropane-1-carboxylate synthase isoforms regulates ethylene biosyntehsis. Genetics 183:979–1003

Tudela D, Primo-Millo E (1992) 1-Aminocyclopropane-1-carboxylic acid transported from roots to shoots promotes leaf abscission in Cleopatra Mandarin (Citrus reshni Hort. ex Tan.) seedlings rehydrated after water stress. Plant Physiol 100:131–137

Van der Straeten D, Zhou HA, Van Onckelen MC, Van Montagu MC (2001) A comparative molecular-physiological study of submergence response in lowland and deepwater rice. Plant Physiol 125:955–968

Vandenbussche F, Vaseva I, Vissenberg K, Van Der Straeten D (2012) Ethylene in vegetative development: a tale with a riddle. New Phytol 194:895–909

Voesenek LA, Colmer TD, Pierik R, Millenaar FF, Peeters AJ (2006) How plants cope with complete submergence. New Phytol 170:213–226

Vriezen WH, Hulzink R, Mariani C, Voesenek LA (1999) 1-aminocyclopropane-1-carboxylate oxidase activity limits ethylene biosynthesis in Rumex palustris during submergence. Plant Physiol 121:189–196

Xu K, Xu X, Fukao T, Canlas P, Maghirang-Rodriguez R, Heuer S, Ismail AM, Bailey-Serres J, Ronald PC, Mackill DJ (2006) Sub1A is an ethylene-response-factor-like gene that confers submergence tolerance to rice. Nature 442:705–708

Yaish MW, El-Kereamy A, Zhu T, Beatty PH, Good AG, Bi YM, Rothstein SJ (2010) The APETALA-2-like transcription factor OsAP2-39 controls key interactions between abscisic acid and gibberellin in rice. PLoS Genet 6(9):e1001098

Yang S-H, Choi D (2006) Characterization of genes encoding ABA 8′-hydroxylase in ethylene-induced stem growth of deepwater rice (Oryza sativa L.). Biochem Biophys Res Commun 350:685–690

Yang SF, Hoffman NE (1984) Ethylene biosynthesis and its regulation in higher plants. Annu Rev Plant Physiol 35:155–189

Yasumura Y, Pierik R, Fricker MD, Voesenek LA, Harberd NP (2012) Studies of Physcomitrella patens reveal that ethylene-mediated submergence responses arose relatively early in land-plant evolution. Plant J 72:947–959

Yip W-K, Jiao X-Z, Yang SF (1988) Dependence of in vivo ethylene production rate on 1-aminocyclopropane-1-carboxylic acid content and oxygen concentrations. Plant Physiol 88:553–558

Yoshida H, Nagata M, Saito K et al (2005) Arabidopsis ETO1 specifically interacts with and negatively regulates type-2 1-aminocyclopropane-1-carboxylate synthases. BMC Biol 5:14

Zarembinski TI, Theologis A (1997) Expression characteristics of Os-ACS1 and Os-ACS2, two members of the 1-aminocyclopropane-1-carboxylate synthase family in rice (Oryza sativa L., cv. Habiganj Aman II) during partial submergence. Plant Mol Biol 33:71–77

Zhu Y, Nomura T, Xu Y, Zhang Y, Peng Y, Mao B, Hanada A, Zhou H, Wang R, Li P, Zhu X, Mander LN, Kamiya Y, Yamaguchi S, He Z (2006) ELONGATED UPPERMOST INTERNODE encodes a cytochrome P450 monooxygenase that epoxidizes gibberellins in a novel deactivation reaction in rice. Plant Cell 18:442–456

Part III
Metabolic Responses

Insights into Algal Fermentation

Wenqiang Yang, Claudia Catalanotti, Matthew C. Posewitz, Jean Alric, and Arthur R. Grossman

Abstract *Chlamydomonas reinhardtii* (Chlamydomonas throughout) is a metabolically versatile, soil-dwelling photosynthetic alga, capable of modulating its metabolism as environmental conditions change; this flexibility enables sustained energy generation and maintenance of cellular redox balance. Because of its many unique features, Chlamydomonas has become a model organism for understanding a range of biological, cellular, molecular, and physiological processes, including fermentation metabolism, which occur in both photosynthetic and non-photosynthetic organisms. Genes encoding proteins associated with multiple fermentation pathways have been identified on the Chlamydomonas genome. Controlled regulation of these pathways has allowed this alga to satisfy its energy requirements when O_2 levels become low, and studies of Chlamydomonas mutant strains have demonstrated substantial metabolic flexibility in the use of the various available fermentation pathways. Furthermore, when Chlamydomonas ferments polysaccharides under anoxic conditions, it has the ability to eliminate reducing power (which sustains glycolysis) through the production of H_2, a molecule that can be developed as a source of sustainable energy. H_2 is produced by three different pathways; direct biophotolysis, indirect biophotolysis, and dark fermentative metabolism. To date, with respect to the algae there is little known about the specific role(s) of the different branches of fermentation metabolism, how they sense oxic conditions, and the mechanisms involved in controlling these responses, at both the transcriptional and posttranscriptional levels. More is known

W. Yang (✉) • C. Catalanotti • A.R. Grossman
Department of Plant Biology, The Carnegie Institution for Science, 260 Panama Street, Stanford, CA 94305, USA
e-mail: wqyang@stanford.edu

M.C. Posewitz
Department of Chemistry and Geochemistry, Colorado School of Mines, Golden, CO 80401, USA

J. Alric
CEA, IBEB, Lab Bioenerget Biotechnol Bacteries & Microalgues, Saint-Paul-lez-Durance, F-13108, France

J.T. van Dongen and F. Licausi (eds.), *Low-Oxygen Stress in Plants*, Plant Cell Monographs 21, DOI 10.1007/978-3-7091-1254-0_8, © Springer-Verlag Wien 2014

about acclimation to anoxic conditions in vascular plants, and recently it was demonstrated that the responses of plants to anoxia depend upon specific ethylene response factor transcriptional elements that are destabilized under oxygen-replete conditions (Trends Plant Sci 17:129–138, 2012; Curr Opin Plant Biol 13:489–494, 2010). In the present review, we focus on fermentation metabolism in Chlamydomonas and other green algae, with some discussion of plants when relevant. We also explore different methods that can be used to probe the physiology of cells exposed to hypoxic/anoxic conditions.

1 Introduction

1.1 Initial Remarks on Anoxia

Maintaining cellular viability during anoxia requires the production and conservation of ATP as well as the recycling of NAD(P)H. Anaerobic pathways generate ~2–3 ATP per molecule of glucose metabolized compared to over 30 ATP generated by the oxidative metabolism of glucose. Under some conditions, a "Pasteur effect" is triggered that leads to accelerated carbohydrate catabolism, which would increase the rate of ATP production (Magneschi and Perata 2009); a two- to threefold increase in glycolysis can occur under anoxia relative to oxic conditions (Gibbs et al. 2000). Other sources of energy may also be exploited to maintain cells exposed to hypoxic/anoxic conditions. For example, pyrophosphate may contribute to the energy balance and survivability of plants during anoxia (Huang et al. 2008), with a decrease in pyrophosphate levels in transgenic potato plants leading to decreased viability under hypoxic conditions (Mustroph et al. 2005). Interestingly, while various metabolites, including O_2 itself, may impact regulatory processes associated with acclimation of organisms to anoxia, other molecules, including nitrate and its reduction to nitric oxide (NO) (and subsequent oxidation by oxyhaemoglobin), may be involved in controlling plant cell cytoplasmic pH homeostasis (Stoimenova et al. 2007) and energetics during hypoxia (Igamberdiev et al. 2004). While this chapter will primarily focus on the ways in which algae, primarily *Chlamydomonas reinhardtii*, respond to low O_2 conditions, we also briefly discuss work with plants; many other contributions in this volume detail the responses of plants to hypoxic conditions.

1.2 Chlamydomonas as a Model Organism

Chlamydomonas, a soil-dwelling photosynthetic alga that is a flagship model organism for the Department of Energy (USA), has several metabolic features similar to those of vascular plants, although it also has structures and activities (e.g., flagella) that were lost during vascular plant evolution. One important feature of Chlamydomonas, that is absent in vascular plants, is its ability to grow in the dark on a fixed carbon source while maintaining the photosynthetic apparatus during that growth; this feature has allowed the dissection of photosynthesis through characterizations of photosynthetic mutants. Over the last 20 years many tools were developed to probe processes in Chlamydomonas at the molecular, genetic, biochemical, and physiological levels. Mutants can be rapidly generated, phenotypically characterized, and the site of the lesion on the genome localized by PCR (Dent et al. 2005) and map-based cloning (Rymarquis et al. 2005). In addition to forward genetic screens, strategies have been developed to generate specific mutant strains using reverse genetics (Gonzalez-Ballester et al. 2011). The sequencing of all three Chlamydomonas genomes (nuclear, chloroplast, and mitochondrion) (Maul et al. 2002; Merchant et al. 2007) has facilitated the capture of information about gene and genome structure, potential regulatory sequences, including promoter regions, 3′ and 5′ UTRs, and intron–exon junctions. The genome sequence has also (1) supported RNA-seq data for exploring differential gene expression and building gene models since reads can be directly mapped to the genome, (2) facilitated the development of RNAi constructs for suppressing the accumulation of specific proteins, and (3) made it routine to identify the sites of insertion in mutant strains. Many of these features have transformed Chlamydomonas into one of the most attractive reference systems for analyzing a range of biological processes including flagellar function and assembly (King and Kamiya 2009; Yang and Smith 2009), the catalytic activities and biogenesis of chloroplasts (Eberhard et al. 2008; Finazzi et al. 2009; Minagawa 2009; Niyogi 2009; Rochaix 2009), regulation of gene expression and posttranscriptional processes in chloroplasts (Choquet and Wollman 2009; Eberhard et al. 2008; Goldschmidt-Clermont 2009; Herrin 2009; Klein 2009; Zerges and Hauser 2009), phototaxis and photoperception (Wagner et al. 2008), and the activities and regulatory circuits associated with the assimilation of macronutrients (Camargo et al. 2007; Fernández et al. 2009; González-Ballester and Grossman 2009; González-Ballester et al. 2008; Moseley et al. 2009; Moseley and Grossman 2009) and micronutrients (Long et al. 2008; Tejada-Jimenez et al. 2007). Moreover, this green alga can synthesize relatively high levels of lipids under nitrogen deprivation conditions and evolve H_2 under anoxic conditions; both of these capabilities suggest the potential use of this alga for biofuel production (Ghirardi et al. 2005; Melis and Happe 2004; Wang et al. 2009).

2 Fermentation Metabolism

2.1 Fermentation in Algae

In a soil environment many algae, including Chlamydomonas, experience hypoxic and anoxic conditions, mainly during the night when photosynthetic O_2 evolution stops and environmental O_2 levels dramatically decline as a consequence of respiration (by both the algae and heterotrophic organisms). Under these conditions the algae generate "maintenance energy" by fermentation metabolism. Such adaptive strategies can also be used by algae in the light under anoxic conditions when respiratory O_2 consumption exceeds photosynthetic O_2 evolution. Furthermore, various products of fermentation may be excreted from algal cells, providing heterotrophic microbes with organic compounds and reductant. Recently, several different genes encoding proteins associated with a diverse set of fermentative pathways have been identified on the Chlamydomonas genome (Grossman et al. 2007, 2011; Merchant et al. 2007), while a number of studies have revealed various fermentation circuits (Atteia et al. 2006; Burgess et al. 2012; Catalanotti et al. 2012; Dubini et al. 2009; Gfeller and Gibbs 1984, 1985; Gibbs et al. 1986; Kreuzberg 1984; Magneschi et al. 2012; Meuser et al. 2012; Mus et al. 2007; Ohta et al. 1987; Philipps et al. 2011). This metabolic flexibility allows Chlamydomonas to satisfy its energy requirements under low O_2 conditions.

During dark fermentation in the evening, the carbohydrate reserves (starch) that accumulated in cells during the day are catabolized to generate ATP, while the reduced pyridine nucleotides NADH and NADPH that is co-produced with the ATP must be re-oxidized to sustain the activity of the fermentation pathways. Based on past studies, pyruvate, a key metabolite generated by the glycolytic breakdown of starch/hexose, serves as the substrate for a number of fermentation circuits used by Chlamydomonas. These pathways, which may also occur in other photosynthetic microbes under anoxic conditions, could support the secretion of organic acids (formate, lactate, malate, acetate, and succinate) and alcohols (ethanol, glycerol), and the evolution of H_2 and CO_2 (Catalanotti et al. 2012; Dubini et al. 2009; Gfeller and Gibbs 1984; Grossman et al. 2011; Kosourov et al. 2003; Kreuzberg 1984; Magneschi et al. 2012; Mus et al. 2007; Ohta et al. 1987; Philipps et al. 2011). A diagram depicting fermentation pathways that operate in Chlamydomonas is presented in Fig. 1. Two prominent enzymes that consume pyruvate when the cells become micro-oxic/anoxic are pyruvate formate lyase (PFL1) and pyruvate: ferredoxin:oxidoreductase (PFR1). In Chlamydomonas, PFL1 appears to be located in both mitochondria and chloroplasts (Atteia et al. 2006; Kreuzberg et al. 1987). The reaction mediated by this enzyme results in the cleavage of pyruvate to acetyl-CoA and formate, while PFR1 converts pyruvate to acetyl-CoA, CO_2, and reduced ferredoxin (FDX). The reduced FDX can be re-oxidized by Chlamydomonas hydrogenases, generating H_2 (Müller 2003), or by reactions that enzymatically reduce nitrite and sulfate/sulfite.

The acetyl-CoA produced by the PFL1 and PFR1 reactions (Fig. 1) is either reduced to ethanol by alcohol/aldehyde dehydrogenase (ADH1) (Atteia et al. 2006; Dubini et al. 2009; Hemschemeier and Happe 2005; Magneschi et al. 2012) or metabolized to acetate by phosphate acetyltransferase (PAT) and acetate kinase (ACK) (Atteia et al. 2006). These latter reactions appear to occur both in mitochondria (PAT1, ACK2) and chloroplasts (PAT2, ACK1) (Bailey-Serres and Voesenek 2008). More questionable is the localization of ADH1, which has been reported to be in the mitochondrion by Atteia et al. (2006) and in chloroplasts by Terashima et al. (2010) (a dual localization cannot be excluded). Support for dual localization also comes from the data of Kreuzberg et al. (1987). An alternative pathway for ethanol production may be the direct decarboxylation of pyruvate to

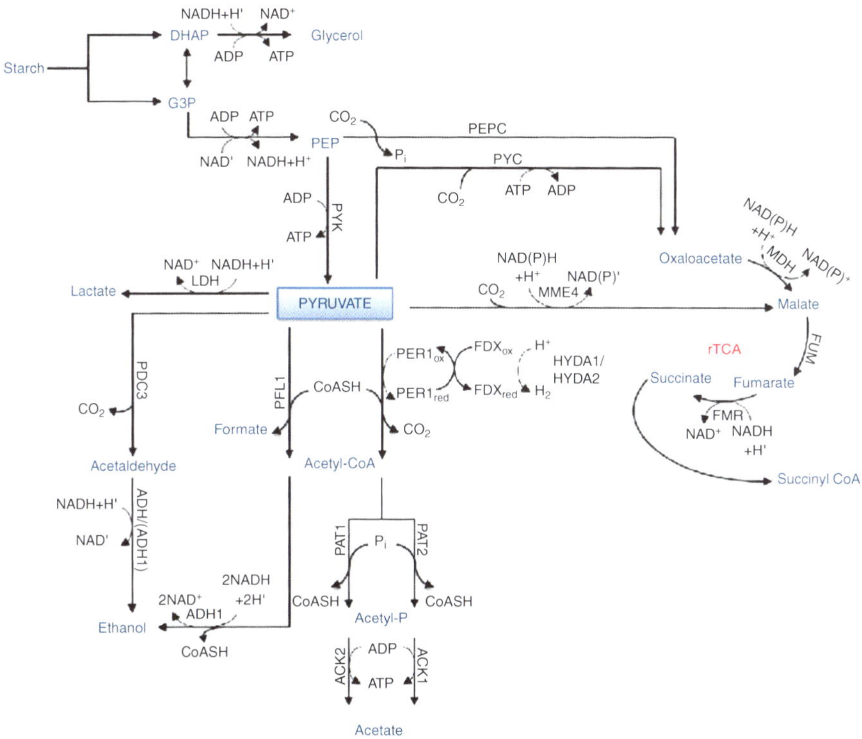

Fig. 1 Fermentation pathways in Chlamydomonas. In wild-type Chlamydomonas cells, the major fermentative products are formate, acetate, and ethanol, with CO_2 and H_2 generated as minor products. Note, the pathway leading to fermentative succinate production is not readily detected in wild-type cells, but becomes prominent in the *hydEF-1* mutant (Dubini et al. 2009). An increase in lactate production, which is almost undetectable in fermenting wild-type cells, has been observed in the *pfl1* mutants (Burgess et al. 2012; Catalanotti et al. 2012; Philipps et al. 2011). Glycerol accumulation was observed in the *adh1* single and the *pfl1-1adh1* double mutants (Catalanotti et al. 2012). The gene designations in this figure are *PYK* for pyruvate kinase, *PEPC* for phosphoenolpyruvate carboxylase, *PYC* for pyruvate carboxylase, *MME4* for malic enzyme, *MDH* for malate dehydrogenase, *FUM* for fumarase, *FMR* for fumarate reductase, *LDH* for lactate dehydrogenase, *PDC3* for pyruvate decarboxylase, *ADH/ADH1* for acetaldehyde-alcohol dehydrogenase, *PFL1* for pyruvate formate lyase, *PFR1* for pyruvate:ferredoxin:oxidoreductase, *PAT1* and *PAT2* for phosphate acetyltransferase, *ACK1* and *ACK2* for acetate kinase, *FDX* for ferredoxin, and *HYDA1* and *HYDA2* for the hydrogenases

CO_2 and acetaldehyde through the action of pyruvate decarboxylase (PDC3). The acetaldehyde generated in this reaction can be reduced to ethanol by ADH1, which requires oxidation of one NADH; the conversion of acetyl-CoA to ethanol recycles two NADH molecules.

2.2 Plant Fermentation Metabolism

The majority of plants are not tolerant to anaerobiosis, but some species, including rice, have developed specialized mechanisms to cope with low O_2 conditions (Magneschi and Perata 2009). One of the main factors that has promoted studies of anoxia in plants is flooding, a natural disaster that occurs frequently in many regions of the world and can dramatically reduce crop yield and quality (Bailey-Serres et al. 2012; Fukao et al. 2006; Kennedy et al. 1992). In recent years, there have been many reviews describing the ways in which plants acclimate to anoxic/hypoxic conditions (Agarwal and Grover 2006; Bailey-Serres et al. 2012; Bailey-Serres and Voesenek 2008, 2010; Bell and Chandel 2007; Pucciariello and Perata 2012; Fukao et al. 2009; Geigenberger 2003; Gibbs et al. 2000; Licausi 2011; Licausi et al. 2010; Magneschi and Perata 2009; Noctor et al. 2007; Perata et al. 2011; Sasidharan and Mustroph 2011; Voesenek et al. 2006). Plant submergence restricts diffusion of O_2 and CO_2 by over 100-fold, which has a dramatic impact on biochemical processes such as aerobic respiration and photosynthesis (Fukao et al. 2006). Several acclimation responses are triggered by low O_2 conditions, including petiole or internode elongation, altered anatomy and cell ultrastructure in both leaves and roots, the development of lateral or adventitious roots, the formation of aerenchyma cells, and a switch from aerobic to anaerobic respiration (Bailey-Serres et al. 2012; Fukao et al. 2006). As hypoxic/anoxic conditions develop, plants elicit specific signaling pathways, activate fermentative enzymes that facilitate the re-oxidization NADH produced by glycolysis, and accumulate organic compounds such as lactate and ethanol (Licausi 2011; Magneschi and Perata 2009). Most of the fermentative enzymes are inactive when O_2 is present, but become rapidly activated when environmental O_2 is depleted (Davies et al. 1974). In certain rice plants, nearly 92 % of the pyruvate is converted into ethanol, 1 % into lactate, and 7 % into alanine, which means that in these plants ethanolic fermentation accounts for most pyruvate consumption and NAD^+ regeneration, which minimizes acidosis of the cytoplasm of cells (Kato-Noguchi 2006). Enzymes associated with anoxia, including alcohol dehydrogenase (ADH), lactate dehydrogenase (LDH), and pyruvate decarboxylase (PDC), have been characterized in many plants including soybean (Komatsu et al. 2011), rice (Tsuji et al. 2003), wheat (Shingaki-Wells et al. 2011), barley (Good and Crosby 1989; Kato-Noguchi et al. 2010), grasses (Kennedy et al. 1987), grape (Mugnai et al. 2011), Arabidopsis and poplar (Narsai et al. 2011), and oat (Kato-Noguchi et al. 2010).

Metabolite accumulation may vary between plant species as well as between plant organs. In addition to ethanol, lactate, and formate, several plants accumulate succinate, acetate, GABA, and alanine upon exposure to low O_2 conditions (Magneschi and Perata 2009; Miyashita and Good 2008). Under hypoxic/anoxic conditions, the accumulation of acetaldehyde is harmful since it can form adducts with cellular proteins (Perata et al. 1992). Two pathways were identified that metabolize acetaldehyde; one is through the activity of ADH which converts the acetaldehyde to ethanol and the other is through the activity of ALDH, which converts acetaldehyde to acetate (Kirch et al. 2004).

3 Hydrogenase, Hydrogen Production, and Mutant Analyses

3.1 Hydrogenase in Algae

Genes encoding the hydrogenases of Chlamydomonas (and various microbes) are induced when the cells experience anoxic conditions. Significant interest in the structure and regulation of hydrogenases arises from the ability of these enzymes to convert reductant plus protons into H_2, a molecule that can be exploited as a fuel (Beer et al. 2009; Esper et al. 2006; Ghirardi et al. 2005, 2007, 2009; Grossman et al. 2011; Hemschemeier and Happe 2005; Hemschemeier et al. 2009; Kosourov and Seibert 2009; Kruse and Hankamer 2010; Radakovits et al. 2010; Rosenberg et al. 2008; Rupprecht 2009; Seibert et al. 2008). Although the precise physiological function of hydrogenase and H_2 synthesis in algal cells is not completely resolved, it is likely to have significant impacts on redox poise, photoprotection, and fermentative energy production.

Green algae have [Fe–Fe] hydrogenases (Ghirardi et al. 2007). In Chlamydomonas these hydrogenases are designated HYDA1 and HYDA2. These enzymes are typically O_2-sensitive (Ghirardi 1997; King et al. 2006) and are irreversibly inactivated within minutes of exposure to atmospheric levels of O_2 (Cohen et al. 2005). The two Chlamydomonas [Fe–Fe] hydrogenases (68 % identical and 74 % similar at the amino acid level) (Forestier et al. 2003; Happe and Naber 1993) have molecular masses of ~49 kDa and N-terminal transit peptides facilitate their transport into chloroplasts (Forestier et al. 2003; Happe et al. 1994; Happe and Naber 1993). The active site of the enzymes is complex and contains Fe–S, CO, and CN ligands. The maturation of microbial [Fe–Fe] hydrogenases (Bock et al. 2006) (associated with the attachment of specific ligands) involves the activities of three auxiliary proteins; two of these proteins, HYDE and HYDG, belong to the radical *S*-adenosylmethionine (radical SAM) superfamily, while the third (HYDF) has a GTPase domain (Mulder et al. 2010; Posewitz et al. 2004a, b). In Chlamydomonas, the HYDE and HYDF proteins are fused (Posewitz et al. 2004a, b).

3.2 Hydrogen Production

Generally, H_2 is known to be produced under three different conditions, all associated with O_2 deprivation: these are (1) direct biophotolysis, (2) indirect biophotolysis, and (3) dark fermentative metabolism (Fig. 2). Direct biophotolysis involves light-dependent oxidation of water by photosystem II (PSII), the transfer of electrons from PSII to photosystem I (PSI), light-dependent excitation of PSI with the concomitant reduction of FDX (gene designation *PETF*), and the subsequent transfer of electrons from FDX to the hydrogenases (Beer et al. 2009; Ghirardi et al. 2007; Greenbaum 1982; Happe and Naber 1993; Winkler et al. 2010). This pathway operates under conditions in which the cells have low PSII activity (e.g., sulfur deprivation) (Melis et al. 2000) and relatively high respiration rates, facilitating the generation of hypoxic/anoxic conditions in the cultures (Chochois et al. 2009; Kosourov et al. 2003). The second photoproduction pathway, indirect biophotolysis, involves non-photochemical plastoquinone (PQ) pool reduction (NPQR) by NAD(P)H that is generated as a consequence of carbohydrate catabolism, followed by light-dependent FDX reduction by PSI and the subsequent transfer of electrons from FDX to the hydrogenases (Kosourov et al. 2003; Mus et al. 2005). In Chlamydomonas, this pathway depends on NAD(P)H-plastoquinone oxidoreductase (NDA2) activity (Jans et al. 2008; Mus et al. 2005) and is independent of PSII activity (Chochois et al. 2009; Kosourov et al. 2003). Finally, the third pathway for H_2 production involves dark fermentative metabolism; the catabolism of fixed carbon provides electrons to the hydrogenases through PFR1 activity (Figs. 1 and 2) (Dubini et al. 2009; Gfeller and Gibbs 1984; Ghirardi 1997; Happe et al. 1994; Kreuzberg 1984; Mus et al. 2007; Ohta et al. 1987; Posewitz et al. 2004a, b). Hydrogenases also function in H_2 uptake, with two distinct uptake pathways in Chlamydomonas (Gaffron 1944; Maione and Gibbs 1986). The ability to take up H_2 may give the organism a competitive edge in the natural environment by providing it with low potential reductant.

3.3 *Mutations That Impact Hydrogenase Activity*

Several lesions in Chlamydomonas genes that either directly or indirectly impact fermentation metabolism have been described.

3.3.1 *hydA* Mutants

Mutants were obtained in each of the two *HYDA* genes of Chlamydomonas; these mutants were generated in both forward and reverse genetic screens (Meuser et al. 2012). The phenotypes of both single (*hydA1* and *hydA2*) and double (*hydA1hydA2*) mutants were analyzed (Meuser et al. 2012). The *hydA1* and *hydA2* mutants could both catalyze H_2 production from reductant derived from either fermentative or photosynthetic metabolism. However, the contribution of the HYDA2 enzyme to H_2 photoproduction under the conditions tested was approximately 25 % of that of HYDA1 (Godman et al. 2010; Meuser et al. 2012). Interestingly, some *hydA1*-rescued strains (transformed with the wild-type *HYDA1* gene) exhibited enhanced in vitro and fermentative H_2 production, while H_2 photoproduction rates never exceeded those of the wild-type strain. In addition,

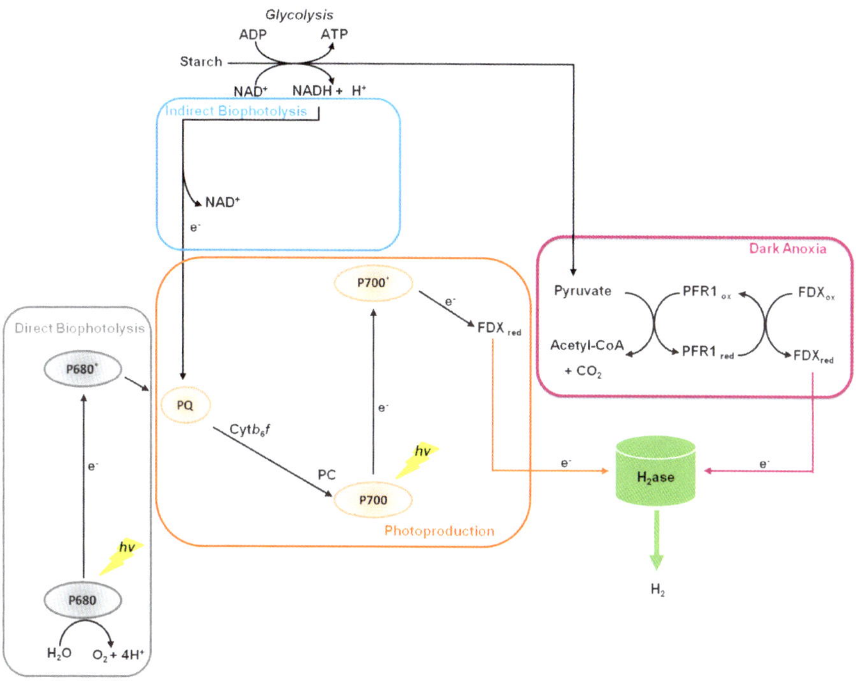

Fig. 2 H_2 production pathways in Chlamydomonas. The two pathways associated with the photoproduction of H_2 are the direct biophotolysis (*gray rectangle*) and indirect biophotolysis (NPQR-dependent; *blue rectangle*) pathways. Both pathways require the generation of a reduced PQ pool, PSI-dependent reduction of FDX (*orange arrow* and *orange rectangle*), and the transfer of electrons from FDX to the hydrogenases (*green*). In addition, during dark fermentation (*purple rectangle*), pyruvate oxidation is coupled to FDX reduction through PFR1. The electrons from FDX are transferred to the hydrogenases (*green*)

Godman et al. (2010) generated a knockdown strain of Chlamydomonas HYD3 (homologue of HYDA1/2) that was impaired in xanthine dehydrogenase activity and may also function in the biogenesis/maintenance of cytosolic FeS proteins.

3.3.2 *hydEF-1* Mutant

The *hydEF-1* strain is unable to assemble the inorganic constituents of the hydrogenase active site, and consequently cannot catalyze H_2 synthesis (Dubini et al. 2009; Posewitz et al. 2004a, b). Anoxic cultures of this mutant also exhibit lower CO_2 evolution, and reduced extracellular formate, acetate, and ethanol accumulation. Interestingly, increased levels of extracellular succinate are observed; the accumulation of this metabolite was not detected in the parental strain (Dubini et al. 2009), indicating activation of a fermentative pathway not typically operating at significant levels in wild-type cells under our standard growth conditions. Transcript and metabolite analyses suggest that in the *hydEF*-1 mutant the carboxylation of pyruvate leads to the generation of either malate or oxalo-acetate, which is converted to succinate through activation of a section of the TCA cycle that operates in the reverse direction (a partial reverse TCA cycle) (Fig. 1 and Dubini et al. 2009). Hence, in the absence of hydrogenase activity, the generation of succinate from pyruvate helps the cells satisfy their need to oxidize NADH. These results suggest that when the hydrogenase is unable to accept electrons from FDX and there are no alternative electron acceptors available (e.g., nitrite, sulfate are potential alternate electron acceptors), cellular redox increases, which might lead to activation of the thioredoxin system through coupling with reduced FDX. These regulatory molecules might in turn control the activities of genes/proteins associated with the various metabolic pathways that facilitate NAD(P)H oxidation; one such pathway results in the generation of succinate. Although speculative at this point, these results raise the possibility that both *MME4* (encoding the malic enzyme, which catalyzes the formation of malate from pyruvate with concomitant CO_2 uptake; see Fig. 1) and *FMR* (encoding fumarate reductase, which reduces fumarate to succinate; see Fig. 1) genes, which are both significantly up-regulated specifically in the *hydEF-1* mutant during anoxia (Dubini et al. 2009), may be under the control of a FDX-thioredoxin regulatory system.

3.3.3 *sta* Mutants

Two Chlamydomonas mutants that do not accumulate starch, *sta6* (Zabawinski et al. 2001) and *sta7* (Posewitz et al. 2004a, b), have reduced hydrogenase activity under dark, anaerobic conditions and the levels of the *HYDA1* and *HYDA2* transcripts are also reduced (Posewitz et al. 2004a, b). The phenotypes associated with these mutants may be indirect, resulting from altered metabolite pools and/or changes in cellular redox and energy status. Starch degradation under anaerobic conditions can influence intracellular NAD(P)H levels and/or the oxidation state of the PQ pool, both of which impact transcriptional processes (Escoubas et al. 1995; Pfannschmidt and Liere 2005; Rutter et al. 2001). It was recently demonstrated that the Chlamydomonas *sta6* and *sta7* mutants have near-wild-type levels of hydrogenase activity under conditions of anoxia and sulfur starvation in the light (Chochois et al. 2009). Furthermore, although DCMU severely attenuated H_2 photoproduction in these cells, uninhibited and DCMU-inhibited cells exhibited a similar level of in vitro hydrogenase activity (Chochois et al. 2009), suggesting that the flow of reductant to the hydrogenases does not significantly alter expression of the *HYDA* genes. Moreover, starch accumulation is not absolutely necessary for high levels of H_2-photoproduction. Feeding the *sta* mutants with acetate allowed them to achieve high respiratory rates, which hastened anaerobiosis and H_2 generation (Chochois et al. 2009); a similar result was demonstrated for immobilized, wild-type cells (Kosourov and Seibert 2009).

3.3.4 *stm6* Mutant

This mutant, disrupted for a homologue of the human mitochondrial transcription termination factor, has a variety of phenotypes including inhibition of cyclic electron transport under anaerobic conditions, hyper-accumulation of starch, a decreased number of active PSII reaction centers, an increased rate of respiration, and an elevated rate of H_2 photoproduction during sulfur deprivation compared to the parental strain (Kruse et al. 2005; Rupprecht et al. 2006). Increased H_2 photoproduction in the mutant probably reflects its aggregate phenotype; attenuated cyclic electron transport would eliminate competition between the hydrogenase and PSI-dependent, cyclic electron transport, overaccumulation of starch would provide additional reductant to the PQ pool, and reduced O_2 evolution by PSII would delay the build-up of O_2 in the culture (Kruse et al. 2005). Additional genetic modifications generated in the *stm6* genetic background have also been examined. The *stm6* mutant transformed with a glucose transporter, *stm6Glc4*, has more efficient glucose uptake and improved H_2 photoproduction (Doebbe et al. 2010). The *stm6Glc4* mutant with reduced antenna size further boosts H_2 production (Beckmann et al. 2009).

3.4 *The Generation of Mutants in Specific Branches of Fermentative Metabolism Using Reverse Genetics*

Reverse genetic approaches have proven extraordinarily effective in generating mutants specifically inactivated for genes relevant to anoxic metabolism in Chlamydomonas (Gonzalez-Ballester et al. 2011). This technique has yielded mutants with lesions in genes encoding pyruvate formate lyase (PFL1: ID 146801), fumarate reductase (FMR: ID 145357), malic enzyme (MME4: ID 196831), acetate kinase (ACK1: ID 129982; ACK2: ID 128476), acetate phosphotransferase (PAT2: ID 191051), alcohol dehydrogenase (ADH1: ID 133318), pyruvate decarboxylase (PDC1: ID 193810), the hydrogenases (HYDA1: ID 183963; HYDA2: ID 24189), prolyl-4-hydroxylases (P4H2: ID 106683; P4H3: ID 206687; P4H4: ID 111255; P4H5: ID 182877), ferredoxin 5 (FDX5: ID 156833), the transhydrogenase (NTH1: ID 139758), serine endopeptidase (SEP: ID 147682), phospholipid-translocating P-type ATPase (ALA2: ID 190292) and a proton ATPase (H^+-ATPase: ID 190023), hybrid cluster proteins (HCP1: ID148253; HCP2: ID 148898; HCP3: ID 191403; HCP4: ID 148255), pyruvate dehydrogenase kinase (PDK1: ID 196270), MAP kinase (MAK1: ID 58843), glyoxal galactose oxidases (GOX8: ID 196818; GOX15: ID 196829), and glycoxylate-induced protein (GIP1: ID 34332). We were unable to identify a mutant with an insertion in the *PFR1* gene, but have subsequently used TILLING (targeted induced local lesions in genomes) to identify strains with point mutations in this gene (Catalanotti, unpublished). Below we present a description of some of the mutants that were generated by our targeted screen and that have been characterized in some detail.

3.4.1 *pfl1* Mutants

pfl1 mutants were isolated by both Catalanotti et al. (2012) and Philipps et al. (2011). Both groups obtained strains that were unable to accumulate the PFL1 protein. When the *pfl1* mutants were subjected to dark fermentative conditions, they displayed a marked accumulation of extracellular lactate, elevated pyruvate decarboxylation, and extracellular ethanol accumulation, and, in one case (Catalanotti et al. 2012), lower H_2 production. The significant level of lactate that accumulates in the medium of *pfl1* mutants allows for recycling of NADH as a consequence of pyruvate reduction by LDH. Catalanotti et al. (2012) also showed that the *pfl1* mutant accumulates high intracellular levels of lactate, succinate, alanine, malate, and fumarate, suggesting that reverse TCA reactions were used to recycle NADH. Additionally, a *pfl1* knockdown strain accumulates 3-hydroxybutyrate during the transition from hypoxic to anoxic conditions in the light after sulfur deprivation (Burgess et al. 2012).

3.4.2 *adh1* Mutant

In the *adh1* mutant both CO_2 and ethanol production were completely abolished; the strain appeared to be completely null for ADH1 activity (Magneschi et al. 2012). The inability of the *adh1* mutant to accumulate ethanol and CO_2, while synthesizing low levels of formate, suggests that the acetaldehyde produced by PDC3 and the acetyl-CoA produced by PFL1 and PFR1 are not readily reduced in mutant cells exposed to anoxic conditions, and that under such conditions the activities of PFL1 and PDC3, the major activities responsible for the accumulation of formate and CO_2, respectively, may markedly decline. These findings also strongly suggest that ADH1 is the only acetaldehyde-alcohol dehydrogenase in Chlamydomonas capable of reducing acetyl-CoA or acetaldehyde under the conditions used in this study.

Changes in fermentative metabolism in the *adh1* mutant would be critical to limit the generation and build-up of substrates normally acted on by ADH1, allowing the mutant to accommodate the complete block in ethanol synthesis. However, it would also be necessary for the mutant to eliminate reducing equivalents generated during the glycolytic production of ATP. The *adh1* strain did show some extracellular lactate accumulation, which was not observed in wild-type cells; however, a significantly larger increase in lactate accumulation was observed in strains lacking PFL1 (Catalanotti et al. 2012). Most interesting was the finding that the medium of the *adh1* mutant maintained under anoxic conditions contained high levels of glycerol relative to wild-type cells; internal metabolite analysis also showed higher intracellular glycerol levels in *adh1* relative to wild-type cells.

Glycerol is synthesized from dihydroxyacetone phosphate (DHAP). This metabolite precedes the formation of pyruvate and the 3C oxidation (NADH formation) step in glycolysis, and glycerol synthesis also effectively recycles one NADH. Hence, glycerol and lactate production in the *adh1* mutant would allow for efficient recycling of NADH, maintenance of redox balance, and sustained glycolytic production of ATP even though the cells are unable to reduce acetaldehyde or acetyl-CoA to ethanol (Magneschi et al. 2012).

To further probe the system, we generated a double mutant (*pfl1-1adh1*) that cannot synthesize either formate or ethanol (Catalanotti et al. 2012). This strain, like the *pfl1* mutants, secreted significant levels of lactate; however, like the *adh1* mutant it also exhibited a marked increase in the levels of extracellular glycerol and acetate, and a decline in dark, fermentative H_2 production. While wild-type Chlamydomonas fermentation primarily produces formate and ethanol, the double mutant exhibits a complete rerouting of glycolytic carbon to lactate and glycerol. Although the metabolic adjustments observed in the single and double mutants facilitate NADH reoxidation and sustain glycolysis under dark, anoxic conditions, the observed changes could not have been predicted, and were only revealed from the biochemical and molecular analyses of the mutants.

4 Regulation of Fermentation Metabolism

4.1 Regulation of Genes Encoding Enzymes Involved in Fermentation Metabolism

Transcripts encoding many enzymes involved in fermentation are elevated in Chlamydomonas during anoxia. The PFR1 transcript and protein both increase dramatically at the onset of anoxia {(Mus et al. 2007), our unpublished data}. In contrast, anoxia triggers an increase in *PFL1* mRNA with little increase in PFL1 protein (Atteia et al. 2006; Catalanotti et al. 2012; Philipps et al. 2011). Interestingly, the molecular mass of PFL1 from anoxic cells was slightly less than that from cells maintained in oxic conditions, consistent with anoxic conditions triggering a posttranslational modification leading to activation of the protein (Atteia et al. 2006; Catalanotti et al. 2012). Transcripts from genes coding for *ADH1*, *PAT1*, *PAT2*, *ACK1*, and *ACK2* are also up-regulated at the onset of anoxia (Mus et al. 2007). Although few studies have been performed to elucidate regulation of genes responsive to anoxic conditions, the region 21–128 upstream of the *HYDA1* transcription start site was shown to be involved in controlling its expression (Stirnberg and Happe 2004).

In plants some studies have revealed that acclimation to anoxic conditions involves the ethylene response factor (EFR) transcriptional elements (Bailey-Serres and Voesenek 2010; Gibbs et al. 2011; Licausi et al. 2011). To date, there is little known about how algae sense oxic conditions. Furthermore, while there are many

EFR-like putative transcriptional elements in Chlamydomonas, none of these have the cysteine at the N terminus (Yang, unpublished), which was shown to be essential for O_2 sensing in plants (See Chap. 1). However, it is likely that the reduction state of the intersystem PQ pool (Antal et al. 2003) and/or the production and detoxification of reactive oxygen species (ROS) play a role in signaling during transitions between oxic and hypoxic/anoxi conditions (Borisjuk and Rolletschek 2008; Licausi et al. 2011). Hydrogen peroxide (H_2O_2), synthesized by a NAD(P)H oxidase, is required for controlled induction of ADH expression in Arabidopsis (Baxter-Burrell et al. 2002). The production of nitric oxide (NO) may be important for regulating mitochondrial activity in plants and animals as they enter hypoxia (Moncada and Erusalimsky 2002). Furthermore, in animal cells prolyl 4-hydroxylases directly sense O_2 and participate in controlling the responses of these cells to hypoxia/anoxia (Guzy and Schumacker 2006).

There have been recent advances in elucidating the ways in which mammalian cells sense hypoxia (Greer et al. 2012; Semenza 2011). A hypoxia inducing factor (HIF) that is constitutively expressed is hydroxylated at conserved proline residues in the presence of O_2. This modification targets HIF for ubiquitin-dependent degradation. In the absence of O_2, HIF hydroxylation ceases, and the protein accumulates and triggers expression of several target genes. The prolyl 4-hydroxylases may have other potential protein targets that accumulate under anoxic conditions, but these are rapidly turned over as the cells transition from anoxic to oxic conditions. Four prolyl 4-hydroxylases encoded on the Chlamydomonas genome are significantly up-regulated in response to anoxic conditions (Mus et al. 2007). Although highly speculative at this point, the activity of one or more of these prolyl hydroxylases in Chlamydomonas may provide a mechanism by which this alga senses O_2, maintains genes in an inactive state when O_2 is present, and targets key proteins involved in fermentation metabolism for destruction as algal cells transition from anoxic to oxic conditions. In Arabidopsis and rice, the levels of transcripts encoding prolyl 4-hydroxylases are markedly induced by O_2 deprivation (Lasanthi-Kudahettige et al. 2007; Vlad et al. 2007), suggesting that their role in animals as sensing elements might be conserved in plants.

Recently, Pape et al. (2012) reported a Chlamydomonas *crr1* mutant impaired in *HYDA1* gene expression. They used a chimeric gene in which the *HYDA1* promoter was fused to *ARS2* (which encodes an arylsulfatase) and the chimeric gene introduced into wild-type Chlamydomonas cells that were then screened for transformants exhibiting arylsulfatase activity. Reporter gene analysis and electrophoretic mobility shift assays have indicated that the SQUAMOSA-PROMOTER BINDING PROTEIN (SBP) domain of CRR1 is involved in regulating *HYDA1* transcription, and that two GTAC motifs are necessary for both CRR1 binding and full promoter activity under hypoxic or Cu-starvation conditions. These combined results demonstrate that CRR1 is involved in *HYDA1* promoter activation (Pape et al. 2012). Whitney et al. (2012) investigated expression of the fermentative genes *PDC3*, *LDH1*, and *ADH2* and showed that the transcripts from these genes were primarily under diurnal regulation, while *PFL1* appeared to be exclusively responsive to diminished O_2. These results suggest that there exist at least three different regulatory mechanisms within the fermentative pathways; therefore, the fermentative pathways are not redundant, but rather enable metabolic versatility that fine tunes metabolic responses to dynamic environmental conditions (Whitney et al. 2012).

5 Probing Cellular Oxic/Redox Conditions

5.1 Flux Balance Analysis and Metabolic Flux Analysis

Recently, several methods have been used to model metabolism, ranging from highly detailed kinetic models to less complex stoichiometric models. A combination of flux balance analysis (FBA) and metabolic flux analysis (MFA) offers a valuable approach for dissecting metabolic pathways and their regulation (Chen et al. 2011) and can be used to help understand anoxic metabolism in photosynthetic organisms. These approaches will become especially potent with improved characterizations of wild-type and mutants of Chlamydomonas that do not properly acclimate to hypoxia/anoxia. FBA attempts to predict fluxes based on reaction stoichiometry, biomass composition, and constraints associated with the thermodynamics of reactions and the limits of uptake/secretion rates. It can also be used to predict theoretical yields and identify bottlenecks and/or metabolic sinks that need to be adjusted to approach maximum yields (Boyle and Morgan 2009; Chen et al. 2011; Orth et al. 2010; Raman and Chandra 2009; Thiele and Palsson 2010). Recently, FBA was used to reconstruct metabolic networks in Chlamydomonas, with almost half of the 458 detected metabolites placed in chloroplasts, a consequence of a large number of plastid-localized reactions (212 out of 484). The cytosol, which contains ~35 % of the metabolites, acts as the "hub" of metabolite transport and the site of synthesis for many macromolecules. A significant fraction of reactions in the model are involved in intracellular transport, suggesting a high interconnectivity between cellular compartments (Boyle and Morgan 2009). FBA combined with knowledge of protein localization (based on predicted target peptide sequences and/or experimental analyses) has begun to reveal compartmentation of metabolic networks in various cellular organelles. These include the synthesis of fatty acids, amino acids, nucleotides, starch, and chlorophyll in chloroplasts; TCA cycle activity and amino acid synthesis in mitochondria; glycolysis and amino acid and fatty acid synthesis in the cytosol (Boyle and Morgan 2009).

In addition to FBA, isotopic labeling-based MFA can be powerfully exploited to elucidate quantitative and regulatory features associated with metabolic networks (Chen et al. 2011). ^{13}C MFA has been used to examine compartment-specific bioprocesses associated with the synthesis of fatty acids, cell walls, protein glycans, and starch; analyses of ^{13}C-labeled metabolites were performed by gas chromatography/mass spectrometry and nuclear magnetic resonance spectroscopy (Allen et al. 2007, 2009; Libourel and Shachar-Hill 2008; Sweetlove and Ratcliffe 2011).

5.2 Sensing O_2 Levels

The ability to noninvasively measure O_2 levels/flux can be used to solve a variety of questions associated with basic metabolism, growth, and stress physiology. Technologies for measuring O_2 concentrations near the surface of cells or tissues include electrochemical and optical techniques. Fiber-optic microprobes (optodes) have been developed for analyzing O_2 fluxes associated with respiration in plant roots, as well as at the boundary layer of phytoplankton mats, while some microsensors have advanced sensing capabilities and can report real-time fluxes (McLamore et al. 2010). Recent applications of O_2-sensitive microsensors have demonstrated steep O_2 gradients in developing seeds of various crops. Porterfield et al. (1999) used miniature glass electrodes to measure the internal O_2 concentration in the siliques of Arabidopsis while others have used O_2 microsensors to detect dynamic changes in tissue O_2 levels in response to environmental factors to evaluate the rate of release of photosynthetic O_2 (Borisjuk and Rolletschek 2008) and to examine differential sensitivity of barley roots to hypoxia (Pang et al. 2006). Other types of sensors, such as the Förster Resonance Energy Transfer (FRET) sensors, may also be adapted to report O_2 redox conditions (Potzkei et al. 2012) during the acclimation of cells to hypoxic/anoxic conditions.

5.3 Redox Measurement Under Anoxic/Hypoxic Conditions

The subcellular localizations of some metabolic pathways differ in green algae and vascular plants. In plants, the chloroplast has differentiated mainly into a photosynthetic organelle, with the end-products of photosynthesis being exported out of chloroplasts and mostly out of leaves (export of sucrose) to fuel whole plant growth and development. In Chlamydomonas, the chloroplast has retained a more complex metabolism that integrates starch synthesis and starch breakdown in the same compartment. This alga can also assimilate external fixed carbon such as acetate, and can synthesize starch in the dark from acetate via gluconeogenesis. Furthermore, under anaerobic conditions many fermentation enzymes become active in chloroplasts (and in mitochondria) while the activities of the mitochondrial cytochrome oxidase, COX (Cardol et al. 2009), the alternative oxidase, AOX (Cardol et al. 2009), and the chloroplast alternative oxidases, PTOX1 and PTOX2 (Houille-Vernes et al. 2011), decline. As a result, electrons accumulate in mitochondrial and chloroplast electron transport chains under dark anaerobic conditions, which strongly impacts chlorophyll fluorescence (associated with photosynthetic electron transport). Therefore, fluorescence measurements can be used as an internal probe to help evaluate chloroplast metabolism and the redox state of cells.

5.3.1 General Principles

The variable component of chlorophyll fluorescence reflects photosystem II (PSII) activity. The fluorescence yield of a cell suspension depends upon the redox state of the primary quinone electron acceptor of PSII, which is designated Q_A; if Q_A is oxidized, it accepts an electron from the primary electron donor of PSII and quenches chlorophyll fluorescence (photochemical quenching). This process leads to a minimum fluorescence yield, designated F_0, when cells are in the dark-adapted state. If Q_A is reduced, chlorophyll fluorescence is not quenched by PSII photochemistry and the maximum level of fluorescence, designated F_M, is attained. Q_A equilibrates thermodynamically with the plastoquinone pool (PQ) via the transfer of electrons to the Q_B site of PSII, making chlorophyll fluorescence also sensitive to the transfer of electrons into and out of the PQ pool. These electron flows are dominated in the light by PSII activity (reduction of PQ) as well as cytochrome b_6f and PSI activity (oxidation of PQ), while in the dark they are dominated by chlororespiration, NDA2 activity (reduction of PQ by NADPH dehydrogenase), PTOX activity (oxidation of PQH_2 by O_2), and fermentation metabolism if the cells become hypoxic. In a mutant devoid of PSI or cytochrome b_6f, the only output for electrons is through the PTOX reaction, and the chlorophyll fluorescence induced by ~1 s of continuous illumination shows a steady rise from F_0 to F_M, as shown in Fig. 3a, representing the progressive filling of the PQ pool with electrons. As demonstrated by others (Forbush and Kok 1968), the area over the fluorescence rise curve plotted against the time for which the cells were exposed to anaerobic conditions (Fig. 3b, blue) is proportional to the number of available PSII electron acceptors (as the area over the rise curve decreases, more PQ is in the reduced state and so less acceptors are available for photochemistry), and allows a direct quantification of the redox state of the PQ pool in the dark in a mutant devoid of cytochrome b_6f. Experiments showing full reduction of the PQ pool under anaerobic conditions (Bennoun 1982) were crucial for demonstrating the existence of a respiratory chain in chloroplasts. Early experiments (Rabinowitch 1951; Shiau and Franck 1947) reported that prolonged anaerobic conditions induced a rise in F_0 and inactivation of photosynthesis. This phenomenon reflects reduction of Q_A in the dark, and therefore it can serve as a precise indicator of the redox state of chloroplasts; E_m (PQ/PQH_2) = +110 mV, E_m (Q_A/Q_A^-) = −30 mV. Also shown by fluorescence measurements (Rabinowitch 1951; Shiau and Franck 1947) is that photosynthesis is slowly reactivated upon illumination of cells experiencing dark anaerobic conditions (see Sect. 5.3.4).

5.3.2 Studies of Mitochondrial Respiration and Chlororespiration

Fluorometric probes have been used to study the metabolic interactions between mitochondria and the chloroplasts. The addition of mitochondrial inhibitors like antimycin A and salicylhydroxamic acid (SHAM) or uncouplers like carbonyl

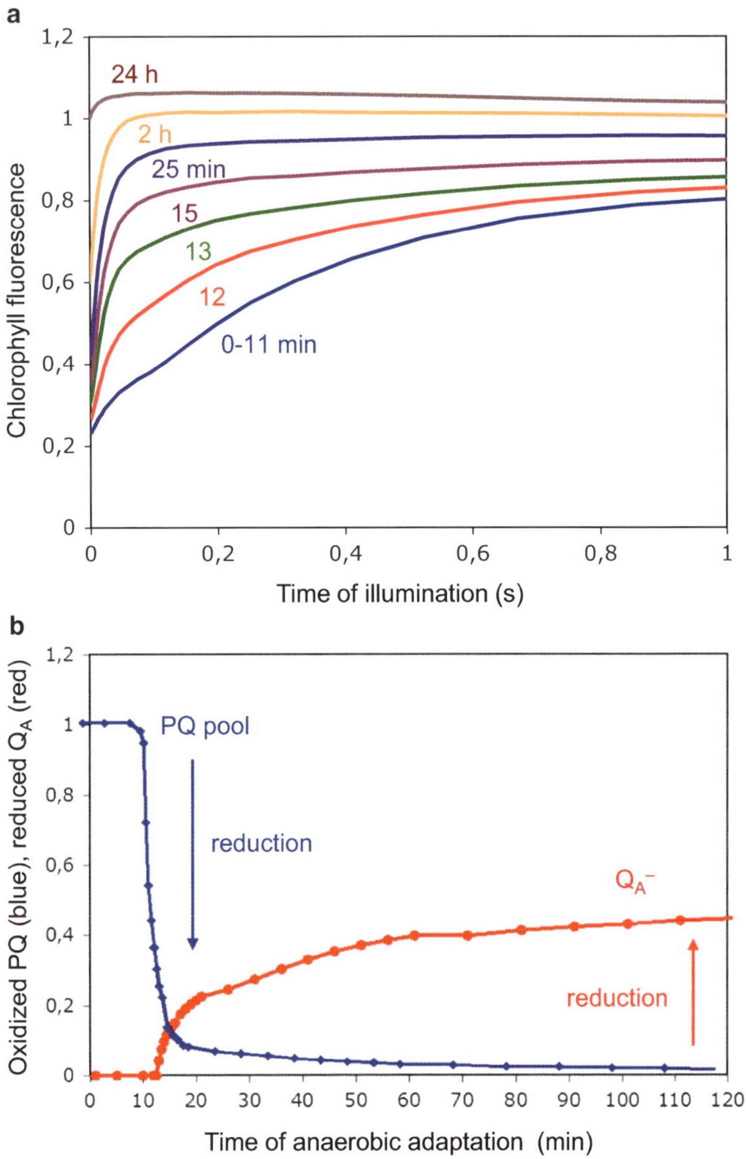

Fig. 3 Chlorophyll fluorescence as an internal probe of the chloroplast redox poise. (**a**) Changes in chlorophyll fluorescence kinetics for Chlamydomonas mutants devoid of the cytochrome b_6f complex when the cells are exposed to dark anaerobic conditions. The area over the fluorescence rise curve, decreasing with increasing exposure to anaerobic conditions, reflects a decline in the size of the pool (PQ and Q_A) able to accept electrons from PSII. (**b**) The area over the fluorescence curve, which decreases with time of anaerobic adaptation (on the *X* axis), represents the concentration of oxidized PQ (*blue line*). An increase in F_0, which indicates elevation in the level of Q_A^- (*red line*), is also shown; the reduction of Q_A increases with increasing exposure to anaerobic conditions

cyanide 3-chlorophenylhydrazone (CCCP) under aerobic conditions were shown to cause a significant reduction of the PQ pool (Alric 2010; Gans and Rebeille 1990; Rebeille and Gans 1988). Inhibition of mitochondrial respiration appears to have at least two major metabolic consequences. First, by blocking electron flow to O_2, NADH accumulates in the mitochondrion and cytosol and likely inhibits the TCA cycle and downstream steps of glycolysis (pyruvate dehydrogenase complex). Second, by depleting ATP in the dark, the upstream steps of glycolysis may be stimulated (ATP acts as an allosteric inhibitor of hexokinase and phosphofructokinase, PFK) (Klock and Kreuzberg 1991); when mitochondria are performing aerobic respiration, the substrate of PFK, fructose-6-phosphate, is more abundant than its product, fructose-1,6-bisphosphate, while this equilibrium is reversed under anaerobic conditions when NAD(P)H is not readily consumed (Klock and Kreuzberg 1991). The consequences of the simultaneous halt in respiratory NAD(P)H oxidation and stimulation of upstream glycolytic steps have an additive effect which would result in a more reducing cellular redox poise. A pronounced reduction of chloroplast electron carriers is observed upon the addition of inhibitors of both mitochondrial respiration and chlororespiration; used individually, these inhibitors (sodium azide, an inhibitor of COX; *n*-propyl gallate, an inhibitor of PTOX) elicit a slight reduction of the PQ pool, while the addition of both inhibitors causes >90 % of the PQ pool to become reduced. These results demonstrate that either respiration or chlororespiration is sufficient to reoxidize most of the NAD(P)H produced by glycolysis (Alric 2010).

5.3.3 Origin of the Reductants

An enzyme likely to have a major impact on NAD(P)H accumulation in the cytosol and chloroplasts during anoxic growth is the glycolytic enzyme glyceraldehyde phosphate dehydrogenase (GAPDH). Under dark aerobic conditions a downstream product of this reaction, 3-phosphoglycerate, is more abundant than GAPDH's substrate (GAP), suggesting rapid oxidation of NAD(P)H in the presence of O_2. This equilibrium is reversed under anaerobic conditions (Klock and Kreuzberg 1991) when cellular NAD(P)H levels can increase. Another reaction likely to impact cellular redox to some extent is glucose-6-phosphate dehydrogenase (G6PDH), a reaction inhibited by NAD(P)H (Lendzian and Bassham 1975). Malate dehydrogenase (MDH), associated with the glyoxylate cycle (in the cytosol), the TCA cycle (in the mitochondrion), and chloroplast metabolism, could also contribute to cellular redox conditions, although in the absence of net acetate assimilation (cells maintained in minimal medium) the net production of NAD(P)H by MDH in the dark should be negligible.

5.3.4 Reactivation of Photosynthesis in the Light

In the preceding section we have described how dark anaerobic adaptation would impact chloroplast metabolism and influences the redox status of photosynthetic electron carriers (dark reduction of Q_A from PSII). Upon light excitation, Q_A is steadily oxidized, and the photochemical yield of PSII increases. Three major pathways have been suggested to play a role in light-induced reactivation of photosynthesis: (1) the activity of hydrogenase, (2) a positive feedback of O_2 production (by PSII) on O_2 evolution (PSII activity), a process known as chlororespiration, and (3) cyclic electron flow around PSI. Each of these mechanisms participates in regenerating electron acceptors under anaerobic conditions. The hydrogenase accepts electrons from FDX, and therefore reactivates the whole linear electron transfer chain (Kessler 1973). The O_2 produced by PSII can serve as an electron acceptor for plastoquinols (Diner and Mauzerall 1973) via PTOX activity. PSII-PTOX electron transfer produces a proton gradient which restores ATP synthesis, and therefore regenerates $NADP^+$ (in the Calvin cycle or other metabolic reactions consuming both ATP and NADPH). A similar ATP dependence of the light anaerobic reoxidation of the PQ pool has been shown in the absence of active PSII, demonstrating the involvement of cyclic electron flow around PSI (Bulte et al. 1990).

5.3.5 Anaerobic Acetate Assimilation and Cyclic Electron Flow

When PSII activity is inhibited with 3(3,4-dichlorophenyl)-1,1-dimethylurea (DCMU) and Chlamydomonas cells do not evolve O_2, they maintain cyclic photophosphorylation (ATP production) under anaerobic conditions (Whatley et al. 1955). The occurrence of cyclic photophosphorylation under anaerobic conditions is important because it sustains anaerobic acetate photoassimilation in *Chlamydomonas mundana* and other green algae (Wiessner 1965). The acetate that is taken up can be incorporated into acetyl-CoA by two possible pathways; direct conversion by acetyl-CoA synthetase (ACS) or a two-step reaction involving ACK (ATP-dependent) and PAT. Both of these pathways consume one ATP molecule per acetyl-CoA synthesized. Acetyl-CoA is a substrate for the glyoxylate cycle, which is related to the TCA cycle except that NADH- and CO_2-producing steps of the TCA cycle are bypassed at the level of isocitrate, which is converted into glyoxylate and succinate. The glyoxylate molecule is condensed with a second acetyl-CoA to form malate, which is then converted to oxaloacetate to reset the cycle. Cyclic photophosphorylation appears to be necessary for anaerobic acetate assimilation in Chlamydomonas (Gibbs et al. 1986); ATP generated by glycolysis in the dark is not sufficient to promote its uptake and activation through the synthesis of acetyl-CoA.

6 Concluding Remarks

Chlamydomonas is a metabolically versatile organism that can perform photosynthetic CO_2 fixation, aerobic respiration, and anaerobic fermentation. Many pathways and enzymes associated with fermentation metabolisms in this organism are just being defined, and there is almost nothing known about the mechanisms by which these pathways are regulated. The generation of lesions that block some of these pathways is providing a wealth of information on the ways in which fermentation metabolism is regulated, compensatory responses that reveal novel mechanisms for balancing the production of ATP with the elimination of reducing equivalents and possible strategies for rerouting electrons through the various branches of fermentation metabolism for H_2 and biofuel production. Furthermore, there are many technologies, including FBA, MFA, time resolve fluorescence measurements, and the use of O_2 microsensors, that can help evaluate the redox/oxic conditions of the cells and correlate those conditions with the activities of both oxic and anoxic metabolisms. An understanding of the various pathways critical for dark metabolism and the ways in which these pathways are controlled constitutes a domain of metabolism that must be fully understood if we are to understand the energy budget of photosynthetic microbes in the environment and potential ways to modify carbon cycling.

References

Agarwal S, Grover A (2006) Molecular biology, biotechnology and genomics of flooding-associated low O_2 stress response in plants. Crit Rev Plant Sci 25:1–21

Allen DK, Shachar-Hill Y, Ohlrogge JB (2007) Compartment-specific labeling information in 13C metabolic flux analysis of plants. Phytochemistry 68:2197–2210

Allen DK, Libourel IG, Shachar-Hill Y (2009) Metabolic flux analysis in plants: coping with complexity. Plant Cell Environ 32:1241–1257

Alric J (2010) Cyclic electron flow around photosystem I in unicellular green algae. Photosynth Res 106:47–56

Antal TK, Krendeleva TE, Laurinavichene TV, Makarova VV, Ghirardi ML et al (2003) The dependence of algal H_2 production on Photosystem II and O_2 consumption activities in sulfur-deprived *Chlamydomonas reinhardtii* cells. Biochim Biophys Acta 1607:153–160

Atteia A, van Lis R, Gelius-Dietrich G, Adrait A, Garin J et al (2006) Pyruvate formate-lyase and a novel route of eukaryotic ATP synthesis in Chlamydomonas mitochondria. J Biol Chem 281: 9909–9918

Bailey-Serres J, Voesenek LA (2008) Flooding stress: acclimations and genetic diversity. Annu Rev Plant Biol 59:313–339

Bailey-Serres J, Voesenek LA (2010) Life in the balance: a signaling network controlling survival of flooding. Curr Opin Plant Biol 13:489–494

Bailey-Serres J, Fukao T, Gibbs DJ, Holdsworth MJ, Lee SC et al (2012) Making sense of low oxygen sensing. Trends Plant Sci 17:129–138

Baxter-Burrell A, Yang Z, Springer PS, Bailey-Serres J (2002) RopGAP4-dependent Rop GTPase rheostat control of Arabidopsis oxygen deprivation tolerance. Science 296:2026–2028

Beckmann J, Lehr F, Finazzi G, Hankamer B, Posten C et al (2009) Improvement of light to biomass conversion by de-regulation of light-harvesting protein translation in *Chlamydomonas reinhardtii*. J Biotechnol 142:70–77

Beer LL, Boyd ES, Peters JW, Posewitz MC (2009) Engineering algae for biohydrogen and biofuel production. Curr Opin Biotechnol 20:264–271

Bell EL, Chandel NS (2007) Mitochondrial oxygen sensing: regulation of hypoxia-inducible factor by mitochondrial generated reactive oxygen species. Essays Biochem 43:17–27

Bennoun P (1982) Evidence for a respiratory chain in the chloroplast. Proc Natl Acad Sci U S A 79:4352–4356

Bock A, King PW, Blokesch M, Posewitz MC (2006) Maturation of hydrogenases. Adv Microb Physiol 51:1–71

Borisjuk L, Rolletschek H (2008) Nitric oxide is a versatile sensor of low oxygen stress in plants. Plant Signal Behav 3:391–393

Boyle NR, Morgan JA (2009) Flux balance analysis of primary metabolism in *Chlamydomonas reinhardtii*. BMC Syst Biol 3:4

Bulte L, Gans P, Rebeille F, Wollman FA (1990) ATP control on state transitions in vivo in *Chlamydomonas reinhardtii*. Biochim Biophys Acta 1020:72–80

Burgess SJ, Tredwell G, Molnar A, Bundy JG, Nixon PJ (2012) Artificial microRNA-mediated knockdown of pyruvate formate lyase (PFL1) provides evidence for an active 3-hydroxybutyrate production pathway in the green alga *Chlamydomonas reinhardtii*. J Biotechnol 162:157–166

Camargo A, Llamas A, Schnell RA, Higuera JJ, Gonzalez-Ballester D et al (2007) Nitrate signaling by the regulatory gene *NIT2* in Chlamydomonas. Plant Cell 19:3491–3503

Cardol P, Figueroa F, Remacle C, Franzén L-G, González-Halphen D (2009) Oxidative phosphorylation: building blocks and related components. In: Harris EH, Stern D, Witman GB (eds) Chlamydomonas sourcebook; organellar and metabolic processes, vol 2. Academic, New York, pp 469–502, Number of 469–502 pp

Catalanotti C, Dubini A, Subramanian V, Yang W, Magneschi L et al (2012) Altered fermentative metabolism in *Chlamydomonas reinhardtii* mutants lacking pyruvate formate lyase and both pyruvate formate lyase and alcohol dehydrogenase. Plant Cell 24:692–707

Chen X, Alonso AP, Allen DK, Reed JL, Shachar-Hill Y (2011) Synergy between (^{13})C-metabolic flux analysis and flux balance analysis for understanding metabolic adaptation to anaerobiosis in *E. coli*. Metab Eng 13:38–48

Chochois V, Dauvillee D, Beyly A, Tolleter D, Cuine S et al (2009) Hydrogen production in Chlamydomonas: photosystem II-dependent and -independent pathways differ in their requirement for starch metabolism. Plant Physiol 151:631–640

Choquet Y, Wollman FA (2009) The CES process. In: Harris EH, Stern D, Witman GB (eds) The Chlamydomonas sourcebook, vol 2. Elsevier, Amsterdam, pp 1027–1063, Number of 1027–1063 pp

Cohen J, Kim K, Posewitz M, Ghirardi ML, Schulten K et al (2005) Molecular dynamics and experimental investigation of H_2 and O_2 diffusion in [Fe]-hydrogenase. Biochem Soc Trans 33:80–82

Davies DD, Grego S, Kenworthy P (1974) The control of the production of lactate and ethanol by higher plants. Planta 118:297–310

Dent RM, Haglund CM, Chin BL, Kobayashi MC, Niyogi KK (2005) Functional genomics of eukaryotic photosynthesis using insertional mutagenesis of *Chlamydomonas reinhardtii*. Plant Physiol 137:545–556

Diner B, Mauzerall D (1973) Feedback controlling oxygen production in a cross-reaction between two photosystems in photosynthesis. Biochim Biophys Acta 305:329–352

Doebbe A, Keck M, La Russa M, Mussgnug JH, Hankamer B et al (2010) The interplay of proton, electron, and metabolite supply for photosynthetic H_2 production in *Chlamydomonas reinhardtii*. J Biol Chem 285:30247–30260

Dubini A, Mus F, Seibert M, Grossman AR, Posewitz MC (2009) Flexibility in anaerobic metabolism as revealed in a mutant of *Chlamydomonas reinhardtii* lacking hydrogenase activity. J Biol Chem 284:7201–7213

Eberhard S, Finazzi G, Wollman FA (2008) The dynamics of photosynthesis. Annu Rev Genet 42:463–515

Escoubas JM, Lomas M, LaRoche J, Falkowski PG (1995) Light intensity regulation of cab gene transcription is signaled by the redox state of the plastoquinone pool. Proc Natl Acad Sci U S A 92:10237–10241

Esper B, Badura A, Rogner M (2006) Photosynthesis as a power supply for (bio-)hydrogen production. Trends Plant Sci 11:543–549

Fernández E, Llamas A, Galván Á (2009) Nitrogen assimilation and its regulation. In: Harris EH, Stern D, Witman GB (eds) The Chlamydomonas sourcebook, vol 2. Elsevier, Amsterdam, pp 69–114, Number of 69–114 pp

Finazzi G, Drapier D, Rappaport F (2009) The CFoF1 ATP-synthase complex of photosynthesis. In: Harris EH, Stern D, Witman GB (eds) The Chlamydomonas sourcebook, vol 2. Elsevier, Amsterdam, pp 639–670, Number of 639–670 pp

Forbush B, Kok B (1968) Reaction between primary and secondary electron acceptors of photosystem II of photosynthesis. Biochim Biophys Acta 162:243–253

Forestier M, King P, Zhang L, Posewitz M, Schwarzer S et al (2003) Expression of two [Fe]-hydrogenases in *Chlamydomonas reinhardtii* under anaerobic conditions. Eur J Biochem 270:2750–2758

Fukao T, Xu K, Ronald PC, Bailey-Serres J (2006) A variable cluster of ethylene response factor-like genes regulates metabolic and developmental acclimation responses to submergence in rice. Plant Cell 18:2021–2034

Fukao T, Harris T, Bailey-Serres J (2009) Evolutionary analysis of the *Sub1* gene cluster that confers submergence tolerance to domesticated rice. Ann Bot 103:143–150

Gaffron H (1944) Photosynthesis, photoreduction and dark reduction of carbon dioxide in certain algae. Biol Rev Camb Philos Soc 19:1–20

Gans P, Rebeille F (1990) Control in the dark of the plastoquinone redox state by mitochondrial activity in *Chlamydomonas reinhardtii*. Biochim Biophys Acta 1015:150–155

Geigenberger P (2003) Response of plant metabolism to too little oxygen. Curr Opin Plant Biol 6:247–256

Gfeller RP, Gibbs M (1984) Fermentative metabolism of *Chlamydomonas reinhardtii*: I. Analysis of fermentative products from starch in dark and light. Plant Physiol 75:212–218

Gfeller RP, Gibbs M (1985) Fermentative metabolism of *Chlamydomonas reinhardtii*: II. Role of plastoquinone. Plant Physiol 77:509–511

Ghirardi ML (1997) Oxygen sensitivity of algal H_2-production. Appl Biochem Biotechnol 63: 141–151

Ghirardi ML, King PW, Posewitz MC, Maness PC, Fedorov A et al (2005) Approaches to developing biological H_2-photoproducing organisms and processes. Biochem Soc Trans 33:70–72

Ghirardi ML, Posewitz MC, Maness PC, Dubini A, Yu J, Seibert M (2007) Hydrogenases and hydrogen photoproduction in oxygenic photosynthetic organisms. Annu Rev Plant Biol 58: 71–91

Ghirardi ML, Dubini A, Yu J, Maness PC (2009) Photobiological hydrogen-producing systems. Chem Soc Rev 38:52–61

Gibbs M, Gfeller RP, Chen C (1986) Fermentative metabolism of *Chlamydomonas reinhardii*: III. Photoassimilation of acetate. Plant Physiol 82:160–166

Gibbs J, Morrell S, Valdez A, Setter TL, Greenway H (2000) Regulation of alcoholic fermentation in coleoptiles of two rice cultivars differing in tolerance to anoxia. J Exp Bot 51:785–796

Gibbs DJ, Lee SC, Isa NM, Gramuglia S, Fukao T et al (2011) Homeostatic response to hypoxia is regulated by the N-end rule pathway in plants. Nature 479:415–418

Godman JE, Molnar A, Baulcombe DC, Balk J (2010) RNA silencing of hydrogenase(-like) genes and investigation of their physiological roles in the green alga *Chlamydomonas reinhardtii*. Biochem J 431:345–351

Goldschmidt-Clermont M (2009) Chloroplast RNA splicing. In: Harris EH, Stern D, Witman GB (eds) The Chlamydomonas sourcebook, vol 2. Elsevier, Amsterdam, pp 915–936, Number of 915–936 pp

González-Ballester D, Grossman AR (2009) Sulfur limitation from the physiological to the genomic. In: Stern D, Harris EH, Witman GB (eds) The Chlamydomonas sourcebook, vol 2. It's Academic Press. p 159–188. Number of 159–188 pp

González-Ballester D, Pollock SV, Pootakham W, Grossman AR (2008) The central role of a SNRK2 kinase in sulfur deprivation responses. Plant Physiol 147:216–227

Gonzalez-Ballester D, Pootakham W, Mus F, Yang W, Catalanotti C et al (2011) Reverse genetics in Chlamydomonas: a platform for isolating insertional mutants. Plant Methods 7:24

Good AG, Crosby WL (1989) Induction of alcohol dehydrogenase and lactate dehydrogenase in hypoxically induced barley. Plant Physiol 90:860–866

Greenbaum E (1982) Photosynthetic hydrogen and oxygen production: kinetic studies. Science 215:291–293

Greer SN, Metcalf JL, Wang Y, Ohh M (2012) The updated biology of hypoxia-inducible factor. EMBO J 31:2448–2460

Grossman AR, Croft M, Gladyshev VN, Merchant SS, Posewitz MC et al (2007) Novel metabolism in Chlamydomonas through the lens of genomics. Curr Opin Plant Biol 10:190–198

Grossman AR, Catalanotti C, Yang W, Dubini A, Magneschi L et al (2011) Multiple facets of anoxic metabolism and hydrogen production in the unicellular green alga *Chlamydomonas reinhardtii*. New Phytol 190:279–288

Guzy RD, Schumacker PT (2006) Oxygen sensing by mitochondria at complex III: the paradox of increased reactive oxygen species during hypoxia. Exp Physiol 91:807–819

Happe T, Naber JD (1993) Isolation, characterization and N-terminal amino acid sequence of hydrogenase from the green alga *Chlamydomonas reinhardtii*. Eur J Biochem 214:475–481

Happe T, Mosler B, Naber JD (1994) Induction, localization and metal content of hydrogenase in the green alga *Chlamydomonas reinhardtii*. Eur J Biochem 222:769–774

Hemschemeier A, Happe T (2005) The exceptional photofermentative hydrogen metabolism of the green alga *Chlamydomonas reinhardtii*. Biochem Soc Trans 33:39–41

Hemschemeier A, Melis A, Happe T (2009) Analytical approaches to photobiological hydrogen production in unicellular green algae. Photosynth Res 102:523–540

Herrin DL (2009) Chloroplast RNA processing and stability. In: Harris EH, Stern D, Witman GB (eds) The Chlamydomonas sourcebook, vol 2. Elsevier, Amsterdam, pp 937–966, Number of 937–966 pp

Houille-Vernes L, Rappaport F, Wollman FA, Alric J, Johnson X (2011) Plastid terminal oxidase 2 (PTOX2) is the major oxidase involved in chlororespiration in Chlamydomonas. Proc Natl Acad Sci U S A 108:20820–20825

Huang X, Guo Z, Zhu W, Xie Y, Tian H (2008) A colorimetric and fluorescent turn-on sensor for pyrophosphate anion based on a dicyanomethylene-4H-chromene framework. Chem Commun (Camb) 41:5143–5145

Igamberdiev AU, Seregélyes C, Manac'h N, Hill RD (2004) NADH-dependent metabolism of nitric oxide in alfalfa root cultures expressing hemoglobin. Planta 219:95–102

Jans F, Mignolet E, Houyoux PA, Cardol P, Ghysels B et al (2008) A type II NAD(P)H dehydrogenase mediates light-independent plastoquinone reduction in the chloroplast of Chlamydomonas. Proc Natl Acad Sci U S A 105:20546–20551

Kato-Noguchi H (2006) Pyruvate metabolism in rice coleoptiles under anaerobiosis. Plant Growth Regul 50:41–46

Kato-Noguchi H, Yasuda Y, Sasaki R (2010) Soluble sugar availability of aerobically germinated barley, oat and rice coleoptiles in anoxia. J Plant Physiol 167:1571–1576

Kennedy RA, Fox TC, Siedow JN (1987) Activities of isolated mitochondria and mitochondrial enzymes from aerobically and anaerobically germinated barnyard grass (Echinochloa) seedlings. Plant Physiol 85:474–480

Kennedy RA, Rumpho ME, Fox TC (1992) Anaerobic metabolism in plants. Plant Physiol 100: 1–6

Kessler E (1973) Effect of anaerobiosis on photosynthetic reactions and nitrogen metabolism of algae with and without hydrogenase. Arch Mikrobiol 93:91–100

King SM, Kamiya R (2009) Axonemal dyneins: assembly, structure and force generation. In: Harris EH, Stern D, Witman GB (eds) The Chlamydomonas sourcebook, vol 3. Elsevier, Amsterdam, pp 131–208, Number of 131–208 pp

King PW, Posewitz MC, Ghirardi ML, Seibert M (2006) Functional studies of [FeFe] hydrogenase maturation in an *Escherichia coli* biosynthetic system. J Bacteriol 188:2163–2172

Kirch H-H, Bartels D, Wei Y, Schnable PS, Wood AJ (2004) The *ALDH* gene superfamily of Arabidopsis. Trends Plant Sci 9:371–377

Klein U (2009) Chloroplast transcription. In: Harris EH, Stern D, Witman GB (eds) The Chlamydomonas sourcebook, vol 2. Elsevier, Amsterdam, pp 893–914, Number of 893–914 pp

Klock G, Kreuzberg K (1991) Compartmented metabolite pools in protoplasts from the green alga *Chlamydomonas reinhardtii*: changes after transition from aerobiosis to anaerobiosis in the dark. Biochim Biophys Acta 1073:410–415

Komatsu S, Deschamps T, Hiraga S, Kato M, Chiba M et al (2011) Characterization of a novel flooding stress-responsive alcohol dehydrogenase expressed in soybean roots. Plant Mol Biol 77:309–322

Kosourov SN, Seibert M (2009) Hydrogen photoproduction by nutrient-deprived *Chlamydomonas reinhardtii* cells immobilized within thin alginate films under aerobic and anaerobic conditions. Biotechnol Bioeng 102:50–58

Kosourov S, Seibert M, Ghirardi ML (2003) Effects of extracellular pH on the metabolic pathways in sulfur-deprived, H_2-producing *Chlamydomonas reinhardtii* cultures. Plant Cell Physiol 44:146–155

Kreuzberg K (1984) Starch fermentation via formate producing pathway in *Chlamydomonas reinhardtii*, *Chlorogonium elongatum* and *Chlorella fusca*. Physiol Plant 61:87–94

Kreuzberg K, Klöck G, Grobheiser D (1987) Subcellular distribution of pyruvate-degrading enzymes in *Chlamydomonas reinhardtii* studied by an improved protoplast fractionation procedure. Physiol Plant 69:481–488

Kruse O, Hankamer B (2010) Microalgal hydrogen production. Curr Opin Biotechnol 21:238–243

Kruse O, Rupprecht J, Bader KP, Thomas-Hall S, Schenk PM et al (2005) Improved photobiological H_2 production in engineered green algal cells. J Biol Chem 280:34170–34177

Lasanthi-Kudahettige R, Magneschi L, Loreti E, Gonzali S, Licausi F et al (2007) Transcript profiling of the anoxic rice coleoptile. Plant Physiol 144:218–231

Lendzian K, Bassham JA (1975) Regulation of glucose-6-phosphate dehydrogenase in spinach chloroplasts by ribulose 1,5-diphosphate and NADPH/NADP+ ratios. Biochim Biophys Acta 396:260–275

Libourel IG, Shachar-Hill Y (2008) Metabolic flux analysis in plants: from intelligent design to rational engineering. Annu Rev Plant Biol 59:625–650

Licausi F (2011) Regulation of the molecular response to oxygen limitations in plants. New Phytol 190:550–555

Licausi F, van Dongen JT, Giuntoli B, Novi G, Santaniello A et al (2010) HRE1 and HRE2, two hypoxia-inducible ethylene response factors, affect anaerobic responses in Arabidopsis thaliana. Plant J 62:302–315

Licausi F, Giorgi FM, Schmalzlin E, Usadel B, Perata P et al (2011) *HRE*-type genes are regulated by growth-related changes in internal oxygen concentrations during the normal development of potato (*Solanum tuberosum*) tubers. Plant Cell Physiol 52:1957–1972

Long JC, Sommer F, Allen MD, Lu SF, Merchant SS (2008) *FER1* and *FER2* encoding two ferritin complexes in *Chlamydomonas reinhardtii* chloroplasts are regulated by iron. Genetics 179: 137–147

Magneschi L, Perata P (2009) Rice germination and seedling growth in the absence of oxygen. Ann Bot 103:181–196

Magneschi L, Catalanotti C, Subramanian V, Dubini A, Yang W et al (2012) A mutant in the ADH1 gene of *Chlamydomonas reinhardtii* elicits metabolic restructuring during anaerobiosis. Plant Physiol 158:1293–1305

Maione TE, Gibbs M (1986) Association of the chloroplastic respiratory and photosynthetic electron transport chains of *Chlamydomonas reinhardii* with photoreduction and the oxy-hydrogen reaction. Plant Physiol 80:364–368

Maul JE, Lilly JW, Cui L, dePamphilis CW, Miller W et al (2002) The *Chlamydomonas reinhardtii* plastid chromosome: islands of genes in a sea of repeats. Plant Cell 14:2659–2679

McLamore ES, Jaroch D, Chatni MR, Porterfield DM (2010) Self-referencing optrodes for measuring spatially resolved, real-time metabolic oxygen flux in plant systems. Planta 232: 1087–1099

Melis A, Happe T (2004) Trails of green alga hydrogen research—from hans gaffron to new frontiers. Photosynth Res 80:401–409

Melis A, Zhang L, Forestier M, Ghirardi ML, Seibert M (2000) Sustained photobiological hydrogen gas production upon reversible inactivation of oxygen evolution in the green alga *Chlamydomonas reinhardtii*. Plant Physiol 122:127–136

Merchant SS, Prochnik SE, Vallon O, Harris EH, Karpowicz SJ et al (2007) The Chlamydomonas genome reveals the evolution of key animal and plant functions. Science 318:245–250

Meuser JE, D'Adamo S, Jinkerson RE, Mus F, Yang W et al (2012) Genetic disruption of both *Chlamydomonas reinhardtii* [FeFe]-hydrogenases: insight into the role of HYDA2 in H_2 production. Biochem Biophys Res Commun 417:704–709

Minagawa J (2009) Light-harvesting proteins. In: Harris EH, Stern D, Witman GB (eds) The Chlamydomonas sourcebook, vol 2. Elsevier, Amsterdam, pp 503–540, Number of 503–540 pp

Miyashita Y, Good AG (2008) Contribution of the GABA shunt to hypoxia-induced alanine accumulation in roots of Arabidopsis thaliana. Plant Cell Physiol 49:92–102

Moncada S, Erusalimsky JD (2002) Does nitric oxide modulate mitochondrial energy generation and apoptosis? Nature reviews. Mol Cell Biol 3:214–220

Moseley J, Grossman AR. 2009. Phosphorus limitation from the physiological to the genomic. In: Harris EH, Stern D, Witman GB (eds) The Chlamydomonas sourcebook, vol 2. It's Academic Press, New York. p 189–215. Number of 189–215 pp

Moseley J, Gonzalez-Ballester D, Pootakham W, Bailey S, Grossman AR (2009) Genetic interactions between regulators of Chlamydomonas phosphorus and sulfur deprivation responses. Genetics 181:889–905

Mugnai S, Marras AM, Mancuso S (2011) Effect of hypoxic acclimation on anoxia tolerance in Vitis roots: response of metabolic activity and K+ fluxes. Plant Cell Physiol 52:1107–1116

Mulder DW, Boyd ES, Sarma R, Lange RK, Endrizzi JA et al (2010) Stepwise [FeFe]-hydrogenase H-cluster assembly revealed in the structure of HydA($^{\Delta EFG}$). Nature 465:248–251

Müller M (2003) Energy metabolism. Part 1: anaerobic protozoa. In: Marr JJ, Nilsen TW, Komuniecki RW (eds) Molecular medical parasitology. Academic, Amsterdam, The Netherlands, pp 125–139, Number of 125–139 pp

Mus F, Cournac L, Cardettini V, Caruana A, Peltier G (2005) Inhibitor studies on non-photochemical plastoquinone reduction and H_2 photoproduction in *Chlamydomonas reinhardtii*. Biochim Biophys Acta 1708:322–332

Mus F, Dubini A, Seibert M, Posewitz MC, Grossman AR (2007) Anaerobic acclimation in *Chlamydomonas reinhardtii*: anoxic gene expression, hydrogenase induction, and metabolic pathways. J Biol Chem 282:25475–25486

Mustroph A, Albrecht G, Hajirezaei M, Grimm B, Biemelt S (2005) Low levels of pyrophosphate in transgenic potato plants expressing *E. coli* pyrophosphatase lead to decreased vitality under oxygen deficiency. Ann Bot 96:717–726

Narsai R, Rocha M, Geigenberger P, Whelan J, van Dongen JT (2011) Comparative analysis between plant species of transcriptional and metabolic responses to hypoxia. New Phytol 190:472–487

Niyogi K (2009) Photoprotection and high light responses. In: Harris EH, Stern D, Witman GB (eds) The Chlamydomonas sourcebook, vol 2. Elsevier, Amsterdam, pp 847–870, Number of 847–870 pp

Noctor G, De Paepe R, Foyer CH (2007) Mitochondrial redox biology and homeostasis in plants. Trends Plant Sci 12:125–134

Ohta S, Miyamoto K, Miura Y (1987) Hydrogen evolution as a consumption mode of reducing equivalents in green algal fermentation. Plant Physiol 83:1022–1026

Orth JD, Thiele I, Palsson BO (2010) What is flux balance analysis? Nat Biotechnol 28:245–248

Pang JY, Newman I, Mendham N, Zhou M, Shabala S (2006) Microelectrode ion and O_2 fluxes measurements reveal differential sensitivity of barley root tissues to hypoxia. Plant Cell Environ 29:1107–1121

Pape M, Lambertz C, Happe T, Hemschemeier A (2012) Differential expression of the Chlamydomonas [FeFe]-hydrogenase encoding *HYDA1* gene is regulated by the Copper Response Regulator 1. Plant Physiol 159:1700–1712

Perata P, Ponzueta-Romero J, Akazawa T, Yamaguchi J (1992) Effect of anoxia on starch breakdown in rice and wheat seeds. Planta 188:611–618

Perata P, Armstrong W, Voesenek LA (2011) Plants and flooding stress. New Phytol 190:269–273

Pfannschmidt T, Liere K (2005) Redox regulation and modification of proteins controlling chloroplast gene expression. Antioxid Redox Signal 7:607–618

Philipps G, Krawietz D, Hemschemeier A, Happe T (2011) A pyruvate formate lyase-deficient *Chlamydomonas reinhardtii* strain provides evidence for a link between fermentation and hydrogen production in green algae. Plant J 66:330–340

Porterfield DM, Kuang A, Smith PJ, Crispi ML, Musgrave ME (1999) Oxygen-depleted zones inside reproductive structures of Brassicaceae: implications for oxygen control of seed development. J Can Bot 77:1439–1446

Posewitz MC, King PW, Smolinski SL, Zhang L, Seibert M, Ghirardi ML (2004a) Discovery of two novel radical S-adenosylmethionine proteins required for the assembly of an active [Fe] hydrogenase. J Biol Chem 279:25711–25720

Posewitz MC, Smolinski SL, Kanakagiri S, Melis A, Seibert M, Ghirardi ML (2004b) Hydrogen photoproduction is attenuated by disruption of an isoamylase gene in *Chlamydomonas reinhardtii*. Plant Cell 16:2151–2163

Potzkei J, Kunze M, Drepper T, Gensch T, Jaeger KE, Buchs J (2012) Real-time determination of intracellular oxygen in bacteria using a genetically encoded FRET-based biosensor. BMC Biol 10:28

Pucciariello P, Perata P (2012) How plants sense low oxygen. Plant Signal Behav 1:7

Rabinowitch E (1951) Photosynthesis. Interscience Publishers, New York

Radakovits R, Jinkerson RE, Darzins A, Posewitz MC (2010) Genetic engineering of algae for enhanced biofuel production. Eukaryot Cell 9:486–501

Raman K, Chandra N (2009) Flux balance analysis of biological systems: applications and challenges. Brief Bioinform 10:435–449

Rebeille F, Gans P (1988) Interaction between chloroplasts and mitochondria in microalgae: role of glycolysis. Plant Physiol 88:973–975

Rochaix J-D (2009) State transitions. In: Harris EH, Stern D, Witman GB (eds) The Chlamydomonas sourcebook, vol 2. Elsevier, Amsterdam, pp 819–846, Number of 819–846 pp

Rosenberg JN, Oyler GA, Wilkinson L, Betenbaugh MJ (2008) A green light for engineered algae: redirecting metabolism to fuel a biotechnology revolution. Curr Opin Biotechnol 19:430–436

Rupprecht J (2009) From systems biology to fuel—*Chlamydomonas reinhardtii* as a model for a systems biology approach to improve biohydrogen production. J Biotechnol 142:10–20
Rupprecht J, Hankamer B, Mussgnug JH, Ananyev G, Dismukes C, Kruse O (2006) Perspectives and advances of biological H_2 production in microorganisms. Appl Microbiol Biotechnol 72: 442–449
Rutter J, Reick M, Wu LC, McKnight SL (2001) Regulation of clock and NPAS2 DNA binding by the redox state of NAD cofactors. Science 293:510–514
Rymarquis LA, Handley JM, Thomas M, Stern DB (2005) Beyond complementation. Map-based cloning in *Chlamydomonas reinhardtii*. Plant Physiol 137:557–566
Sasidharan R, Mustroph A (2011) Plant oxygen sensing is mediated by the N-end rule pathway: a milestone in plant anaerobiosis. Plant Cell 23:4173–4183
Seibert M, King PW, Posewitz MC, Melis A, Ghirardi ML (2008) Photosynthetic water-splitting for hydrogen production. In: Wall J, Harwood CS, Demain A (eds) Bioenergy. ASM Press, Washington, DC, pp 273–291, Number of 273–291 pp
Semenza GL (2011) Oxygen sensing, homeostasis and disease. N Engl J Med 365:537–547
Shiau YG, Franck J (1947) Chlorophyll fluorescence and photosynthesis in algae, leaves and chloroplasts. Arch Biochem 14:253–295
Shingaki-Wells RN, Huang S, Taylor NL, Carroll AJ, Zhou W, Millar AH (2011) Differential molecular responses of rice and wheat coleoptiles to anoxia reveal novel metabolic adaptations in amino acid metabolism for tissue tolerance. Plant Physiol 156:1706–1724
Stirnberg M, Happe T (2004) Identification of a cis-acting element controlling anaerobic expression of the HydA gene from *Chlamydomonas reinhardtii*. In: Miyake J, Igarashi Y, Roegner M (eds) Biohydrogen III. Elsevier Science, Oxford, pp 117–127, Number of 117–127 pp
Stoimenova M, Igamberdiev AU, Gupta KJ, Hill RD (2007) Nitrite-driven anaerobic ATP synthesis in barley and rice root mitochondria. Planta 226:465–474
Sweetlove LJ, Ratcliffe RG (2011) Flux-balance modeling of plant metabolism. Front Plant Sci 2:38
Tejada-Jimenez M, Llamas A, Sanz-Luque E, Galván A, Fernández E (2007) A high-affinity molybdate transporter in eukaryotes. Proc Natl Acad Sci U S A 104:20126–20130
Terashima M, Specht M, Naumann N, Hippler M (2010) Characterizing the anaerobic response of *Chlamydomonas reinhardtii* by quantitative proteomics. Mol Cell Proteomics 9:1514–1532
Thiele I, Palsson BO (2010) A protocol for generating a high-quality genome-scale metabolic reconstruction. Nat Protoc 5:93–121
Tsuji H, Meguro N, Suzuki Y, Tsutsumi N, Hirai A, Nakazono M (2003) Induction of mitochondrial aldehyde dehydrogenase by submergence facilitates oxidation of acetaldehyde during re-aeration in rice. FEBS Lett 546:369–373
Vlad F, Spano T, Vlad D, Daher FB, Ouelhadj A, Kalaitzis P (2007) Arabidopsis prolyl 4-hydroxylases are differentially expressed in response to hypoxia, anoxia and mechanical wounding. Physiol Plant 130:471–483
Voesenek LA, Colmer TD, Pierik R, Millenaar FF, Peeters AJ (2006) How plants cope with complete submergence. New Phytol 170:213–226
Wagner V, Ullmann K, Mollwo A, Kaminski M, Mittag M, Kreimer G (2008) The phosphoproteome of a *Chlamydomonas reinhardtii* eyespot fraction includes key proteins of the light signaling pathway. Plant Physiol 146:772–788
Wang ZT, Ullrich N, Joo S, Waffenschmidt S, Goodenough U (2009) Algal lipid bodies: stress induction, purification, and biochemical characterization in wild-type and starchless *Chlamydomonas reinhardtii*. Eukaryot Cell 8:1856–1868
Whatley FR, Allen MB, Arnon DI (1955) Photosynthetic phosphorylation as an anaerobic process. Biochim Biophys Acta 16:605–606
Whitney LA, Novi G, Perata P, Loreti E (2012) Distinct mechanism regulating gene expression coexist within the fermentative pathways in *Chlamydomonas reinhardtii*. Sci World J. doi:10.1100/2012/565047

Wiessner W (1965) Quantum requirement for acetate assimilation and its significance for quantum measurements in photophosphorylation. Nature 205:56–57

Winkler M, Hemschemeier A, Jacobs J, Stripp S, Happe T (2010) Multiple ferredoxin isoforms in *Chlamydomonas reinhardtii*—their role under stress conditions and biotechnological implications. Eur J Cell Biol 89:998–1004

Yang P, Smith EF (2009) The flagellar radial spokes. In: Harris EH, Stern D, Witman GB (eds) The Chlamydomonas sourcebook, vol 3. Elsevier, Amsterdam, pp 209–234, Number of 209–234 pp

Zabawinski C, Van Den Koornhuyse N, D'Hulst C, Schlichting R, Giersch C et al (2001) Starchless mutants of *Chlamydomonas reinhardtii* lack the small subunit of a heterotetrameric ADP-glucose pyrophosphorylase. J Bacteriol 183:1069–1077

Zerges W, Hauser C (2009) Protein synthesis in the chloroplast. In: Harris EH, Stern D, Witman GB (eds) The Chlamydomonas sourcebook, vol 2. Elsevier, Amsterdam, pp 967–1026, Number of 967–1026 pp

Hypoxic Energy Metabolism and PPi as an Alternative Energy Currency

Angelika Mustroph, Natalia Hess, and Rashmi Sasidharan

Abstract During periods of oxygen-deficiency stress, mitochondrial respiration in plants is strongly inhibited and ATP production decreases. One potential strategy to cope with this energy crisis is the induction of enzymes that use pyrophosphate (PPi) as the energy source as an alternative to ATP-using enzymes. Four enzymatic pathways in plant primary metabolism are discussed here in the context of available gene expression and proteomic studies as well as mutant analyses. One PPi-dependent pathway, sucrose cleavage by sucrose synthase and UDP-glucose pyrophosphorylase, is clearly important for plant hypoxic metabolism, while the impact of the other three enzymes, pyrophosphate: fructose-6-phosphate phosphotransferase, pyruvate-orthophosphate dikinase, and vacuolar proton-transporting pyrophosphorylase in plants remains unclear and needs to be studied further. The latter three enzymes could be potentially involved in the flooding tolerance of species such as *Rorippa sylvestris* or *Oryza sativa* as indicated by gene expression and activity data.

Abbreviations

H^+-PPase	vacuolar PPi-dependent proton pump
PEP	Phosphoenolpyruvate
PEPC	Phosphoenolpyruvate carboxylase
PEPCK	Phosphoenolpyruvate carboxykinase
PF2K	Phosphofructo-2-kinase

A. Mustroph (✉) • N. Hess
Plant Physiology, Bayreuth University, Universitaetsstrasse 30, 95440 Bayreuth, Germany
e-mail: angelika.mustroph@uni-bayreuth.de

R. Sasidharan
Plant Ecophysiology, Institute of Environmental Biology, Utrecht University, Padualaan 8, 3584 CH Utrecht, The Netherlands

J.T. van Dongen and F. Licausi (eds.), *Low-Oxygen Stress in Plants*, Plant Cell Monographs 21, DOI 10.1007/978-3-7091-1254-0_9, © Springer-Verlag Wien 2014

PFK	Phosphofructokinase
PFP	Pyrophosphate:fructose-6-phosphate phosphotransferase
PK	Pyruvate kinase
PPDK	Pyruvate-orthophosphate dikinase
PPi	Inorganic pyrophosphate
SuSy	Sucrose synthase
UGPase	UDP-glucose pyrophosphorylase

1 Introduction

Plants, despite being the main source of oxygen through photosynthesis, are strongly dependent on oxygen. On the one hand, oxygen is required for a number of biosynthetic processes, for example through cytochrome-P450 monooxygenases. On the other hand, oxygen is absolutely necessary for ATP production via mitochondrial respiration, especially in tissues where photosynthesis is not active, for example in roots or during the night. The normal oxygen concentration in air is about 20.9 % and therefore commonly not limiting. However, due to the drastically reduced diffusion rate of oxygen in water and water-containing cells (Armstrong 1979; Armstrong and Armstrong 2014, Chap. 14), it can become limiting within bulky plant tissues such as meristems, seeds, or stems, because they do not possess an oxygen distribution system as animals do. Water surrounding the plant externally, for example under waterlogged or submerged conditions, can also result in oxygen deficiency due to restricted diffusion of oxygen.

While a number of plants can avoid oxygen deficiency through formation of aerenchyma and the elongation of above-ground tissues to gain access to air and enable transport of oxygen into submerged plant parts, they can also withstand this stress for a limited period of time by metabolic acclimatization. The survival time strongly depends on the severity of the stress as well as on the metabolic status and species- or variety-specific differences (Bailey-Serres and Voesenek 2008; Vashisht et al. 2011; Bailey-Serres et al. 2012). *Arabidopsis thaliana*, for example, can survive a period of about 3 weeks under complete submergence under a normal day-night rhythm (Vashisht et al. 2011), while the related *Rorippa sylvestris* can withstand similar conditions for about 100 days (Akman et al. 2012). Among cultured cereals, rice is the most flooding-tolerant species and is therefore an important tool to study adaptative responses to submergence and oxygen-deficiency stress.

2 Hypoxic Plants Have to Cope with Energy Deficiency

Upon oxygen deficiency, one important reaction is the induction of fermentative enzymes in order to allow the glycolytic flux to be maintained, including its production of 2 mole ATP per mole glucose. However, when compared to aerobic respiration, where about 36 mole ATP are produced, the energy outcome is very poor. For this reason, plants have to cope with a severe decrease of ATP very soon after the onset of a decrease in oxygen concentrations (e.g., Kato-Noguchi 2002b; Branco-Price et al. 2008; Mustroph et al. 2009). Since fermentable carbohydrates are limited in most plant tissues, the low ATP yield cannot be compensated by a strongly increased glycolytic rate for a long time. On the contrary, many plants even try to downregulate carbohydrate consumption by a mechanism which is still not completely understood (Geigenberger et al. 2000; Zabalza et al. 2009). A number of highly submergence-tolerant rice varieties switch to a so-called "quiescence" state in order to save carbohydrates and energy, until the submergence period is over. This response is mediated by the transcription factor Sub1A, an ethylene response factor-like gene (Fukao et al. 2006; Xu et al. 2006).

The decrease in ATP causes a number of problems. Growth is strongly repressed during a period of low-oxygen stress, apart from some elongating rice varieties and several wild species, where elongation is actually stimulated under these conditions, for a restricted time (Bailey-Serres and Voesenek 2008). Most of the biosynthetic processes are severely inhibited by oxygen deficiency. This is for example obvious when downregulated genes are compared across various relevant microarray experiments (e.g., van Dongen et al. 2009; Mustroph et al. 2009, 2010; Narsai et al. 2011). These downregulated genes are associated with energetically expensive processes that include cellulose and cell wall biosynthesis, lipid biosynthesis, and nucleotide and protein biosynthesis. The process that consumes the most cellular energy is the translation of proteins (Edwards et al. 2012). The translation of many mRNAs is strongly downregulated in Arabidopsis (Branco-Price et al. 2005, 2008), including housekeeping mRNAs and mRNAs that mediate cell-type specificity (Mustroph et al. 2009), while only a small portion of low-oxygen-stress-related genes is actively translated. The regulation of this selective translation is discussed elsewhere in this book (Chap. 6, Sorenson and Bailey-Serres 2014).

Another mechanism to compensate for the severe ATP deficiency is believed to be the induction of alternative pathways that can use inorganic pyrophosphate (PPi) instead of ATP for phosphorylation reactions. This hypothesis was suggested many years ago, after the discovery of PPi-using enzymes that work in parallel to ATP-using ones (Weiner et al. 1987; Mertens 1991; Stitt 1998). However, the impact of these alternative enzymes is still not absolutely clear for a number of reasons. First, PPi is rather a side product of biosynthetic reactions, formed upon the hydrolysis of energy-rich polyphosphates and needs to be cleaved in order to drive energetically costly reactions. Other sources of PPi in plant metabolism are not yet known. The down-regulation of biosynthetic processes should also downregulate

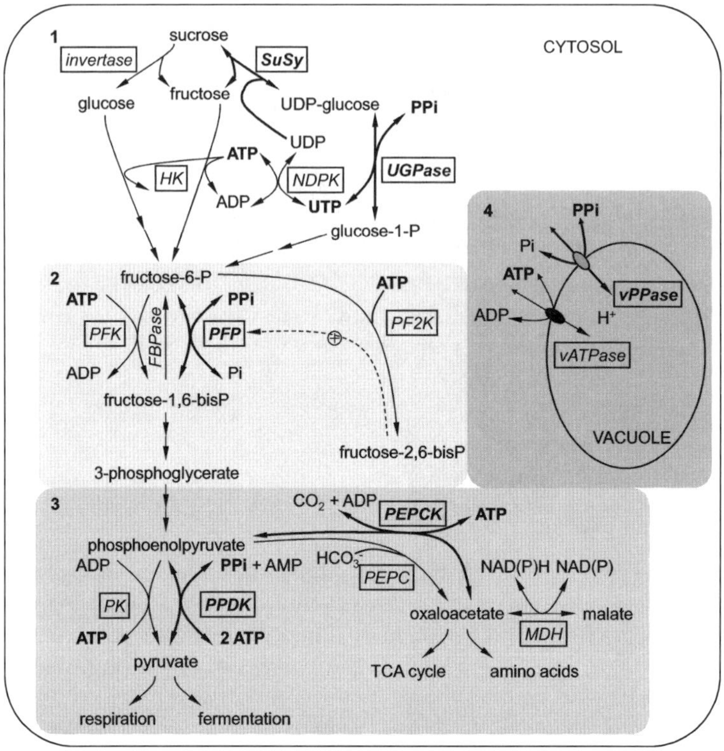

Fig. 1 An overview of plant primary metabolism. The four PPi-using alternative pathways mentioned in the text are presented in *bold* and highlighted within numbered, *shaded boxes*. *FBPase* fructose-1,6-bisphosphatase, *HK* hexokinase, *MDH* malate dehydrogenase, *NDPK* nucleoside diphosphate kinase, *PEPC* phosphoenolpyruvate carboxylase, *PEPCK* phosphoenolpyruvate carboxykinase, *PF2K* phosphofructo-2-kinase, *PFK* phosphofructokinase, *PFP* pyrophosphate: fructose-6-phosphate phosphotransferase, *PK* pyruvate kinase, *PPDK* pyruvate-orthophosphate dikinase, *SuSy* sucrose synthase, *UGPase* UDP-glucose pyrophosphorylase, *vATPase* vacuolar ATP-dependent proton pump, *vPPase* vacuolar PPi-dependent proton pump

PPi production. Accordingly, PPi should be a limited resource during oxygen deficiency, although the content of PPi under these conditions stays mostly stable, unlike ATP, suggesting PPi production from alternative sources (Dancer and ap Rees 1989; Mohanty et al. 1993; Geigenberger et al. 2000; Gibon et al. 2002; Kato-Noguchi 2002a; Mustroph et al. 2005). Second, most of the alternative enzymes perform reversible reactions making it difficult to estimate which direction is the favorable one under stress conditions (Fig. 1). Third, knock-down approaches published so far (see below) for some alternative enzymes have not revealed a severe impairment of growth of different plant species under normal growth conditions.

At least four reactions in the plant primary metabolism have PPi-using alternative enzymes and will be summarized here (Fig. 1). (1) Sucrose cleavage can be catalyzed by invertases, and subsequent action of hexokinases provides phosphorylated hexoses. This ATP-using process can be substituted by a reaction involving sucrose synthase (SuSy) and UDP-glucose pyrophosphorylase (UGPase), also providing hexose-phosphates from sucrose, but getting the energy from PPi. (2) Another versatile reaction is the phosphorylation of fructose-6-phosphate during glycolysis. This can be achieved by phosphofructokinases (PFKs), which are ATP-dependent, but alternatively by pyrophosphate:fructose-6-phosphate phosphotransferase (PFP). (3) The enzyme pyruvate-orthophosphate dikinase (PPDK) is involved in pyruvate metabolism and catalyzes the reversible reaction from phosphoenolpyruvate (PEP) to pyruvate, a reaction that potentially can substitute for pyruvate kinase (Chastain et al. 2011). (4) Vacuolar proton transporters can either use ATP or PPi for their energy supply.

Recent global profiling technologies such as microarray studies or proteome analyses of several flooding-tolerant and sensitive species have now enabled a more complete picture of induction of the respective genes encoding for alternative enzymes and might help in elucidating the role of PPi under oxygen-deficiency stress. This review therefore will summarize evidence from biochemical analyses, transcriptome studies, and transgenic plants for a better understanding of energy metabolism under oxygen-deficiency stress.

3 Sucrose Cleavage by Sucrose Synthase as an Energy-Saving Pathway

In order to degrade sucrose via the glycolytic process, it needs to be cleaved, and the resulting hexoses (glucose and fructose) need to be activated by phosphorylation. In comparison with the invertase-hexokinase pathway, which requires 2 mole of ATP per mole sucrose, the alternative, but reversible pathway via SuSy and UGPase only requires 1 mole of PPi, thus saving 1 mole of energy-rich bonds (Fig. 1, panel 1). This is because the energy of the bond between glucose and fructose is used for the production of an energy-rich sugar, UDP-glucose, in contrast to sucrose cleavage by invertases. While SuSy seems to catalyze the rate-limiting step, the UGPase reaction is the reaction that uses PPi, but is not rate-limiting (Kleczkowski et al. 2004).

SuSy was first discovered as one of the major anaerobic polypeptides in maize (Springer et al. 1986). Further studies revealed its gene induction under oxygen deficiency in rice (Ricard et al. 1991; Lasanthi-Kudahettige et al. 2007; Mustroph et al. 2010), poplar (Kreuzwieser et al. 2009), *Sorghum bicolor* (Chourey et al. 1991), *Potamogeton distinctus* (Harada et al. 2005), soybean (Nanjo et al. 2011), cotton (Christianson et al. 2010), and Arabidopsis (e.g., Loreti et al. 2005; Branco-Price et al. 2008; van Dongen et al. 2009; Mustroph et al. 2009; Lee et al. 2011) (Table 1). The increase in mRNA levels did not result in all species

Table 1 List of genes that are significantly induced by oxygen-deficiency stress in different plant species, based on published microarray data

	Arabidopsis thaliana[a]	*Rorippa spec.*[b]	*Populus trichocarpa*[c]	*Glycine max*[d]	*Gossypium hirsutum*[e]	*Oryza sativa*[f]	*Zea mays*[g]
Sucrose synthase	At5g20830 At3g43190 (6 genes)	At5g20830 At3g43190	Potri.002G202300 (7 genes)	Glyma09g08550 Glyma13g17420 Glyma15g20180 (9 genes)	Gorai.004G142700 Gorai.009G038000 Gorai.013G222400 (8 genes)	LOC_Os03g28330 LOC_Os03g22120 (7 genes)	GRMZM2G089713 (5 genes)
UGPase	Not induced (2 genes)	Not induced	Not induced (3 genes)	Glyma11g33160 (5 genes)	Gorai.007G188400 (3 genes)	Not induced (2 genes)	Not induced (6 genes)
Phosphofructokinase (ATP)	At4g26270 At4g32840 (7 genes)	At4g32840	Potri.006G235100 Potri.018G069200 (6 genes)	Glyma04g09180 Glyma06g09320 (16 genes)	Gorai.001G025000 Gorai.009G029100 Gorai.010G079100 Gorai.003G042800 (9 genes)	LOC_Os01g09570 LOC_Os05g10650 LOC_Os05g44922 (10 genes)	GRMZM5G879882 (10 genes)
PPi-Fru6P-Phosphotransferase (α- and β-subunit)	Not induced (2 + 2 genes)	At1g76550 At4g04040	Potri.002G003100 (2 + 3 genes)	Not induced (2 + 4 genes)	Gorai.005G115600 (1 + 3 genes)	LOC_Os08g25720 LOC_Os06g13810 (4 + 1 genes)	Not induced (4 + 1 genes)
Pyruvate kinase	Not induced (12 genes)	Not induced	Potri.010G233200 Potri.001G001600 (16 genes)	Glyma03g34740 Glyma05g09310 Glyma19g00870 Glyma19g37420 Glyma20g33060 (24 genes)	Gorai.003G154200 Gorai.005G214000 Gorai.006G158900 (17 genes)	LOC_Os01g16960 (10 genes)	Not induced (12 genes)
PPDK	At4g15530[h] (1 gene)	At4g15530	Not induced (1 gene)	Not induced (2 genes)	Not induced (1 gene)	LOC_Os03g31750 (2 genes)	GRMZM2G306345 (2 genes)
PEP carboxylase	Not induced (4 genes)	Not induced	Not induced (5 genes)	Not induced (13 genes)	Gorai.013G173500 (6 genes)	LOC_Os02g14770[i] (6 genes)	Not induced (5 genes)

PEP carboxykinase	At4g37870[h] (2 genes)	At4g37870	Not induced (2 genes)	Not induced (6 genes)	Not induced (3 genes)	LOC_Os10g13700 LOC_Os03g15050 (2 genes)	Not induced (2 genes)
Vacuolar H^+-PPase	Not induced (1 gene)	Not induced	Potri.018G119500 (4 genes)	Glyma07g00350 (5 genes)	Gorai.004G237000 (7 genes)	LOC_Os02g55890 (6 genes)	Not induced (5 genes)

Microarray data have been re-analyzed as described in Mustroph et al. (2010). Numbers in brackets define the size of the gene family in each genome
[a]Microarray data from Branco-Price et al. (2008), Mustroph et al. (2009), Lee et al. (2011)
[b]Microarray data from Sasidharan et al. (2013); annotation according to ATH1 array
[c]Microarray data from Kreuzwieser et al. (2009); annotation according to Tuskan et al. (2006) and the poplar genome version "Populus trichocarpa v3.0" (DOE-JGI, http://www.phytozome.net/poplar)
[d]Microarray data from Nanjo et al. (2011); annotation according to Schmutz et al. (2010) and the soybean genome version "Glycine max v1.0" (DOE-JGI, http://www.phytozome.net/soybean)
[e]Microarray data from Christianson et al. (2010); annotation according to the cotton D genome version "Cotton D v2.1" (DOE-JGI, http://www.phytozome.net/cotton)
[f]Microarray data from Lasanthi-Kudahettige et al. (2007), Narsai et al. (2009), Mustroph et al. (2010); annotation according to Ouyang et al. (2007)
[g]Microarray data from Rajhi et al. (2011); annotation according to Schnable et al. (2009) and the maize genome version "Maize Golden Path B73 RefGen_v2" (DOE-JGI, http://www.phytozome.net/maize)
[h]Induced not in hypoxic, but flooded or submerged Arabidopsis plants (Hsu et al. 2011 and Sasidharan et al. 2013)
[i]Induced only in some experiments

in higher protein content or activity under severe stress conditions (McElfresh and Chourey 1988; Chourey et al. 1991; Nanjo et al. 2010) maybe due to a general decrease in translation efficiency. In other species, SuSy activity and/or protein content were increased upon low-oxygen conditions, for example in potato (Biemelt et al. 1999; Mustroph et al. 2005), wheat (Mustroph and Albrecht 2003), Arabidopsis (Déjardin et al. 1999; Bieniawska et al. 2007), and rice (Ricard et al. 1991; Guglielminetti et al. 1995), while invertase activities usually decreased.

These expression data point to an important function of SuSy during oxygen deficiency. Furthermore, knock-down mutants of SuSy provide more evidence for its role under these conditions. Although mutants of one maize gene for SuSy, *Shrunken1* (*SS1*), were not impaired in anoxic glycolysis, it was shown that the isoenzyme *SUS1* (*SS2*) could compensate for its mutation (Guglielminetti et al. 1996). When a double mutant was created, its survival and ethanolic fermentation capacity were strongly impaired (Ricard et al. 1998). Potato root tips with an antisense inhibition of *SuSy* also performed worse under low oxygen in comparison to the wild-type, although levels of glycolytic intermediates did not change significantly (Biemelt et al. 1999). Also, when potato tubers that are usually hypoxic in the inner core, with genetically increased invertase activity, were observed, their energy status and oxygen requirements were impaired (Bologa et al. 2003).

In this line of evidence, Arabidopsis plants affected in *SUS1* and *SUS4* expression (the two *SuSy* genes which are low-oxygen responsive, Baud et al. 2004, Table 1) also showed reduced survival under flooding (Bieniawska et al. 2007). The latter study also suggests that sucrose cleavage is the main direction of the reversible enzyme, since the *sus1*/*sus4* double knock-out contained more sucrose and hexoses in its leaves than the wild-type. Moreover, even some cyanobacterial species contain hypoxia-inducible *SuSy* genes, and their knock-out also performs worse under low-oxygen conditions (Kolman et al. 2012). Until now, no *SuSy* mutants have been described in rice, which would be very interesting for a complete comparison.

In most datasets obtained so far, genes for the accompanying enzyme UGPase are not induced under conditions of oxygen deficiency (Table 1). However, this enzyme is believed to be not rate-limiting (Kleczkowski et al. 2004) and its activity in wheat or potato plants is about 30- to 50-fold higher than that of SuSy under normal and hypoxic conditions (Mustroph and Albrecht 2003; Mustroph et al. 2005). Interestingly, flooding induced this enzyme in soybean as evidenced by increased gene expression and protein amount (Alam et al. 2010; Komatsu et al. 2010; Nanjo et al. 2011). One UGPase-like gene, the plastidic *UGP3* from Arabidopsis is induced in at least one low-oxygen experiment (9 h hypoxia, Branco-Price et al. 2008), but this enzyme was shown to be involved in sulfolipid biosynthesis, and not in glycolysis (Okazaki et al. 2009).

Besides the role in sucrolysis, SuSy and UGPase are also considered as being important for cellulose and callose synthesis, enabling the production of UDP-glucose, the building block of cellulose. This function under low-oxygen conditions was associated with enhanced cellulose formation in wheat roots under hypoxic conditions, in which aerenchyma are formed and the root system requires strengthening in order to avoid collapsing of the tissue (Albrecht and Mustroph 2003). UGPase was also shown to be important for callose synthesis in pollen (Park et al. 2010); the knock-out of both *UGPase* genes in Arabidopsis caused male sterility. However, the importance of UGPases under oxygen-deficiency conditions remains undetermined.

4 Two Forms of Phosphofructokinase Play a Role During Oxygen Deficiency

The previously mentioned pathway via SuSy and UGPase is truly energy-saving (1 mole PPi instead of 2 mole ATP) in comparison to the invertase-hexokinase pathway, and therefore the importance of SuSy under oxygen deficiency is widely accepted. In contrast, the role of different types of PFKs during hypoxia stress is still a matter of debate. Plants possess two types of PFKs, the irreversible PFK, which uses ATP for the phosphorylation of fructose-6-phosphate (PFK), and the reversible enzyme that can convert fructose-6-phosphate to fructose-1,6-bisphosphate with the help of PPi (PFP) (Fig. 1, panel 2, and Plaxton 1996). In this case, 1 mole of ATP could be replaced by 1 mole of PPi, but both pools of substrates might be limited under low-oxygen conditions.

PFP has been considered as an important low-oxygen stress protein after the discovery that its activity was strongly increased in anoxically germinated rice coleoptiles (Mertens et al. 1990; Kato-Noguchi 2002a) and suspension cells (Mohanty et al. 1993). Following its first description in 1979 (Carnal and Black 1979), different groups have argued about mainly gluconeogenic or glycolytic directions of the enzyme, with no final conclusion yet (e.g., Kruger et al. 1983; Botha and Botha 1991b; Kato-Noguchi 2002a; Costa dos Santos et al. 2003; Groenewald and Botha 2008; van der Merwe et al. 2010). Unlike animals, that only possess PFK isoforms, which are regulated by fructose-2,6-bisphosphate, the plant PFKs are insensitive to this regulatory metabolite, while PFP is responsive to it (Stitt 1990; Nielsen et al. 2004). This finding alone would suggest a glycolytic direction of the enzyme (Mertens et al. 1990). Despite its unclear role with regard to hypoxic stress, PFP has been extensively studied at the biochemical level. PFP is a heterotetramer consisting of 2 alpha- and 2 beta-subunits, with PFP-beta being the active and PFP-alpha being the regulatory domain (Wong et al. 1990; Teramoto et al. 2000). In contrast, less information is available for PFK, but plants appear to contain large families of PFK isoforms which can be located in the chloroplast or cytosol (Mustroph et al. 2007, 2013; Winkler et al. 2007).

Despite the fact that *PFP* was induced in anoxic rice tissue, it was not observed as a hypoxia-induced protein in all species. Microarray studies from Arabidopsis (e.g., Loreti et al. 2005; Branco-Price et al. 2008; van Dongen et al. 2009; Mustroph et al. 2009; Lee et al. 2011), maize (Rajhi et al. 2011), and soybean (Nanjo et al. 2011) did not show a significant induction of *PFP* genes upon oxygen-deficiency stress

(Table 1). Only in rice leaves or coleoptiles, *PFP* genes were significantly induced (Lasanthi-Kudahettige et al. 2007; Narsai et al. 2009; Mustroph et al. 2010), as well as in flooded cotton (Christianson et al. 2010) and poplar roots (Kreuzwieser et al. 2009). Furthermore, PFP and PFK activities changed only slightly in wheat, maize, or potato roots subjected to root hypoxia (Biemelt et al. 1999; Mustroph and Albrecht 2003; Mustroph et al. 2005), and even decreased in anoxic *Citrullus lanatus* cotyledons (Botha and Botha 1991a). In suspension cultures, only the tolerant rice but not the sensitive soybean was able to strongly increase PFP activity (Mohanty et al. 1993). Interestingly, several *PFK* genes were among the hypoxia-induced genes in all plant organisms studied so far (Christianson et al. 2010; Rajhi et al. 2011; Nanjo et al. 2011 and summarized in Mustroph et al. 2010) (Table 1). These data suggest that PFK and not PFP catalyzes the main glycolytic pathway under hypoxic conditions in most organisms.

When *PFP* expression and activity are evaluated, another fact also needs to be considered. The metabolite Fructose-2,6-bisphosphate (Fru2,6BP) is an important regulator of PFP, being absolutely necessary for the activity of PFP (summarized in Stitt 1990; Nielsen et al. 2004). Its synthesis by the enzyme Phosphofructo-2-kinase (PF2K, Fig. 1) is tightly regulated. The transcript level of *PF2K* is generally not induced by oxygen-deficiency stress in plants, although it is induced in humans (Mustroph et al. 2010). However, the metabolite level of Fru2,6BP is enhanced under oxygen-deficiency stress in rice (Mertens et al. 1990; Kato-Noguchi 2002a), carrots (Kato-Noguchi and Watada 1996), and the chloroplast-containing protist *Euglena gracilis* (Enomoto et al. 1990). Since data for Arabidopsis or other hypoxia-sensitive species are not yet available, the significance of this metabolite under oxygen-deficiency stress remains elusive so far.

Experiments with potato plants with an artificially low PPi content indicated that PFP might not function in the glycolytic direction under hypoxic conditions (Mustroph et al. 2005). *PFP*-antisense plants did not perform worse under either aerated or hypoxic conditions compared to their wild-type (Hajirezaei et al. 1994, and Mustroph, unpublished results). Sugar cane plants with reduced PFP activity also did not show an altered growth phenotype, but modification in primary metabolites pointed to a glycolytic role of PFP in younger and a gluconeogenic role in older tissues (Groenewald and Botha 2008; van der Merwe et al. 2010), a result that was also described for PFP-deficient tobacco plants (Paul et al. 1995; Nielsen and Stitt 2001). In contrast, Arabidopsis plants with increased or decreased PFP activity had higher and lower growth rates, respectively, in comparison to the wild-type (Lim et al. 2009), but their survival under oxygen-deficiency stress has not yet been analyzed. These data point to the hypothesis that plant metabolism surrounding PFP is tightly regulated and can be adjusted to a situation where most of the PFP activity is missing without obvious growth defects, for example by modifying levels of Fru2,6BP and PFK. Such regulation most likely also occurs in plants in different developmental and metabolic states.

Although an important role of PFP for metabolism of plants under oxygen deficiency could not yet be demonstrated, one cannot exclude that it might contribute to tolerance of certain plant species such as rice or the Arabidopsis relative

Rorippa. Rice plants show increased *PFP* mRNA levels and/or enzyme activities under conditions of low-oxygen concentrations (Mertens et al. 1990; Kato-Noguchi 2002a; Lasanthi-Kudahettige et al. 2007; Narsai et al. 2009; Mustroph et al. 2010, 2013), and they accumulate Fru2,6BP (Mertens et al. 1990; Kato-Noguchi 2002a). For a full understanding of its role, experiments with rice plants lacking PFP activity need to be performed. Preliminary results from our lab indicate that PFP-knock-out rice lines could be lethal at the 3-week-seedling stage, but this needs to be confirmed in independent experiments (Mustroph and Hess, unpublished results). The flooding-tolerant Arabidopsis relative *Rorippa sylvestris* is also able to induce *PFP* genes under submerged conditions (Sasidharan et al. 2013, Table 1), and future comparative analyses could shed light on this still unsolved question.

5 C4 Metabolism Enzymes Such as PPDK and PEPCK Are Among the Hypoxia-Induced Genes

The enzyme PPDK is recognized as an important enzyme for carbon fixation in C4 plants such as maize or sorghum. However, many years ago its occurrence was already described in C3-species such as wheat (Meyer et al. 1978), spinach (Aoyagi and Bassham 1984), rice (Imaizumi et al. 1997), and Arabidopsis (Parsley and Hibberd 2006). Unexpectedly, it was discovered that one of two rice *PPDK* genes was strongly induced by low-oxygen stress (Moons et al. 1998), making it an interesting target to study in this context. This was later confirmed in rice micro-array experiments where *OsPPDKA*, LOC_Os03g31750, was one of the most strongly upregulated genes under oxygen-deficiency conditions (Lasanthi-Kudahettige et al. 2007; Narsai et al. 2009; Mustroph et al. 2010).

PPDK catalyzes the reversible reaction of PEP, AMP, and PPi to pyruvate and ATP (Fig. 1, panel 3). In C4 plants, its function is to provide PEP as a substrate for CO_2 fixation by PEP carboxylase or PEP carboxykinase. In C3 plants, its role is still unclear. For rice *OsPPDKB*, a function in seed filling was described (Kang et al. 2005), while the function of the anoxia-responsive *OsPPDKA* is unknown. It is, however, hypothesized that it might participate in the energy-saving PPi metabolism (Huang et al. 2008). Hypothetically, PPDK can replace pyruvate kinase (PK). The latter enzyme transforms PEP and ADP to pyruvate and ATP, while the first performs the following reaction: PEP + AMP + PPi $\leftrightarrow$ pyruvate + ATP + Pi. In conjunction with the reaction of adenylate kinase (2 ADP $\leftrightarrow$ ATP + AMP), the PPDK reaction yields 2 mole ATP out of 2 mole ADP and 1 mole PPi per PEP, a reaction slightly more effective than the usual pyruvate kinase (Mertens 1993).

The enzyme PPDK is not the only C4 enzyme induced in rice organs under oxygen-deficient conditions. Another strongly induced gene is the phosphoenol-pyruvate carboxykinase (PEPCK) (Lasanthi-Kudahettige et al. 2007; Narsai et al. 2009; Mustroph et al. 2010), catalyzing the reversible reaction PEP + CO_2 +

ADP ↔ oxaloacetate + ATP (Fig. 1, panel 3) (Leegood and Walker 2003). Although this enzyme seems not to be involved in PPi-dependent metabolism, its co-regulation with PPDK and other PPi-using enzymes in rice, such as PFP or H^+-PPase, suggests an important function under oxygen-deficiency stress. This was furthermore confirmed by activity assays of submerged rice tissue for PPDK and PEPCK (Moons et al. 1998). Although these experiments also demonstrated enhanced malate dehydrogenase activity, an enzyme also involved in C4 metabolism, this induction could not be confirmed by microarray expression analyses.

While *PPDK* and *PEPCK* are among the most strongly induced genes under oxygen-deficiency stress in rice, they are commonly not induced in Arabidopsis, poplar (summarized in Mustroph et al. 2010), soybean (Nanjo et al. 2011), or cotton (Christianson et al. 2010) and were also not reported as anaerobic polypeptides in maize, although the transcript of *PPDK* was found to be slightly upregulated in submerged maize roots (Rajhi et al. 2011) (Table 1). Under some circumstances, PPDK and PEPCK can be also induced in Arabidopsis, for example during root flooding (Hsu et al. 2011), a treatment which is not as severe as hypoxic stress. Furthermore, there are hints that both enzymes can be induced by submergence in the flooding-tolerant Arabidopsis relative *Rorippa sylvestris* (Sasidharan et al. 2013, Table 1). Interestingly, also another wild species, *Rumex acetosa*, strongly increases *PEPCK* transcript amount and activity in the petioles of submerged leaves (van Veen et al. 2013). From these data one could hypothesize that this metabolic pathway could be one reason for higher tolerance of rice and other species to oxygen deficiency. This needs to be confirmed by further experiments.

Although not induced in Arabidopsis, the function of these enzymes is best studied in this model organism. *PPDK* over-expressing plants revealed a role of PPDK in the formation of the transport amino acid glutamine and functions in nitrogen remobilization during senescence (Taylor et al. 2010). Unfortunately, no knock-out mutants have been described so far. For PEPCK, a role in amino acid and/or malate metabolism has been suggested, since knock-out mutants show modified amino acid content and enhanced stomata opening (Brown et al. 2010; Penfield et al. 2012). It is therefore likely that these enzymes are also involved in amino acid metabolism under oxygen-deficiency stress, although the direct function remains to be elucidated. Indeed, plants do show changes in metabolite content under oxygen-deficiency stress, for example the accumulation of succinate, GABA, and Alanine (e.g., Branco-Price et al. 2008; van Dongen et al. 2009; Kreuzwieser et al. 2009; Narsai et al. 2009). These metabolites are suggested as alternative fermentative products besides lactate and ethanol, but their formation and significance under oxygen deficiency is not fully understood. Alternatively, the enzymes could be involved in the regulation of cytosplasmic pH under oxygen-deficiency stress, as suggested for rice (Kulichikhin et al. 2009). This pathway would also involve the observed induction of one gene for the malic enzyme, LOC_Os05g09440 (Lasanthi-Kudahettige et al. 2007; Narsai et al. 2009), but further experiments are required to verify this hypothesis.

6 Vacuolar Proton Transport Using PPi

Although not directly involved in primary metabolism, the vacuolar PPi-dependent proton pump (H^+-PPase) might also be involved in an alternative energy-consuming pathway. The enzyme can replace the ATP-dependent proton pump in the vacuolar membrane (H^+-ATPase), allowing proton exchange between cytosol and the vacuole (Fig. 1, panel 4). Proton transport into the vacuole is of major importance under oxygen-deficiency stress since the initial production of lactic acid after the onset of the stress period, as well as the inhibition of H^+-ATPases due to ATP deficiency, rapidly leads to a decrease in cytoplasmic pH (Roberts et al. 1984; Gout et al. 2001; Kulichikhin et al. 2007). This cytoplasmic acidosis can inhibit enzyme activities in the cytosol. Since rice roots are able to restore their cytoplasmic pH after about 6 h of anoxia (Kulichikhin et al. 2007), it could be hypothesized that induction of H^+-PPases could be part of the adaptive response.

Indeed, *H^+-PPase* is strongly induced in oxygen-deficient rice tissue (Lasanthi-Kudahettige et al. 2007; Narsai et al. 2009; Mustroph et al. 2010), and to a lower extent in cotton (Christianson et al. 2010), poplar (Kreuzwieser et al. 2009), and soybean (Nanjo et al. 2011) (Table 1). However, no induction has been observed in Arabidopsis plants under any oxygen-deficiency-related condition (e.g., Loreti et al. 2005; Branco-Price et al. 2008; van Dongen et al. 2009; Mustroph et al. 2009; Lee et al. 2011), or in maize (Rajhi et al. 2011). Similar results have been confirmed independently during a comparison of rice and maize, where H^+-PPase was induced at the transcript, protein and activity level in rice, but not in maize roots (Carystinos et al. 1995). This induction was found to be mainly due to one *H^+-PPase* gene, *OVP3* (LOC_Os02g55890), while five other *OVP* members of the gene family were not induced (Liu et al. 2010). Anoxic gene induction was confirmed by promoter:GUS fusion constructs. Interestingly, there were variations among rice varieties, with the most tolerant variety Amaroo inducing the transcript to the highest level. Accordingly, the submergence-tolerant transgenic line M202 (Sub1) induced *OVP3* fourfold higher than the wild-type M202 (Mustroph et al. 2010).

Until now, transgenic plants lacking expression of H^+-PPase are only described for Arabidopsis and show an interesting germination phenotype. Mutant seedlings only showed defects while germinating without sucrose, whereas sucrose addition rescued their growth (Ferjani et al. 2012). The knock-out of the only Arabidopsis vacuolar *H^+-PPase* gene, At1g15690, led to accumulation of PPi, and the authors suggested that this PPi accumulation might inhibit gluconeogenesis from fatty acids stored in the seed. They therefore concluded that the main function of H^+-PPase is the removal of PPi from the cytosol. However, rice (as well as other plant species) has not just one but six vacuolar *H^+-PPase* genes, of which one is oxygen-responsive. Its function under oxygen-deficiency stress might be different from the single Arabidopsis gene and could be involved in energy metabolism. Indeed, it is still debatable whether this enzyme is possibly able to catalyze the reversible transport of protons to synthesize PPi under normal or hypoxic growth conditions (Rocha Facanha and de Meis 1998; Serrano et al. 2007; Gaxiola et al. 2012).

7 Conclusions and Open Questions

Despite a growing collection of gene expression data from different plant species, major questions regarding plant responses to low-oxygen stress remain partially unsolved. Here, the role of alternative PPi-using pathways was discussed. From the available data it became obvious that the SuSy pathway should have a general energy-saving function in all plant species under low oxygen, while other PPi-using enzymes are not ubiquitously induced. However, induction of most of the genes encoding PPi-utilizing alternative enzymes has been observed in flooding-tolerant rice plants (Table 1), and this could potentially be part of the metabolic acclimation response of rice and maybe other tolerant species to low oxygen.

Besides the importance of genes involved in the PPi-dependent energy metabolism, it is also still unclear what a continuous source of PPi could be in order to be a true alternative to replace ATP. It could be hypothesized that a number of the pathways mentioned above generate PPi, while others consume it, thereby maintaining a tight balance of the cytoplasmic PPi levels. From the proposed energy balances one could suggest that the SuSy/UGPase pathway might function in the direction of PPi consumption while PFP could generate PPi for this reaction, but the precise role of H^+-PPase and PPDK is rather unclear.

Evidence that the alternative pathways are indeed important for anoxic survival does come from a very different side: A potential role for PFP and PPDK in energy-saving metabolism under oxygen deficiency has not only been suggested for plants, but also for other organisms (Mertens 1993; Slamovits and Keeling 2006). For example, *Entamoeba histolytica*, *Trichomonas vaginalis,* and *Giardia lamblia* are obligate fermenting protists and contain *PFP* as well as *PPDK* genes. Also other organisms contain sequences for both enzymes, for example the bacteria *Caldicellulosiruptor saccharolyticus* (Bielen et al. 2010) and *Propionibacterium shermanii* (O'Brien et al. 1975). PPDK is an essential enzyme in *Giardia lamblia*, because its knock-down leads to severe ATP deficiency (Feng et al. 2008). These data suggest a link between these enzymes and survival under oxygen-deficiency conditions. But like in plants, collection of more data on these genes, for example in transgenic bacteria and protists, is required and will help to elucidate the role of these pathways in future.

From the evidence presented here it is obvious that PPi-using alternative enzymes play a role in primary metabolism of plants under oxygen-deficiency stress. It remains to be determined how important this role is in general, and what impact the enzymes have in plant species with different tolerance to the stress. Studies with transgenic plants impaired in those enzymes, or with over-expressing Arabidopsis plants are absolutely required for completely solving this question.

Acknowledgments The authors thank Alex Boonman (Monsanto Vegetable Seeds Division, Bergschenhoek, The Netherlands) for providing data on *Rorippa* species, and Maria Bongartz, Willi Riber, and Philipp Gasch (University Bayreuth, Germany) for critical reading of the manuscript.

References

Akman M, Bhikharie AV, McLean EH, Boonman A, Visser EJ, Schranz ME, van Tienderen PH (2012) Wait or escape? Contrasting submergence tolerance strategies of *Rorippa amphibia*, *Rorippa sylvestris* and their hybrid. Ann Bot 109:1263–1276

Alam I, Lee DG, Kim KH, Park CH, Sharmin SA, Lee H, Oh KW, Yun BW, Lee BH (2010) Proteome analysis of soybean roots under waterlogging stress at an early vegetative stage. J Biosci 35:49–62

Albrecht G, Mustroph A (2003) Localization of sucrose synthase in wheat roots: increased in situ activity of sucrose synthase correlates with cell wall thickening by cellulose deposition under hypoxia. Planta 217:252–260

Aoyagi K, Bassham JA (1984) Pyruvate orthophosphate dikinase of C(3) seeds and leaves as compared to the enzyme from maize. Plant Physiol 75:387–392

Armstrong W (1979) Aeration in higher plants. In: Woolhouse HW (ed) Advances in botanical research. Academic, London, pp 226–328

Armstrong W, Armstrong J (2014) Plant internal oxygen transport (diffusion and convection) and measuring and modelling oxygen gradients. In: van Dongen JT, Licausi F (eds) Low-oxygen stress in plants. Springer, Heidelberg

Bailey-Serres J, Voesenek LACJ (2008) Flooding stress: acclimations and genetic diversity. Annu Rev Plant Biol 59:313–339

Bailey-Serres J, Fukao T, Gibbs DJ, Holdsworth MJ, Lee SC, Licausi F, Perata P, Voesenek LACJ, van Dongen JT (2012) Making sense of low oxygen sensing. Trends Plant Sci 17:129–138

Baud S, Vaultier MN, Rochat C (2004) Structure and expression profile of the sucrose synthase multigene family in Arabidopsis. J Exp Bot 55:397–409

Bielen AA, Willquist K, Engman J, van der Oost J, van Niel EW, Kengen SW (2010) Pyrophosphate as a central energy carrier in the hydrogen-producing extremely thermophilic *Caldicellulosiruptor saccharolyticus*. FEMS Microbiol Lett 307:48–54

Biemelt S, Hajirezaei MR, Melzer M, Albrecht G, Sonnewald U (1999) Sucrose synthase activity does not restrict glycolysis in roots of transgenic potato plants under hypoxic conditions. Planta 210:41–49

Bieniawska Z, Paul Barratt DH, Garlick AP, Thole V, Kruger NJ, Martin C, Zrenner R, Smith AM (2007) Analysis of the sucrose synthase gene family in Arabidopsis. Plant J 49:810–828

Bologa KL, Fernie AR, Leisse A, Loureiro ME, Geigenberger P (2003) A bypass of sucrose synthase leads to low internal oxygen and impaired metabolic performance in growing potato tubers. Plant Physiol 132:2058–2072

Botha AM, Botha FC (1991a) Effect of anoxia on the expression and molecular form of the pyrophosphate dependent phosphofructokinase. Plant Cell Physiol 32:1299–1302

Botha AM, Botha FC (1991b) Pyrophosphate dependent phosphofructokinase of *Citrullus lanatus*: molecular forms and expression of subunits. Plant Physiol 96:1185–1192

Branco-Price C, Kawaguchi R, Ferreira RB, Bailey-Serres J (2005) Genome-wide analysis of transcript abundance and translation in Arabidopsis seedlings subjected to oxygen deprivation. Ann Bot 96:647–660

Branco-Price C, Kaiser KA, Jang CJ, Larive CK, Bailey-Serres J (2008) Selective mRNA translation coordinates energetic and metabolic adjustments to cellular oxygen deprivation and reoxygenation in *Arabidopsis thaliana*. Plant J 56:743–755

Brown NJ, Palmer BG, Stanley S, Hajaji H, Janacek SH, Astley HM, Parsley K, Kajala K, Quick WP, Trenkamp S, Fernie AR, Maurino VG, Hibberd JM (2010) C acid decarboxylases required for C photosynthesis are active in the mid-vein of the C species *Arabidopsis thaliana*, and are important in sugar and amino acid metabolism. Plant J 61:122–133

Carnal NW, Black CC (1979) Pyrophosphate dependent 6-phosphofructokinase, a new glycolytic enzyme in pineapple leaves. Biochem Biophys Res Commun 86:20–26

Carystinos GD, Heather MR, Monroy AF, Rajinder DS, Poole RJ (1995) Vacuolar H^+-translocating pyrophosphatase is induced by anoxia or chilling in seedlings of rice. Plant Physiol 108:641–649

Chastain CJ, Failing CJ, Manandhar L, Zimmerman MA, Lakner MM, Nguyen TH (2011) Functional evolution of C(4) pyruvate, orthophosphate dikinase. J Exp Bot 62:3083–3091

Chourey PS, Taliercio EW, Kane EJ (1991) Tissue-specific expression and anaerobically induced posttranscriptional modulation of sucrose synthase genes in *Sorghum bicolor* M. Plant Physiol 96:485–490

Christianson JA, Llewellyn DJ, Dennis ES, Wilson IW (2010) Global gene expression responses to waterlogging in roots and leaves of cotton (*Gossypium hirsutum* L.). Plant Cell Physiol 51:21–37

Costa dos Santos A, da Silva WS, de Meis L, Galina A (2003) Proton transport in maize tonoplasts supported by fructose-1,6-bisphosphate cleavage. Pyrophosphate-dependent phosphofructokinase as a pyrophosphate-regenerating system. Plant Physiol 133:885–892

Dancer JE, ap Rees T (1989) Effects of 2,4-dinitrophenol and anoxia on the inorganic pyrophosphate content of the spadix of *Arum maculatum* and the root apices of *Pisum sativum*. Planta 178:421–424

Déjardin A, Sokolov LN, Kleczkowski LA (1999) Sugar/osmoticum levels modulate differential abscisic acid-independent expression of two stress-responsive sucrose synthase genes in Arabidopsis. Biochem J 344(Pt 2):503–509

Edwards JM, Roberts TH, Atwell BJ (2012) Quantifying ATP turnover in anoxic coleoptiles of rice (*Oryza sativa*) demonstrates preferential allocation of energy to protein synthesis. J Exp Bot 63(12):4389–402

Enomoto T, Ohyama H, Inui H, Miyatake K, Nakano Y, Kitaoka S (1990) Roles of pyrophosphate: D-fructose 6-phosphate 1-phosphotransferase in the regulation of glycolysis during acclimation of aerobic *Euglena gracilis* to anaerobiosis. Plant Sci 67:161–167

Feng XM, Cao LJ, Adam RD, Zhang XC, Lu SQ (2008) The catalyzing role of PPDK in *Giardia lamblia*. Biochem Biophys Res Commun 367:394–398

Ferjani A, Segami S, Horiguchi G, Sakata A, Maeshima M, Tsukaya H (2012) Regulation of pyrophosphate levels by H^+-PPase is central for proper resumption of early plant development. Plant Signal Behav 7:38–42

Fukao T, Xu K, Ronald PC, Bailey-Serres J (2006) A variable cluster of ethylene response factor-like genes regulates metabolic and developmental acclimation responses to submergence in rice. Plant Cell 18:2021–2034

Gaxiola RA, Sanchez CA, Paez-Valencia J, Ayre BG, Elser JJ (2012) Genetic manipulation of a "vacuolar" H^+-PPase: from salt tolerance to yield enhancement under phosphorus-deficient soils. Plant Physiol 159:3–11

Geigenberger P, Fernie AR, Gibon Y, Christ M, Stitt M (2000) Metabolic activity decreases as an adaptive response to low internal oxygen in growing potato tubers. Biol Chem 381:723–740

Gibon Y, Vigeolas H, Tiessen A, Geigenberger P, Stitt M (2002) Sensitive and high throughput metabolite assays for inorganic pyrophosphate, ADPGlc, nucleotide phosphates, and glycolytic intermediates based on a novel enzymic cycling system. Plant J 30:221–235

Gout E, Boisson A, Aubert S, Douce R, Bligny R (2001) Origin of the cytoplasmic pH changes during anaerobic stress in higher plant cells. Carbon-13 and phosphorous-31 nuclear magnetic resonance studies. Plant Physiol 125:912–925

Groenewald JH, Botha FC (2008) Down-regulation of pyrophosphate: fructose 6-phosphate 1-phosphotransferase (PFP) activity in sugarcane enhances sucrose accumulation in immature internodes. Transgenic Res 17:85–92
Guglielminetti L, Perata P, Alpi A (1995) Effect of anoxia on carbohydrate metabolism in rice seedlings. Plant Physiol 108:735–741
Guglielminetti L, Alpi A, Perata P (1996) Shrunken-1-encoded sucrose synthase is not required for the sucrose-ethanol transition in maize under anaerobic conditions. Plant Sci 119:1–10
Hajirezaei MR, Sonnewald U, Viola R, Carlise S, Dennis D, Stitt M (1994) Transgenic potato plants with strongly decreased expression of pyrophosphate:fructose-6-phosphate phosphotransferase show no visible phenotype and only minor changes in metabolic fluxes in their tubers. Planta 192:16–33
Harada T, Satoh S, Yoshioka T, Ishizawa K (2005) Expression of sucrose synthase genes involved in enhanced elongation of pondweed (*Potamogeton distinctus*) turions under anoxia. Ann Bot 96:683–692
Hsu FC, Chou MY, Peng HP, Chou SJ, Shih MC (2011) Insights into hypoxic systemic responses based on analyses of transcriptional regulation in Arabidopsis. PLoS One 6:e28888
Huang S, Colmer TD, Millar AH (2008) Does anoxia tolerance involve altering the energy currency towards PPi? Trends Plant Sci 13:221–227
Imaizumi N, Ku MS, Ishihara K, Samejima M, Kaneko S, Matsuoka M (1997) Characterization of the gene for pyruvate, orthophosphate dikinase from rice, a C3 plant, and a comparison of structure and expression between C3 and C4 genes for this protein. Plant Mol Biol 34:701–716
Kang HG, Park S, Matsuoka M, An G (2005) White-core endosperm floury endosperm-4 in rice is generated by knockout mutations in the C-type pyruvate orthophosphate dikinase gene (OsPPDKB). Plant J 42:901–911
Kato-Noguchi H (2002a) The catalytic direction of pyrophosphate:fructose 6-phosphate 1-phosphotransferase in rice coleoptiles in anoxia. Physiol Plant 116:345–350
Kato-Noguchi H (2002b) Hypoxic acclimation to anoxia in Avena roots. Plant Growth Regul 38:1–5
Kato-Noguchi H, Watada AE (1996) Low-oxygen atmosphere increases fructose-2,6-bisphosphatase in fresh-cut carrots. J Am Soc Hortic Sci 121:307–309
Kleczkowski LA, Geisler M, Ciereszko I, Johansson H (2004) UDP-Glucose pyrophosphorylase. An old protein with new tricks. Plant Physiol 134:912–918
Kolman MA, Torres LL, Martin ML, Salerno GL (2012) Sucrose synthase in unicellular cyanobacteria and its relationship with salt and hypoxic stress. Planta 235:955–964
Komatsu S, Sugimoto T, Hoshino T, Nanjo Y, Furukawa K (2010) Identification of flooding stress responsible cascades in root and hypocotyl of soybean using proteome analysis. Amino Acids 38:729–738
Kreuzwieser J, Hauberg J, Howell KA, Carroll A, Rennenberg H, Millar AH, Whelan J (2009) Differential response of gray poplar leaves and roots underpins stress adaptation during hypoxia. Plant Physiol 149:461–473
Kruger NJ, Kombrink E, Beevers H (1983) Pyrophosphate:fructose-6-phosphate phosphotransferase in germinating castor bean seedlings. FEBS Lett 153:409–412
Kulichikhin KY, Aitio O, Chirkova TV, Fagerstedt KV (2007) Effect of oxygen concentration on intracellular pH, glucose-6-phosphate and NTP content in rice (*Oryza sativa*) and wheat (*Triticum aestivum*) root tips: in vivo 31P-NMR study. Physiol Plant 129:507–518
Kulichikhin KY, Chirkova TV, Fagerstedt KV (2009) Activity of biochemical pH-stat enzymes in cereal root tips under oxygen deficiency. Russ J Plant Physiol 56:377–388
Lasanthi-Kudahettige R, Magneschi L, Loreti E, Gonzali S, Licausi F, Novi G, Beretta O, Vitulli F, Alpi A, Perata P (2007) Transcript profiling of the anoxic rice coleoptile. Plant Physiol 144:218–231
Lee SC, Mustroph A, Sasidharan R, Vashisht D, Pedersen O, Oosumi T, Voesenek LACJ, Bailey-Serres J (2011) Molecular characterization of the submergence response of the *Arabidopsis thaliana* ecotype Columbia. New Phytol 190:457–471

Leegood RC, Walker RP (2003) Regulation and roles of phosphoenolpyruvate carboxykinase in plants. Arch Biochem Biophys 414:204–210
Lim H, Cho MH, Jeon JS, Bhoo SH, Kwon YK, Hahn TR (2009) Altered expression of pyrophosphate: fructose-6-phosphate 1-phosphotransferase affects the growth of transgenic Arabidopsis plants. Mol Cell 27:641–649
Liu Q, Zhang Q, Burton RA, Shirley NJ, Atwell BJ (2010) Expression of vacuolar H^+-pyrophosphatase (OVP3) is under control of an anoxia-inducible promoter in rice. Plant Mol Biol 72:47–60
Loreti E, Poggi A, Novi G, Alpi A, Perata P (2005) A genome-wide analysis of the effects of sucrose on gene expression in Arabidopsis seedlings under anoxia. Plant Physiol 137:1130–1138
McElfresh KC, Chourey PS (1988) Anaerobiosis induces transcription but not translation of sucrose synthase in maize. Plant Physiol 87:542–546
Mertens E (1991) Pyrophosphate-dependent phosphofructokinase, an anaerobic glycolytic enzyme? FEBS Lett 285:1–5
Mertens E (1993) ATP versus pyrophosphate: glycolysis revisited in parasitic protists. Parasitol Today 9:122–126
Mertens E, Laroundelle Y, Hers H-G (1990) Induction of pyrophosphate:fructose 6-phosphate 1-phosphotransferase by anoxia in rice seedlings. Plant Physiol 93:584–587
Meyer AO, Kelly GJ, Latzko E (1978) Pyruvate ortho-phosphate dikinase of immature wheat grains. Plant Sci Lett 12:35–40
Mohanty B, Wilson PM, ap Rees T (1993) Effects of anoxia on growth and carbohydrate metabolism in suspension cultures of soybean and rice. Phytochemistry 34:75–82
Moons A, Valcke R, Van Montagu M (1998) Low-oxygen stress and water deficit induce cytosolic pyruvate orthophosphate dikinase (PPDK) expression in roots of rice, a C3 plant. Plant J 15:89–98
Mustroph A, Albrecht G (2003) Tolerance of crop plants to oxygen deficiency stress: fermentative activity and photosynthetic capacity of entire seedlings under hypoxia and anoxia. Physiol Plant 117:508–520
Mustroph A, Albrecht G, Hajirezaei MR, Grimm B, Biemelt S (2005) Low levels of pyrophosphate in transgenic potato plants expressing *E. coli* pyrophosphatase lead to decreased vitality under oxygen deficiency. Ann Bot 96:717–726
Mustroph A, Sonnewald U, Biemelt S (2007) Characterisation of the ATP-dependent phosphofructokinase gene family from *Arabidopsis thaliana*. FEBS Lett 581:2401–2410
Mustroph A, Zanetti ME, Jang CJ, Holtan HE, Repetti PP, Galbraith DW, Girke T, Bailey-Serres J (2009) Profiling translatomes of discrete cell populations resolves altered cellular priorities during hypoxia in Arabidopsis. Proc Natl Acad Sci U S A 106:18843–18848
Mustroph A, Lee SC, Oosumi T, Zanetti ME, Yang H, Ma K, Yaghoubi-Masihi A, Fukao T, Bailey-Serres J (2010) Cross-kingdom comparison of transcriptomic adjustments to low-oxygen stress highlights conserved and plant-specific responses. Plant Physiol 152:1484–1500
Mustroph A, Stock J, Hess N, Aldous S, Dreilich A, Grimm B (2013) Characterization of the phosphofructokinase gene family in rice and its expression under oxygen deficiency stress. Front Plant Sci 4:125
Nanjo Y, Skultety L, Ashraf Y, Komatsu S (2010) Comparative proteomic analysis of early-stage soybean seedlings responses to flooding by using gel and gel-free techniques. J Proteome Res 9:3989–4002
Nanjo Y, Maruyama K, Yasue H, Yamaguchi-Shinozaki K, Shinozaki K, Komatsu S (2011) Transcriptional responses to flooding stress in roots including hypocotyl of soybean seedlings. Plant Mol Biol 77:129–144
Narsai R, Howell KA, Carroll A, Ivanova A, Millar AH, Whelan J (2009) Defining core metabolic and transcriptomic responses to oxygen availability in rice embryos and young seedlings. Plant Physiol 151:306–322

Narsai R, Rocha M, Geigenberger P, Whelan J, van Dongen JT (2011) Comparative analysis between plant species of transcriptional and metabolic responses to hypoxia. New Phytol 190:472–487

Nielsen TH, Stitt M (2001) Tobacco transformants with strongly decreased expression of pyrophosphate:fructose-6-phosphate expression in the base of their young growing leaves contain much higher levels of fructose-2,6-bisphosphate but no major changes in fluxes. Planta 214:106–116

Nielsen TH, Rung JH, Villadsen D (2004) Fructose-2,6-bisphosphate: a traffic signal in plant metabolism. Trends Plant Sci 9:556–563

O'Brien WE, Bowien S, Wood HG (1975) Isolation and characterization of a pyrophosphate-dependent phosphofructokinase from *Propionibacterium shermanii*. J Biol Chem 250:8690–8695

Okazaki Y, Shimojima M, Sawada Y, Toyooka K, Narisawa T, Mochida K, Tanaka H, Matsuda F, Hirai A, Hirai MY, Ohta H, Saito K (2009) A chloroplastic UDP-glucose pyrophosphorylase from Arabidopsis is the committed enzyme for the first step of sulfolipid biosynthesis. Plant Cell 21:892–909

Ouyang S, Zhu W, Hamilton J et al (2007) The TIGR rice genome annotation resource: improvements and new features. Nucleic Acids Res 35:D883–D887

Park JI, Ishimizu T, Suwabe K, Sudo K, Masuko H, Hakozaki H, Nou IS, Suzuki G, Watanabe M (2010) UDP-glucose pyrophosphorylase is rate limiting in vegetative and reproductive phases in *Arabidopsis thaliana*. Plant Cell Physiol 51:981–996

Parsley K, Hibberd JM (2006) The Arabidopsis PPDK gene is transcribed from two promoters to produce differentially expressed transcripts responsible for cytosolic and plastidic proteins. Plant Mol Biol 62:339–349

Paul M, Sonnewald U, Hajirezaei M, Dennis D, Stitt M (1995) Transgenic tobacco plants with strongly decreased expression of pyrophosphate:fructose-6-phosphate 1-phosphotranferase do not differ significantly from wild type in photosynthate partitioning, plant growth or their ability to cope with limiting phosphate, limiting nitrogen and suboptimal temperatures. Planta 196:277–283

Penfield S, Clements S, Bailey KJ, Gilday AD, Leegood RC, Gray JE, Graham IA (2012) Expression and manipulation of phosphoenolpyruvate carboxykinase 1 identifies a role for malate metabolism in stomatal closure. Plant J 69:679–688

Plaxton WC (1996) The organization and regulation of plant glycolysis. Annu Rev Plant Physiol Plant Mol Biol 48:185–214

Rajhi I, Yamauchi T, Takahashi H, Nishiuchi S, Shiono K, Watanabe R, Mliki A, Nagamura Y, Tsutsumi N, Nishizawa NK, Nakazono M (2011) Identification of genes expressed in maize root cortical cells during lysigenous aerenchyma formation using laser microdissection and microarray analyses. New Phytol 190:351–368

Ricard B, Rivoal J, Spiteri A, Pradet A (1991) Anaerobic stress induces the transcription of sucrose synthase in rice. Plant Physiol 95:669–674

Ricard B, Van Toai T, Chourey P, Saglio P (1998) Evidence for the critical role of sucrose synthase for anoxic tolerance of maize roots using a double mutant. Plant Physiol 116:1323–1331

Roberts JKM, Callis J, Jardetzky O, Walbot V, Freeling M (1984) Cytoplasmic acidosis as a determinant of flooding intolerance in plants. Proc Natl Acad Sci U S A 81:6029–6033

Rocha Facanha A, de Meis L (1998) Reversibility of H^+-ATPase and H^+-pyrophosphatase in tonoplast vesicles from maize coleoptiles and seeds. Plant Physiol 116:1487–1495

Sasidharan R, Mustroph A, Boonman A, Akman M, Ammerlaan AM, Breit T, Schranz ME, Voesenek LA, van Tienderen P (2013) Root transcript profiling of two Rorippa (Brassicaceae) species reveals gene clusters associated with extreme submergence tolerance. Plant Physiol 163:1277–1292

Schmutz J, Cannon SB, Schlueter J et al (2010) Genome sequence of the palaeopolyploid soybean. Nature 463:178–183

Schnable PS, Ware D, Fulton RS et al (2009) The B73 maize genome: complexity, diversity, and dynamics. Science 326:1112–1115

Serrano A, Pérez-Castiñeira JR, Baltscheffsky M, Baltscheffsky H (2007) H^+-PPases: yesterday, today and tomorrow. IUBMB Life 59:76–83

Slamovits CH, Keeling PJ (2006) Pyruvate-phosphate dikinase of oxymonads and parabasalia and the evolution of pyrophosphate-dependent glycolysis in anaerobic eukaryotes. Eukaryot Cell 5:148–154

Sorenson R, Bailey-Serres J (2014) Selective mRNA translation tailors low oxygen energetics. In: van Dongen JT, Licausi F (eds) Low-oxygen stress in plants. Springer, Heidelberg

Springer B, Werr W, Starlinger P, Bennet DC, Zokolica M, Freeling M (1986) The shrunken gene on chromosome 9 of *Zea mays* L. is expressed in various plant tissues and encodes an anaerobic protein. Mol Gen Genet 205:461–468

Stitt M (1990) Fructose-2,6-bisphosphate as a regulatory molecule in plants. Annu Rev Plant Physiol Plant Mol Biol 41:153–185

Stitt M (1998) Pyrophosphate as an energy donor in the cytosol of plant cells: an enigmatic alternative to ATP. Bot Acta 111:167–175

Taylor L, Nunes-Nesi A, Parsley K, Leiss A, Leach G, Coates S, Wingler A, Fernie AR, Hibberd JM (2010) Cytosolic pyruvate, orthophosphate dikinase functions in nitrogen remobilization during leaf senescence and limits individual seed growth and nitrogen content. Plant J 62:641–652

Teramoto M, Koshiishi C, Ashihara H (2000) Wound-induced respiration and pyrophosphate: fructose-6-phosphate phosphotransferase in potato tubers. Z Naturforsch 55:953–956

Tuskan GA, Difazio S, Jansson S et al (2006) The genome of black cottonwood, *Populus trichocarpa* (Torr. & Gray). Science 313:1596–1604

van der Merwe MJ, Groenewald JH, Stitt M, Kossmann J, Botha FC (2010) Downregulation of pyrophosphate: D-fructose-6-phosphate 1-phosphotransferase activity in sugarcane culms enhances sucrose accumulation due to elevated hexose-phosphate levels. Planta 231:595–608

van Dongen JT, Frohlich A, Ramirez-Aguilar SJ, Schauer N, Fernie AR, Erban A, Kopka J, Clark J, Langer A, Geigenberger P (2009) Transcript and metabolite profiling of the adaptive response to mild decreases in oxygen concentration in the roots of Arabidopsis plants. Ann Bot 103:269–280

van Veen H, Mustroph A, Barding GA, Vergeer-van Eijk M, Welschen-Evertman RA, Pedersen O, Visser EJ, Larive CK, Pierik R, Bailey-Serres J, Voesenek LA, Sasidharan R (2013) Two Rumex species from contrasting hydrological niches regulate flooding tolerance through distinct mechanisms. Plant Cell [Epub ahead of print]. doi: http://dx.doi.org/10.1105/tpc.113.119016

Vashisht D, Hesselink A, Pierik R, Ammerlaan JM, Bailey-Serres J, Visser EJ, Pedersen O, van Zanten M, Vreugdenhil D, Jamar DC, Voesenek LACJ, Sasidharan R (2011) Natural variation of submergence tolerance among *Arabidopsis thaliana* accessions. New Phytol 190:299–310

Weiner H, Stitt M, Heldt HW (1987) Subcellular compartmentation of pyrophosphate and alkaline pyrophosphatase in leaves. Biochim Biophys Acta 893:13–21

Winkler C, Delvos B, Martin W, Henze K (2007) Purification, microsequencing and cloning of spinach ATP-dependent phosphofructokinase link sequence and function for the plant enzyme. FEBS J 274:429–438

Wong JH, Kiss F, Wu M-X, Buchanan BB (1990) Pyrophosphate fructose-6-P 1-phosphotransferase from tomato fruit. Evidence for change during ripening. Plant Physiol 94:499–506

Xu K, Xu X, Fukao T, Canlas P, Maghirang-Rodriguez R, Heuer S, Ismail AM, Bailey-Serres J, Ronald PC, Mackill DJ (2006) Sub1A is an ethylene-response-factor-like gene that confers submergence tolerance to rice. Nature 442:705–708

Zabalza A, van Dongen JT, Froehlich A, Oliver SN, Faix B, Gupta KJ, Schmälzlin E, Igal M, Orcaray L, Royuela M, Geigenberger P (2009) Regulation of respiration and fermentation to control the plant internal oxygen concentration. Plant Physiol 149:1087–1098

Oxygen Consumption Under Hypoxic Conditions

Carola Päpke, Santiago Ramirez-Aguilar, and Carla Antonio

Abstract Even during optimal growth conditions, the availability of oxygen to plant cells can get limited due to the diffusion resistance of the tissue for oxygen. However, plants can probably regulate the rate of respiratory oxygen consumption which is thought to play a role during periods of low-oxygen stress as it postpones oxygen to become limiting as respiratory substrate. At present, limitation of oxygen becomes more and more important in the context of climate change and concomitant changed environmental stress conditions such as flooding. One mechanism by which plants can react to survive periods of low-oxygen stress is to save on oxygen by downregulating energy requiring processes such as storage metabolism. For this, a series of drastic metabolic adaptations is initiated in plants upon hypoxia. In this chapter, it is discussed how plants adapt to low-oxygen availability via metabolic responses and by regulating respiratory oxygen consumption as a function of the actual oxygen concentration.

C. Päpke (✉)
Max-Planck Institute of Molecular Plant Physiology, Am Mühlenberg 1,
14476 Potsdam-Golm, Germany
e-mail: paepke@mpimp-golm.mpg.de

S. Ramirez-Aguilar
Max-Planck Institute of Molecular Plant Physiology, Am Mühlenberg 1,
14476 Potsdam-Golm, Germany

Metanomics GmbH, Tegeler Weg 33, 10589 Berlin, Germany

C. Antonio
Max-Planck Institute of Molecular Plant Physiology, Am Mühlenberg 1,
14476 Potsdam-Golm, Germany

Instituto de Tecnologia Química e Biológia (ITQB-UNL), Av. da República, Estação
Agronómica Nacional, 2780-157 Oeiras, Portugal

J.T. van Dongen and F. Licausi (eds.), *Low-Oxygen Stress in Plants*, Plant Cell Monographs 21, DOI 10.1007/978-3-7091-1254-0_10, © Springer-Verlag Wien 2014

1 Introduction

Oxygen is an essential substrate of respiratory energy metabolism in aerobic organisms. Plants are the main source for oxygen production on earth, yet little is understood about oxygen as a potential limiting substrate for cellular respiration in plants when exposed to suboptimal oxygen conditions. Plants are obligate aerobic organisms, but even under optimal growth conditions they have to cope with highly variable oxygen concentrations inside their different tissues. These tissues can vary from very thin and oxygen-saturated tissue, such as leaves, to bulky or dense tissues, like fruits and stems of trees. In the first case, oxygen can reach all cells with ease, whereas in the latter example oxygen diffusion to the central cells and tissues can be strongly hampered due to the high resistance for oxygen diffusion into the tissue (Geigenberger et al. 2000; Rolletschek et al. 2002; van Dongen et al. 2003). Together with cellular oxygen consumption through mitochondrial respiration, the high diffusion resistance leads to relatively low oxygen concentrations inside the tissue (Geigenberger et al. 2000; van Dongen et al. 2003; Borisjuk and Rolletschek 2009). In contrast, in photosynthetically active parts of the plant, oxygen evolution through photosynthesis can raise the internal oxygen concentration above ambient levels (Rolletschek et al. 2004; Schmälzlin et al. 2005), which leads to a decrease of the photosynthetic efficiency due to photorespiration and an increase of the production of reactive oxygen species (ROS; Hütter et al. 2002; Puntarulo and Cederbaum 1988). The latter might damage cellular compounds including DNA, lipids, and proteins (Apel and Hirt 2004).

The availability of oxygen and concomitant metabolic regulation eventually has a great impact on plant production and storage metabolism and thus on the expected yield. In this chapter, we discuss the occurrence of oxygen gradients in plants and their effect on respiratory oxygen consumption rates along with mechanisms on how plants might regulate and adapt to low oxygen availability by metabolic responses and regulation of enzyme activity.

2 Oxygen Gradients Inside Plant Tissues

It was shown that in bulky plant organs the diffusion resistance of the tissue leads to steep oxygen gradients, being lowest towards the center of the plant organs (see also Chap. 14; Armstrong and Armstrong 2014). Probably, this affects the availability of oxygen as substrate for respiration which thus would affect the energy status of the plant tissue (Gupta et al. 2009). However, one should be aware that different plant tissues display various rates of respiratory oxygen uptake and diffusion resistance parameters. In trees, the vascular cambium is considered to have a very low oxygen concentration (Kimmerer and Stringer 1988), but also root steles with embedded vascular tissue are hypoxic (Thomson and Greenway 1991; Ober and Sharp 1996; Drew 1997). Phloem tissue, for instance, is a very dense and metabolically active tissue with high respiration rates, which was shown to correlate with local hypoxic conditions (van Dongen et al. 2003). These highly metabolically active plant tissues contribute - together with the local resistance for oxygen diffusion - to low internal oxygen levels in plant tissues.

It was observed that local oxygen concentrations in plant tissue can vary over time or due to environmental changes. During development, changes in the internal oxygen concentration were observed depending on the metabolic activity of the tissue (van Dongen et al. 2003; Benamar et al. 2008; Licausi et al. 2011). Furthermore, alterations in the environment of plant tissues, e.g., due to the day(light)-night(dark) cycle (Rolletschek et al. 2002), cause changes in oxygen availability inside tissues. Moreover, temporal flooding or waterlogging of soil is a major cause of environmental stress which can lead to a strong decrease in the availability of oxygen (Bailey-Serres and Voesenek 2008; van Bodegom et al. 2008).

Initial attempts to determine in situ oxygen concentrations in plant tissues using oxygen microsensors were made and in such studies a steep decrease of the oxygen concentration was determined from the outside to the inside of the plant tissue. In maize roots (*Zea mays*) a steep oxygen gradient from the cortex (10 % oxygen) to the stele (5 % oxygen) could be observed (Armstrong et al. 1994). Moreover, when the shoot was exposed to 10 % oxygen only, the oxygen gradient resulted in very hypoxic concentrations in the stele. In potato tuber, an oxygen gradient through the tissue from 11–15 % oxygen under the periderm to 2–5 % oxygen in the center was found (Geigenberger et al. 2000). In *Ricinus communis* phloem tissue an oxygen profile with a gradient from 21 % down to 7 % in the vascular region and in the hollow center an increase of oxygen up to 15 % was reported (van Dongen et al. 2003). In *Pisum sativum* roots an internal oxygen concentration of 8 % was shown (Zabalza et al. 2009). Furthermore, it was demonstrated that a low oxygen tension is also present within seeds, e.g., seeds of *Arabidopsis thaliana* (Porterfield et al. 1999; Gibon et al. 2002), *Vicia faba,* and *Pisum sativum* (Rolletschek et al. 2002). All these examples show that even under normal growth conditions, the availability of oxygen to plant cells can be restricted due to the resistance for diffusion of oxygen through the tissue. Unfortunately, the technology to measure oxygen concentrations inside cells or cellular compartments of plant tissues is still limited and our understanding of oxygen uptake processes is thus based on interpretation of indirect measurements of oxygen concentrations around the cells and miniaturized invasive sensors (see also Chap. 21; Ast and Draaijer 2014).

3 Oxidative Phosphorylation

The oxygen consuming process of respiration is the mitochondrial electron transport chain (mETC), which is located in the inner mitochondrial membrane (IMM). In most eukaryotes, the classical operation of the mETC involves electron passage from reducing equivalents such as NADH or succinate to molecular oxygen via four integral membrane oxidoreductases: NADH dehydrogenase (complex I), succinate dehydrogenase (complex II), cytochrome c reductase (complex III), and cytochrome c oxidase (complex IV, COX). The ATP synthase (complex V) uses the

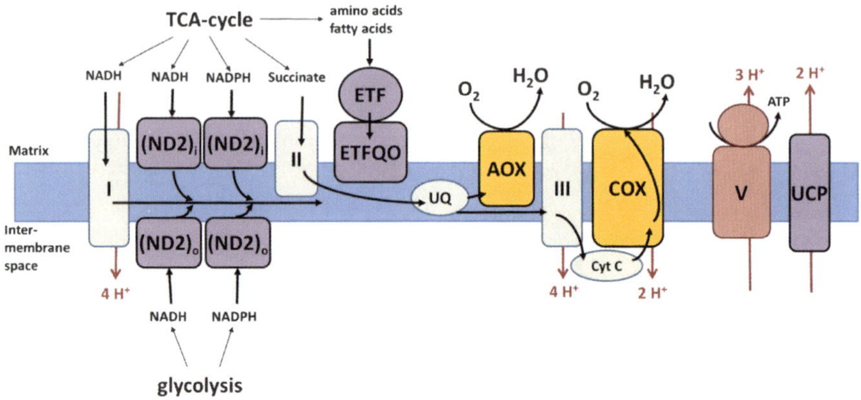

Fig. 1 Schematic overview of the components involved in conventional and alternative reactions of the mitochondrial oxidative phosphorylation. Complexes I–III, that is NADH dehydrogenase, succinate dehydrogenase, cytochrome c reductase, respectively; COX, cytochrome c oxidase; V, ATP synthase; AOX, alternative oxidase; cyt c, cytochrome c; ETF, electron transfer flavoprotein; ETFQO, electron transfer flavoprotein quinine oxidoreductase; ND2i, type II NAD(P)H dehydrogenase, located at the matrix side of the mitochondrial inner membrane; ND2o, type II NAD(P)H dehydrogenase, located at the outside of the mitochondrial inner membrane; UCP, plant uncoupling protein; UQ, ubiquinone pool. *Black arrows* show the electron flow; *red arrows* show the proton flow

potential energy of the electrochemical H^+ gradient across the IMM, which is build up by the activity of the complexes I, III, and IV to generate ATP (Fig. 1, for review see Dudkina et al. 2006). The mitochondria of plants, fungi, and protists contain several additional enzymes for electron transport from NADH and NADPH to oxygen (Vanlerberghe and McIntosh 1997; Rasmusson et al. 2004) without pumping H^+ across the IMM. These systems are indicated as alternative pathways (Review van Dongen et al. 2011).

The alternative NAD(P)H dehydrogenases (ND2) are located on the inner and outer side of the IMM and reduce ubiquinone (UQ). These enzymes enable plant mitochondria to oxidize NADH and NADPH directly from the cytosol and the matrix in a non-energy-conserving way (Rasmusson et al. 2004). Under different physiological conditions or stress situations which could affect the redox status of the cell, the electron input into the mETC through ND2 can change. Also a decrease in substrate specificity at low pH was observed (Felle 2005), which might happen upon cellular acidification during anoxic stress, and could therefore support the oxidation of cytosolic NADH and thereby keeping NAD^+ available as substrate for glycolysis (van Dongen et al. 2011).

The alternative oxidase (AOX) oxidizes ubiquinol (UQH_2) and transfers the electrons directly to oxygen, thereby bypassing complexes III and IV. Gene expression of this AOX was shown to be upregulated upon hypoxia (Clifton et al. 2005). However, it was found that the protein amount and enzyme capacity did not change during hypoxia (Skutnik and Rychter 2009), although the capacity increased after reoxygenation (Blokhina et al. 2001). AOX activity plays an important role in preventing and reducing ROS (McDonald and Vanlerberghe 2006; Maxwell et al. 1999) by avoiding an "over-reduction" of UQ and probably also by decreasing the oxygen concentration inside the mitochondria. AOX can also buffer oscillations of energy production (Rasmusson et al. 2009) and help to endure natural inhibitors of the COX pathway like nitric oxide (NO), a potent reversible respiratory inhibitor of COX (as shown by in vitro experiments; Gupta et al. 2009). Neither AOX nor COX is active at full capacity under non-stress conditions and both were shown to compete for the distribution of electrons (Ribas-Carbo et al. 1995). However, fast regulatory changes affect the activity of both enzymes, which in turn influences the distribution of electrons between both these terminal oxidases.

Another potential regulator of respiratory metabolism is the uncoupling protein (UCP), which belongs to the mitochondrial anion carrier superfamily and transports protons across the IMM following the electrochemical gradient, thereby uncoupling the activity of the mETC from ATP production while the potential energy is dissipated as heat (Vercesi et al. 2006). Its activity is dependent on purine nucleotide di- and triphosphate concentrations (ADP, GDP, ATP, and GTP) as well as on the pH, the redox status of UQ pool, and on Mg^{2+} concentrations (Borecký et al. 2001). Furthermore, a decrease in pH enhances the inhibitory effect of ATP (Borecký et al. 2001), which could be especially important under hypoxic conditions, where the cytosolic pH decreases (Felle 2005) and the enhanced inhibition of UCP by purine nucleotide di- and triphosphate increases the ATP production efficiency.

The electron transfer flavoprotein (ETF)/electron transfer flavoprotein quinone oxidoreductase (ETFQO) complex is another additional electron donor of the mETC (Heazlewood and Millar 2005; Araújo et al. 2011). It was suggested that the ETF/ETFQO pathway is involved in lysine degradation as well as in the breakdown of aromatic acids and phytol. By doing so, it provides alternative substrates to feed electrons into the mETC (van Dongen et al. 2011). However, the understanding of the regulation and integration of the mitochondrial alternative pathways of the mETC in plants is still limited and further analysis of this aspect of mitochondrial metabolism is necessary.

4 Rate of Respiratory Oxygen Consumption

The mETC requires oxygen as a final acceptor and the rate of respiratory oxygen consumption decreases when the oxygen availability goes down. At present, it is under discussion if this decrease of respiratory activity is due to oxygen being limiting as substrate or because of some kind of proactive regulatory mechanism that fine-tunes the oxygen consumption rate to the oxygen concentration whereby both suggested mechanisms are not mutually exclusive.

It was observed that in response to low internal oxygen concentrations, the metabolic flux through glycolysis slows down, respiratory oxygen consumption decreases, and ATP levels drop (Geigenberger et al. 2000; Bologa et al. 2003). Several studies evaluated how oxygen availability affects the respiratory oxygen consumption rate. Tucker and Laties (1985) found in avocado fruits a biphasic relation for the mitochondrial respiration (dependent on the timescale). These authors suggested that the down-regulation of respiration cannot be explained by a limitation of oxygen as substrate for the mETC and they assumed that a change in the mobilization of carbohydrates within the cells could be an explanation. Laisk et al. (2007), however, found in a short-time experiment (60 s) with sunflower and aspen leaves - in which optimal oxygen diffusion is provided by the stomatal openings and spongy parenchyma - that a Michaelis–Menten kinetic (MMK) fitted well with an apparent Km for oxygen of 0.3–1 μM. One explanation for the different respiratory profiles could be that biochemical adaptations affect the respiratory behavior.

A different approach was adopted by Zabalza et al. (2009). They used *Pisum sativum* roots in a tightly closed cuvette containing air saturated buffer to prevent oxygen diffusion from the outside environment in to the buffer and measured the oxygen consumption through time. The respiratory profiles obtained with these experiments show a change of respiratory oxygen consumption in a biphasic manner, with a slow and linear decrease between air saturated buffer and approximately 20 μM oxygen and a subsequent sharp drop of the respiratory activity below 20 μM. A similar measurement with soybean roots is shown in Fig. 2. This response is observed in several plant tissues, namely roots (Mancuso and Boselli 2002; Maricle and Lee 2007; Gupta et al. 2009), fruits (Tucker and Laties 1985; Ho et al. 2010), and seeds (van Dongen et al. 2004).

In addition, it was shown that the addition of pyruvate strongly enhanced the oxygen consumption rate of plant tissues (Zabalza et al. 2009). This effect is probably explained by the allosteric activation of the AOX by alpha-ketoacids (such as pyruvate, a-ketoglutarate, succinate, and malate) and an increased capacity of AOX (Millar et al. 1996; more detail in Sect. 6.3). Based on these observations, it is assumed that control of cellular pyruvate levels during hypoxic conditions, while the TCA cycle activity is reduced, is of crucial importance to prevent oxygen being "wasted" in alternative respiratory pathways that would only provide a limited ATP gain.

5 Metabolic Responses of Plants to Hypoxic Conditions

Certain plant species and ecotypes have developed a number of metabolic strategies to adapt to a low-oxygen environment by avoiding or delaying the depletion of oxygen to concentrations that would limit oxidative phosphorylation. In this section, studies on plant responses to low oxygen stress at the metabolite level will be discussed taking into account the various factors that can contribute to a decrease of the oxygen concentration within a plant cell (Fig. 3).

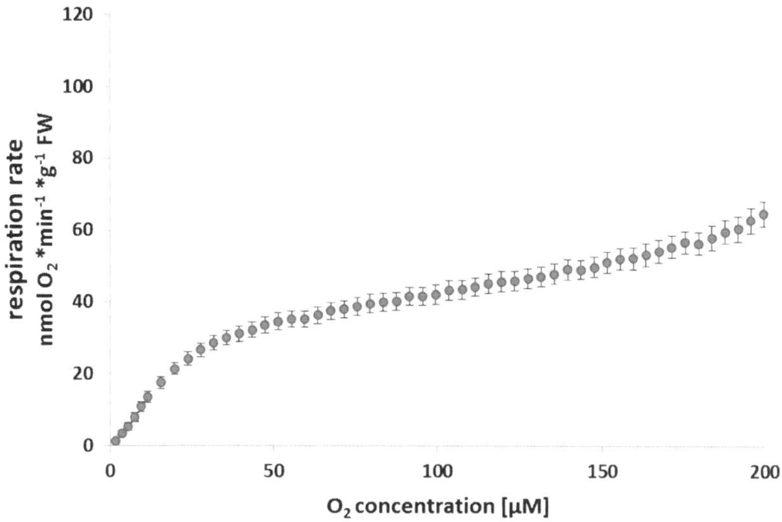

Fig. 2 Respiratory oxygen consumption rates of soybean roots. Values are mean ±SE from at least 20 independent measurements (of freshly cut roots 50 min in 100 mM HEPES buffer, pH 7.4) plotted against the oxygen concentration (μM). Measurements were performed in a closed vial at 25 °C connected to an OXY-4 multi-channel optical oxygen sensor (PreSens, Regensburg, Germany)

5.1 Metabolic Responses of Plants to Flooding, Waterlogging, and Submergence

Abiotic stresses like flooding, waterlogging, and complete submergence are different scenarios that can induce low-oxygen stress in plant cells. In a broad sense, the term *flooding* is used to describe the situation in which the inundation by water affects the whole plant or part of a plant. *Waterlogging* corresponds to the full saturation of the soil with water. Hence, under waterlogged conditions, only the root system of the plant is exposed to low-oxygen stress, while the shoot is exposed to aerobic conditions. The term *submergence* describes the situation where most or all aerial tissue is under water (Bailey-Serres et al. 2012). These common environmental stress conditions greatly affect plant growth and agricultural productivity by reducing the entry of oxygen into the soil so that roots and other organs are restricted in respiratory energy production. However, it is important to highlight that these stresses do not necessarily lead to oxygen limitation, especially when photosynthesis remains active (Colmer and Pedersen 2008; Mommer and Visser 2005).

Kreuzwieser et al. (2009) investigated the molecular and physiological responses of waterlogged poplar (*Populus* x *canescens*) in roots and leaves using transcript and metabolite profiling. Metabolite profiling revealed that hypoxia induced rapid changes in the levels of several amino acids in the roots: strong

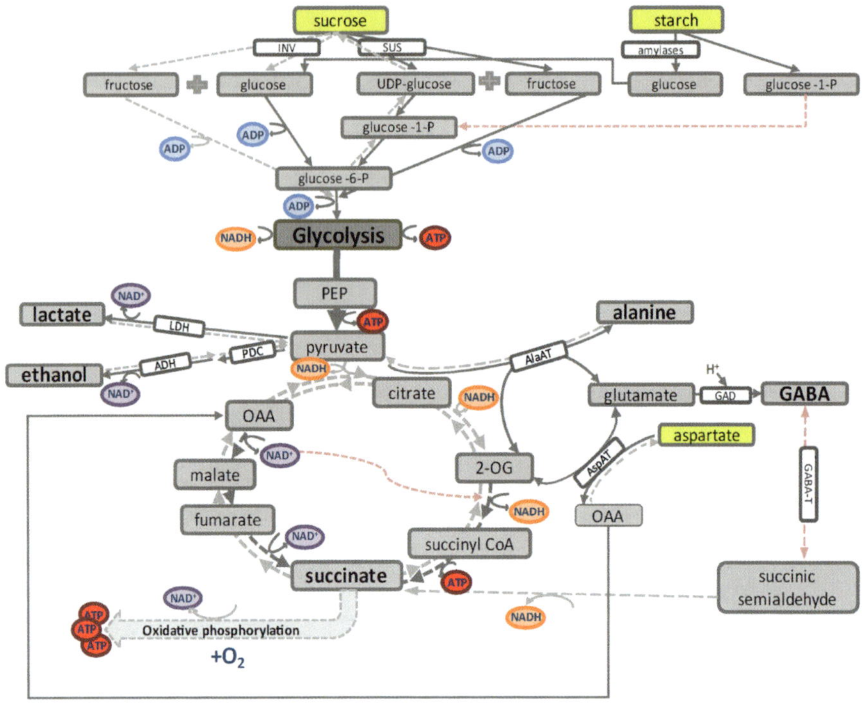

Fig. 3 Metabolic responses of plants to low oxygen availability. In order to maximize ATP production under reduced oxygen availability a number of metabolic changes occur. Various processes, such as a sucrose–starch metabolism, glycolysis, fermentation, the tricarboxylic acid (TCA) flow (shown in a modified form), an alanine and 2-oxoglutarate (2-OG) shunt and a γ-aminobutyric acid (GABA) shunt, are known to undergo metabolic alterations. During low oxygen stress some pathways are enhanced (indicated by *gray lines*), activated (*red dashed lines*) or inhibited (*gray dashed lines*). Increased metabolite levels during stress are indicated by a *larger, black font* while decreased metabolite levels are depicted in *yellow*. Abbreviations: *2-OG* 2-oxoglutarate, *OAA* oxaloacetate, *PEP* phosphoenolpyruvate, *ADH* alcohol dehydrogenase, *AlaAT* alanine aminotransferase, *AspAT* aspartate aminotransferase, *GABA-T* γ-aminobutyric acid transaminase, *GAD* glutamic acid decarboxylase, *INV* invertase, *LDH* lactate dehydrogenase, *PDC* pyruvate decarboxylase, *SUS* sucrose synthase

accumulating was observed for alanine, valine, leucine, glycine, serine, and tyrosine. The amino acids glutamate, glutamine, aspartate, and asparagine, which are derived from TCA cycle intermediates, were all shown to decrease. This observation supports the hypothesis that hypoxia led to a decrease of the carbon-flux into the TCA cycle, resulting in a redirection of glycolytic carbon into glycolytic-derived amino acids and a decrease of flux into TCA cycle intermediate-derived amino acids. Additionally, changes in amino acids were frequently associated with dynamic changes in the levels of transcripts encoding enzymes involved in amino acid metabolism. Interestingly, changes in metabolite levels were described for both roots and leaves, whereas changes in transcript abundance were only observed in the roots. It was concluded that the coordinated physiological and molecular responses in leaves and roots, coupled with the transport of metabolites, revealed important stress adaptations to ensure survival during long periods of root hypoxia (Kreuzwieser et al. 2009).

A metabolite profiling analysis of waterlogged roots of the flooding-tolerant *Lotus japonicus* confirmed the accumulation of alanine as well as succinate under low oxygen conditions (Rocha et al. 2010a, b). In this study, a model for the metabolic regulation during hypoxia was described. When the oxygen availability goes down, pyruvate from glycolysis will be used as substrate of the enzyme alanine amino transferase to produce alanine and 2-oxoglutarate (2OG). Alanine accumulates while 2OG reacts further to produce succinate via individual reactions of the TCA-cycle. This interpretation is in agreement with the observed steady state levels of primary metabolites measured by GC-MS. The observed accumulation of succinate is explained by the inhibition of the enzyme succinate dehydrogenase (SDH) during hypoxia. As a consequence, the TCA-cycle will not operate as a net cycle but rather as two parallel branches; one oxidative and one reductive. This model explains both alanine and succinate accumulation during hypoxia via reorganization of the TCA cycle, thereby preventing pyruvate accumulation also when the glycolytic activity is upregulated during waterlogging (Rocha et al. 2010a, b). Moreover, the parallel action of an oxidative and reductive branch of the TCA-cycle would drastically affect the net production of mitochondrial NADH, which influences the substrate-dependent activity of the mETC and thus the mitochondrial oxygen consumption.

Work on submergence stress has been extensively investigated in rice (*Oryza sativa*) using the submergence tolerance regulator gene SUBMERGENCE1A (SUB1A), conferred by the presence of the *SUB1A-1* allele of the *SUB1A* gene from the group VII ethylene responsive transcription factors (Xu et al. 2006; Fukao and Bailey-Serres 2008; Fukao et al. 2012). Although the genetic mechanism of submergence survival of rice varieties containing the *SUB1A* gene is known for some years now already, its effect at the metabolite level has only recently been revealed (Barding et al. 2011; 2013). The metabolomes of two different rice genotypes, namely *Oryza sativa ssp. japonica* cv. M202 (*Sub1*) and *Oryza sativa ssp. japonica* cv. M202, were profiled using NMR-technology to compare the metabolic effect of submergence and subsequent recovery of rice in the presence or absence of *SUB1A* (Barding et al. 2011). Under submergence conditions, higher levels of leucine, isoleucine, threonine, valine, glutamine, and glutamate were observed in M202 (without SUB1A expression) as compared to M202 (*Sub1*) (with SUB1A expression), but alanine showed the largest difference between the two varieties. Additionally, Barding and coworkers reported the identification and characterization using ^{1}H NMR of alanylglycine (AlaGly) in rice; however, no significant changes were observed between the rice varieties following submergence. The response of the same varieties of rice plants subjected to submergence stress was later investigated by the same authors using a combined GC-MS and NMR metabolite profiling study (Barding et al. 2013). Due to its high sensitivity as compared to NMR-based methods, metabolites such as phenyalanine, proline, citrate, malate, succinate, or GABA could only be detected using GC-MS. However, glutamine, AlaGly, and *S*-methyl-methionine (SMM) could only be detected using NMR, showing the complementary use of this technique.

A combination of protein and metabolite profiling analysis was used to compare the differences in response to the total absence of oxygen (anoxia) between rice (*Oryza sativa*) and wheat (*Triticum aestivum*) coleoptiles (Shingaki-Wells et al. 2011). This study revealed that rice coleoptiles responded strongly to anoxia, both at the protein and metabolite level, whereas wheat responded to anoxia in a moderate fashion. Significant increases were observed in the abundance of proteins in rice coleoptiles related to protein translation and antioxidant defense, and an accumulation of a set of enzymes involved in serine, glycine, and alanine biosynthesis which correlates with an observed accumulation of these amino acids in anoxic rice. In addition to changes in amino acids, variations in intermediates in the TCA cycle were also observed. In the TCA cycle, the step converting succinate to fumarate by SDH requires the operation of the mETC and the consumption of oxygen. Without oxygen, the TCA cycle will be interrupted at SDH and succinate will accumulate, as observed in both rice and wheat coleoptiles. The consequences and metabolic advantages of the accumulation of succinate under anoxia have been widely discussed in the context of the extra ATP production that can result (Bailey-Serres and Voesenek 2008; Rocha et al. 2010a, b).

5.2 Metabolic Responses to Hypoxia in Bulky Plant Tissues and Developing Plants

Another example of metabolic adaptation that occurs in metabolically active, dense, or bulky plant tissues is the decrease of oxygen consumption in response to changes in the oxygen concentration. This response involves a restriction of respiration and, at the same time, a decrease in ATP consumption that results from the inhibition of a wide range of biosynthetic processes. Such metabolic responses were investigated in detail by Geigenberger et al. (2000) using potato tuber discs incubated in different oxygen concentrations. When the oxygen concentration in the potato tuber discs decreased from 21 to 4 %, extensive inhibition of several biosynthetic processes was observed, such as partial inhibition of sucrose breakdown, glycolysis, and respiration. The inhibition of sucrose breakdown and glycolysis was followed by a small increase in sugars (sucrose, fructose), glycolic intermediates (3-phosphoglycerate, phosphoenolpyruvate, pyruvate), and a small increase in TCA cycle intermediates (isocitrate, 2-oxoglutarate). Under anoxic conditions, the opposite response generally known as the Pasteur Effect was observed, in which glycolysis increased twofold to support fermentation, starch synthesis was strongly inhibited, while the level of lactate, and the ratios lactate:pyruvate and triose-phosphate:glycerate-3-phosphate increased dramatically. Based on these studies, it was concluded that declining internal oxygen leads to: (1) a restriction of glycolysis and respiration which decreases the adenylate energy status; (2) a widespread decrease in biosynthetic activity which decreases ATP consumption; and (3) a switch to pathways that consume less ATP (Geigenberger et al. 2000; Bologa et al. 2003).

Similar conclusions were drawn by van Dongen et al. (2003) in a study of the consequences of endogenous limitations in oxygen delivery for phloem transport in the castor oil plant *Ricinus communis*. Using an oxygen microsensor to measure in situ oxygen tensions, it was observed that decreasing oxygen concentrations led to a progressive increase in the amino acids alanine, GABA, methionine, isoleucine, and to an increase in the succinate to malate ratio in the phloem. They concluded that oxygen concentrations are low inside the phloem leading to metabolic changes in phloem metabolism and function.

Fruits are also considered an interesting model to analyse the metabolic changes due to hypoxic or anoxic conditions, since gradients occur inside the fruit during normal ripening. However, only a few studies have addressed this issue. A metabolomics approach combining ^{1}H NMR and GC-MS metabolite profiling was employed to study the spatial changes of metabolism in melon fruit (*Cucumis melo* L.) in response to hypoxia (Biais et al. 2010). Direct ^{1}H NMR profiling of fresh fruit collected from different sites within the fruit revealed several gradients of metabolites, for example sucrose, alanine, valine, GABA, and the fermentation product ethanol were shown to increase in their concentration from the periphery to the center of the fruit. GC-MS metabolite profiling offered the advantage of revealing gradients of metabolites not detected using ^{1}H NMR, including the organic acids pyruvate and fumarate. The quantification of adenine nucleotides highlighted a strong decrease in both ATP/ADP ratios and the adenylate energy charge from the periphery to the center of the fruit. These concentration patterns are consistent with an increase in ethanol fermentation due to oxygen limitation and were confirmed by observed changes in alanine and GABA concentrations. It was suggested that these gradients may be related to in situ hypoxia in the central part of the fruit during normal ripening. Similar results were previously obtained in pear fruit during hypoxia (Pedreschi et al. 2009), and later in peach fruit in response to anoxia (Lara et al. 2011). The main metabolic changes observed in fruits included induction of the fermentative pathway, glycolysis, and enzymes involved in both sucrose synthesis and degradation.

Developing seeds are metabolically highly active, and a high respiration rate, combined with a limited gas-exchange capability, results in depletion of oxygen inside the seed. Oxygen-sensitive microsensors were used to study seed hypoxic metabolism, and the occurrence of steep gradients in developing seeds was observed (Borisjuk and Rolletschek 2009). An interesting study by Rolletschek et al. (2011) combined noninvasive imaging and modeling approaches that revealed compartmentation of metabolic activity within the internal environment of barley (*Hordeum vulgare*) endosperm. By applying a noninvasive assay based on $^{13}C/^{1}H$-magnetic resonance imaging (MRI), it was shown that during grain filling the primary site of alanine synthesis was the central region of the endosperm and that the caryopsis experienced the highest level of hypoxia. Furthermore, using flux balance analysis it was predicted that in the central region of the endosperm the TCA cycle shifts to a noncyclic mode followed by an increased glycolytic flux and subsequently, accumulation of alanine.

Changes in overall seed metabolite levels during development and maturation and in response to a stepwise decrease of the external oxygen concentration were investigated by Gibon et al. (2002). Decreased external oxygen concentration led to a dramatic decrease of ATP, a small decrease of ADP, and a marked decrease of the ATP/ADP ratio in seeds. This contrasts with the silique wall, where the ATP/ADP ratio did not decrease until the external oxygen concentration was very low. Evidence that seed metabolism becomes oxygen-limited at a relatively high external oxygen concentration was provided by the response of metabolites produced in pathways that recycle NADH. A decrease of external oxygen concentration in the range between 21 and 8 % led to a decrease of aspartate and glutamate and an increase of alanine. Accumulation of lactate and GABA started when the external oxygen concentration fell below about 8 %. In contrast to the seeds, alanine, lactate, and GABA did not increase in the silique walls until external oxygen concentration was below 1 % (Gibon et al. 2002).

In order to better understand the molecular mechanisms underlying hypoxic responses, van Dongen et al. (2008) used combined transcript and metabolite profiling studies to investigate the genomic response of *Arabidopsis thaliana* roots to a mild decrease in oxygen concentrations. Root growth was inhibited and transcript and metabolite profiles were significantly altered in response to a moderate decrease in oxygen concentrations. The levels of many metabolites increased at low oxygen concentrations, such as the amino acids alanine, proline, and GABA, and the phosphorylated intermediates glucose-6-P and glycerol-3-P.

In summary, without oxygen, the glycolytic pathway is the main source of non-photosynthetic energy production in plants, and lactate and alanine accumulate in oxygen-deprived plant tissues (Drew 1997; Fukao and Bailey-Serres 2004; Ricoult et al. 2006; Limami et al. 2008; Bailey-Serres et al. 2012). Metabolites such as succinate and γ-amino butyric acid (GABA) are also observed to accumulate in hypoxic plant tissues. The relative abundance of specific end products varies according to the plant species, genotype, and tissue as well as the duration and severity of oxygen limitation (Bailey-Serres and Voesenek 2008; Narsai et al. 2011).

6 Regulation of Enzyme Activity Under Low Oxygen Availability

Metabolic changes are an important part of the physiological adaptations that plants are able to perform under low oxygen conditions. Many studies have shown the rearrangements in the steady state levels of metabolic intermediates that take place under hypoxic or anoxic stress in plants (Narsai et al. 2011). However, the changes in metabolite levels result from modifications in the fluxes of primary metabolic pathways, which derive from increased or decreased enzymatic activity rates. These reaction rates depend on several factors, like substrate/product ratios, the specific kinetic characteristics of each enzyme isoform, the sum of all active enzymes present in a cellular compartment, and different kinds of posttranslational regulatory mechanisms (Brown et al. 2005), such as phosphorylation, acetylation, glutathionylation, disulfide bonding, allosteric regulation, modification of subunit composition, etc.

6.1 *Enzymatic Adaptation of Primary Metabolism During Oxygen Limitation*

The best known mechanism of enzymatic adaptation of primary metabolism to low oxygen availability is the activation of lactate dehydrogenase, pyruvate decarboxylase, and alcohol dehydrogenase. This biochemical adaptation of metabolism contributes to glycolytic ATP synthesis when the tissue is impeded in NAD^+-regeneration, which is needed to maintain the glycolytic flux (Bailey-Serres and Voesenek 2008; Roberts et al. 1984). The activity of this pathway becomes relevant, when the glycolytic flux increases and pyruvate starts to accumulate due to the decrease of the pyruvate dehydrogenase activity, which is inhibited by the high reduction level of the $NADH/NAD^+$ pool and the accumulation of TCA-cycle intermediates (Drew 1997). When pyruvate levels start to increase, the activity of the enzyme lactate dehydrogenase is stimulated as compared to the enzyme pyruvate dehydrogenase, since the affinity of lactate dehydrogenase to pyruvate is much lower than the affinity of the pyruvate dehydrogenase (Dat et al. 2004). However, under continuous oxygen limitation, an accumulation of lactic acid contributes to the overall accumulation of organic acids in the cell. The increased levels of organic acids and the characteristic reduction of the ATP levels under hypoxia/anoxia limit the activity of the vacuolar and plasma membrane H^+-ATPases (Felle 2005) and lead to a drop in cellular pH. This acidification inhibits the lactate dehydrogenase and activates pyruvate decarboxylase (Tihanyi et al. 1989; Gibbs and Greenway 2003; Bailey-Serres and Voesenek 2008). This shift from lactic fermentation to ethanolic fermentation varies among plant species, while hypoxia-tolerant plants have often higher affinities of pyruvate decarboxylase to pyruvate and a higher pH optimum of this enzyme compared to hypoxia-sensitive plants (Lee and Langston-Unkefer 1985; Rivoal et al. 1990). This adaptation helps the hypoxia-tolerant plants to shift from lactic to ethanolic fermentation earlier during development of hypoxic conditions and to avoid or reduce the deleterious effects of cytosolic acidification.

Another pH-regulated step in primary catabolic metabolism is the conversion of PEP into pyruvate by the cytosolic pyruvate kinase enzyme. Plaxton (1996) and Podestá and Plaxton (1991) suggested that the activity of pyruvate kinase in *Ricinus communis* seedlings increased during hypoxic germination caused by the drop in cytosolic pH. The acidification made the enzyme less sensitive to allosteric inhibitors, such as ATP. An increase in the activity of this enzyme would be important to maintain the ATP production derived from glycolysis and it is in agreement with several studies that show an increased glycolytic flux when oxygen availability is reduced (Geigenberger 2003).

The pyruvate dehydrogenase complex is the main entrance of substrates into the TCA cycle during normoxic conditions and therefore a very important regulatory point of respiratory metabolism. It has been shown that this enzyme complex is strongly inhibited by the concentration of its products NADH and acetyl-CoA (Tovar-Méndez et al. 2003). For this reason, a decrease in oxygen and concomitant mitochondrial electron transport lead to an accumulation of NADH which diminishes the rate of reaction catalyzed by the pyruvate dehydrogenase complex.

Among the regulation points in plant energy metabolism, the TCA-cycle enzyme activities are also modulated by various components. It is known that during oxygen-limiting conditions, the metabolic activity of the TCA-cycle is reduced, and that specific enzymatic steps of the cycle are slowed down or even blocked in an oxygen concentration depending manner (Kennedy et al. 1992). One important regulatory mechanism in the TCA cycle is the adaptation of the metabolic flux by the redox regulation of specific enzymes. All enzymes involved in the reactions of the TCA-cycle have been documented to be inhibited by high concentrations of NADH, although not all of them use NAD^+ as substrate for their reaction (Araújo et al. 2012). During hypoxic conditions, the NADH consumption by the mETC is reduced and the concomitant accumulation of NADH in the cell will inhibit the metabolic flux through the TCA-cycle, thereby diminishing the NADH production via a negative-feedback loop. In this way, the activity of the TCA-cycle is inherent to the activity of the mETC and to the oxygen availability in the cell.

6.2 Reactive Oxygen Species

Several reactions of respiratory metabolism can be regulated by the redox status of the cell and the production of ROS. It is therefore highly interesting that a decrease of the oxygen availability can lead to the production in ROS. Pucciariello et al. (2012) and Baxter-Burrell et al. (2002) showed that slow decreases of the oxygen concentration led to the activation of NADPH oxidases, thus inducing an oxidative burst just before oxygen became limiting as substrate of energy metabolism. The hydrogen peroxide that is formed as a result of this oxidative burst could be a regulatory signal in the TCA-cycle. It has been already shown that the TCA-cycle enzyme aconitase is highly sensitive towards ROS, and that its activity is strongly inhibited by the presence of hydrogen peroxide (Verniquet et al. 1991). By this mechanism, aconitase would be one of the enzymes that become less active when the oxygen concentration decreases.

In a study by Sweetlove et al. (2002) it was reported that the amount of an important set of mitochondrial proteins diminished after treatment of Arabidopsis cell cultures with hydrogen peroxide, menadione, or antimycin A. Among these proteins different enzymes of primary catabolic metabolism were found, like lipoic acid containing subunits of the pyruvate dehydrogenase complex, the 2-oxoglutarate dehydrogenase complex, aconitase, malate dehydrogenase, succinyl-CoA ligase, and several subunits of the ATP synthase. This supports the idea that the regulation of respiration by changes in metabolic fluxes can be regulated at the onset of hypoxia/anoxia by the formation of ROS.

6.3 Regulation of the Mitochondrial Electron Transport Chain

Among the several adaptation mechanisms that take place during low oxygen stress, it was found that the electron transport chain in mitochondria can undergo changes in the activity of the different complexes and other proteins of the alternative pathways. These changes have not been completely elucidated; however, some modifications in the complexes were already found. For example, the exchange of subunits within complexes has been documented; many proteins of the complexes of the mETC have more than one isoform and the expression level of the genes coding for these subunits can vary under developmental and environmental conditions (Millar et al. 2004). In most of the cases, the physiological function of these changes is not clear yet, although a regulatory function for metabolic activity seems likely.

Posttranslational modifications of the proteins of the mETC have also been observed. These modifications, like disulfide-bond formation, separation (Millar et al. 1993; Holtzapffel et al. 2003), or phosphorylation of specific subunits (Bykova et al. 2003; Ito et al. 2009) of the complexes lead to alterations in the kinetic properties of the enzyme complexes and are examples of regulatory mechanisms of respiratory metabolism. Some examples of protein phosphorylation were shown for subunits of complex II, complex III, and complex V (Bykova et al. 2003; Ito et al. 2009). The function of this is still unclear, although phosphorylation of subunits of complex IV is a probable regulatory mechanism for the rate of electron transfer across the mETC: Fang et al. (2007) found that the phosphorylation status of specific complex IV subunits from rabbit heart changed by the treatment with ischemia/reperfusion, which lead to a reduction of the oxygen levels of the treated cells. Similar changes in activity rate or substrate affinity due to posttranslational modifications in complexes of the mETC in plants are very well possible though yet to be elucidated.

Some further mechanisms for regulation of the activity of the mETC are well described. For instance, the accumulation of pyruvate levels as a product of an enhanced glycolytic flux under hypoxic conditions could activate the AOX. Pyruvate is an alpha-ketoacids which can bind to the reduced form of a cysteine residue in the N-terminal part of the protein (Rhoads et al. 1998; Berthold et al. 2000).

This binding changes the covalent interaction between the subunits of the AOX homo-dimer. This interaction is crucial for the redox(de-)activation of AOX (Umbach et al. 2006) and can be further enhanced by the overall higher reduction levels in the redox equivalent-pools, NAD^+, and ubiquinone, which would facilitate the reduction of the disulfide bridge between the monomers of the AOX homo-dimer (Umbach et al. 2006). However, in most plant species the affinity of AOX to oxygen is one to two orders of magnitude lower compared to the affinity of Cytochrome c oxidase (COX) to oxygen. For this reason, it is not expected that AOX competes with COX for oxygen as substrate during low-oxygen conditions. Therefore, it is unlikely that the ATP production efficiency of the mETC is compromised at low oxygen levels by AOX activity.

Other possible regulatory mechanisms of respiration activity under low oxygen conditions include the formation of nitric oxide (NO) due to its inhibitory effect on the activity of several enzymes of glycolysis and the TCA cycle (Navarre et al. 2000), as well as on complexes of the mETC, especially on COX. Under strong hypoxic conditions (near anoxia) nitrite can be used by the mETC as an alternative electron acceptor and as a result of this reduction NO is formed (Planchet et al. 2005). Additionally, NO can be produced by nitrate reductase. This reaction is favored by the accumulation of nitrite and a drop in the cellular pH (Gupta et al. 2011). These conditions occur especially when the oxygen availability is low. Under these circumstances, the inhibition of the activity of the mETC (especially by the inhibition of COX and of several key enzymes in the primary catabolic metabolism) (Rocha et al. 2010a, b; Gardner et al. 1997; Navarre et al. 2000; Millar and Day 1997; Yamasaki et al. 2001) could contribute to the decrease of the metabolic flux and consequently leads to the reduction of the rate of oxygen consumption by respiration under reduced oxygen conditions.

Another regulatory mechanism of the action of the mETC are respiratory supercomplexes (Eubel et al. 2004; Schäfer et al. 2007). The complexes complex I, III, and IV of the mETC can be ordered in protein superstructures with specific configuration and stoichiometry to form so-called respirasomes. Modifications in the composition of the different supercomplexes in the mETC were suggested to result in modifications of the electron pathway through the different complexes thereby affecting the respiratory rate and ATP production efficiency via electron partition. For this reason, it is suggested that the decrease in the respiration rate, occurring during a drop in the oxygen levels, could be further facilitated by separation of the respirasomes, as shown by Ramirez-Aguilar et al. (2011). Supercomplex dissociation would shift the electron partitioning from a more cytochrome c-pathway-based respiration into an increased activity of the proteins of the alternative pathways (compare Fig. 1). An increase in the activity of the type II NAD(P)H dehydrogenases is based on the reduced specificity against NADH and NADPH among the different dehydrogenases and to an increased activity at lower pH levels (Rasmusson et al. 2008). Additionally, the shift in electron partitioning towards the alternative pathways during hypoxia is further supported by the inhibition of complex I via lower pH levels (Hatefi 1985). Furthermore, the separation of the respirasomes, caused by the pH drop, opens the binding sites for ubiquinol within the supercomplex, increasing the probability of ubiquinol to be used as substrate by the AOX. The most probable advantage of the modification in the supercomplex composition for the plant is to diminish oxidative damage caused by the production of ROS when oxygen is reintroduced after a period of hypoxic stress (Turrens 2004). Moreover, the activation of the external type II NADH dehydrogenases supports the NAD^+ resupply in the cytosol, which is important to maintain the ATP production resulting from the increased glycolytic flux present under hypoxia.

7 Hypotheses of Regulatory Mechanism for Respiratory Oxygen Consumption

While the previous section described the research studies showing how oxygen availability can affect the respiratory oxygen consumption rate, this section focuses on the ongoing debate about whether the oxygen-consumption rate is proactively regulated or not in response to hypoxia. Two hypotheses were suggested. One postulates that kinetic parameters of respiratory metabolism (components of mETC like COX and AOX, but also of glycolysis and TCA cycle) can change depending on the availability of oxygen. The second supposes that just Michaelis-Menten kinetics linked to diffusive oxygen transport inside the tissue are sufficient to describe the respiratoty activity as a function of the environmental oxygen concentration. Both hypothesis have their merit, but a combination of both could be likely as well.

Studies in yeast (Burke and Poyton 1998) and mammals (Chandel et al. 1996; Gnaiger et al. 1998; Fukuda et al. 2007; Galkin et al. 2009; Solaini et al. 2010) provide evidence for the first hypothesis, e.g., it was shown in isolated mitochondria from animal tissue, that the affinity for oxygen as well as the maximum flux of oxygen can alter as a function of the metabolic state (Gnaiger et al. 1998).

A main problem in this debate still is that due to current technical limitations, it is not possible to measure oxygen concentrations inside plant cells or mitochondria. This hinders the exact correlation between the oxygen affinity constant (Km) of plant oxidases (COX and AOX) and the subcellular, in vivo oxygen concentration at the reaction site. Therefore, these values can only be estimated using indirect measurements, such as invasive sensor methods for the determination of oxygen concentration inside tissues (Armstrong et al. 1994; van Dongen et al. 2003) and the rate of oxygen consumption from the buffer solution that surrounds the tissue (Zabalza et al. 2009; Gupta et al. 2009).

In order to demonstrate the value of the second hypothesis, Armstrong and Beckett (2011) used a mathematical model to describe the data from Zabalza et al. (2009), which links a MMK for oxygen consumption and diffusive oxygen uptake. This study of Armstrong and Beckett (2011) provides a first and essential step towards an accurate model of oxygen consumption in plant tissues. With this model, oxygen consumption is described using partial differential equations (PDEs) based on Michaelis-Menten kinetics with variable phenomenological parameters across eight considered spatial gradients. In an attempt to verify one of both existing hypotheses, Armstrong and Beckett fit the experimental data to a model based on the distribution of respiratory demand and oxygen diffusion rates through various empirically determined zones within the root in combination with a fixed value for the oxygen affinity of COX. The mathematical formula with which the authors come up with describes the experimental data well. The authors therefore conclude that there is no need for any other explanation anymore. However, various critical components such as calibrating the model through parameter-estimation, measuring the stability through perturbation tests, and validation in different environments and tissue types have not been sufficiently addressed. Further theoretical and experimental analysis remains required therefore.

Another modeling approach was provided by Ho et al. (2011). They used a multiscale gas-exchange model for carrying out in silico experiments in fruits by combining the microscale geometry of the tissue and local concentrations in the cells from macroscopic gas concentration profiles. This approach allows the cell metabolism in any plant organ, during hypoxic and anoxic conditions, to be analyzed efficiently. They demonstrated in their model, that in apple (*Malus* x *domestica*) parenchyma tissue, the cellular gas exchange was bit influenced by the gas-exchange rates through the cell wall and cell membranes. This observation provides strong evidence that carbondioxide is imported via alternative mechanisms in the cells; however, oxygen uptake occurs through diffusion from intercellular spaces. Furthermore, the gas concentration gradient is predicted to be steeper in the cells than in the surrounding air, which indicates passive transport mainly through intercellular spaces and not through cells.

The debate is still not finished and alternative approaches to model oxygen uptake and consumption are summarized and discussed by Nikoloski and van Dongen (2011). However, the results discussed demonstrate that respiration is quite complex and understanding the mechanism by which respiration is regulated under oxygen limitations remains limited, but a combination of oxygen gradients and substrate limitation with adaptations in respiratory metabolism seems to be the most probable.

8 Conclusion

Even under optimal growth conditions, the availability of oxygen to plant cells can be restricted due to high resistances to diffusion of oxygen through the tissue. The ability of plants to regulate their respiratory oxygen consumption upon hypoxic condition to avoid or postpone the depletion of oxygen is therefore most important. Some mechanisms to survive prolonged periods of low-oxygen stress by conserving carbon within the plant were discussed. Upon hypoxia, a series of drastic metabolic adaptations are initiated in plants. The best known is the activation of fermentation and up-regulation of glycolytic activity to increase the yield of ATP from the glycolytic pathway. Alternative carbon flux that would explain the accumulation of alanine, GABA, and succinate upon hypoxia via pathways mediated by an Ala- and GABA-shunt was suggested. Additionally, a decrease of the production of redox equivalents in the mitochondria could explain the down-regulation of respiratory oxygen consumption during hypoxic conditions.

References

Apel K, Hirt H (2004) Reactive oxygen species: metabolism, oxidative stress, and signal transduction. Annu Rev Plant Biol 55:373–399

Araújo WL, Ishizaki K, Nunes-Nesi A, Tohge T, Larson TR, Krahnert I, Balbo I, Witt S, Dörmann P, Graham IA, Leaver CJ, Fernie AR (2011) Analysis of a range of catabolic mutants provides evidence that phytanoyl-coenzyme A does not act as a substrate of the electron-transfer flavoprotein/electron-transfer flavoprotein:ubiquinone oxidoreductase complex in Arabidopsis during dark-induced senescence. Plant Physiol 157:55–69

Araújo WL, Nunes-Nesi A, Nikoloski Z, Sweetlove LJ, Fernie AR (2012) Metabolic control and regulation of the tricarboxylic acid cycle in photosynthetic and heterotrophic plant tissues. Plant Cell Environ 35:1–21

Armstrong W, Armstrong J (2014) Plant internal oxygen transport (diffusion and convection) and measuring and modelling oxygen gradients. In: van Dongen JT, Licausi F (eds) Low oxygen stress in plants. Springer, Heidelberg

Armstrong W, Beckett PM (2011) Experimental and modelling data contradict the idea of respiratory down-regulation in plant tissues at an internal [O2] substantially above the critical oxygen pressure for cytochrome oxidase. New Phytol 190:431–441

Armstrong W, Strange ME, Cringle S, Beckett PM (1994) Microelectrode and modelling study of oxygen distribution in roots. Ann Bot 74:287–299

Ast C, Draaijer A (2014) Oxygen sensing in plants. In: van Dongen JT, Licausi F (eds) Low oxygen stress in plants, vol 21, Plant cell monographs. Springer, Heidelberg

Bailey-Serres J, Voesenek LA (2008) Flooding stress: acclimations and genetic diversity. Annu Rev Plant Biol 59:313–339

Bailey-Serres J, Fukao T, Gibbs DJ, Holdsworth MJ, Lee SC, Licausi F, Perata P, Voesenek L, van Dongen JT (2012) Making sense of low oxygen sensing. Trends Plant Sci 17:129–138

Barding GA Jr, Fukao T, Béni S, Bailey-Serres J, Larive CK (2011) Differential metabolic regulation governed by the rice *SUB1A* gene during submergence stress and identification of alanylglycine by ^{1}H NMR spectroscopy. J Proteome Res 11:320–330

Barding GA Jr, Béni S, Fukao T, Bailey-Serres J, Larive CK (2013) Comparison of GC-MS and NMR for metabolite profiling of rice subjected to submergence stress. J Proteome Res 12:898–909

Baxter-Burrell A, Yang Z, Springer PS, Bailey-Serres J (2002) RopGAP4-dependent Rop GTPase Rheostat control of Arabidopsis oxygen deprivation tolerance. Science 296(5575):2026–2028

Benamar A, Rolletschek H, Borisjuk L, Avelange-Macherel MH, Curien G, Mostefai HA, Andriantsitohaina R, Macherel D (2008) Nitrite-nitric oxide control of mitochondrial respiration at the frontier of anoxia. Biochim Biophys Acta 1777(10):1268–1275

Berthold DA, Andersson ME, Nordlund P (2000) New insights into the structure and function of the alternative oxidase. Biochim Biophys Acta 1460:241–254

Biais B, Beauvoit B, Allwood J, Deborde C, Maucourt M, Goodacre R, Rolin D, Moing A (2010) Metabolic acclimation to hypoxia revealed by metabolite gradients in melon fruit. J Plant Physiol 167:242–245

Blokhina OB, Chirkova TV, Fagerstedt KV (2001) Anoxic stress leads to hydrogen peroxide formation in plant cells. J Exp Bot 52:1179–1190
Bologa KL, Fernie AR, Leisse A, Loureiro ME, Geigenberger P (2003) A bypass of sucrose synthase leads to low internal oxygen and impaired metabolic performance in growing potato tubers. Plant Physiol 132:2058–2072
Borecký J, Maia IG, Costa ADT, Jezek P, Chaimovich H, de Andrade PBM, Vercesi AE, Arruda P (2001) Functional reconstitution of Arabidopsis thaliana plant uncoupling mitochondrial protein (AtPUMP1) expressed in Escherichia coli. FEBS Lett 505:240–244
Borisjuk L, Rolletschek H (2009) The oxygen status of the developing seed. New Phytol 182 (1):17–30
Brown GC, Hafner RP, Brand MD (2005) A 'top-down' approach to the determination of control coefficients in metabolic control theory. Eur J Biochem 188:321–325
Burke PV, Poyton RO (1998) Structure/function of oxygen-regulated isoforms in cytochrome c oxidase. J Exp Biol 201:1163–1175
Bykova NV, Egsgaard H, Møller IM (2003) Identification of 14 new phosphoproteins involved in important plant mitochondrial processes. FEBS Lett 540(1–3):141–146
Chandel NS, Budinger GR, Schumacker PT (1996) Molecular oxygen modulates cytochrome c oxidase function. J Biol Chem 271:18672–18677
Clifton R, Lister R, Parker KL, Sappl PG, Elhafez D, Millar AH, Day DA, Whelan J (2005) Stress-induced co-expression of alternative respiratory chain components in Arabidopsis thaliana. Plant Mol Biol 58:193–212
Colmer TD, Pedersen O (2008) Oxygen dynamics in submerged rice (Oryza sativa). New Phytol 178:326–334
Dat JF, Capelli N, Folzer H, Bourgeade P, Badot PM (2004) Sensing and signalling during plant flooding. Plant Physiol Biochem 42:273–282
Drew MC (1997) Oxygen deficiency and root metabolism: injury and acclimation under hypoxia and anoxia. Annu Rev Plant Physiol Plant Mol Biol 48:223–250
Dudkina NV, Heinemeyer J, Sunderhaus S, Boekema EJ, Braun HP (2006) Respiratory chain supercomplexes in the plant mitochondrial membrane. Trends Plant Sci 11(5):232–240
Eubel H, Heinemeyer J, Sunderhaus S, Braun HP (2004) Respiratory chain supercomplexes in plant mitochondria. Plant Physiol Biochem 42(12):937–942
Fang JK, Prabu SK, Sepuri NB, Raza H, Anandatheerthavarada HK, Galati D, Spear J, Avadhani NG (2007) Site specific phosphorylation of cytochrome c oxidase subunits I, IVi1 and Vb in rabbit hearts subjected to ischemia/reperfusion. FEBS Lett 581(7):1302–1310
Felle HH (2005) pH regulation in anoxic plants. Ann Bot 96(4):519–532
Fukao T, Bailey-Serres J (2004) Plant responses to hypoxia—is survival a balancing act? Trends Plant Sci 9:449–456
Fukao T, Bailey-Serres J (2008) Submergence tolerance conferred by *Sub1A* is mediated by SLR1 and SLRL1 restriction of gibberellin responses in rice. Proc Natl Acad Sci U S A 105:16814–16819
Fukao T, Yeung E, Bailey-Serres J (2012) The submergence tolerance gene *SUB1A* delays leaf senescence under prolonged darkness through hormonal regulation in rice. Plant Physiol 160:1795–1807
Fukuda R, Zhang H, Kim J, Shimoda L, Dang CV, Semenza GL (2007) HIF-1 regulates cytochrome oxidase subunits to optimize efficiency of respiration in hypoxic cells. Cell 129:111–122
Galkin A, Abramov AY, Frakich N, Duchen MR, Moncada S (2009) Lack of oxygen deactivates mitochondrial complex I: implications for ischemic injury? J Biol Chem 284:36055–36061
Gardner PR, Costantino G, Szabó C, Salzman AL (1997) Nitric oxide sensitivity of the aconitases. J Biol Chem 272:25071–25076
Geigenberger P (2003) Response of plant metabolism to too little oxygen. Curr Opin Plant Biol 6:247–256

Geigenberger P, Fernie AR, Gibon Y, Christ M, Stitt M (2000) Metabolic activity decreases as an adaptive response to low internal oxygen in growing potato tubers. Biol Chem 381:723–740

Gibbs J, Greenway H (2003) Review: mechanisms of anoxia tolerance in plants. I. Growth, survival and anaerobic catabolism. Funct Plant Biol 30(1):1–47

Gibon Y, Vigeolas H, Tiessen A, Geigenberger P, Stitt M (2002) Sensitive and high throughput metabolite assays for inorganic pyrophosphate, ADPGlc, nucleotide phosphates, and glycolytic intermediates based on a novel enzymic cycling system. Plant J 30:221–235

Gnaiger E, Lassnig B, Kuznetsov AV, Margreiter R (1998) Mitochondrial respiration in the low oxygen environment of the cell. Effect of ADP on oxygen kinetics. Biochim Biophys Acta 1365:249–254

Gupta KJ, Zabalza A, van Dongen JT (2009) Regulation of respiration when the oxygen availability changes. Physiol Plant 137:383–391

Gupta KJ, Fernie AR, Kaiser WM, van Dongen JT (2011) On the origins of nitric oxide. Trends Plant Sci 16(3):160–168

Hatefi Y (1985) The mitochondrial electron transport and oxidative phosphorylation system. Annu Rev Biochem 54:1015–1069

Heazlewood JL, Millar AH (2005) AMPDB: the Arabidopsis mitochondrial protein database. Nucleic Acids Res 33:605–610

Ho QT, Verboven P, Verlinden BE, Schenk A, Delele MA, Rolletschek H, Vercammen J, Nicolaï BM (2010) Genotype effects on internal gas gradients in apple fruit. J Exp Bot 61:2745–2755

Ho QT, Verboven P, Verlinden BE, Herremans E, Wevers M, Carmeliet J, Nicolaï BM (2011) A three-dimensional multiscale model for gas exchange in fruit. Plant Physiol 155(3):1158–1168

Holtzapffel RC, Castelli J, Finnegan PM, Millar AH, Whelan J, Day DA (2003) A tomato alternative oxidase protein with altered regulatory properties. Biochim Biophys Acta 1606 (1–3):153–162

Hütter E, Renner K, Jansen-Dürr P, Gnaiger E (2002) Biphasic oxygen kinetics of cellular respiration and linear oxygen dependence of antimycin A inhibited oxygen consumption. Mol Biol Rep 29:83–87

Ito J, Taylor NL, Castleden I, Weckwerth W, Millar AH, Heazlewood JL (2009) A survey of the Arabidopsis thaliana mitochondrial phosphoproteome. Proteomics 9(17):4229–4240

Kennedy RA, Rumpho ME, Fox TC (1992) Anaerobic metabolism in plants. Plant Physiol 100 (1):1–6

Kimmerer TW, Stringer MA (1988) Alcohol dehydrogenase and ethanol in the stems of trees: evidence for anaerobic metabolism in the vascular cambium. Plant Physiol 87:693–697

Kreuzwieser J, Hauberg J, Howell KA, Carroll A, Rennenberg H, Millar AH, Whelan J (2009) Differential response of gray poplar leaves and roots underpins stress adaptation during hypoxia. Plant Physiol 149:461–473

Laisk A, Oja V, Eichelmann H (2007) Kinetics of leaf oxygen uptake represent in planta activities of respiratory electron transport and terminal oxidases. Physiol Plant 131:1–9

Lara MV, Budde CO, Porrini L, Borsani J, Murray R, Andreo CS, Drincovich MF (2011) Peach (Prunus persica) fruit response to anoxia: reversible ripening and biochemical changes. Plant Cell Physiol 52(2):392–403

Lee TC, Langston-Unkefer PJ (1985) Pyruvate decarboxylase from Zea mays L.: I. Purification and partial characterization from mature kernels and anaerobically treated roots. Plant Physiol 79(1):242–247

Licausi F, Giorgi FM, Schmälzlin E, Usadel B, Perata P, van Dongen JT, Geigenberger P (2011) HRE-type genes are regulated by growth-related changes in internal oxygen concentrations during the normal development of potato (Solanum tuberosum) tubers. Plant Cell Physiol 52 (11):1957–1972

Limami AM, Glévarec G, Ricoult C, Clique J-B, Planchet E (2008) Concerted modulation of alanine and glutamate metabolism in young Medicago truncatula seedlings under hypoxic stress. J Exp Bot 59:2325–2335

Mancuso S, Boselli M (2002) Characterisation of the oxygen fluxes in the division, elongation and mature zones of Vitis roots: influence of oxygen availability. Planta 214:767–774
Maricle BR, Lee RW (2007) Root respiration and oxygen flux in salt marsh grasses from different elevational zones. Mar Biol 151:413–423
Maxwell DP, Wang Y, McIntosh L (1999) The alternative oxidase lowers mitochondrial reactive oxygen production in plant cells. Proc Natl Acad Sci U S A 14:8271–8276
McDonald AE, Vanlerberghe GC (2006) Origins, evolutionary history, and taxonomic distribution of alternative oxidase and plastoquinol terminal oxidase. Comp Biochem Physiol Part D Genomics Proteomics 1:357–364
Millar AH, Day DA (1997) Nitric oxide inhibits the cytochrome oxidase but not the alternative oxidase of plant mitochondria. FEBS Lett 398(2–3):155–158
Millar AH, Wiskich JT, Whelan J, Day DA (1993) Organic acid activation of the alternative oxidase of plant mitochondria. FEBS Lett 329:259–262
Millar AH, Hoefnagel MHN, Day DA, Wiskich JT (1996) Specificity of the organic acid activation of alternative oxidase in plant mitochondria. Plant Physiol 111:613–618
Millar A, Eubel H, Jänsch L, Kruft V, Heazlewood J, Braun HP (2004) Mitochondrial cytochrome c oxidase and succinate dehydrogenase complexes contain plant specific subunits. Plant Mol Biol 56(1):77–90
Mommer L, Visser EJ (2005) Underwater photosynthesis in flooded terrestrial plants: a matter of leaf plasticity. Ann Bot 96:581–589
Narsai R, Rocha M, Geigenberger P, Whelan J, van Dongen JT (2011) Comparative analysis between plant species of transcriptional and metabolic responses to hypoxia. New Phytol 190 (2):472–487
Navarre DA, Wendehenne D, Durner J, Noad R, Klessig DF (2000) Nitric oxide modulates the activity of tobacco aconitase. Plant Physiol 122(2):573–582
Nikoloski Z, van Dongen JT (2011) Modeling alternatives for interpreting the change in oxygen-consumption rates during hypoxic conditions. New Phytol 190(2):273–276; author reply 276–278
Ober ES, Sharp RE (1996) A microsensor for direct measurement of O2 partial pressure within plant tissues. J Exp Bot 47:447–457
Pedreschi R, Franck C, Lammertyn J, Erban A, Kopka J, Hertog M, Verlinden B, Nicolai B (2009) Metabolic profiling of 'Conference' pears under low oxygen stress. Postharvest Biol Technol 51:123–130
Planchet E, Gupta JK, Sonoda M, Kaiser WM (2005) Nitric oxide emission from tobacco leaves and cell suspensions: rate limiting factors and evidence for the involvement of mitochondrial electron transport. Plant J 41(5):732–743
Plaxton WC (1996) The organization and regulation of plant glycolysis. Annu Rev Plant Physiol Plant Mol Biol 47:185–214
Podestá FE, Plaxton WC (1991) Kinetic and regulatory properties of cytosolic pyruvate kinase from germinating castor oil seeds. Biochem J 279(Pt 2):495–501
Porterfield DM, Kuang A, Smith PJS, Crispi ML, Musgrave ME (1999) Oxygen-depleted zones inside reproductive structures of Brassicaceae: implications for oxygen control of seed development. Can J Bot 77:1439–1446
Pucciariello C, Parlanti S, Banti V, Novi G, Perata P (2012) Reactive oxygen species-driven transcription in Arabidopsis under oxygen deprivation. Plant Physiol 159(1):184–196
Puntarulo S, Cederbaum AI (1988) Effect of oxygen concentration on microsomal oxidation of ethanol and generation of oxygen radicals. Biochem J 251:787–794
Ramirez-Aguilar SJ, Keuthe M, Rocha M, Fedyaev VV, Kramp K, Gupta KJ, Rasmusson AG, Schulze WX, van Dongen JT (2011) The composition of plant mitochondrial supercomplexes changes with the oxygen availability. J Biol Chem 286:43045–43053
Rasmusson AG, Sool KL, Ethon TE (2004) Alternative NAD(P)H dehydrogenases of plant mitochondria. Annu Rev Plant Biol 55:23–39

Rasmusson AG, Geisler DA, Møller IM (2008) The multiplicity of dehydrogenases in the electron transport chain of plant mitochondria. Mitochondrion 8(1):47–60, Unique aspects of plant mitochondria

Rasmusson AG, Fernie AR, van Dongen JT (2009) Alternative oxidase: a defence against metabolic fluctuations? Physiol Plant 137:371–382

Rhoads DM, Umbach AL, Sweet CR, Lennon AM, Rauch GS, Siedow JN (1998) Regulation of the cyanide-resistant alternative oxidase of plant mitochondria. Identification of the cysteine residue involved in alpha-keto acid stimulation and intersubunit disulfide bond formation. J Biol Chem 13:30750–30756

Ribas-Carbo M, Berry JA, Yakir D, Giles L, Robinson SA, Lennon AM, Siedow JN (1995) Electron partitioning between the cytochrome and alternative pathways in plant mitochondria. Plant Physiol 109:829–837

Ricoult C, Echeverria LO, Cliquet JB, Limami AM (2006) Characterization of alanine aminotransferase (AlaAT) multigene family and hypoxic response in young seedlings of the model legume Medicago truncatula. J Exp Bot 57:3079–3089

Rivoal J, Ricard B, Pradet A (1990) Purification and partial characterization of pyruvate decarboxylase from Oryza sativa L. Eur J Biochem 194(3):791–797

Roberts JKM, Callis J, Wemmer D, Walbot V, Jardetzky O (1984) Mechanism of cytoplasmic pH regulation in hypoxic maize root tips and its role in survival under hypoxia. Proc Natl Acad Sci U S A 81:3379–3383

Rocha M, Sodek L, Licausi F, Hameed MW, Dornelas MC, van Dongen JT (2010a) Analysis of alanine aminotransferase in various organs of soybean (*Glycine max*) and in dependence of different nitrogen fertilisers during hypoxic stress. Amino Acids 39(4):1043–1053

Rocha M, Licausi F, Araujo WL, Nunes-Nesi A, Sodek L, Fernie AR, van Dongen JT (2010b) Glycolysis and the TCA-cycle are linked by Alanine aminotransferase during hypoxia induced by waterlogging of Lotus japonicus. Plant Physiol 152(3):1501–1513

Rolletschek H, Borisjuk L, Koschorreck M, Wobus U, Weber H (2002) Legume embryos develop in a hypoxic environment. J Exp Bot 53(371):1099–1107

Rolletschek H, Weschke W, Weber H, Wobus U, Borisjuk L, Rosenfeld E, Beauvoit B (2004) Energy state and its control on seed development: starch accumulation is associated with high ATP and steep oxygen gradients within barley grains. J Exp Bot 55:1351–1359

Rolletschek H, Melkus G, Grafahrend-Belau E, Fuchs J, Heinzel N, Schreiber F, Jakob PM, Borisjuk L (2011) Combined noninvasive imaging and modeling approaches reveal metabolic compartmentation in the barley endosperm. Plant Cell 23(8):3041–54

Schäfer E, Dencher NA, Vonck J, Parcej DN (2007) Three-dimensional structure of the respiratory chain supercomplex I1III2IV1 from bovine heart mitochondria. Biochemistry 46:12579–12585

Schmälzlin E, van Dongen JT, Klimant I, Marmodée B, Steup M, Fisahn J, Geigenberger P, Löhmannsröben HG (2005) An optical multifrequency phase-modulation method using microbeads for measuring intracellular oxygen concentrations in plants. Biophys J 89:1339–1345

Shingaki-Wells RN, Huang S, Taylor NL, Carroll AJ, Zhou W, Millar AH (2011) Differential molecular responses of rice and wheat coleoptiles to anoxia reveal novel metabolic adaptations in amino acid metabolism for tissue tolerance. Plant Physiol 156:1706–1724

Skutnik M, Rychter AM (2009) Differential response of antioxidant systems in leaves and roots of barley subjected to anoxia and post-anoxia. J Plant Physiol 166(9):926–937

Solaini G, Baracca A, Lenaz G, Sgarbi G (2010) Hypoxia and mitochondrial oxidative metabolism. Biochim Biophys Acta Bioenerg 1797:1171–1177

Sweetlove LJ, Heazlewood JL, Herald V, Holtzapffel R, Day DA, Leaver CJ, Millar AH (2002) The impact of oxidative stress on Arabidopsis mitochondria. Plant J 32:891–904

Thomson CJ, Greenway H (1991) Metabolic evidence for stelar anoxia in maize roots exposed to low O2 concentrations. Plant Physiol 96:1294–1301

Tihanyi K, Talbot B, Brzezinski R, Thirion JP (1989) Purification and characterization of alcohol dehydrogenase from soybean. Phytochemistry 28(5):1335–1338
Tovar-Méndez A, Miernyk JA, Randall DD (2003) Regulation of pyruvate dehydrogenase complex activity in plant cells. Eur J Biochem 270(6):1043–1049
Tucker ML, Laties GG (1985) The dual role of oxygen in avocado fruit respiration; kinetic analysis and computer modelling of diffusion-affected respiratory oxygen isotherms. Plant Cell Environ 8:117–127
Turrens JF (2004) Mitochondrial formation of reactive oxygen species. J Physiol 552(2):335–344
Umbach AL, Ng VS, Siedow JN (2006) Regulation of plant alternative oxidase activity: a tale of two cysteines. Biochim Biophys Acta 1757(2):135–142
van Bodegom PM, Sorrell BK, Oosthoek A, Bakker C, Aerts R (2008) Separating the effects of partial submergence and soil oxygen demand on plant physiology. Ecology 89(1):193–204
van Dongen JT, Schurr U, Pfister M, Geigenberger P (2003) Phloem metabolism and function have to cope with low internal oxygen. Plant Physiol 131:1529–1543
van Dongen JT, Roeb GW, Dautzenberg M, Fröhlich A, Vigeolas H, Minchin PEH, Geigenberger P (2004) Phloem import and storage metabolism are highly coordinated by the low oxygen concentrations within developing wheat seeds. Plant Physiol 135:1809–1821
van Dongen JT, Fröhlich A, Ramírez-Aguilar SJ, Schauer N, Fernie AR, Erban A, Kopka J, Clark J, Langer A, Geigenberger P (2008) Transcript and metabolite profiling of the adaptive response to mild decreases in oxygen concentration in the roots of arabidopsis plants. Ann Bot 103:269–280
van Dongen JT, Gupta KJ, Ramírez-Aguilar SJ, Araújo WL, Nunes-Nesi A, Fernie AR (2011) Regulation of respiration in plants: a role for alternative metabolic pathways. J Plant Physiol 168(12):1434–1443
Vanlerberghe GC, McIntosh L (1997) Alternative oxidase: from gene to function. Annu Rev Plant Physiol Plant Mol Biol 48:703–734
Vercesi AE, Borecký J, Maia Ide G, Arruda P, Cuccovia IM, Chaimovich H (2006) Plant uncoupling mitochondrial proteins. Annu Rev Plant Biol 57:383–404
Verniquet F, Gaillard J, Neuburger M, Douce R (1991) Rapid inactivation of plant aconitase by hydrogen-peroxide. Biochem J 276:643–648
Xu K, Xu X, Fukao T, Canlas P, Maghirang-Rodriguez R, Heuer S, Ismail AM, Bailey-Serres J, Ronald PC, Mackill DJ (2006) Sub1A is an ethylene-response-factor-like gene that confers submergence tolerance to rice. Nature 442(7103):705–708
Yamasaki H, Shimoji H, Ohshiro Y, Sakihama Y (2001) Inhibitory effects of nitric oxide on oxidative phosphorylation in plant mitochondria. Nitric Oxide 5(3):261–270
Zabalza A, van Dongen JT, Fröhlich A, Oliver SN, Faix B, Gupta KJ, Schmälzlin E, Igal M, Orcaray L, Royuela M, Geigenberger P (2009) Regulation of respiration and fermentation to control the plant internal oxygen concentration. Plant Physiol 149:1087–1098

Adaptations of Nitrogen Metabolism to Oxygen Deprivation in Plants

Anis M. Limami

Abstract Acclimation of plants to O_2 deprivation depends on their ability to mitigate detrimental effects related to energy crisis and cytosolic acidosis. Accordingly, lactic and ethanol fermentative pathways are activated under low oxygen stress in order to regenerate NAD^+ to maintain a high glycolysis rate that becomes the major route for ATP production. Paradoxically lactic acid worsens cytosolic acidosis and ethanol fermentation drains carbon for the production of a metabolically useless dead-end product. Nitrogen metabolism is profoundly affected by O_2 deprivation. Interestingly hypoxic N metabolism not only contributes to tolerate O_2 deprivation but also mitigates negative effects of lactic and ethanol fermentation. The most salient event is the concerted modulation of alanine and glutamate pathways that allow for the substitution of ATP-dependent enzymes glutamine synthetase (GS) and asparagine synthetase (AS) by alanine aminotransferase (AlaAT) and glutamate oxoglutarate aminotransferase (NADH-GOGAT) as essential enzymes of N assimilation. This adaptation saves ATP, regenerates NAD^+, and saves carbon in the form of alanine, a C/N storage form readily remobilized upon recovery. As for acidosis amelioration, nitrogen metabolism participates in the cellular pH-stat through GABA and putrescine pathways. Alanine accumulation contributes indirectly to pH homeostasis by using pyruvate competitively with lactate dehydrogenase.

1 Introduction

Flooding of the root system is known as waterlogging and is a major cause of O_2 deprivation to plants (Bailey-Serres et al. 2012; Bailey-Serres and Voesenek 2008). When soils are saturated with water, the root environment becomes hypoxic or

A.M. Limami (✉)
Institut de Recherche en Horticulture et Semences (IRHS), UMR 1345, INRA/Agrocampus Ouest/Université d'Angers, 2 Bd Lavoisier, 49045 Angers cedex, France
e-mail: anis.limami@univ-angers.fr

J.T. van Dongen and F. Licausi (eds.), *Low-Oxygen Stress in Plants*, Plant Cell Monographs 21, DOI 10.1007/978-3-7091-1254-0_11, © Springer-Verlag Wien 2014

anoxic due to the insufficient diffusion of O_2 through water and the competition for O_2 with respiring microorganisms. Gases diffuse approximately 10,000 times slower in water than in air (Drew 1997; Jackson 1985) and as a consequence the flux of oxygen into plants becomes too slow to support respiration, resulting in energy deficits and, eventually, death of cells and tissues in non-adapted plants (Gout et al. 2001; Jackson and Armstrong 1999). Major disorders caused by hypoxia are related to (1) an energy crisis due to the inhibition of mitochondrial oxidative phosphorylation and subsequent reduction of the cellular ATP/ADP ratio and the adenylate energy charge ([ATP + 0.5ADP]/[ATP + ADP + AMP]) (Greenway and Gibbs 2003) and (2) cytoplasm acidification, a major determinant in intolerance to O_2 deficiency (Roberts et al. 1989) caused by the release of H^+ accompanying hydrolysis of the pools of Mg-nucleoside triphosphate (NTP) and sugar phosphates, impaired functioning of the plasma membrane H^+-pumping ATPase (Gout et al. 2001), accumulation of non-processed acidic intermediates like glycolytic compounds (Felle 1996), and a poor CO_2 removal (Saglio et al. 1999). The initial cellular reaction to cope with this energy crisis in both tolerant and intolerant species relies on the acceleration of glycolysis and lactate and ethanol fermentation to generate ATP and regenerate NAD^+. Paradoxically the onset of fermentation worsens cytoplasm acidification due to lactate synthesis by lactate dehydrogenase (LDH) (Davies 1987; Davies and Patil 1974) and the production of acetaldehyde by pyruvate decarboxylase (PDC) a highly reactive chemical affecting cellular damage by forming acetaldehyde–protein adducts (Braun et al. 1995; Jackson 1985; Jackson and Armstrong 1999). Furthermore higher rates of glycolysis and ethanol fermentation induce a faster depletion of sugar stores and thus carbon-starvation stress. Ethanol produced by alcohol dehydrogenase (ADH) is a dead-end product that either accumulates or leaks out of the tissue representing in both cases a net loss of carbon skeletons. Nevertheless it seems that the regeneration of NAD^+ by fermentative enzymes ADH and LDH is vital for hypoxia/anoxia tolerance because in the absence of NAD^+ glycolysis ceases (Albrecht et al. 2004; Ismond et al. 2003; Kursteiner et al. 2003).

The difference between tolerant and intolerant plants to O_2 deprivation is dependent on their ability to mitigate damaging effects of energy crisis and acidosis by reducing energy requirement for tissue maintenance and setting effective control of cytoplasmic pH (Greenway and Gibbs 2003). The reduction of energy consumption includes the reduction of storage compound (proteins, lipids, and starch) synthesis, preferential use of PPi-dependent enzymes like pyruvate orthophosphate dikinase (PPDK) instead of pyruvate kinase (PK) for pyruvate synthesis, and sucrose synthase pathway instead of invertase pathway for sucrose catabolism (Bailey-Serres and Voesenek 2008). Major functions involved in the control of cytoplasmic pH involve the extrusion of protons through H^+-ATPases located at the plasma membrane and through H^+-ATPases and H^+-PPiases located at the tonoplast allowing for the generation of free energy gradient for uptake of the strong cation K^+ (Greenway and Gibbs 2003): decarboxylation of organic acids contributes also to protons removal (Gout et al. 2001; Greenway and Gibbs 2003; Roberts et al. 1989). In *Oryza sativa* var arborio, a species highly resistant to O_2 deprivation, acidification in shoots (pH 7.4–7.0) stabilized after 10 min of anoxia and alkalinization of both cytoplasm and vacuole followed thereafter. In contrast in O_2-deprivation-intolerant wheat (*Triticum aestivum* var MEK shoots), the same treatment caused a sharper and progressive cytoplasmic acidification (pH 7.4–6.6) during the anoxia period and there was no vacuolar alkalinization comparable to the one observed in the rice species.

Cellular acclimation to O_2 deprivation is an important issue, to which nitrogen metabolism may contribute. In the present chapter; the contribution of several aspects of nitrogen metabolism, including nitrate reduction and amino acids as well as polyamine metabolism, to the cellular response to hypoxia in plants is presented and discussed.

2 Modulation of Nitrogen Metabolism Upon O_2 Deprivation

2.1 *Nitrate, Nitrite, and Nitric Oxide*

Nitrate reductase (NR) is a cytosolic enzyme that catalyzes NADH-dependent nitrate reduction into nitrite. NR gene expression and enzyme activity were shown to increase significantly in response to O_2 deprivation in several plant species, e.g., *Arabidopsis thaliana*, *Medicago truncatula*, and *Oryza sativa* (Allègre et al. 2004; Botrel and Kaiser 1997; Lasanthi-Kudahettige et al. 2007; Loreti et al. 2005; Stoimenova et al. 2007). In *Cucumis sativus* increased NR activity correlated with an increase in transcript levels of NR and its cofactor-binding domain genes *FAD* (*FAD binding*) and *CYP51G1* (*Heme binding*). Furthermore, it was proposed that the activation of NR activity was probably induced by both new enzyme synthesis and dephosphorylation of the phosphor-NR protein because the expression of the *PP2A* gene that encodes the phosphor-NR phosphatase increased several-fold under hypoxic stress (Shi et al. 2008). It has also been proposed that cytosolic pH acidification may increase NR activity because of the low pH optimum of this enzyme (Botrel and Kaiser 1997; Stoimenova et al. 2007).

Altogether these findings point out to an important role played by NR under hypoxic conditions with species exhibiting greater ability to tolerate O_2 deprivation showing higher nitrate reductase activity. Indeed, tobacco plants with low NR activity displayed several metabolic disorders linked to hypoxia stress with enhanced ethanol and lactate production and increased acidification of the cytosol. Conversely, supplying maize seedlings with nitrate during anoxia maintains a slightly higher cytosolic pH than that in the control seedlings (Libourel et al. 2006). In *Cucumis sativus*, increased NR activity upon hypoxia was accompanied by decrease in NO_3^- and increase in NO_2^- contents indicating that nitrite reduction under hypoxic conditions does not match the increased nitrate reduction process and that the benefit of nitrate supply was not due to a metabolic effect (Shi et al. 2008). Nitrate reduction may rather contribute to cellular acclimation to low O_2 deprivation by regenerating NAD^+ from NADH.

The effect of nitrite was also investigated. For this aim maize seedling roots were fed nitrite under hypoxic condition. It appeared that the provision of micromolar levels of nitrite was effective on the cytosolic pH adjustment indicating that nitrite is also implicated in cellular acclimation to O_2 deprivation. Since very low levels of nitrite effected cellular response to hypoxia its role is unlikely to be limited to the regeneration of $NAD(P)^+$ and might rather be linked to a regulatory mechanism (Libourel et al. 2006) such as NO emission that may activate mitochondrial ROS production and Ca^{2+} release (Ma et al. 2012; Talwar et al. 2012; Zhang et al. 2007). During O_2 deficit electrons generated by the oxidation of NAD(P)H by the Ca^{2+}-sensitive NAD(P)H dehydrogenases on the inner mitochondrial membrane surface are proposed to be accepted by nitrite at complex III (ubiquinone:cytochrome *c* reductase) or IV (COX) of the mitochondrial electron transport chain, producing the signal molecule NO and contributing to ATP synthesis due to proton pumping at the sites of complex III or COX (Planchet et al. 2005; Stoimenova et al. 2007). Alternatively nitrite may participate to NO emission by being reduced in the cytosol by nitrate reductase (Kaiser et al. 2002; Sakihama et al. 2002; Yamasaki et al. 1999). Due to the high K_m (100 mM) of nitrate reductase for nitrite, emission of NO is competitively inhibited by nitrate (Rockel et al. 2002). However, the rate of NO production by nitrate reductase increases each time nitrate reduction exceeded nitrite reduction such that nitrite accumulated; typically this is the case under low oxygen condition, e.g., NO emission could be established in leaves exposed to hypoxia/anoxia in the dark and in detached leaves fed nitrite through the petiole and maintained in the dark (Kaiser et al. 2002). Accordingly the inactivation of nitrate reductase by maintaining the enzyme in its phosphorylated state by the inhibitors of PP2A decreased the rate of NO emission (Kaiser et al. 2002). Nonsymbiotic hemoglobin, which genes expression was found to be greatly induced by low oxygen in flooded roots, was proposed to scavenge NO in hypoxic tissues by catalyzing its turnover to nitrate (Dordas et al. 2003a, b; Igamberdiev and Hill 2004; Igamberdiev et al. 2004). When hemoglobin (Hb) is coupled with nitrite reductase activity in hypoxic cells, this forms the Hb/NO cycle, in which excess NAD(P)H is oxidized (Hebelstrup et al. 2012). Additional positive effect of this reaction is the limitation of nitrogen loss under NO emission (Hebelstrup et al. 2012). Over-expression of Hb in alfalfa maintained ATP levels, ATP/ADP ratio, and increased survival during hypoxia compared to wild type and Hb-silenced plants in which ATP and ATP/ADP ratio declined under low oxygen condition (Dordas et al. 2003a). NO emission was 2.5-fold higher in Hb-silenced plants compared to Hb-over-expressers suggesting the involvement of Hb in plant response to hypoxia through the modulation of NO emission and the loss of nitrogen (Dordas et al. 2003a).

2.2 Alanine and GABA

Nitrogen assimilation and amino acids metabolism are profoundly affected by O_2 deprivation and energy shortage (Fig. 1). One of the most salient effects of hypoxia is the accumulation of alanine. In *Medicago truncatula* seedlings it has been shown that these changes were related to the stimulation of the expression of the mitochondrial alanine aminotransferase (*mAlaAT*) isogene and the accumulation of the encoded protein mAlaAT (Ricoult et al. 2005, 2006). Expression of the gene encoding the mitochondrial alanine glyoxylate aminotransferase (*AGT*) was inhibited by hypoxia while the gene encoding the cytosolic *AlaAT* was not expressed in young seedlings. AlaAT activity, as determined in vivo by using $^{15}NH_4$ labeling, increased in hypoxic seedlings (Limami et al. 2008; Ricoult et al. 2006). Alanine metabolism was investigated more thoroughly by feeding seedlings with either ^{15}N-glutamate or ^{15}N-alanine upon normoxic and hypoxic conditions (Ricoult et al. 2006). Feeding embryo axis with ^{15}N-glutamate or ^{15}N-alanine under normoxic conditions showed that mAlaAT catalyzed a reversible reaction allowing for synthesis of alanine with glutamate as amino donor and synthesis of glutamate with alanine as amino donor.

The same experiment showed that glycine synthesis occurred at the expense of either glutamate or alanine indicating that besides alanine glyoxylate aminotranferase (AGT) a glutamate glyoxylate aminotranferase was also operating. Feeding seedlings either ^{15}N-glutamate or ^{15}N-alanine under hypoxia showed that mAlaAT activity was directed towards alanine synthesis using glutamate as amino donor while the reaction of glutamate synthesis using alanine as amino donor was inhibited (Ricoult et al. 2006). The results indicate that mAlaAT isoform is regulated at both transcriptional and posttranslational levels. This dual mode of regulation by hypoxia allowed for an increase in the enzyme content through an increase in the expression of the coding gene and at the posttranslational level for the orientation of the equilibrium of the catalyzed reaction towards alanine synthesis. As a result ^{15}N-alanine was 4 times higher in hypoxic seedlings than in the control and alanine accumulated as the major amino acid instead of asparagine (Limami et al. 2008).

In vivo AGT activity in the direction of glycine synthesis was inhibited by hypoxia as shown by the fact that almost no labeled glycine was detected in seedlings when they were fed ^{15}N-alanine (Ricoult et al. 2006). Total absence of glycine (both labeled and unlabeled) in hypoxic seedlings fed ^{15}N-alanine means that glutamate glyoxylate transaminase did not compensate for the lack of AGT activity probably because glutamate was competitively recruited for alanine and GABA synthesis by mAlaAT and GDC (glutamate decarboxylase). Labeling experiment showed also that there was a dramatic decrease in de novo synthesis of glutamine and particularly asparagine which is the most abundant amino acid in *Medicago truncatula* (Glevarec et al. 2004). Both the enzymes glutamine synthetase (GS) and asparagine synthetase (AS) are ATP-dependent; inhibition of their activities is probably related to the energy crisis. Consistently, more than just a

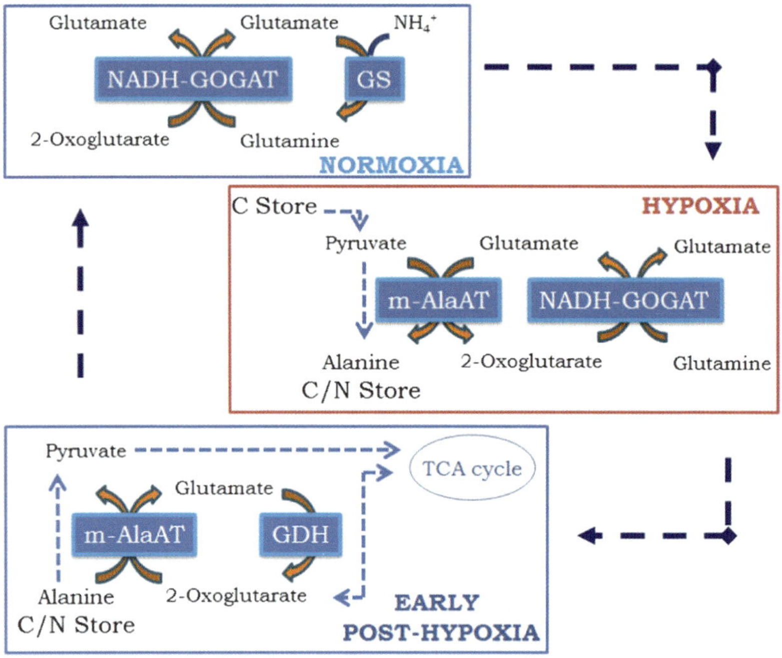

Fig. 1 Schematic representation of the central role of alanine metabolism during hypoxia and early post-hypoxia periods as revealed by ^{15}N labeling experiments in young *Medicago truncatula* seedlings. *Dashed lines* in Hypoxia and Early Post-hypoxia boxes represent fluxes of carbon from carbon storage compounds to alanine and further mobilization into Krebs cycle during post-hypoxia recovery period

change in *AlaAT* expression and activity significant changes were induced by hypoxia in amino acids metabolism that resulted in the accumulation of alanine as the major amino acid instead of asparagine in *Medicago truncatula*.

Changes in amino acids metabolism with alanine accumulation as the culminant event may contribute to mitigate the most damaging consequences of O_2 deprivation; i.e., an energy crisis and acidosis. Synthesis of alanine is accompanied by the generation of 2-oxoglutarate which can be further metabolized to succinate, via the TCA cycle enzyme succinate CoA ligase, thereby providing additional ATP per molecule of sucrose metabolized (Rocha et al. 2010). Accumulation of alanine upon hypoxia as a C/N storage compound saves ATP that otherwise is needed for the synthesis of asparagine or glutamine and saves C3 skeletons avoiding a shortage in carbon availability. Ethanol is a dead-end product that either accumulates or leaks out of the tissue representing in both cases a net loss of carbon. As a matter of fact the increase in the flow of carbon to ethanol by overexpressing *PDC* in *Arabidopsis thaliana* was effective in improving anoxia tolerance only in plants growing on a full nutrient Murashige and Skoog medium containing 3 % (w/v) of sugar (Ismond et al., 2003; Albrecht et al., 2004). Under sugar-limiting conditions as in the field, due to higher rates of glycolysis and ethanol fermentation, a faster depletion of sugar stores leads to decreased survival upon O_2 deprivation (Ismond et al. 2003). Finally, although cytoplasmic pH homeostasis relies on the cellular pH-stat, synthesis of alanine contributes to a limitation of cytoplasmic acidification by lactate because of the competition for pyruvate by both lactate fermentation and alanine synthesis pathways (Menegus et al. 1991; Reggiani et al. 2000; Ricoult et al. 2005, 2006).

Alanine synthesis and accumulation is thought to occur upon O_2 deprivation through another metabolic pathway. Alanine may accumulate as a byproduct of the GABA shunt that involves three reactions catalyzed by glutamate decarboxylase (GDC), GABA transaminase (GABA-T), and succinic semialdehyde dehydrogenase (SSADH). GABA is derived from the decarboxylation of glutamate by GDC thereby contributing to cellular pH-stat under hypoxia stress as glutamate decarboxylation is a proton consuming reaction that reduces the weak acid content and increases pH (Drew 1997; Greenway and Gibbs 2003). GDC is regulated by H^+ and Ca^{2+} which levels increases under hypoxia. Consistently GDC activity increases as the cytoplasmic pH declines and decreases as the pH recovers (Carroll et al. 1994). Ca^{2+} with calmodulin as Ca/CAM complex binds to the enzyme, thereby relieving it from autoinhibition (Bouché et al. 2004, 2005). Succinic semialdehyde (SSA) is produced from GABA via GABA-T that can use either pyruvate or 2-oxoglutarate as amino acceptor. It seems however that under hypoxic conditions the enzyme uses preferentially pyruvate thus leading to alanine synthesis. In favor of this hypothesis alanine accumulation was observed under hypoxic conditions in Arabidopsis T-DNA knockout mutants affected in *AlaAT* (Miyashita et al. 2007; Miyashita and Good 2008) indicating that the production of alanine might depend on another metabolic pathway. However GABA-T null mutants accumulated only slightly less alanine upon hypoxia compared with wild-type plants (Miyashita et al. 2007; Miyashita and Good 2008) suggesting that AlaAT and GABA-T pathways might be redundant at least under hypoxic conditions.

2.3 Glutamate

Enzymes of primary nitrogen assimilation involved in glutamate synthesis or using glutamate as amino donor were shown to be differently affected by O_2 deprivation in *Medicago truncatula* seedlings (Limami et al. 2008). Expression of genes encoding the ATP-consuming enzymes GS and AS and activities of these enzymes were inhibited by hypoxia stress. NADH-dependent glutamate synthase (*NADH-GOGAT*) expression was inhibited by hypoxia stress, while NADH-GOGAT activity increased. Conversely glutamate dehydrogenase (*GDH*) expression was up-regulated by hypoxic stress, while GDH activity, determined either in vitro or by native PAGE staining, was down-regulated. In vivo $^{15}NH_4^+$ labeling in the presence and absence of the GS-inhibitor methionine sulfoximine (MSX), used in combination with GC–MS amino acids analyses, indicated that the residual GDH activity was not contributing to glutamate synthesis upon hypoxic conditions.

In parallel to these investigations a blend of metabolic experiments using the incorporation of $^{15}NH_4^+$ and ^{15}N-amino acids during hypoxic stress showed that the pools of newly synthesized glutamate (^{15}N-glutamate) in normoxic and hypoxic *Medicago truncatula* seedlings were very similar indicating that the glutamate content was subjected to a very tight control. Consistently it is suggested that rather than just an activation of alanine synthesis, the adaptive reaction of the plant to hypoxic stress consists of a concerted modulation of nitrogen flux through both glutamate and alanine synthesis pathways.

It appears as if the decrease in glutamate utilization by GS and AS—probably due to the lack of ATP—was compensated for by increased NADH-GOGAT and AlaAT activities, as revealed by increased amounts of newly synthesized alanine (^{15}N-alanine) during hypoxic stress. Therefore, it is likely that the reductive amination of 2-oxoglutarate by NADH-GOGAT during hypoxic stress fulfills two major roles. The first is the synthesis of glutamate, the substrate of AlaAT, and the second is the oxidation of NADH when oxidative phosphorylation is totally or partially inhibited by the lack of oxygen, thus making NAD^+ available to enable glycolysis to proceed (Limami et al. 2008).

Finally the discrepancy between the levels of *GDH* gene expression and enzyme activity questions the significance of the up-regulation of the expression of *GDH* in *Medicago truncatula* immediately following hypoxic stress? It has been suggested that the return to aerobic conditions is anticipated in plants subjected to hypoxic stress by expressing genes whose products have functions during the subsequent recovery period (Drew 1997). Consistently, it is proposed that *GDH1* was up-regulated by hypoxia-induced carbon stress, in anticipation that the product of its transcription would regenerate 2-oxoglutarate by deaminating glutamate during the subsequent post-hypoxic recovery period (Fig. 1) (Limami et al. 2008). Higher rates of glycolysis and ethanolic fermentation are known to lead to carbon stress due to the faster depletion of sugar stores in hypoxic tissues (Ismond et al. 2003). Up-regulation of the expression of *GDH* genes has been observed under various conditions associated with carbon stress, i.e., senescing leaves, low light and dark, and C/N imbalance due to excess ammonium nutrition (Masclaux-Daubresse et al. 2005; Melo-Oliveira et al. 1996; Skopelitis et al. 2006). The function of GDH in these conditions is assumed to be the oxidative deamination of glutamate that provides C skeletons (2-oxoglutarate) and reducing power (for review, see Forde and Lea 2007). The same pattern of regulation was observed for the hypoxia-inducible *AlaAT1* in *Arabidopsis thaliana* (Miyashita et al. 2007). *AlaAT1* was up-regulated at the transcriptional level during hypoxic stress while the major role of the encoded enzyme was shown to be the conversion of alanine into glutamate during the post-hypoxic period.

2.4 Polyamines

The first step in polyamines biosynthesis in higher plants is the decarboxylation of either arginine by arginine decarboxylase (ADC) or ornithine by ornithine decarboxylase (ODC). Ultimately, both reactions lead to putrescine, the diamine precursor of spermidine and spermine. The latter are formed by sequential addition of an aminopropyl moiety onto putrescine and spermidine in reactions catalyzed by spermidine synthase and spermine synthase (Shelp et al. 2012).

Accumulation of putrescine, rather than spermidine or spermine, was associated in several plant species to hypoxia tolerance. Greater capacity of putrescine accumulation was observed in species like rice and barnyard grass which are well adapted to hypoxic environment than that in anoxia-intolerant species (Reggiani et al. 1989). In several graminae species acclimation to O_2 deprivation was associated to putrescine accumulation as a result of the induction of ADC and ODC and concomitant inhibition of degradation of putrescine by diamine oxidase (DAO). Furthermore, ethylene—which synthesis is boosted by O_2 deprivation (Bailey-Serres and Voesenek 2008)—is known to inhibit spermidine and spermine synthesis by competition for a common precursor, the aminopropyl moiety donor *S*-adenosylmethionine (SAM) (Amir 2010). Hypoxia-induced shoot elongation in the flood-tolerant grassweed *Scirpus mucronatus* coincided with increased putrescine content. Alternatively the inhibition of putrescine synthesis through the inhibition of ADC and ODC resulted in an inhibition of shoot elongation upon O_2 deprivation. This inhibitory effect was reversed by exogenous putrescine treatment pointing out to the important role this diamine may play in acclimation to O_2 deprivation (Lee and Kende 2001, 2002). Similarly rice coleoptiles elongation under anoxia was also associated with putrescine accumulation, inhibited by an ADC inhibitor and reestablished by exogenous putrescine treatment (Reggiani et al. 1989, 2000). Root hypoxia tolerance was increased and injury linked to O_2 deprivation was alleviated by putrescine application to tomato plants (Nada et al. 2004). Altogether these findings strongly suggest that putrescine may play a protective role in acclimation to O_2 in plants (Lee and Kende 2001, 2002). Putrescine and to a very lesser extent spermidine but not spermine accumulated in waterlogged *Medicago truncatula* roots as well as the precursors ornithine and arginine (Fig. 2, Diab and Limami, unpublished data).

The beneficial role of putrescine is still poorly understood. As a cation and one of the compounds thought to produce basic equivalents it was suggested that the diamine can contribute to balancing the anoxic production of organic acids and to the homeostatic buffering mechanism for stabilizing intracellular pH. Putrescine accumulated in anoxic rice coleoptiles to concentrations comparable in magnitude to the sum of concentrations of lactic acid and succinic acid (Reggiani et al. 1989).

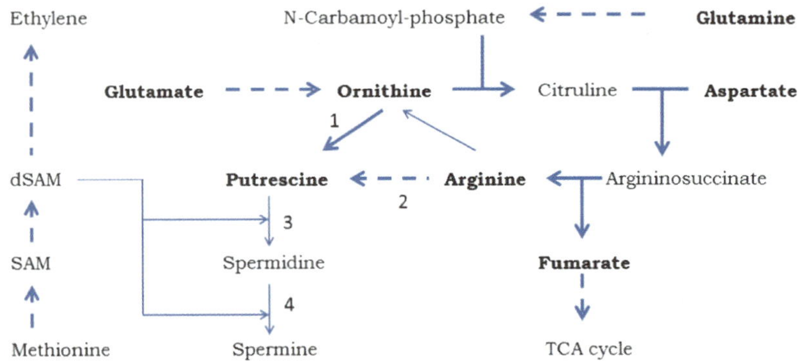

Fig. 2 Schematic representation of putrescine, spermidine, and spermine biosynthesis pathway and its relation with ethylene biosynthesis through the common precursor, decarboxylated *S*-adenosylmethionine (dSAM). In *bold characters* are shown the compounds that accumulated in waterlogged roots of *Medicago truncatula*. Putrescine is synthesized directly by ornithine decarboxylase (1) or through agmatine and *N*-carbamoylputrescine by arginine decarboxylase (2). Reactions 3 and 4 are catalyzed, respectively, by spermidine synthase and spermine synthase

3 Conclusion

Modulation of nitrogen metabolism is an important component of the acclimation of plants to waterlogging-induced O_2 deprivation. Hypoxic N metabolism not only contributes to energy crisis and acidosis amelioration but also counteracts detrimental effects of lactic and ethanol fermentation. A large bulk of information was gathered during the last decades on the adaptive response of each of N and C primary metabolisms to O_2 deprivation. However, a broad picture of the interaction between both metabolisms under hypoxic conditions including secondary metabolism is still needed for a full understanding of the metabolic adaptation of plants to low O_2 stress. For this aim an integrative approach including metabolomic and transcriptomic analysis would be suitable.

Acknowledgments To Claudie Ricoult *in memoriam*.

References

Albrecht G, Mustroph A, Theodore C (2004) Sugar and fructan accumulation during metabolic adjustment between respiration and fermentation under low oxygen conditions in wheat roots. Physiol Plant 120:93–104

Allègre A, Silvestre J, Morard P, Kallerhoff J, Pinelli E (2004) Nitrate reductase regulation in tomato roots by exogenous nitrate: a possible role in tolerance to long-term root anoxia. J Exp Bot 55:2625–2634

Amir R (2010) Current understanding of the factors regulating methionine content in vegetative tissues of higher plants. Amino Acids 39:917–931

Bailey-Serres J, Voesenek LA (2008) Flooding stress: acclimations and genetic diversity. Annu Rev Plant Biol 59:313–339
Bailey-Serres J, Fukao T, Gibbs DJ, Holdsworth MJ, Lee SC, Licausi F, Perata P, Voesenek LA, van Dongen JT (2012) Making sense of low oxygen sensing. Trends Plant Sci 17:129–138
Botrel A, Kaiser WM (1997) Nitrate reductase activation state in barley roots in relation to the energy and carbohydrate status. Planta 201:496–501
Bouché N, Fait A, Zik M, Fromm H (2004) The root-specific glutamate decarboxylase (GAD1) is essential for sustaining GABA levels in Arabidopsis. Plant Mol Biol 55:315–325
Bouché N, Yellin A, Snedden WA, Fromm H (2005) Plant-specific calmodulin-binding proteins. Annu Rev Plant Biol 56:435–466
Braun KP, Cody RB, Jones JDR, Peterson CM (1995) A structural assignment for a stable acetaldehyde-lysine adduct. J Biol Chem 270:11263–11266
Carroll AD, Fox GG, Laurie S, Phillips R, Ratcliffe RG, Stewart GR (1994) Ammonium assimilation and the role of [gamma]-aminobutyric acid in pH homeostasis in carrot cell suspensions. Plant Physiol 106:513–520
Davies DD (1987) The role of lactate dehydrogenase isozymes in controlling the cytosolic pH of plant cells. Isozymes Curr Top Biol Med Res 16:193–207
Davies DD, Patil KD (1974) Regulation of 'malic' enzyme of Solanum tuberosum by metabolites. Biochem J 137:45–53
Dordas C, Hasinoff BB, Igamberdiev AU, Manac'h N, Rivoal J, Hill RD (2003a) Expression of a stress-induced hemoglobin affects NO levels produced by alfalfa root cultures under hypoxic stress. Plant J 35:763–770
Dordas C, Rivoal J, Hill RD (2003b) Plant haemoglobins, nitric oxide and hypoxic stress. Ann Bot 91 Spec No: 173–178
Drew MC (1997) Oxygen deficiency and root metabolism: injury and acclimation under hypoxia and anoxia. Annu Rev Plant Physiol Plant Mol Biol 48:223–250. doi:10.1146/annurev.arplant.48.1.223
Felle HH (1996) Control of cytoplasmic pH under anoxic conditions and its implication for plasma membrane proton transport in Medicago sativa root hairs. J Exp Bot 47:967–973. doi:10.1093/jxb/47.7.967
Forde BG, Lea PJ (2007) Glutamate in plants: metabolism, regulation, and signalling. J Exp Bot 58:2339–2358
Glevarec G, Bouton S, Jaspard E, Riou MT, Cliquet JB, Suzuki A, Limami AM (2004) Respective roles of the glutamine synthetase/glutamate synthase cycle and glutamate dehydrogenase in ammonium and amino acid metabolism during germination and post-germinative growth in the model legume Medicago truncatula. Planta 219:286–297
Gout E, Boisson A, Aubert S, Douce R, Bligny R (2001) Origin of the cytoplasmic pH changes during anaerobic stress in higher plant cells. Carbon-13 and phosphorous-31 nuclear magnetic resonance studies. Plant Physiol 125:912–925
Greenway H, Gibbs J (2003) Mechanisms of anoxia tolerance in plants. II. Energy requirements for maintenance and energy distribution to essential processes. Funct Plant Biol 30:37
Hebelstrup KH, van Zanten M, Mandon J, Voesenek LA, Harren FJ, Cristescu SM, Møller IM, Mur LA (2012) Haemoglobin modulates NO emission and hyponasty under hypoxia-related stress in Arabidopsis thaliana. J Exp Bot 63:5581–5591
Igamberdiev AU, Hill RD (2004) Nitrate, NO and haemoglobin in plant adaptation to hypoxia: an alternative to classic fermentation pathways. J Exp Bot 55:2473–2482
Igamberdiev AU, Seregélyes C, Manac'h N, Hill RD (2004) NADH-dependent metabolism of nitric oxide in alfalfa root cultures expressing barley hemoglobin. Planta 219:95–102
Ismond KP, Dolferus R, De Pauw M, Dennis ES, Good AG (2003) Enhanced low oxygen survival in Arabidopsis through increased metabolic flux in the fermentative pathway. Plant Physiol 132:1292–1302
Jackson M (1985) Ethylene and responses of plants to soil waterlogging and submergence. Annu Rev Plant Physiol Plant Mol Biol 36:145–174

Jackson M, Armstrong W (1999) Formation of aerenchyma and the processes of plant ventilation in relation to soil flooding and submergence. Plant Biol 1:274–287

Kaiser WM, Weiner H, Kandlbinder A, Tsai CB, Rockel P, Sonoda M, Planchet E (2002) Modulation of nitrate reductase: some new insights, an unusual case and a potentially important side reaction. J Exp Bot 53:875–882

Kursteiner O, Dupuis I, Kuhlemeier C (2003) The pyruvate decarboxylase1 gene of Arabidopsis is required during anoxia but not other environmental stresses. Plant Physiol 132:968–978

Lasanthi-Kudahettige R, Magneschi L, Loreti E, Gonzali S, Licausi F, Novi G, Beretta O, Vitulli F, Alpi A, Perata P (2007) Transcript profiling of the anoxic rice coleoptile. Plant Physiol 144:218–231

Lee Y, Kende H (2001) Expression of beta-expansins is correlated with internodal elongation in deepwater rice. Plant Physiol 127:645–654

Lee Y, Kende H (2002) Expression of alpha-expansin and expansin-like genes in deepwater rice. Plant Physiol 130:1396–1405

Libourel IG, van Bodegom PM, Fricker MD, Ratcliffe RG (2006) Nitrite reduces cytoplasmic acidosis under anoxia. Plant Physiol 142:1710–1717

Limami AM, Glevarec G, Ricoult C, Cliquet J-B, Planchet E (2008) Concerted modulation of alanine and glutamate metabolism in young Medicago truncatula seedlings under hypoxic stress. J Exp Bot 59:2325–2335. doi:10.1093/jxb/ern102

Loreti E, Poggi A, Novi G, Alpi A, Perata P (2005) A genome-wide analysis of the effects of sucrose on gene expression in Arabidopsis seedlings under anoxia. Plant Physiol 137:1130–1138

Ma F, Lu R, Liu H, Shi B, Zhang J, Tan M, Zhang A, Jiang M (2012) Nitric oxide-activated calcium/calmodulin-dependent protein kinase regulates the abscisic acid-induced antioxidant defence in maize. J Exp Bot 63:4835–4847

Masclaux-Daubresse C, Carrayol E, Valadier MH (2005) The two nitrogen mobilisation- and senescence-associated GS1 and GDH genes are controlled by C and N metabolites. Planta 221:580–588

Melo-Oliveira R, Oliveira IC, Coruzzi GM (1996) Arabidopsis mutant analysis and gene regulation define a nonredundant role for glutamate dehydrogenase in nitrogen assimilation. Proc Natl Acad Sci U S A 93:4718–4723

Menegus F, Cattaruzza L, Mattana M, Beffagna N, Ragg E (1991) Response to anoxia in rice and wheat seedlings: changes in the pH of intracellular compartments, glucose-6-phosphate level, and metabolic rate. Plant Physiol 95:760–767

Miyashita Y, Good AG (2008) Contribution of the GABA shunt to hypoxia-induced alanine accumulation in roots of Arabidopsis thaliana. Plant Cell Physiol 49(1):92–102

Miyashita Y, Dolferus R, Ismond KP, Good AG (2007) Alanine aminotransferase catalyses the breakdown of alanine after hypoxia in Arabidopsis thaliana. Plant J 49:1108–1121

Nada K, Iwatani E, Doi T, Tachibana S (2004) Effect of putrescine pretreatment to roots on growth and lactate metabolism in the root of tomato (Lycopersicum esculentum Mill.) under root-zone hypoxia. J Jpn Soc Hortic Sci 73:3

Planchet E, Jagadis Gupta K, Sonoda M, Kaiser WM (2005) Nitric oxide emission from tobacco leaves and cell suspensions: rate limiting factors and evidence for the involvement of mitochondrial electron transport. Plant J 41:732–743

Reggiani R, Hochkoeppler A, Bertani A (1989) Polyamines in rice seedlings under oxygen-deficit stress. Plant Physiol 91:1197–1201

Reggiani R, Nebuloni M, Mattana M, Brambilla I (2000) Anaerobic accumulation of amino acids in rice roots: role of the glutamine synthetase/glutamate synthase cycle. Amino Acids 18:207–217

Ricoult C, Cliquet J-B, Limami AM (2005) Stimulation of alanine amino transferase (AlaAT) gene expression and alanine accumulation in embryo axis of the model legume Medicago truncatula contribute to anoxia stress tolerance. Physiol Plant 123:30–39

Ricoult C, Echeverria LO, Cliquet JB, Limami AM (2006) Characterization of alanine aminotransferase (AlaAT) multigene family and hypoxic response in young seedlings of the model legume Medicago truncatula. J Exp Bot 57:3079–3089
Roberts JKM, Chang K, Webster C, Callis J, Walbot V (1989) Dependence of ethanolic fermentation, cytoplasmic pH regulation, and viability on the activity of alcohol dehydrogenase in hypoxic maize root tips. Plant Physiol 89:1275–1278
Rocha M, Licausi F, Araujo WL, Nunes-Nesi A, Sodek L, Fernie AR, van Dongen JT (2010) Glycolysis and the tricarboxylic acid cycle are linked by alanine aminotransferase during hypoxia induced by waterlogging of Lotus japonicus. Plant Physiol 152:1501–1513. doi:10.1104/pp. 109.150045
Rockel P, Strube F, Rockel A, Wildt J, Kaiser WM (2002) Regulation of nitric oxide (NO) production by plant nitrate reductase in vivo and in vitro. J Exp Bot 53:103–110
Saglio P, Germain V, Richard B (1999) The response of plants to oxygen deprivation : role of enzyme induction in the improvement of tolerance to anoxia. In: Lerner HR (ed) Plant responses to environmental stresses. Marcel Dekker, New York, pp 373–393
Sakihama Y, Nakamura S, Yamasaki H (2002) Nitric oxide production mediated by nitrate reductase in the green alga Chlamydomonas reinhardtii: an alternative NO production pathway in photosynthetic organisms. Plant Cell Physiol 43:290–297
Shelp BJ, Bozzo GG, Trobacher CP, Zarei A, Deyman KL, Brikis CJ (2012) Hypothesis/review: contribution of putrescine to 4-aminobutyrate (GABA) production in response to abiotic stress. Plant Sci 193–194:130–135
Shi K, Ding X-T, Don D-K, Zhou Y-H, Yu JQ (2008) Putrescine enhancement of tolerance to root-zone hypoxia in Cucumis sativus: a role in increased nitrate reductase. Funct Plant Biol 35:48
Skopelitis DS, Paranychianakis NV, Paschalidis KA, Pliakonis ED, Delis ID, Yakoumakis DI, Kouvarakis A, Papadakis AK, Stephanou EG, Roubelakis-Angelakis KA (2006) Abiotic stress generates ROS that signal expression of anionic glutamate dehydrogenases to form glutamate for proline synthesis in tobacco and grapevine. Plant Cell 18:2767–2781
Stoimenova M, Igamberdiev AU, Gupta KJ, Hill RD (2007) Nitrite-driven anaerobic ATP synthesis in barley and rice root mitochondria. Planta 226:465–474
Talwar PS, Gupta R, Maurya AK, Deswal R (2012) Brassica juncea nitric oxide synthase like activity is stimulated by PKC activators and calcium suggesting modulation by PKC-like kinase. Plant Physiol Biochem 60:157–164
Yamasaki H, Sakihama Y, Takahashi S (1999) An alternative pathway for nitric oxide production in plants: new features of an old enzyme. Trends Plant Sci 4:128–129
Zhang A, Jiang M, Zhang J, Ding H, Xu S, Hu X, Tan M (2007) Nitric oxide induced by hydrogen peroxide mediates abscisic acid-induced activation of the mitogen-activated protein kinase cascade involved in antioxidant defense in maize leaves. New Phytol 175:36–50

Adaptation of Storage Metabolism to Oxygen Deprivation

Peter Geigenberger

Abstract Recent studies document that oxygen deprivation is not only restricted to stress conditions such as flooding, but is also associated with the normal development and growth of storage organs such as tubers, fruits, and seeds in well-oxygenated surroundings. The decrease in internal oxygen concentrations in these tissues is attributable to their active metabolism and internal restrictions in oxygen entry during their normal development. Short-term balancing of internal oxygen concentrations has been shown to involve an oxygen-sensing system that regulates respiration in response to normal fluctuations in internal oxygen concentrations. The resulting changes in adenylate energy state lead to a widespread regulation of ATP-consuming processes which are involved in storage and growth. Developmental changes in internal oxygen concentrations also have been found to be linked to global changes in gene transcription networks, leading to longer-term adaptive responses in storage metabolism and growth. This involves a switch to more energy-conserving pathways of (1) sucrose degradation via sucrose synthase, (2) nucleotide synthesis via salvage pathways, and (3) respiratory metabolism via alanine aminotransferase. Group VII ethylene-response-factor transcription factors have been identified to function as oxygen sensors in the regulation of these processes. They were shown to respond to a decrease in internal oxygen that occurs during the normal development of potato tubers, suggesting a role of these oxygen-sensor proteins in the regulation of sucrose and starch metabolism in response to tuber development. Possible implications for strategies to improve crop yield are discussed.

P. Geigenberger (✉)
Department of Biology I, Ludwig-Maximilians-Universität München (LMU), Grosshaderner Str. 2-4, 82152 Martinsried, Germany
e-mail: geigenberger@bio.lmu.de

J.T. van Dongen and F. Licausi (eds.), *Low-Oxygen Stress in Plants*, Plant Cell Monographs 21, DOI 10.1007/978-3-7091-1254-0_12, © Springer-Verlag Wien 2014

1 Introduction

During photosynthesis, plants have the ability to convert inorganic carbon dioxide to organic sugars which are used for storage and growth. In leaves, a part of the reduced carbon is transiently stored as starch or exported as sucrose to non-photosynthetic tissues via the phloem. In heterotrophic storage organs such as developing tubers, fruits, or seeds, sucrose is unloaded from the phloem and converted to glycolytic intermediates to support the synthesis of reserve products and growth (Weber et al. 1997; Geigenberger et al. 2004). Glycolysis, respiration, and the oxidative pentose phosphate pathway provide energy, reducing equivalents and carbon compounds for various biosynthetic pathways, including the biosynthesis of lipids, amino acids, secondary products, and nucleic acids. Carbon skeletons required for amino acid biosynthesis, one prerequisite for storage protein accumulation, are mainly generated in the tricarboxylic acid (TCA) cycle. The key metabolite of the starch biosynthesis pathway, ADP-glucose, is generated from hexose phosphates by ADP-glucose pyrophosphorylase (AGPase).

Starch is the major storage product in higher plants. In developing tubers, fruits, stems, and seeds, starch serves as a long-term carbon store, which is remobilized later in development to support phases of reproductive growth (Geigenberger 2011). Also lipids are important reserve products, preferentially accumulating in oil-storing seeds or fruits. In addition to their important functions in plant physiology, both, starch and lipids, are of great economical importance, representing a major resource of our diet and feedstock for many industrial applications, including bio-fuel production (Geigenberger and Fernie 2006; Vigeolas et al. 2007). Starch and lipid metabolism have therefore been the subject of intense research in the past, leading to an advanced knowledge of the pathways and their regulation. The interested reader is referred to recent reviews on the regulation of starch (Geigenberger 2011; Stitt and Zeeman 2012) and storage lipid metabolism (Chapman and Ohlrogge 2012).

While most of the previous studies have analyzed the regulation of these pathways in response to physiological inputs, such as nutrient supply, less is known about their regulation in response to environmental constraints. High temperature, drought, or flooding have been identified as important stress conditions, usually leading to strong detrimental effects on seed and tuber yield (Geigenberger et al. 2004; Tiessen et al. 2006; Bailey-Serres and Voesenek 2008). The underlying mechanisms, however, are difficult to resolve since many aspects of plant physiology are being affected in response to these stress conditions. Transient flooding or waterlogging, for example, leads to a rapid depletion of soil oxygen which impairs root function and overall plant growth (Drew 1997; Bailey-Serres and Voesenek 2008; Ferner et al. 2012). Lack of oxygen in the rhizosphere leads to a shift from oxidative phosphorylation to fermentation within the roots, leading to a dramatic decrease in ATP formation, accumulation of harmful fermentation products, and cytosolic acidification. On the whole plant level, this has adverse effects on mineral nutrient uptake, water relations, photosynthesis, and carbohydrate translocation, leading to impaired plant growth and storage product accumulation also in aerial storage organs not directly affected by the low oxygen concentrations in the soil.

As it will be outlined in the present chapter, oxygen deprivation is not only restricted to stress conditions such as flooding, but is also associated with the normal development and growth of storage organs such as tubers, fruits, and seeds in well-oxygenated surroundings (21 % v/v external oxygen, corresponding to a partial pressure of 21 kPa or an oxygen concentration of 250 μM). The decrease in internal oxygen concentrations in these tissues is attributable to their active metabolism and internal restrictions in oxygen entry during their normal development. Low internal oxygen leads to adaptive changes in storage metabolism to save ATP and decrease oxygen consumption, to prevent a fall into anoxia (Geigenberger 2003). In addition to this, low internal oxygen could be an important factor to regulate the correct progression of seed and tuber development (Licausi et al. 2011a), as it has been found for embryo development in mammals (Forristal et al. 2010). The underlying oxygen-sensing and signalling mechanisms as well as possible implications for strategies to improve crop yield are discussed.

2 Dynamics of Internal Oxygen Concentrations in Storage Tissues During Their Development

Historically, the first measurements of internal oxygen concentrations have been performed with bulky storage organs such as tubers and fruits (Gerber 1896; Devaux 1891). The large size of these organs allowed the application of methods to sample the internal atmosphere from intercellular spaces either directly via a syringe or indirectly from a chamber that was allowed to equilibrate with the internal atmosphere before sampling (Magness 1920; Wardlaw and Leonard 1939; Smith 1947). From the 1960th onwards, miniaturized Clarke-type oxygen electrodes and microelectrodes were developed and used for fine-scale in situ measurements of internal oxygen concentrations in fruits (Brändle and Wanner 1965), roots (Armstrong et al. 1994), siliques (Porterfield et al. 1999), and growing potato tubers (Geigenberger et al. 2000). More recently, the development of optical oxygen microsensors allowed comprehensive analyses of oxygen concentrations within developing seeds (Rolletschek et al. 2002). Compared to electrochemical microelectrodes, optoelectronic microsensors are usually more robust and easier to use (Rolletschek et al. 2009). Moreover, the latter are more precise as they do not disturb internal oxygen concentrations by consuming oxygen molecules while measuring (Rolletschek et al. 2009; Ast et al. 2012). While the diameter of commercially available optodes (approx. 50 μm) usually limits the spatial resolution of the measurements, the development of ultrafine fiberoptic sensors with tip diameters down to 10 μm allowed oxygen gradients to be measured in roots with a comparable resolution as provided by microelectrodes (Zabalza et al. 2009).

Table 1 Dynamics of internal oxygen concentrations in storage tissues at ambient external oxygen concentrations (21 % v/v, corresponding to a partial pressure of 21 kPa or a concentration of 250 μM)

Tissue/Factor	Internal oxygen concentration (% v/v)																	Reference
	0	0.1	0.2	0.5	1	2	3	4	5	6	8	12	16	21	30	40	60	80	
Potato tuber																			
-transect							■	■	■	■	■	■	■						Geigenberger et al. (2000)
-development/size						■	■	■	■	■	■	■	■						Licausi et al. (2011)
-genotype					■	■	■	■	■	■	■	■	■						Bologa et al. (2003)
-storage/dormant												■	■	●					Geigenberger et al. (2000)
Apple fruit																			
-transect						■	■	■	■	■	■	■	■						Ho et al. (2010)
-genotype						■	■	■	■	■	■								Ho et al. (2010)
Maize seed																			
-transect				■	■	■	■	■	■	■	■	■	■						Rolletschek et al. (2005b)
Sunflower seed																			
-transect		■	■	■	■	■	■	■	■	■	■	■	■						Rolletscheck et al. (2007)
-development		■	■	■															Rolletschek et al. (2007)
-temperature	■	■	■																Rolletscheck et al. (2007)
Barley seed																			
-transect (dark)			■	■	■	■	■	■	■	■	■	■	■	●					Rolletschek et al. (2004)
-dark/light			■	■	■	■	■	■	■	■	■	■	■	●	▲	▲	▲		Rolletschek et al. (2004)
Soybean seed																			
-transect (dark)			■	■	■	■	■	■	■	■	■								Rolletschek et al (2005a)
-dark/light			■	■	■	■	■	■	■	■	■	■	■						Rolletschek et al. (2005a)
-NO injection				■	■	■	■	■	■										Borisjuk et al. (2007)
Broad bean																			
-transect (dark)			■	■	■	■	■	■	■	■	■	■							Rolletschek et al. (2002)
-development		■	■	■	■	■	■	■	■	■	■	■							Rolletschek et al. (2002)
-dark/light			■	■	■	■	■	■	■	■	■	■							Rolletschek et al. (2002)
Pea seed																			
-transect (dark)			■	■	■	■	■	■	■	■	■	■	■						Rolletschek et al. (2003)
-dark/light							■	■	■	■	■	■							Rolletschek et al, (2003)
-germination			■	■	■	■	■	■	■	■	■	■	■	●					Benamar et al. (2008); Rolletschek et al. (2009)
Oilseed-rape seed																			
-transect (dark)		■	■	■	■	■	■	■	■	■	■	■	■						Vigeolas et al. (2003); Borisjuket al.(2009)
-dark/light		■	■	■	■	■	■	■	■	■	■	■	■	●	▲	▲	▲	▲	Borisjuk et al. (2009); Waldeck (unpublished)
-DCMU injection						■	■	■	■	■	■	■	■	●					Waldeck & Geigenberger, (unpublished)

Internal oxygen concentrations in potato tubers, fruits, and diverse seeds show large dynamics in internal oxygen concentrations along spatial transects, during development, across genotypes, and in response to environmental and experimental alterations. In seeds showing a full (broad bean, pea, soybean, and oil-seed rape) or partial photosynthetic capacity (barley pericarp, but not endosperm), dynamics of internal oxygen concentrations in response to light/dark changes are shown. In this case, spatial transects are referred to dark conditions. Injection of the signalling molecule nitric oxide (NO), which inhibits respiration, leads to an increase in internal oxygen in the dark. Injection of the inhibitor of photosynthetic electron transport 3-(3,4-dichlorophenyl)-1,1-dimethylurea (DCMU) leads to a decrease in internal oxygen in the light

In diverse fruits such as bananas, apples, or melon, large internal oxygen gradients down to 2–4 % v/v (compared to 21 % v/v in air) have been monitored and found to be mainly determined by the rate of respiration and the resistance to gaseous diffusion provided by intercellular spaces and the anatomy of the tissue (Magness 1920; Wardlaw and Leonard 1940; Brändle 1968; Banks 1983). These metabolic and morphological properties are strongly affected by developmental processes, leading to large changes in internal oxygen concentrations during ripening. More recently, the use of optical oxygen microsensors allowed changes in internal oxygen gradients between different apple cultivars to be resolved and models on the dynamics of internal oxygen concentrations in apples to be tested (Ho et al. 2010; see Table 1).

Large gradients in internal oxygen concentrations also evolve during the normal development of starch storing potato tubers (Licausi et al. 2011a). Oxygen concentrations around 18 % (v/v) were found in stolons before the onset of tuberisation. With tuber development and enlargement, internal oxygen concentrations progressively decreased to approx. 2 % (v/v) in the centre of 25-day-old tubers. There were also oxygen gradients along transects within a single tuber, with oxygen falling to 16–18 % (v/v) in the tuber periphery and 2–8 % (v/v) in the tuber centre (Geigenberger et al. 2000; Table 1). These oxygen gradients were found to be determined (1) by the increased resistance to gas diffusion as tubers grow, since the gradients were getting larger when the tubers were growing bigger and (2) by metabolic activity, since harvested tubers remaining in a dormant metabolic state revealed much smaller internal oxygen gradients than developing tubers of the same size actively storing starch (Geigenberger et al. 2000; Table 1). Moreover, transgenic tubers with increased respiration rates were characterized by steeper internal oxygen gradients, compared to the wild-type (Bologa et al. 2003; Table 1).

Steep oxygen gradients have also been found in developing seeds of various crops, such as broad bean (Rolletschek et al. 2002), pea (Rolletschek et al. 2003), oil-seed rape (Vigeolas et al. 2003; Rolletschek et al. 2009; Waldeck and Geigenberger, unpublished), wheat (van Dongen et al. 2004), barley (Rolletschek et al. 2004), soybean (Rolletschek et al. 2005a), maize (Rolletschek et al. 2005b), and sunflower (Rolletscheck et al. 2007). The results of these studies have been reviewed by Borisjuk and Rolletschek (2009) and are summarized in Table 1, providing an overview on the dynamics of internal oxygen concentrations in diverse tissues and conditions. In addition to developmental parameters that lead to changes in metabolic and morphological properties, seed oxygen gradients have been found to be strongly determined by environmental factors, such as light and temperature (see Table 1 and references therein). When oxygen gradients are compared in the dark, lipid storing seed such as sunflower or oil-seed rape showed a stronger decrease in internal oxygen concentrations (down to 1 μM) compared to starch storing seeds such as maize or barley (down to 2–5 μM). This is most likely attributable to the higher energy demand of lipid biosynthesis compared to starch biosynthesis, leading to higher rates of oxygen consumption within lipid-storing tissues. Depending on the photosynthetic capacity of developing seeds, hypoxic conditions are relieved or even turned into hyperoxia in the light. There is higher light supply to outer versus inner regions of developing seeds, leading to gradients in photosynthetic activity across seeds (Borisjuk and Rolletschek 2009). In developing seeds of oil-seed rape showing a high capacity of oil production in the field, large and progressive increases in internal oxygen concentrations have been found in response to increased light intensities up to 800 μE (Borisjuk and Rolletschek 2009; Geigenberger and Waldeck, unpublished results), which can be prevented by injection of the inhibitor of photosynthetic electron transport 3-(3,4-dichlorophenyl)-1,1-dimethylurea (DCMU; Waldeck and Geigenberger, unpublished results).

3 Low Oxygen Signals Lead to Adaptive Changes in Storage Metabolism to Prevent Internal Anoxia in Developing Storage Organs

3.1 Low Internal Oxygen Leads to an Adaptive Suppression of Storage Metabolism

While the data visualized in Table 1 show large dynamics of internal oxygen concentrations in diverse storage tissues during their normal development, they do not provide any evidence for the occurrence of internal anoxia. Moreover, typical fermentation products such as lactate or ethanol were found to be rather low in these tissues, providing additional evidence that anoxia is obviously avoided (Geigenberger 2003; Borisjuk and Rolletschek 2009). This requires regulatory mechanisms for balancing of internal oxygen concentrations to prevent a fall into anoxia. There is evidence for rapid adjustments of mitochondrial respiration and global storage metabolism in response to experimental perturbations of internal oxygen concentrations by changing (1) external oxygen supply in growing potato tubers (Geigenberger et al. 2000) seeds of wheat (van Dongen et al. 2004), maize (Rolletschek et al. 2005b), and oil-seed rape (Vigeolas et al. 2003) or (2) light conditions in soybean (Rolletschek et al. 2005a; Borisjuk et al. 2005). Work with growing potato tubers provided direct evidence that a fall in internal oxygen leads to an inhibition of respiration, a decrease in the adenylate energy state, and a global decrease in biosynthetic activities to diminish oxygen consumption and prevent the tissue to fall into anoxia (Geigenberger et al. 2000). Strong correlations have been found between changes in oxygen concentrations and the energy state in diverse tissues, including growing potato tubers (Geigenberger et al. 2000; Geigenberger 2003; Bologa et al. 2003), developing seeds (Vigeolas et al. 2003; van Dongen et al. 2004; Rolletschek et al. 2004, 2005a, b), and fruits (Biais et al. 2010), providing evidence that this metabolic adaptation to low oxygen is a general phenomenon in plants.

Geigenberger et al. (2000) proposed that the inhibition of respiration in response to low oxygen requires the existence of an oxygen-sensing and signalling system in plants. This hypothesis is supported by several independent lines of experimental evidence as follows:

First, inhibition of respiration by low oxygen occurs at internal oxygen concentrations that are several magnitudes higher than the published Km (oxygen) of cytochrome oxidase (0.14 μM, equivalent to 0.01 % v/v oxygen, see Drew 1997).

This has been found in studies with developing potato tubers (Geigenberger et al. 2000; Geigenberger 2003), seeds (Vigeolas et al. 2003; van Dongen et al. 2004; Borisjuk and Rolletschek 2009), and fruits (Ho et al. 2010) as summarized in Table 1. It seems therefore to be unlikely that oxygen is so low in the centre of these tissues that it limits electron transport directly as a substrate. This view was recently supported by simulations using a multi-scale model, which incorporates the actual 3-D microstructure of the tissue, showing local intracellular oxygen concentrations near the core of apple fruits to be well above the reported Km values of cytochrome oxidase (Ho et al. 2011). While these fine-scale and advanced modeling data are in full support of the hypothesis initially proposed by Geigenberger et al. (2000), more work is needed to develop methods for intracellular oxygen measurements in plant tissues. This may involve microinjection of oxygen-sensing nanobeads into cells (Schmälzlin et al. 2005) or expression of genetically encoded FRET-based oxygen biosensors, which recently have been demonstrated to allow real-time determination of intracellular oxygen in bacteria in a noninvasive manner (Potzkei et al. 2012).

Second, the decrease in adenylate energy state as oxygen concentration falls reflects a progressive inhibition of respiration that occurs at oxygen concentrations that are clearly higher than those that lead to fermentation (Geigenberger 2003; van Dongen et al. 2003; Vigeolas et al. 2003; Borisjuk and Rolletschek 2009). In various metabolic studies, the proactive inhibition of respiration by oxygen as a signal could be clearly distinguished from direct inhibition by oxygen as a substrate, with the former depressing and the latter accelerating glycolysis and fermentation (Geigenberger et al. 2000; Geigenberger 2003; Vigeolas et al. 2003; Borisjuk and Rolletschek 2009).

Third, the signal metabolite nitric oxide (NO) has recently been identified to be part of an oxygen-sensing system that regulates respiration in response to normal changes in internal oxygen concentrations in developing seeds of pea and soybean (Borisjuk et al. 2007). A fall in internal oxygen leads to an increase in the level of NO, which in turn inhibits respiration, resulting in a rise of internal oxygen concentrations in the seeds to prevent fermentation (Benamar et al. 2008; Borisjuk et al. 2007; Table 1; Fig. 1).

Concomitant to the decrease in respiration and adenylate energy state, low internal oxygen also leads to a widespread inhibition of ATP-consuming processes, which are involved in biosynthesis and growth. This includes the synthesis of sucrose, amino acids, protein, lipids, starch, secondary products, and nucleotide cofactors (Geigenberger 2003). Protein, starch, and lipid biosynthesis have been found to be strongly limited by the decrease in internal oxygen concentrations during the normal development of potato tubers (Geigenberger 2003; Bologa et al. 2003; Vigeolas, van Dongen, and Geigenberger, unpublished) and diverse seeds (Vigeolas et al. 2003; van Dongen et al. 2004; Rolletschek et al. 2005a, b). This is most likely attributable to a decrease in the levels of adenine and uridine nucleotides, which serve as important cofactors to drive key reactions in these pathways. In many experiments with developing tubers and seeds, a strong correlation has been found between internal oxygen, adenylate energy state, and

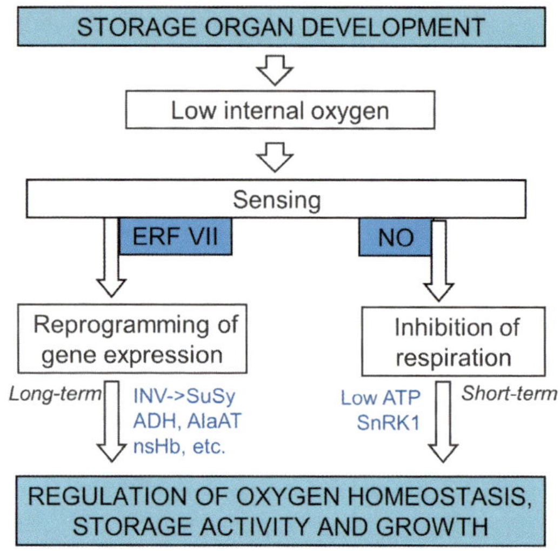

Fig. 1 *Oxygen-sensing and signalling pathways regulating storage metabolism in response to a fall in internal oxygen during the normal development of storage organs, such as potato tubers and seeds*. Short-term balancing of internal oxygen concentrations involves an oxygen-sensing system that regulates respiration within minutes in response to fluctuations in internal oxygen concentrations (Geigenberger 2003). The signal metabolite nitric oxide (NO) has been suggested to be involved in this response, since it is increased under low oxygen and inhibits respiration (Borisjuk et al. 2007). The resulting decrease in adenylate energy state leads to a widespread inhibition of ATP-consuming processes which are involved in storage and growth, possibly via SnRK1 signalling (Geigenberger 2003; Borisjuk and Rolletschek 2009; Geigenberger et al. 2010). Developmental changes in internal oxygen concentrations are also linked to global changes in gene transcription networks, leading to longer-term adaptive responses in storage metabolism and growth that require several hours or days to develop (van Dongen et al. 2009; Narsai et al. 2011). This involves a switch to more energy-conserving pathways of (1) sucrose degradation via sucrose synthase, (2) nucleotide synthesis via salvage pathways, and (3) respiratory metabolism via alanine aminotransferase (AlaAT). Moreover, non-symbiotic hemoglobins will allow NO levels to be decreased (Thiel et al. 2011) and oxygen delivery to be facilitated (Vigeolas et al. 2011). Group VII ethylene-response-factor (ERF VII) transcription factors are stabilized under hypoxia and degraded upon reoxygenation via the N-end rule pathway of targeted proteolysis, functioning as an oxygen-sensing mechanism in the regulation of transcriptional responses in plants (Licausi et al. 2011c; Gibbs et al. 2011). ERF VII genes were also identified to respond to a decrease in internal oxygen that occurs during the normal development of potato tubers (Licausi et al. 2011a). Co-expression analysis suggests a role of potato ERF VII genes in the regulation of sucrose and starch metabolism during tuber development (Licausi et al. 2011a)

biosynthetic activities of storage product synthesis (reviewed in Geigenberger 2003; Borisjuk and Rolletschek 2009). Interestingly, syntheses of different storage products revealed different sensitivities towards inhibition by low oxygen. Lipid and protein synthesis have been found to be more strongly affected than starch synthesis, which is probably attributable to the different energy requirements and regulation of these processes (Geigenberger 2003; Vigeolas et al. 2003; Rolletschek et al. 2005a). In developing seeds of broad bean, protein synthesis has been found to be mainly located in the outer region, where oxygen and ATP are relatively high, while starch synthesis dominates in the inner region of the seed, where oxygen and ATP are lowest (Rolletschek et al. 2003; Borisjuk et al. 2003).

The studies above provide extensive correlative evidence for a role of adenylates in the regulation of storage activities in response to low internal oxygen. To analyze the role of a decrease in adenylate energy state more directly without a concomitant decrease in oxygen supply, the ATP/ADP ratio was manipulated in growing potato tubers by over-expression of an ATP diphosphohydrolase (apyrase) using an inducible promoter (Riewe et al. 2008a). Inducing apyrase over-expression for a 24-h period resulted in a decrease in the ATP/ADP ratio that was accompanied by a decrease in the intermediates of sucrose-to-starch conversion and respiration, indicating a general depression of tuber metabolism. More long-term expression of apyrase using a constitutive tuber-specific promoter revealed a strong inhibition of tuber starch accumulation, a shift to the production of amylopectin instead of amylose, alterations in tuber morphology, and a decreased tuber size (Riewe et al. 2008a). There were corresponding changes in the rate of tuber starch synthesis when synthesis (Loef et al. 2001; Oliver et al. 2008), equilibration (Regierer et al. 2002; Oliver et al. 2008), salvaging (Riewe et al. 2008b, 2008c), or transport of adenylates (Geigenberger et al. 2001) were changed across various genetic and physiological manipulations in growing potato tubers. A similar relationship between ATP/ADP ratio and cotton fiber elongation has been found in response to manipulating mitochondrial ATP synthase (Pang et al. 2010). More recently, cofactome analysis in developing oat seeds provided evidence that an elevated supply of nucleotide cofactors also supports carbon partitioning into the production of storage oil and protein (Hayden et al. 2011), while an increase in nucleotide cofactors by injecting nucleotide precursors into seeds resulted in an increased flux to storage lipids in seeds of oil-seed rape (Vigeolas and Geigenberger, unpublished).

The results presented above provide cumulative evidence that the widespread inhibition of biosynthetic processes in response to low oxygen is most likely due to a decrease in the adenylate energy status, rather than to direct oxygen signalling effects. Regulation of biosynthetic activities can be achieved by (1) direct interaction of ATP with the respective enzymes in the biosynthetic pathways or by (2) ATP signalling (Geigenberger et al. 2010). The latter may involve the conserved SNF1-related kinase (SnRK1), which has been found to inhibit a number of biosynthetic enzymes by reversible protein phosphorylation (Sugden et al. 1999) and to affect hypoxic gene expression (Baena-González et al. 2007) and flooding tolerance in plants (Lee et al. 2009). Interestingly, SnRK1 is also involved in sugar signalling pathways, allowing sugar and hypoxic responses to be coordinated in plants. In developing potato tubers (Tiessen et al. 2003; McKibbin et al. 2006) and pea seeds (Radchuck et al. 2010), manipulation of SnRK1 has been found to affect starch accumulation. Analysis of the key enzyme of starch synthesis, AGPase, revealed a central role of this enzyme in the adaptation of starch biosynthesis to low oxygen and low energy in many of these studies. This involves regulation of

AGPase by changes in the concentration of ATP as a substrate (Geigenberger et al. 2004), posttranslational redox-regulation by thiol-disulfide modulation (Oliver et al. 2008; Riewe et al. 2008a), and transcriptional regulation (reviewed in Geigenberger 2011).

3.2 Low Internal Oxygen Leads to a Switch to Pathways that Consume Less Energy

A complementary strategy to a depression of metabolism would be to prioritize metabolic pathways that conserve energy and utilize oxygen more efficiently. One of the unusual features of plant metabolism is the existence of alternative pathways for the same metabolic reaction. While this has previously been attributed to increase the flexibility of metabolism in a changing environment, recent studies indicate this feature to be linked to the need to decrease oxygen consumption in developing storage tissues.

One example is the use of inorganic pyrophosphate (PPi) as alternative energy donor to ATP in many reactions of central metabolism and transport (Stitt 1998). While PPi is produced as a waste-product in many biosynthetic reactions, recycling of this waste energy to fuel important central metabolic and cellular functions will save ATP and ultimately lead to a decrease in oxygen consumption in storage tissues. There is physiological and genetic evidence for an important role of PPi under low internal oxygen conditions in developing tubers and seeds. While ATP falls to low levels in response to low oxygen in these tissues, PPi concentrations are maintained (Geigenberger et al. 2000; Gibon et al. 2002; Vigeolas et al. 2003). Moreover, genetic manipulations of PPi concentrations in transgenic tubers (Sonnewald 1992; Geigenberger et al. 1998) and seeds (Meyer et al. 2012) lead to corresponding changes in storage activities under the prevailing low internal oxygen concentrations. Intriguingly, storage lipid accumulation is increased in developing seeds lacking a cytosolic inorganic pyrophosphatase (Meyer et al. 2012).

A second example is the existence of two alternative pathways for sucrose degradation to hexose phosphates, which differ in their energy requirements (Bologa et al. 2003). While the breakdown of a molecule of sucrose by invertase and hexokinase requires two molecules of ATP, its breakdown by sucrose synthase and UDP-glucose pyrophosphorylase requires only one molecule of PPi. A main feature of low oxygen responses in many tissues, including developing tubers and seeds, is an up-regulation of sucrose synthase, while invertase is repressed (Bologa et al. 2003; see also Sect. 4). This represents an adaptation to low oxygen, since it provides a shift to a more energy-saving pathway of sucrose degradation that consumes less ATP and utilize oxygen more efficiently. Invertase is expressed early and sucrose synthase later in the development of potato tubers (Appeldoorn et al. 1997), seeds (Weber et al. 1997), and sugar cane stalks (Watt 2005), as storage activity increases and internal oxygen concentration decreases in these organs. This implies that the developmental switch from invertase to sucrose synthase is most likely activated by falling internal oxygen concentrations in these tissues, to allow higher storage activities as would otherwise be possible.

Genetic evidence confirms this interpretation in growing potato tubers. A transgenic bypass of the sucrose synthase pathway through ectopic over-expression of invertase or sucrose phosphorylase in growing potato tubers led to a decrease in internal oxygen concentrations, adenylate energy state, and storage starch accumulation, resulting in impaired tuber development (Bologa et al. 2003). Conversely, over-expression of sucrose synthase in growing potato tubers led to increased accumulation of tuber starch (Baroja-Fernández et al. 2009), while over-expression of sucrose synthase in cotton seeds led to increased biomass, fiber yield, and quality (Jiang et al. 2012). This contrasts with studies showing that sucrose synthase is redundant in vegetative tissues that reveal no oxygen problems (Barratt et al. 2009), implying a specific role of sucrose synthase in the response to low internal oxygen in developing storage tissues to allow higher rates of biosynthetic activities.

A third example is the existence of multiple routes of nucleotide biosynthesis in plants. This holds specifically true for the synthesis of uridine nucleotides, which are important cofactors in the use of sucrose for various biosynthetic processes, including starch and cell wall biosynthesis (Loef et al. 1999). In addition to the de-novo pathway of uridine nucleotide synthesis, several salvage pathways exist that allow cells to use preformed nucleotides as precursors, thereby avoiding the high energy costs of de-novo biosynthesis. This energy efficient means of uridine nucleotide synthesis is specifically important in developing storage organs such as tubers, which can use preformed nucleosides and/or nucleobases delivered by the phloem. In developing potato tubers, low oxygen concentrations lead to an induction of genes involved in salvaging of nucleotides, while genes involved in the de-novo pathway remained unaffected (Langer, van Dongen, and Geigenberger, unpublished results). A shift to more energy-conserving pathways of nucleotide synthesis will allow more nucleotides to be made and thus more energy to be available to fuel biosynthetic processes under low oxygen conditions. This has been confirmed in a reverse-genetic approach, where induction of uridine salvage pathways relative to the de-novo pathway led to increased uridine nucleotide levels and increased accumulation of starch and cell wall components in transgenic potato tubers (Geigenberger et al. 2005).

3.3 *Low Internal Oxygen Leads to an Inhibition of Phloem Transport and Unloading*

For several plant species it has been shown that low oxygen concentrations lead to impaired growth and development of seeds (Porterfield et al. 1999; Kuang et al. 1998; Akita and Tanaka 1973; Quebedeaux and Hardy 1975), tubers (Langer, van Dongen, and Geigenberger, unpublished), and roots (van Dongen et al. 2009). This will ultimately save ATP and decrease oxygen consumption in these tissues. The inhibition of growth under low oxygen concentrations could be a direct consequence of the inhibition of storage metabolism under these conditions (see above). In developing potato tubers (Geigenberger et al. 2004) and barley seeds (Faix et al. 2012), it has been found that a specific block in starch synthesis by inactivating AGPase will lead to an inhibition of growth and development, rather than inducing other metabolic pathways. A second possibility is that low oxygen leads to an inhibition of phloem transport (van Dongen et al. 2003) and unloading of assimilates in these tissues (Thorne 1982; van Dongen et al. 2004; Ferner et al. 2012). The low oxygen concentrations found within the phloem of castor bean plants (around 5–6 % v/v) are indeed in a range at which they limit phloem energy metabolism and sucrose transport under normal growth conditions (van Dongen et al. 2003). Conversely, the prevailing low oxygen concentrations in developing wheat seeds (approx. 2 % v/v) were found to be limiting for energy metabolism and sucrose unloading from the phloem (van Dongen et al. 2004). This was revealed by in-planta labeling studies using the carbon isotope C-11, showing that manipulation of seed oxygen concentration affects the rate of carbon translocation from source leaves to the developing seeds (van Dongen et al. 2004). Overall, these studies show that low internal oxygen leads to a coordinated inhibition of phloem import and storage metabolism in developing seeds, allowing growth rates to be decreased without large changes in metabolite levels.

In developing seeds, there are no symplastic connections between maternal and filial tissues, where storage metabolism takes place. Uptake of sucrose and amino acids therefore occurs via an apoplastic route, which requires ATP to energize the transport proteins at the plasmalemma of the sink cells (Weber et al. 1997). Down-regulation of these transport steps in response to low oxygen will save ATP and allow oxygen consumption to be decreased. During the development of potato tubers, there is a switch in the mechanism of phloem unloading from an apoplastic to a more energy-conserving symplastic mode (Viola et al. 2001). This switch in the unloading mechanism occurs in the early stages of tuber development when internal oxygen concentrations are starting to decrease. It is paralleled by an analogous switch to a more energy-conserving pathway of sucrose degradation via sucrose synthase (see above). While symplastic unloading via plasmodesmata is probably less efficient as apoplastic unloading, it will allow growing tubers to save energy and to keep higher internal oxygen concentrations as would otherwise be possible.

4 Reprogramming of Storage Metabolism at the Transcriptional Level in Response to Low-Oxygen Signals

During the last 10 years, post-genomic tools have been applied to study the molecular mechanisms underlying low oxygen responses in plants. Transcript and metabolite profiling of hypoxic responses were mainly applied to roots and seedlings of diverse species, analyzing the effect of very low external oxygen concentrations close to anoxia (reviewed by Narsai et al. 2011). In van Dongen et al. (2009), the response of transcript and metabolite profiles to mild changes in oxygen concentrations (1, 4, 8, 12, and 21 % v/v) were studied in Arabidopsis roots, which resemble the normal alterations in internal oxygen concentrations in developing tubers and seeds (see Table 1). Low oxygen led to a preferential up-regulation of genes that are potentially important to trigger adaptive responses in the plants. These included genes encoding (1) sucrose synthase providing an energy-conserving pathway of sucrose degradation to save oxygen and prevent anoxia (see above), (2) pyruvate decarboxylase and alcohol dehydrogenase involved in the induction of fermentation to cope with anoxia once it may arrive, as well as (3) transcription factors (LOB domain and group VII ethylene-response factors, see also Licausi et al. 2011b) and non-symbiotic hemoglobins potentially involved in low-oxygen signalling. Genes that were downregulated mainly encoded proteins involved in energy-consuming processes, such as transport, signalling, lipid metabolism, secondary metabolism, and redox-regulation. This is in line with the inhibition of biosynthetic activities, transport, and growth in response to low oxygen as observed in metabolic studies (see Sect. 3) which will ultimately save ATP and decrease oxygen consumption.

In developing storage organs of crops, detailed studies on transcript and metabolite profiling of the low oxygen response are largely lacking. In citrus fruits, gene expression profiling allowed the identification of low oxygen-regulated genes involved in plant development, carbohydrate, and amino acid metabolism, as well as in the biosynthesis of brassinosteroids, vitamins, cofactors, and starch (Pasentsis et al. 2007). Expression profiling in growing potato tubers using a POCI array (Kloosterman et al. 2008) was used to analyze changes in transcript levels in response to a decrease in oxygen to 4 % (Langer, van Dongen, and Geigenberger, unpublished). Results were similar to the study of van Dongen et al. (2009) with Arabidopsis roots. A number of genes with potential importance to trigger adaptive responses were upregulated, encoding proteins involved in central metabolism (sucrose synthase, alanine aminotransferase, isocitrate lyase, 1-aminocyclopropane-1-carboxylate oxidase), fermentation (pyruvate decarboxylase, alcohol dehydrogenase, lactate dehydrogenase), and signalling (calmodulin-binding, heat shock, LOB domain proteins). Interestingly, there was also an up-regulation of genes involved in the metabolism of the signal metabolite trehalose-6-phosphate (Tre6P). Tre6P has been found to be an important sugar signal regulating storage metabolism and development in various plant tissues, including developing seeds (Gomez et al. 2006; Radchuck et al. 2010; Martínez-Barajas et al. 2011) and potato tubers (Kolbe et al. 2005; Debast et al. 2011). Under the low internal oxygen conditions in these tissues, Tre6P could be important to coordinate sugar and hypoxic responses during development, similar to SnRK1 (see above). This is underlined by recent studies showing interaction of Tre6P and SnRK1 in growing tissues that are prone to internal hypoxia such as seeds (Martínez-Barajas et al. 2011) and tubers (Debast et al. 2011), but not in photosynthesizing leaves (Zhang et al. 2009). Interestingly, perturbation of Tre6P levels in growing potato tubers not only affected starch accumulation, but also had strong effects on respiration rates (Debast et al. 2011).

Genes that were downregulated by low oxygen in growing potato tubers encoded enzymes involved in energy-consuming processes, including the synthesis of secondary metabolites, such as flavonoids and terpens (Langer, van Dongen, and Geigenberger, unpublished). This will support the inhibition of metabolic processes to save ATP and oxygen (see above).

Metabolite profiling of growing potato tubers by gas chromatography-mass-spectrometry (GC-MS) was used to analyze global changes in metabolite levels in response to low oxygen (Langer, Faix, van Dongen, Fernie, and Geigenberger, unpublished). First, there were characteristic changes in glycolytic and TCA-cycle intermediates, indicating pyruvate kinase, aconitase, and succinate dehyrogenase as potential sites for the hypoxic regulation of respiration. The latter two have also been implicated to be regulated by the low-oxygen signal NO (Besson-Bard et al. 2008). Second, a large number of amino acids have been found to be strongly increased, indicating an inhibition of protein synthesis at the level of translation to conserve energy in a manner as it has been found previously in hypoxic Arabidopsis plants (see Branco-Price et al. 2005). Third, alanine, succinate, and gamma-aminobutyrate were strongly increased, indicating induction of an energy-conserving pathway linking glycolysis with the TCA cycle via alanine aminotransferase (see Narsai et al. 2011). This is *in-line* with the strong increase in alanine aminotransferase in the potato transcript profiles at low oxygen (see above). Fourth, there was a strong decrease in chlorogenic acid, the main secondary compound accumulating in potato tubers, confirming that energy-consuming biosynthetic processes have been inhibited.

Non-symbiotic class 1 hemoglobins (Hb1) have been found to be induced under hypoxic conditions in many plant tissues, including seeds (Gupta et al. 2011). They act as NO scavenger, maintaining the energy and redox status under low oxygen conditions (Sowa et al. 1998; Thiel et al. 2011). While Hb1 over-expression has been found to promote survival under severe hypoxia, its role in regulating storage product formation remains unclear (Gupta et al. 2011). Also a second class of non-symbiotic hemoglobins (Hb2) has been implicated in low-oxygen responses, although it is not directly induced by low oxygen concentrations (Gupta et al. 2011). Recent results provide genetic evidence for a specific function of Hb2 in seed oil production and in promoting the accumulation of poly-unsaturated omega-3 fatty acids by facilitating oxygen supply in developing Arabidopsis seeds (Vigeolas et al. 2011). The 40 % increase in oil content was due to a threefold higher energy state leading to a stimulation of the rate of triacylglycerol synthesis during normal seed development. Under low external oxygen, Hb2 over-expression maintained an up to fivefold higher energy state and prevented fermentation. This shows Hb1 and Hb2 to have different functions in plants, with the former regulating NO and energy metabolism under hypoxia and the latter facilitating oxygen delivery in developing storage organs to allow higher rates of storage product formation than would otherwise be possible (Vigeolas et al. 2011).

5 A Role of ERF Transcription Factors in Oxygen-Sensing and Signalling Pathways Linking Storage Metabolism with the Development of Potato Tubers

The results presented above show that developmental changes in internal oxygen concentrations are linked to global changes in gene transcription networks, leading to adaptive responses in storage metabolism and growth in tubers and seeds. From the discussion presented in Sect. 3.1 of the present chapter it becomes evident that this requires oxygen-sensing and signalling systems to be operating in plants. There has been rapid progress in our understanding of these low oxygen signalling pathways in the last years (see Fig.1). Recent reports indicate group VII ethylene-response-factor (ERF VII) transcription factors such as HRE1, HRE2 (Licausi et al. 2010), RAP2.2, and RAP2.12 (Hinz et al. 2010) as important regulators of hypoxic gene expression and survival in Arabidopsis plants. Intriguingly, it was shown that group VII ERFs are stabilized under hypoxia and degraded upon reoxygenation via the N-end rule pathway of targeted proteolysis, functioning as an oxygen-sensing mechanism in Arabidopsis (Licausi et al. 2011c; Gibbs et al. 2011).

Three hypoxia-responsive members of the ERF group VII were also identified in growing potato tubers, StHRE1, StHRE2a, and StHRE2b (Licausi et al. 2011a). Intriguingly, these StHRE genes did not only respond to external hypoxia applied to the plant, but were also strongly affected by self-imposed low oxygen conditions that are part of normal tuber development and growth. Specifically, the expression of these genes was regulated by the gradual decrease in oxygen tension that occurs when potato tubers grow larger during their development. Co-expression analysis suggests a role of potato HREs in the regulation of sucrose and starch metabolism during tuber development (Licausi et al. 2011a). This implicates potato HREs to be involved in oxygen-sensing and signalling mechanisms linking storage metabolism with tuber development (Fig. 1). Since potato HREs are very similar to the Arabidopsis ERFs RAP2.2 and RAP2.12, oxygen-sensing via the N-end rule pathway is most likely also operating in developing potato tubers in a similar manner as it has been discovered in Arabidopsis (see above). More studies will be required to provide genetic evidence for the role played by HREs in potato tuber development and to investigate whether results can be extended to seeds.

Intriguingly, recent studies show that tuber development is impaired by over-expression of a heterologous ERF VII gene in transgenic potato tubers (Youm et al. 2008).

6 Implications of the Normal Occurrence of Oxygen Limitations for Crop Yield

In developing potato tubers and diverse seeds, low internal oxygen concentrations have been found to limit sucrose import rates (van Dongen et al. 2004) and the conversion of carbon to storage starch (Geigenberger 2003; Bologa et al. 2003; van Dongen et al. 2004), proteins (Geigenberger 2003; Borisjuk et al. 2003; Vigeolas et al. 2003), and lipids (Vigeolas et al. 2003; Rolletschek et al. 2005a). While these studies clearly demonstrate internal oxygen to be a limiting factor for storage metabolism, much less is known on its implications for crop productivity. In previous studies with potato, oxygen availability in the soil has been found to affect tuber yield and quality (Holder and Cary 1984; Cary 1986). To investigate whether tuber yield can be increased by increasing oxygen supply, growing potato tubers were exposed to super-ambient oxygen concentrations for 6 weeks, while roots and other parts of the plants remained at ambient oxygen concentrations (Langer, van Dongen, and Geigenberger, unpublished results). Increased oxygen supply led to an increase in tuber yield per plant and to an increase in starch content per tuber at the time of harvest. This was accompanied by an increase in ATP levels, providing evidence that increased tuber yield at elevated oxygen supply was attributable to an improved energy metabolism.

The relationship between oxygen supply and crop yield has clear implications for potential strategies to increase plant productivity. First, molecular strategies could be employed to increase oxygen availability and/or transport within the storage tissue. This could involve an increase of the photosynthetic capacity and/or over-expression of oxygen transport proteins. Intriguingly, over-expression of Hb2 improved oxygen availability in developing Arabidopsis seeds, resulting in an increase in oil quantity and quality under normal growth conditions (Vigeolas et al. 2011). Second, potential regulators of low-oxygen responses could be manipulated in developing potato tubers or seeds, such as ERF transcription factors or NO. While over-expression of class-1 hemoglobin has been found to decrease NO levels under hypoxic stress, its effect on seed yield and crop productivity under normal growth conditions still has to be determined (Thiel et al. 2011). Third, direct genetic manipulations could be applied to specifically improve supply, conversion, or transport of nucleotide cofactors in storage tissues. In this respect, tuber-specific antisense repression of plastidial adenylate kinase resulted in a strong increase in ATP levels leading to a 30–40 % increase of potato yield in several field trials (Regierer et al. 2002). A similar increase in potato yield was observed when uridine nucleotide levels were increased in transgenic potato tubers providing a more energy-saving pathway of uridine nucleotide synthesis (Geigenberger et al. 2005).

7 Concluding Remarks

Here, we have summarized recent studies documenting that oxygen deprivation is not only restricted to stress conditions such as flooding, but is also associated with the normal development and growth of storage organs such as tubers, fruits, and seeds in well-oxygenated surroundings. The decrease in internal oxygen concentrations is sensed in these tissues, leading to adaptive responses in storage metabolism and growth. ERF VII transcription factors have been identified to function as oxygen sensors in the transcriptional regulation of these processes, while well-known metabolic signals such as SnRK1 and NO have been implied in posttranslational regulation. More work is needed to fully elucidate the oxygen-sensing and signalling system in developing storage organs, its coordination with other signalling systems (specifically sugar signalling), and its significance to regulate storage metabolism, development, and yield.

Acknowledgments Work from the author's laboratory was supported by the Deutsche Forschungsgemeinschaft.

References

Akita S, Tanaka I (1973) Studies on the mechanism of differences in photosynthesis among species III. Influence of low oxygen concentration on dry matter production and grain fertility of rice. Proc Crop Sci Soc Jpn 42:18–23

Appeldoorn NJG, de Bruijn SM, Koot-Gronsveld EAM, Visser RGF, Vreugdenhil D, van der Plas LHW (1997) Developmental changes of enzymes involved in conversion of sucrose to hexose-phosphate during early tuberisation of potato. Planta 202:220–226

Armstrong W, Strange ME, Cringle S, Beckett PM (1994) Microelectrode and modelling study of oxygen distribution in roots. Ann Bot 74:287–299

Ast C, Schmälzlin E, Löhmannsröben H-G, van Dongen JT (2012) Optical oxygen micro- and nanosensors for plant applications. Sensors (Basel) 12:7015–7032

Baena-González E, Rolland F, Thevelein JM, Sheen J (2007) A central integrator of transcription networks in plant stress and energy signalling. Nature 448:938–942

Bailey-Serres J, Voesenek LACJ (2008) Flooding stress: acclimations and genetic diversity. Annu Rev Plant Biol 59:313–339

Banks NH (1983) Evaluation of methods for determining internal gases in banana fruit. J Exp Bot 34:871–879

Baroja-Fernández E, Muñoz FJ, Montero M, Etxeberria E, Sesma MT, Ovecka M, Bahaji A, Ezquer I, Li J, Prat S, Pozueta-Romero J (2009) Enhancing sucrose synthase activity in transgenic potato (Solanum tuberosum L.) tubers results in increased levels of starch, ADPglucose and UDPglucose and total yield. Plant Cell Physiol 50:1651–1662

Barratt DHP et al (2009) Normal growth of Arabidopsis requires cytosolic invertase but not sucrose synthase. Proc Natl Acad Sci U S A 106:13124–13129

Benamar A, Rolletschek H, Borisjuk L, Avelange-Macherel MH, Curien G, Mostefai HA, Andriantsitohaina R, Macherel D (2008) Nitrite-nitric oxide control of mitochondrial respiration at the frontier of anoxia. Biochim Biophys Acta 1777:1268–1275

Besson-Bard A, Pugin A, Wendehenne D (2008) New insights into nitric oxide signaling in plants. Annu Rev Plant Biol 59:21–39

Biais B, Beauvoit B, William Allwood J, Deborde C, Maucourt M, Goodacre R, Rolin D, Moing A (2010) Metabolic acclimation to hypoxia revealed by metabolite gradients in melon fruit. J Plant Physiol 167:242–245

Bologa KL, Fernie AR, Leisse A, Ehlers Loureiro M, Geigenberger P (2003) A bypass of sucrose synthase leads to low internal oxygen and impaired metabolic performance in growing potato tubers. Plant Physiol 132:2058–2072

Borisjuk L, Rolletschek H (2009) The oxygen status of the developing seed. New Phytol 182: 17–30

Borisjuk L, Rolletschek H, Walenta S, Panitz R, Wobus U, Weber H (2003) Energy status and its control on embryogenesis of legumes: ATP distribution within Vicia faba embryos is developmentally regulated and correlated with photosynthetic capacity. Plant J 36:318–329

Borisjuk L, Nguyen TH, Neuberger T, Rutten T, Tschiersch H, Claus B, Feussner I, Webb AG, Jacob P, Weber H et al (2005) Gradients of lipid storage, photosynthesis and plastid differentiation in developing soybean seeds. New Phytol 167:761–776

Borisjuk L, Macherel D, Benamar A, Wobus U, Rolletschek H (2007) Low oxygen sensing and balancing in plant seeds: a role for nitric oxide. New Phytol 176:813–823

Branco-Price C, Kawaguchi R, Ferreira RB, Bailey-Serres J (2005) Genome-wide analysis of transcript abundance and translation in Arabidopsis seedlings subjected to oxygen deprivation. Ann Bot 96:647–660

Brändle (1968) Die Verteilung der Sauerstoffkonzentrationen in fleischigen Speicherorganen (Äpfel, Bananen und Kartoffelknollen). Berichte der Schweizerischen botanischen Gesellschaft 78:330–364

Brändle R, Wanner H (1965) Die Verteilung der Sauerstoffkonzentration in reifenden Äpfeln. Planta 65:65–72

Cary JW (1986) Effects of relative humidity, oxygen, and carbon dioxide on initiation and early development of stolons and tubers. Am J Potato Res 63:619–627

Chapman KD, Ohlrogge JB (2012) Compartmentation of triacylglycerol accumulation in plants. J Biol Chem 287:2288–2294

Debast S, Nunes-Nesi A, Hajirezaei MR, Hofmann J, Sonnewald U, Fernie AR, Börnke F (2011) Altering trehalose-6-phosphate content in transgenic potato tubers affects tuber growth and alters responsiveness to hormones during sprouting. Plant Physiol 156:1754–1771

Devaux H (1891) Etude experimentale sur l'aeration des tissus massifs. Annales des sciences naturelles (Botanique) 7 serie 14:297–308

Drew MC (1997) Oxygen deficiency and root metabolism: injury and acclimation under hypoxia and anoxia. Annu Rev Plant Physiol Plant Mol Biol 48:223–250

Faix B, Radchuk V, Nerlich A, Hümmer C, Radchuk R, Emery RJN, Keller H, Götz K-P, Weschke W, Geigenberger P, Weber H (2012) Barley grains, deficient in cytosolic small subunit of ADP-glucose pyrophosphorylase, reveal coordinate adjustment of C:N metabolism mediated by an overlapping metabolic-hormonal control. Plant J 69:1077–1093

Ferner E, Rennenberg H, Kreuzwieser J (2012) Effect of flooding on C-metabolism of flood tolerant (*Quercus robur*) and non-tolerant (*Fagus sylvatica*) tree species. Tree Physiol 32: 135–145

Forristal CE, Wright KL, Hanley NA, Oreffo RO, Houghton FD (2010) Hypoxia inducible factors regulate pluripotency and proliferation in human embryonic stem cells cultured at reduced oxygen tensions. Reproduction 139:85–97

Geigenberger P (2003) Response of plant metabolism to too little oxygen. Curr Opin Plant Biol 6:247–256

Geigenberger P (2011) Regulation of starch biosynthesis in response to a fluctuating environment. Plant Physiol 155:1566–1577
Geigenberger P, Fernie AR (2006) Starch synthesis in the potato tuber. In: Hui YH, Corke H, DeLeyn I, Nip W-K, Cross N (eds) Food biochemistry and food processing. Blackwell, Ames, Iowa, pp 253–270
Geigenberger P, Hajirezaei M, Geiger M, Deiting U, Sonnewald U, Stitt M (1998) Overexpression of pyrophosphatase leads to increased sucrose degradation and starch synthesis, increased activities of enzymes involved in sucrose-starch interconversion, and increased levels of nucleotides in growing potato tubers. Planta 205:428–434
Geigenberger P, Fernie AR, Gibon Y, Christ M, Stitt M (2000) Metabolic activity decreases as an adaptive response to low internal oxygen in growing potato tubers. Biol Chem 381:723–740
Geigenberger P, Stamme C, Tjaden J, Schulz A, Quick PW, Betsche T, Kersting HJ, Neuhaus HE (2001) Tuber physiology and properties of starch from tubers of transgenic potato plants with altered plastidic adenylate transporter activity. Plant Physiol 125:1667–1678
Geigenberger P, Stitt M, Fernie AR (2004) Metabolic control analysis and regulation of the conversion of sucrose to starch in growing potato tubers. Plant Cell Environ 27:655–673
Geigenberger P, Regierer B, Nunes-Nesi A, Leisse A, Urbanczyk-Wochniak E, Springer F, van Dongen JT, Kossmann J, Fernie AR (2005) Inhibition of de novo pyrimidine synthesis in growing potato tubers leads to a compensatory stimulation of the pyrimidine salvage pathway and a subsequent increase in biosynthetic performance. Plant Cell 17:2077–2088
Geigenberger P, Riewe D, Fernie AR (2010) The central regulation of plant physiology by adenylates. Trends Plant Sci 15:98–105
Gerber C (1896) Recherches sur la maturation des fruits charnus. Ann Sci Bot VIII 4:1–280
Gibbs DJ et al (2011) Homeostatic response to hypoxia is regulated by the N-end rule pathway in plants. Nature 479:415–418
Gibon Y, Vigeolas H, Tiessen A, Geigenberger P, Stitt M (2002) Sensitive and high throughput metabolite assays for inorganic pyrophosphate, ADPGlc, nucleotide phosphates, and glycolytic intermediates based on a novel enzymic cycling system. Plant J 30:221–235
Gomez LD, Baud S, Gilday A, Li Y, Graham IA (2006) Delayed embryo development in the Arabidopsis trehalose 6-phosphate synthase 1 mutantis associated with altered cell wall structure, decreased cell division and starch accumulation. Plant J 46:69–84
Gupta KJ, Hebelstrup KH, Mur LA, Igamberdiev AU (2011) Plant hemoglobins: important players at the crossroads between oxygen and nitric oxide. FEBS Lett 585:3843–3849
Hayden DM, Rolletschek H, Borisjuk L, Corwin J, Kliebenstein DJ, Grimberg A, Stymne S, Dehesh K (2011) Cofactome analyses reveal enhanced flux of carbon into oil for potential biofuel production. Plant J 67:1018–1028
Hinz M, Wilson IW, Yang J, Buerstenbinder K, Llewellyn D, Dennis ES et al (2010) Arabidopsis RAP2.2: an ethylene response transcription factor that is important for hypoxia survival. Plant Physiol 144:218–231
Ho QT, Verboven P, Verlinden BE, Schenk A, Delele MA, Rolletschek H, Vercammen J, Nicolaï BM (2010) Genotype effects on internal gas gradients in apple fruit. J Exp Bot 61:2745–2755
Ho QT, Verboven P, Verlinden BE, Herremans E, Wevers M, Carmeliet J, Nicolaï BM (2011) A three-dimensional multiscale model for gas exchange in fruit. Plant Physiol 155:1158–1168
Holder CB, Cary JW (1984) Soil oxygen and moisture in relation to Russet Burbank potato yield and quality. Am J Potato Res 61:67–75
Jiang Y, Guo W, Zhu H, Ruan YL, Zhang T (2012) Overexpression of GhSusA1 increases plant biomass and improves cotton fiber yield and quality. Plant Biotechnol J 10:301–312
Kloosterman B, De Koeyer D, Griffiths R, Flinn B, Steuernagel B, Scholz U, Sonnewald S, Sonnewald U, Bryan GJ, Prat S, Bánfalvi Z, Hammond JP, Geigenberger P, Nielsen KL, Visser RGF, Bachem CWB (2008) Genes driving potato tuber initiation and growth: identification based on transcriptional changes using the POCI array. Funct Integr Genomics 8:329–340

Kolbe A, Tiessen A, Schluepmann H, Paul M, Ulrich S, Geigenberger P (2005) Trehalose-6-phosphate regulates starch synthesis via post-translational redox-activation of ADP-glucose pyrophosphorylase. Proc Natl Acad Sci U S A 102:11118–11123

Kuang A, Crispi M, Musgrave ME (1998) Control of seed development in Arabidopsis thaliana by atmospheric oxygen. Plant Cell Environ 21:71–78

Lee KW, Chen PW, Lu C-A, Chen S, Ho T-H, Yu S-M (2009) Coordinated responses to oxygen and sugar deficiency allow rice seedlings to tolerate flooding. Sci Signal 2:ra61

Licausi F, van Dongen JT, Giuntoli B, Novi G, Santaniello A, Geigenberger P, Perata P (2010) HRE1 and HRE2, two hypoxia-inducible ethylene response factors, affect anaerobic responses in Arabidopsis thaliana. Plant J 62:302–315

Licausi F, Giorgi F, Schmälzlin E, Usadel B, Perata P, van Dongen J, Geigenberger P (2011a) HRE-type genes are regulated by growth-related changes in internal oxygen concentrations during the normal development of potato (Solanum tuberosum) tubers. Plant Cell Physiol 52:1957–1972

Licausi F, Weits DA, Pant BD, Scheible W-R, Geigenberger P, van Dongen JT (2011b) Hypoxia responsive gene expression is mediated by various subsets of transcription factors and miRNAs that are determined by the actual oxygen availability. New Phytol 190:442–456

Licausi F, Kosmacz M, Weits DA, Giuntoli B, Giorgi FM, Voesenek LACJ, Perata P, van Dongen JT (2011c) Oxygen sensing in plants is mediated by an N-end rule pathway for protein destabilization. Nature 479:419–423

Loef I, Stitt M, Geigenberger P (1999) Feeding orotate leads to a specific increase in uridine nucleotide levels, resulting in a stimulation of sucrose degradation and starch synthesis in discs of growing potato tubers. Planta 209:314–323

Loef I, Stitt M, Geigenberger P (2001) Increased adenine nucleotide levels modify the interaction between respiration and starch synthesis when adenine is fed to discs of growing potato tubers. Planta 212:782–791

Magness JR (1920) Composition of gases in intercellular spaces of apples and potatoes. Bot Gaz 70:308–316

Martínez-Barajas E, Delatte T, Schluepmann H, de Jong GD, Somsen GW, Nunes C, Primavesi LF, Coello P, Mitchell RAC, Paul MJ (2011) Wheat grain development is characterized by remarkable trehalose 6-phosphate accumulation pregrain filling: tissue distribution and relationship to SNF1-related protein kinase1 activity. Plant Physiol 156:373–381

McKibbin RS, Muttucumarub N, Paul MJ, Powers SJ, Burrell MM, Coates S, Purcell PC, Tiessen A, Geigenberger P, Halford NG (2006) Production of high starch, low glucose potatoes through over-expression of the metabolic regulator, SnRK1. Plant Biotechnol J 4:409–418

Meyer K, Stecca KL, Ewell-Hicks K, Allen SM, Everard JD (2012) Oil and protein accumulation in developing seeds is influenced by the expression of a cytosolic pyrophosphatase in Arabidopsis. Plant Physiol 159:1221–1234

Narsai R, Rocha M, Geigenberger P, Whelan J, van Dongen JT (2011) Comparative analysis between plant species of transcriptional and metabolic responses to hypoxia. New Phytol 190: 472–487

Oliver SN, Tiessen A, Fernie AR, Geigenberger P (2008) Decreased expression of plastidial adenylate kinase in potato tubers results in an enhanced rate of respiration and a stimulation of starch synthesis that is attributable to post-translational redox-activation of ADP-glucose pyrophosphorylase. J Exp Bot 59:315–325

Pang Y, Wang H, Song WQ, Zhu YX (2010) The cotton ATP synthase delta1 subunit is required to maintain a higher ATP/ADP ratio that facilitates rapid fiber elongation. Plant Biol (Stuttg) 12:903–909

Pasentsis K, Falara V, Pateraki I, Gerasopoulos D, Kanellis AK (2007) Identification and expression profiling of low oxygen regulated genes from *Citrus flavedo* tissues using RT-PCR differential display. J Exp Bot 58:2203–2216

Porterfield D, Kuang A, Smith P, Crispi M, Musgrave ME (1999) Oxygen-depleted zones inside reproductive structures of Brassicaceae: implications for oxygen control of seed development. Can J Biochem Physiol 77:1439–1446

Potzkei J, Kunze M, Drepper T, Gensch T, Jaeger KE, Büchs J (2012) Real-time determination of intracellular oxygen in bacteria using a genetically encoded FRET-based biosensor. BMC Biol 10:28

Quebedeaux B, Hardy RWF (1975) Reproductive growth and dry matter production of *Glycine max* (L.) Merr. In response to oxygen concentration. Plant Physiol 55:102–107

Radchuck R, Emery NEJ, Weier D, Vigeolas H, Geigenberger P, Lunn JE, Feil R, Weschke W, Weber H (2010) Sucrose non-fermenting kinase 1 (SnRK1) coordinates metabolic and hormonal signals during pea cotyledon growth and differentiation. Plant J 61:324–338

Regierer B, Fernie AR, Springer F, Perez-Melis A, Leisse A, Koehl K, Willmitzer L, Geigenberger P, Kossmann J (2002) Starch content and yield increase as a result of altering adenylate pools in transgenic plants. Nat Biotechnol 20:1256–1260

Riewe D, Grosman L, Zauber H, Wucke C, Fernie AR, Geigenberger P (2008a) Metabolic and developmental adaptations of growing potato tubers in response to specific manipulations of the adenylate energy status. Plant Physiol 146:1579–1598

Riewe D, Grosman L, Fernie AR, Wucke C, Geigenberger P (2008b) The potato-specific apyrase is apoplastically localized and has influence on gene expression, growth and development. Plant Physiol 147:1092–1109

Riewe D, Grosman L, Fernie AR, Zauber H, Wucke C, Geigenberger P (2008c) A cell wall-bound adenosine nucleosidase is involved in the salvage of extracellular ATP in Solanum tuberosum. Plant Cell Physiol 49:1572–1579

Rolletscheck H, Borisjuk L, Sanchez-Garcia A, Gotor C, Romero LC, Martinez-Rivas JM, Mancha M (2007) Temperature-dependent endogenous oxygen concentration regulates microsomal oleate desaturase in developing sunflower seeds. J Exp Bot 58:3171–3181

Rolletschek H, Borisjuk L, Koschorreck M, Wobus U, Weber H (2002) Legume embryos develop in a hypoxic environment. J Exp Bot 53:1099–1107

Rolletschek H, Weber H, Borisjuk L (2003) Energy status and its control on embryogenesis of legumes: embryo photosynthesis contributes to oxygen supply and is coupled to biosynthetic fluxes. Plant Physiol 132:1196–1203

Rolletschek H, Weschke W, Weber H, Wobus U, Borisjuk L (2004) Energy state and its control on seed development: starch accumulation is associated with high ATP and steep oxygen gradients within barley grains. J Exp Bot 55:1351–1359

Rolletschek H, Radchuk R, Klukas C, Schreiber F, Wobus U, Borisjuk L (2005a) Evidence of a key role for photosynthetic oxygen release in oil storage in developing soybean seeds. New Phytol 167:777–786

Rolletschek H, Koch K, Wobus U, Borisjuk L (2005b) Positional cues for the starch/lipid balance in maize kernels and resource partitioning to the embryo. Plant J 42:69–83

Rolletschek H, Stangelmayer A, Borisjuk L (2009) Methodology and significance of microsensor-based oxygen mapping in plant seeds—an overview. Sensors (Basel) 9:3218–3227

Schmälzlin E, van Dongen JT, Klimant I, Marmodée B, Steup M, Fisahn J, Geigenberger P, Löhmannsröben H-G (2005) An optical multi-frequency phase modulation method using microbeads for measuring intracellular oxygen concentrations in plants. Biophys J 89:1339–1345

Smith WH (1947) A new method for the determination of the composition of the internal atmosphere of fleshy plant organs. Ann Bot 11:363–368

Sonnewald U (1992) Expression of E. coli pyrophosphatase in transgenic plants alters photoassimilate partitioning. Plant J 2:571–581

Sowa AW, Duff SMG, Guy PA, Hill RD (1998) Altering hemoglobin levels changes energy status in maize cells under hypoxia. Proc Natl Acad Sci U S A 95:10317–10321

Stitt M (1998) Pyrophosphate as an alternative energy donor in the cytosol of plant cells: an enigmatic alternative to ATP. Bot Acta 111:167–175

Stitt M, Zeeman SC (2012) Starch turnover: pathways, regulation and role in growth. Curr Opin Plant Biol 15:282–292

Sugden C, Crawford RM, Halford NG, Hardie DG (1999) Regulation of spinach SNF1-related (SnRK1) kinases by protein kinases and phosphatases is associated with phosphorylation of the T loop and is regulated by 5′-AMP. Plant J 19:433–439

Thiel J, Rolletschek H, Friedel S, Lunn JE, Nguyen TH, Feil R, Tschiersch H, Müller M, Borisjuk L (2011) Seed-specific elevation of non-symbiotic hemoglobin AtHb1: beneficial effects and underlying molecular networks in Arabidopsis thaliana. BMC Plant Biol 11:48

Thorne JH (1982) Temperature and oxygen effects on ^{14}C-photosynthate unloading and accumulation in developing soybean seeds. Plant Physiol 69:48–53

Tiessen A, Prescha K, Branscheid A, Palacios N, McKibbin R, Halford NG, Geigenberger P (2003) Evidence that SNF1-related kinase and hexokinase are involved in separate sugar-signaling pathways modulating post-translational redox activation of ADP-glucose pyrophosphorylase in potato tubers. Plant J 35:490–500

Tiessen A, Lunn JE, Geigenberger P (2006) Carbohydrate metabolism under water-limited conditions. In: Ribaut J-M (ed) Drought adaptation in cereals. The Haworth Press, Binghamton, NY, pp 449–503

van Dongen JT, Schurr U, Pfister M, Geigenberger P (2003) Phloem metabolism and function have to cope with low internal oxygen. Plant Physiol 131:1529–1543

van Dongen JT, Roeb GW, Dautzenberg M, Froehlich A, Vigeolas H, Minchin PE, Geigenberger P (2004) Phloem import and storage metabolism are highly coordinated by the low oxygen concentrations within developing wheat seeds. Plant Physiol 135:1809–1821

van Dongen JT, Fröhlich A, Ramirez-Aguilar S, Schauer N, Fernie AR, Erban A, Kopka J, Clarke J, Langer A, Geigenberger P (2009) Transcript and metabolite profiling of the adaptive response to mild decreases in oxygen concentration in the roots of Arabidopsis plants. Ann Bot 103:269–280

Vigeolas H, van Dongen JT, Waldeck P, Hühn D, Geigenberger P (2003) Lipid storage metabolism is limited by the prevailing low oxygen concentrations within developing seeds of oilseed rape. Plant Physiol 133:2048–2060

Vigeolas H, Waldeck P, Zank T, Geigenberger P (2007) Increasing seed oil content in oil-seed rape (*Brassica napus* L.) by overexpression of a yeast glycerol-3-phosphate dehydrogenase under the control of a seed-specific promoter. Plant Biotechnol J 5:431–441

Vigeolas H, Hühn D, Geigenberger P (2011) Non-symbiotic hemoglobin-2 leads to an elevated energy state and to a combined increase in polyunsaturated fatty acids and total oil content when over-expressed in developing seeds of transgenic Arabidopsis plants. Plant Physiol 155: 1435–1444

Viola R, Roberts AG, Haupt S, Gazzani S, Hancock RD, Marmiroli N, Machray GC, Oparka KJ (2001) Tuberization in potato involves a switch from apoplastic to symplastic phloem unloading. Plant Cell 13:385–398

Wardlaw CW, Leonard ER (1939) Studies in tropical fruits. IV. Methods in the investigation of respiration with special reference to the banana. Ann Bot 3:27–42

Wardlaw CW, Leonard ER (1940) Studies in tropical fruits. IX. The respiration of bananas during ripening at tropical temperatures. Ann Bot 4:269–315

Watt DA (2005) Do external oxygen levels influence sucrose metabolism in the sugarcane stalk? Proc S Afr Sug Technol Ass 79:517–520

Weber H, Borisjuk L, Wobus U (1997) Sugar import and metabolism during seed development. Trends Plant Sci 2:169–174

Youm JW, Jeon JH, Choi D, Yi SY, Joung H, Kim HS (2008) Ectopic expression of pepper CaPF1 in potato enhances multiple stresses tolerance and delays initiation of in vitro tuberization. Planta 228:701–708

Zabalza A, van Dongen JT, Froehlich A, Oliver SN, Faix B, Kapuganti JG, Schmälzlin E, Igal M, Orcaray L, Koehl K, Royuela M, Geigenberger P (2009) Regulation of respiration and fermentation to control the plant internal oxygen concentration. Plant Physiol 149:1087–1098

Zhang Y, Primavesi LF, Jhurreea D, Andralojc PJ, Mitchell RAC, Powers SJ, Schluepmann H, Delatte T, Wingler A, Paul MJ (2009) Inhibition of SNF1-related protein kinase1 activity and regulation of metabolic pathways by trehalose-6-phosphate. Plant Physiol 149:1860–1871

Part IV
Morphological Adaptations

Aerenchyma Formation in Plants

Hirokazu Takahashi, Takaki Yamauchi, Timothy David Colmer, and Mikio Nakazono

Abstract Aerenchyma enhances internal aeration between, and within, shoots and roots. Aerenchyma formation is therefore important for the adaptation of plants in environments with excess water, such as plants with roots in waterlogged soils or submerged shoots. Aerenchyma can form in primary tissues (primary aerenchyma) and in secondary tissues (secondary aerenchyma). Primary tissues have two main types of aerenchyma: schizogenous aerenchyma and lysigenous aerenchyma. Both types provide enlarged spaces for gas-phase diffusion. Schizogenous aerenchyma is formed by the separation of adjacent files (radial rows) of cortical cells and by enlargement of existing intercellular spaces through cell division and differential cell enlargement. By contrast, lysigenous aerenchyma results from the collapse and lysis of files of cortical cells via programmed cell death. Secondary aerenchyma differentiates from phellogen, cambium, and pericycle in stems, hypocotyls, or roots of some dicots to form a gas-filled and low-resistance pathway for gas movement. Presently, the mechanisms of schizogenous and secondary aerenchyma formation are less well understood than the mechanisms of lysigenous aerenchyma formation. Here, we summarize the characteristics of primary aerenchyma (schizogenous and lysigenous aerenchymas) and secondary aerenchyma types, and present recent advances in understanding the mechanisms of lysigenous aerenchyma formation.

H. Takahashi • T. Yamauchi • M. Nakazono (✉)
Laboratory of Plant Genetics and Breeding, Graduate School of Bioagricultural Sciences, Nagoya University, Furo-cho, Chikusa, Nagoya 464-8601, Japan
e-mail: nakazono@agr.nagoya-u.ac.jp

T.D. Colmer
School of Plant Biology (M084), The University of Western Australia, 35 Stirling Highway, Crawley, WA 6009, Australia

J.T. van Dongen and F. Licausi (eds.), *Low-Oxygen Stress in Plants*, Plant Cell Monographs 21, DOI 10.1007/978-3-7091-1254-0_13, © Springer-Verlag Wien 2014

1 Introduction

The formation of aerenchyma tissue is an anatomical adaptation to excess water stress. Aerenchyma consists of longitudinally interconnected gas spaces that enable the rapid transport of gases (O_2, CO_2, ethylene, and methane) between, and within, shoots and roots (Armstrong 1979; Colmer 2003; Evans 2003; Jackson and Armstrong 1999). The gas movements in roots occur by diffusion (Armstrong 1979), but in the rhizome and shoot aerenchyma of some wetland species, the gases can be transported by Venturi- and humidity-induced pressure flows (Armstrong et al. 1992, 1996) (see also Chapter 14 of this volume). In some wetland plant species, aerenchyma may promote photosynthesis by delivering CO_2 produced by root respiration (Constable and Longstreth 1994) or CO_2 that enters from waterlogged soil (Greenway et al. 2006) to the leaf intercellular spaces. Aerenchyma can also provide a conduit for CO_2 movement from the rhizosphere to the shoots when the shoots are completely submerged (Winkel and Borum 2009).

In many wetland and aquatic plants, aerenchyma is developed in the shoots as well as the roots. For example, in rice (*Oryza sativa*), aerenchyma is well developed in the roots, the internodes, and the leaf sheaths and mid-ribs (Colmer and Pedersen 2008; Matsukura et al. 2000; Parlanti et al. 2011; Steffens et al. 2011) and contributes to effective internal aeration between the shoots and roots (Armstrong 1971; Colmer 2003; Colmer and Pedersen 2008). Aerenchyma takes a variety of forms (Arber 1920; Justin and Armstrong 1987; Jung et al. 2008; Seago et al. 2005; Smirnoff and Crawford 1983) and can form in primary tissues (primary aerenchyma) or can form as a secondary tissue (secondary aerenchyma) (Jackson and Armstrong 1999). In roots, primary aerenchyma forms in the primary cortex and can be broadly classified into two types: schizogenous aerenchyma and lysigenous aerenchyma (Arber 1920; Armstrong and Armstrong 1994; Jackson and Armstrong 1999; Jung et al. 2008; Justin and Armstrong 1987; Seago et al. 2005; Smirnoff and Crawford 1983). Schizogenous aerenchyma is formed by the creation of gas spaces between cells as a result of cell separation and expansion of existing spaces by differential cell divisions and/or cell expansion in particular directions (Fig. 1a, b; Arber 1920; Jackson and Armstrong 1999; Justin and Armstrong 1987). Lysigenous aerenchyma is formed by the creation of gas spaces as a result of the death and subsequent lysis causing collapse of files of cells (e.g., root cortical cells; Fig. 1c, d; Arber 1920; Evans 2003; Jackson and Armstrong 1999; Justin and Armstrong 1987). In contrast, secondary aerenchyma differentiates from the phellogen (cork cambium), cambium, or pericycle that can produce either a porous secondary cortex (Fig. 1e, f) or an aerenchymatous phellem (Fig. 1g, h) in stems, hypocotyls, and roots of some dicots (Arber 1920; Jackson and Armstrong 1999; Stevens et al. 2002; Shimamura et al. 2003).

Aerenchyma formation has been studied both anatomically and physiologically. Lysigenous aerenchyma has been particularly well studied, and a molecular understanding of the regulation of its formation is emerging. Here, we summarize the mechanisms of formation of primary aerenchyma (schizogenous aerenchyma and

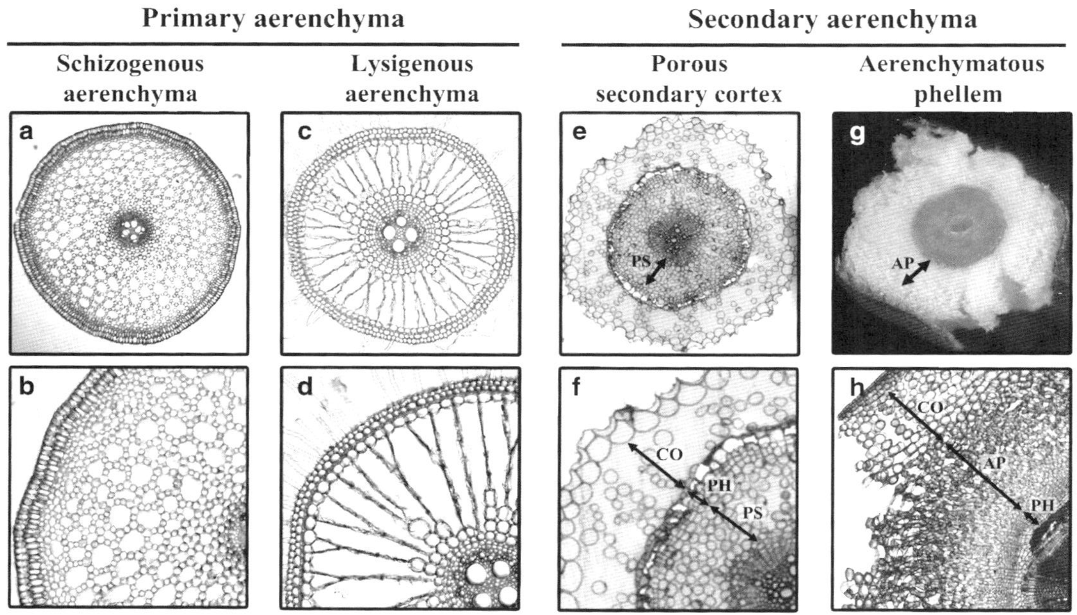

Fig. 1 Examples of primary aerenchyma (schizogenous and lysigenous types) and secondary aerenchyma (porous secondary cortex and aerenchymatous phellem) in plants. (**a**, **b**) Schizogenous aerenchyma in *Acorus calamus* root. (**c**, **d**) Lysigenous aerenchyma in rice root. (**e**, **f**) Porous secondary cortex in *Plantago maritima* root. (**g**, **h**) Aerenchymatous phellem in soybean hypocotyl. *AP* aerenchymatous phellem, *CO* primary cortex, *PH* phellogen, *PS* porous secondary cortex. Photographs in panels (**e**) and (**f**) from Justin and Armstrong (1987), New Phytologist, with permission from John Wiley and Sons (photographs in (**e**) and (**f**) were kindly provided by William Armstrong)

lysigenous aerenchyma) and secondary aerenchyma, and the functioning of these aerenchymas, under conditions of excessive water.

2 Primary Aerenchyma

Justin and Armstrong (1987) classified the types of primary cortex and their associated gas-space features into three major groups with 14 sub-categories based on cellular configuration and aerenchyma patterns found in the root primary cortex. More recently, Seago et al. (2005) proposed to distinguish "expansigeny" from "schizogeny" and categorized root aerenchyma into seven groups (honeycomb expansigeny, radial expansigeny, schizogeny, schizo-lysigeny, packet lysigeny, tangential lysigeny, and radial lysigeny) based on the pattern of origin of aerenchymas and their systematic distributions. According to Seago et al. (2005), the aerenchymas of some plants (e.g., *Acorus calamus*), which were previously considered schizogenous aerenchyma (Armstrong and Armstrong 1994; Jackson and Armstrong 1999), could be reclassified as expansigenous aerenchyma. However, in this chapter we continue to use the more broad term of schizogenous aerenchyma, and thus do not distinguish between schizogeny and expansigeny types [see Seago et al. (2005) and Jung et al. (2008) for detailed characterizations of these types], as that level of distinction of aerenchyma types is not required for our present purposes. Both schizogeny and expansigeny types of aerenchyma form without cell death, which occurs only in lysigenous aerenchyma formation (Bailey-Serres and Voesenek 2008; Jung et al. 2008; Seago et al. 2005).

2.1 Schizogenous Aerenchyma

2.1.1 Schizogenous Aerenchyma Formation in Roots

Schizogenous aerenchyma is constitutively formed not only in wetland plants but also in some non-wetland species (Evans 2003; Jackson and Armstrong 1999). In the genus *Rumex*, schizogenous aerenchyma is formed in the taproot and adventitious roots, and the aerenchyma develops from the tip towards the base of adventitious roots (Colmer et al. 2004; Justin and Armstrong 1987). In *R. crispus* and *R. palustris*, waterlogging enhances the formation of schizogenous aerenchyma along the entire length of adventitious roots, thereby increasing root porosity (Colmer et al. 2004; Laan et al. 1989; Visser et al. 2000). The enhanced formation of schizogenous aerenchyma in *R. palustris* (Colmer et al. 2004) and in *R. hydrolapathum* (Justin and Armstrong 1987) has been suggested to result from the promotion of cell division in the files (radial rows) of cortical cells in adventitious roots.

2.1.2 Schizogenous Aerenchyma Formation in Petioles

In *Sagittaria lancifolia*, aerenchyma in the roots is formed by lysigeny, but aerenchyma in the petioles is formed by schizogeny (Schussler and Longstreth 1996). Cylinders of the petiole aerenchyma of *S. lancifolia* and *Sagittaria trifolia* are formed in a developmental pattern based on cell separation, division, and expansion, rather than on cell lysis. Perpendicular and parallel division of the cells of the cylinders leads to the formation of gas-filled compartments within the cylinders during development (Liang et al. 2008; Schussler and Longstreth 1996). Diaphragm cells form single-cell layers across the cylinder and partition the cylinder. The space between these partitions increases as the cylinder cells divide and enlarge (Evans 2003; Liang et al. 2008; Schussler and Longstreth 1996). Observation of the petiole aerenchyma in *S. trifolia* at different developmental stages has revealed that cell division at the early stage and cell enlargement in the later stages contribute to the increase in gas space (Liang et al. 2008). An immunofluorescence study has shown that cortical microtubules are localized at the boundary between the diaphragm cells and the edge of the intercellular spaces (Liang et al. 2008). Moreover, the cell walls at the fringes of the intercellular spaces are clearly thicker than normal. These results suggest that cortical microtubule arrays play an important role in intercellular space formation, cell deformation, expansion, and deposition of secondary cell walls in diaphragm cells (Liang et al. 2008).

2.2 Lysigenous Aerenchyma

2.2.1 Lysigenous Aerenchyma Formation in Roots

Lysigenous aerenchyma is formed by the selective death and subsequent lysis of cortical cells in the roots of some wetland plants and many gramineous species [e.g., *Phragmites australis* (Justin and Armstrong 1987); *Juncus effusus* (Smirnoff and Crawford 1983); rice (Jackson et al. 1985); maize (*Zea mays*; Drew et al. 1981); wheat (*Triticum aestivum*; Trought and Drew 1980); and barley (*Hordeum vulgare*; Arikado and Adachi 1955)]. In general, lysigenous aerenchyma is formed by packet lysigeny, tangential lysigeny, or radial lysigeny (Seago et al. 2005). *Hydrocharis morsus-ranae* is an example of a species with the packet lysigeny pattern in roots, in which anticlinal and periclinal divisions occur in cortical cells and are followed by cell lysis (Seago et al. 1999, 2005). *Cyperus alternifolius*, *Cladium jamaicense*, and *Carex nigra* are examples of plants with tangential lysigeny patterns, in which cell separation and collapse occur in tangential sectors between intact radial files of root cortical cells (Justin and Armstrong 1987; Seago et al. 2005). Rice is a species with a radial lysigeny pattern, in which cell death begins at the cells in the mid-cortex and then spreads out radially to the surrounding cortical cells of the roots (Kawai et al. 1998). In rice roots, the contents of the cortical cells that are lysed are completely digested, leaving only remnants of collapsed cell walls, whereas the stele and endodermis, along with the outer cell layers (i.e., sclerenchyma, hypodermis/exodermis, and epidermis), are unaffected (Fig. 1c, d). Fully developed aerenchyma, which is generally observed in the basal parts of the rice roots, separates the stele from the outer cell layers with only a few files/strands of remaining cells and cell wall remnants forming radial bridges across the gas spaces in the cortex (Fig. 1c, d; Ranathunge et al. 2003; Shiono et al. 2011). These radial cell files and strands of cell walls are important for the structural integrity of the root and for both apoplastic and symplastic movements of ions (transport of nutrients) and water (Drew and Fourcy 1986). Recently, using X-ray computed tomography, Karahara et al. (2012) visualized lysigenous aerenchyma in rice roots and highlighted the well-connected channels in a longitudinal direction along the root axis (also with radial connections), supporting earlier measurements and models that aerenchyma provides an interconnected longitudinal gas-filled pathway in roots (Armstrong 1979).

In wetland species, such as rice, *J. effusus*, and *Zea nicaraguensis*, lysigenous aerenchyma is constitutively formed even under aerobic conditions (Abiko et al. 2012; Jackson et al. 1985; Mano et al. 2006; Mano and Omori 2013; Visser and Bögemann 2006), and its formation can be further induced during waterlogging (Colmer et al. 2006; Shiono et al. 2011; Visser and Bögemann 2006). Aerenchyma formed under aerobic conditions is designated "constitutive" lysigenous aerenchyma, whereas that formed during waterlogging is designated "inducible" lysigenous aerenchyma (Colmer and Voesenek 2009). In non-wetland plant species such as maize, wheat, and barley, lysigenous aerenchyma does not typically form under well-drained soil conditions but is induced by poor aeration (McDonald et al. 2001; McPherson 1939; Trought and Drew 1980). Interestingly, aerenchyma in some non-wetland plant species can also be induced during nutrient deficiency or drought. The purpose does not appear to be facilitation of gas transport but reduction of the costs of resource investment per unit root length by converting a proportion of living cortical tissue to a gas-filled volume (Konings and Verschuren 1980; Postma and Lynch 2011; Zhu et al. 2010). Induction of aerenchyma formation in non-wetland plants usually takes several hours after the start of the waterlogging/low O_2 treatment (Haque et al. 2010; Malik et al. 2003; Rajhi et al. 2011). In addition, aerenchyma formation is less extensive in non-wetland species than in wetland species (Armstrong 1979; Justin and Armstrong 1987; Colmer and Voesenek 2009). Differences in capacity to form aerenchyma may be a major factor explaining why non-wetland plants are less tolerant to waterlogging than wetland plants (Justin and Armstrong 1987). Hence, the capacity to provide adequate aerenchyma, with low radial O_2 loss, for O_2 diffusion from the above-ground parts to the root tips, is important for increasing tolerance to waterlogging in dryland crops (Malik et al. 2011; Setter and Waters 2003).

2.2.2 Lysigenous Aerenchyma Formation in Shoots

Lysigenous aerenchyma forms not only in roots but also in shoots (e.g., leaf sheath, internodes, and leaf mid-rib in rice; Colmer and Pedersen 2008; Matsukura et al. 2000; Parlanti et al. 2011; Steffens et al. 2011). In the rice leaf sheath, lysigenous aerenchyma is formed developmentally by cell death in the middle layers of the parenchymal cells, which are located between the vascular bundles (Matsukura et al. 2000). Leaf sheath aerenchyma formation is enhanced by submergence (Parlanti et al. 2011). In the internodes of deepwater rice, lysigenous aerenchyma is formed constitutively and is also further enhanced by partial submergence or by ethylene exposure (Steffens et al. 2011). Pre-aerenchymal cells contain less starch and have larger vacuoles and thinner cell walls, than do other parenchymal cells (Steffens et al. 2011).

2.2.3 Mechanisms of Lysigenous Aerenchyma Formation

Ethylene, a gaseous plant hormone, rapidly accumulates in roots surrounded by water (Visser and Voesenek 2004) and is involved in inducible lysigenous aerenchyma formation in maize and rice (Drew et al. 1979, 2000; Justin and Armstrong 1991). Evans (2003) proposed that cell death in the root cortex during lysigenous aerenchyma formation can be divided into five stages: (1) perception of hypoxia and initiation of ethylene biosynthesis (note: if ethylene production remained the same as occurs in roots in drained soil, impeded outwards diffusion owing to the surrounding water would result in elevated levels of ethylene within roots in waterlogged soil or stagnant solution cultures, so the ethylene signal can occur without "perception of hypoxia"); (2) perception of an ethylene signal by cells of the mid-cortex; (3) initiation of cell death with loss of ions to the surrounding cells, initiation of plasma membrane invagination, and the formation of small vesicles; (4) chromatin condensation, increased activities of cell wall hydrolytic enzymes, and the surrounding of organelles by membranes; and (5) cell-wall degradation, cell lysis, and absorption of cell contents and water by the surrounding cells.

So far, most studies of lysigenous aerenchyma formation have been conducted from an anatomical and/or a physiological perspective. In contrast, the molecular mechanisms of lysigenous aerenchyma formation in roots are poorly understood. Rajhi et al. (2011) used a microarray analysis combined with laser microdissection to identify genes associated with lysigenous aerenchyma formation in maize roots. They found that genes associated with Ca^{2+} signaling and cell-wall modification were induced in the root cortical cells under waterlogged conditions. Inductions of expression of Ca^{2+} signaling-related genes or cell-wall modification-related genes were suppressed by treatment with an ethylene perception inhibitor, 1-methylcyclopropene (Rajhi et al. 2011). These findings support the previously proposed mechanism of ethylene-mediated lysigenous aerenchyma formation (Drew et al. 2000; Evans 2003; Jackson and Armstrong 1999).

Interestingly, a respiratory burst oxidase homolog (RBOH), a plant homolog of gp91^{phox} in mammalian NADPH oxidase, which has a key role in ROS (reactive oxygen species) generation was upregulated strongly in the cortical cells and slightly less strongly in the stelar cells and outer cell layers of maize roots under O_2-deficient conditions (Rajhi et al. 2011; Yamauchi et al. 2011). Waterlogging-inducible aerenchyma formation was inhibited by diphenyleneiodonium (DPI), an NADPH oxidase inhibitor (Yamauchi et al. 2011). RBOH, whose activity is controlled by Ca^{2+} signaling, plays an important role in ROS-mediated signaling in plants, for example, in the defense response, programmed cell death (PCD), and plant development (Torres and Dangl 2005). Similarly, an interaction between Ca^{2+} signaling and RBOH-mediated H_2O_2 production might also be involved in PCD in the root cortical cells during lysigenous aerenchyma formation. In addition, metallothionein (MT), which plays a role in ROS scavenging (Wong et al. 2004; Xue et al. 2009), is constitutively expressed in all cortical cells, stelar cells, and outer cell layers of maize roots under aerobic conditions, whereas in waterlogged conditions the *MT* gene was hardly expressed in the cortical cells, but it was still highly expressed in the stelar cells and the outer cell layers (Rajhi et al. 2011; Yamauchi et al. 2011).

MTs are low-molecular-weight, cysteine-rich metal-binding proteins that have roles in metal homeostasis, general stress responses, and ROS scavenging (Coyle et al. 2002; Sato and Bremner 1993). Many cysteine residues in MT proteins, which are highly reactive to oxidizing agents, are likely to play a central role in the scavenging of ROS (Chae et al. 1994; Coyle et al. 2002). Although the antioxidative capacity of MTs against H_2O_2 has not been directly evaluated, animal Zn- or Cd-binding MTs have more than 100 times the antioxidant activity of reduced glutathione against hydroxyl radicals (Sato and Bremner 1993; Thornalley and Vašák 1985). Rice MT2b and cotton MT3a also have high antioxidative capacity against hydroxyl radicals in vitro (Wong et al. 2004; Xue et al. 2009). This suggests that H_2O_2 and other ROS are scavenged by MT constitutively expressed in stelar cells and the outer cell layers, whereas in the cortical cells, decreased *MT* expression prevents ROS scavenging under waterlogged conditions, thereby possibly leading to greater ROS accumulation, which activates the subsequent processes of PCD (and thus lysigenous aerenchyma formation) in maize roots (Fig. 2; Nishiuchi et al. 2012; Yamauchi et al. 2013).

Submergence-induced ROS may also be involved in aerenchyma formation in the leaf sheath of rice, because leaf sheath aerenchyma formation is inhibited by treatment with DPI (Parlanti et al. 2011). Ethylene-promoted down-regulation of expression of the gene encoding MT2b, which acts as an ROS scavenger, enhances the accumulation of H_2O_2 produced by NADPH oxidase and thus induces aerenchyma formation in rice internodes (Steffens et al. 2011) and epidermal cell death in rice stem nodes prior to adventitious root appearance (Steffens and Sauter 2009). Furthermore, a deficiency of MT2b promotes internode aerenchyma formation in rice (Steffens et al. 2011). These results suggest that ROS accumulation is involved in aerenchyma formation in the leaf sheaths and internodes.

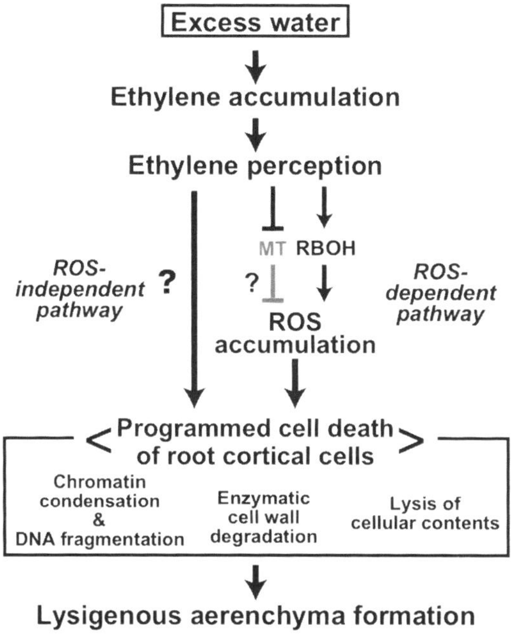

Fig. 2 Processes involved in lysigenous aerenchyma formation. Drawn according to data and schemes in Evans (2003), Drew et al. (2000), He et al. (1996a), Rajhi et al. (2011), Shiono et al. (2008), and Yamauchi et al. (2011). This simplified scheme provides a general overview and does not show all factors that might be involved, such as cytosolic Ca^{2+} signaling (e.g., He et al. 1996b; Subbaiah et al. 1994; Subbaiah and Sachs 2003) and some cytoplasmic events described in Evans (2003) and in Sect. 2.2.3 of this chapter. *RBOH* respiratory burst oxidase homolog, *ROS* reactive oxygen species, *MT* metallothionein. For a full explanation, see the text

3 Secondary Aerenchyma

Secondary aerenchyma usually arises from the phellogen, but sometimes it arises from the cambium or pericycle (Arber 1920; Armstrong, personal communication). Secondary aerenchyma is generally formed radially inwards or outwards from phellogen. Here, we refer to the former as porous secondary cortex (Fig. 1e, f) and to the latter as aerenchymatous phellem (Fig. 1g, h). In each case, the newly formed tissues (secondary aerenchyma) have much gas-filled space and are important for gas transport through these secondary thickened tissues (Arber 1920; Jackson and Armstrong 1999). On the other hand, a gas-filled pith-like structure for internal aeration is observed in Lodgepole pine roots (Coutts and Philipson 1978). This may be categorized as one of the secondary aerenchyma types, but there is no direct evidence that it is secondary aerenchyma. Thus, in this section, we focus on only the secondary aerenchyma (porous secondary cortex and aerenchymatous phellem) derived from the phellogen.

3.1 Porous Secondary Cortex (Phelloderm)

Secondary cortex (phelloderm) is formed radially inwards from phellogen by secondary thickening (i.e., secondary growth). Some wetland plant species, such as *Plantago maritima* and *Trifolium resupinatum*, have well-developed porous secondary cortex in roots (Justin and Armstrong 1987; Gibberd et al. 2001). These plants seem to form secondary cortex from the phellogen, but *Limonium vulgare* shows anomalous secondary cortex formation. The secondary cortex of *L. vulgare* is formed by the differentiation of both phellogen cells and proliferating pericycle cells (Armstrong, personal communication). The aerenchyma in the secondary cortex of these plants appears to have developed by radial schizogeny and is similar to the schizogenous aerenchyma in the primary cortex (Justin and Armstrong 1987; Gibberd et al. 2001).

Some *Rumex* species have "honeycomb-type" schizogenous aerenchyma in the adventitious roots as described in the section of "Schizogenous aerenchyma." In addition, *R. palustris* and *R. maritimus* develop a secondary aerenchyma in the basal part of their taproot (Colmer et al. 2004; Laan et al. 1989). In roots of *R. palustris*, *α-expansin* genes, whose products mediate cell-wall loosening, are expressed in the cells of secondary aerenchyma near the vascular cambium (Colmer et al. 2004). This suggests that α-expansins are involved in cell elongation during the formation of this porous secondary cortex in *R. palustris*.

3.2 Aerenchymatous Phellem

3.2.1 Formation of Aerenchymatous Phellem

Phellogen forms phellem (cork tissue), which is an impervious tissue that covers the surface of plants. It is ruptured by lenticels which allow the exchange of gases between the outside and inside of a plant. In some plant species, secondary aerenchyma is formed radially outwards from phellogen. This type of secondary aerenchyma is homologues to phellem, and thus it is referred to as aerenchymatous phellem (Lempe et al. 2001; Stevens et al. 1997). Phellem usually consists of dead cells with suberized cell walls. In contrast, aerenchymatous phellem consists of living cells with non-suberized cell walls and is thus distinguishable from cork tissues (Arber 1920; Fraser 1931; Jackson and Armstrong 1999). Thus, nuclei and cytoplasm are observed in the cells of aerenchymatous phellem (Angeles 1992; Scott and Wager 1888; Shimamura et al. 2007).

Aerenchymatous phellem is highly porous tissue (Fig. 1g, h) that is formed by successive cell divisions of phellogen followed by expansion of these cells (Arber 1920; Fraser 1931; Jackson and Armstrong 1999; Scott and Wager 1888). The cells that constitute aerenchymatous phellem can be elongated in the radial direction, but generally they remain unbranched, like spongy tissue in soybean (*Glycine max*;

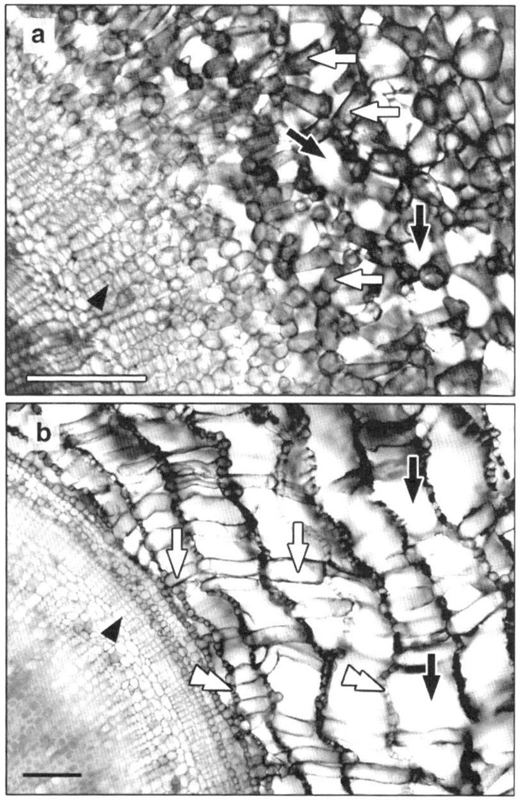

Fig. 3 Two types of aerenchymatous phellem. (**a**) Aerenchymatous phellem in soybean hypocotyl. *Black arrowhead* (i.e., *triangle*) indicates phellogen. *White arrows* indicate expanded cells in the aerenchymatous phellem and the *black arrows* indicate some of the gas spaces between the cells—the resulting high porosity tissue functions as aerenchyma for internal gas diffusion. (**b**) Aerenchymatous phellem in *Lythrum salicaria* hypocotyl. *Black arrowhead* (i.e., *triangle*) indicates phellogen. *White double arrowheads* indicate small isodiametric cells and *white arrows* indicate radially elongated cells. *Black arrows* indicate some of the gas spaces in this tissue of high porosity which functions as aerenchyma for internal gas diffusion. Scale bar in both (**a**) and (**b**): 100 μm

Fig. 3a; Shimamura et al. 2003; Thomas et al. 2005), *Melilotus siculus* (Teakle et al. 2011), *Sesbania aculeata* (Scott and Wager 1888), and *Viminaria juncea* (Walker et al. 1983). In contrast, the aerenchymatous phellem of *Lythrum salicaria* consists of alternating bands of radially elongated cells and small isodiametric cells, which are adjacent to the phellogen and a band of non-elongated cells (Fig. 3b). Under well-drained soil conditions, the stem base of *L. salicaria* has several layers of small isodiametric cells lateral to the phellogen. These cell layers are separated by radially elongated cells (Fig. 3b; Lempe et al. 2001; Stevens et al. 1997). The cell elongations increase the size of the gas spaces in secondary aerenchyma.

Aerenchymatous phellem can form in several organs such as the base of the stem, hypocotyls, adventitious roots, taproots, and nodules in many dicotyledonous plant species [e.g., *S. aculeata* (Scott and Wager 1888); *Sesbania rostrata* (Saraswati et al. 1992; Shiba and Daimon 2003); *V. juncea* (Walker et al. 1983); *L. salicaria* (Stevens et al. 1997, 2002); *M. siculus* (Teakle et al. 2011; Verboven et al. 2012); *Lotus uliginosus* (James and Crawford 1998; James and Sprent 1999); and soybean (Shimamura et al. 2002, 2003, 2010; Thomas et al. 2005)]. In soybean,

L. salicaria, and *V. juncea*, aerenchymatous phellem is not formed under well-drained soil conditions, but it is induced under waterlogged soil conditions (Shimamura et al. 2003; Stevens et al. 1997; Walker et al. 1983). Interestingly, in soybean hypocotyls, phellogen is also newly formed in response to waterlogged conditions (Shimamura et al. 2003).

3.3 Physiological Roles of Aerenchymatous Phellem

Aerenchymatous phellem contains a large volume of gas space. For example, the porosity of aerenchymatous phellem tissues exceeds 25 % in soybean hypocotyls (Shimamura et al. 2003) and *M. siculus* roots (Teakle et al. 2011), and is about 60–70 % in the stem of *L. salicaria* (Stevens et al. 1997). Recently, using micro-computed tomography, Verboven et al. (2012) demonstrated that the gas spaces in *M. siculus* aerenchymatous phellem form a well-connected network. Removal of stem aerenchymatous phellem caused a decline in the O_2 concentration in the roots of *M. siculus* (Teakle et al. 2011) and *L. salicaria* (Stevens et al. 2002). In soybean, O_2 diffusion to flooded roots and nodules was inhibited when the part of the stem with aerenchymatous phellem was completely submerged in water (Shimamura et al. 2010). The aerenchymatous phellem along these basal stem parts usually protrudes above the water surface, and epidermal "cracking" and/or lenticels likely provide an entry path for O_2 from the surrounding air (Shimamura et al. 2003). In woody plants, lenticels are important sites for O_2 entry into stems (Armstrong 1968; Dittert et al. 2006; Hook et al. 1971). Thus, the stem aerenchymatous phellem has an important function in internal aeration, facilitating the entry and longitudinal transport of O_2 from shoots to roots under waterlogged conditions.

For legumes, aerenchymatous phellem plays a role in facilitation of nitrogen fixation in nodules under waterlogged soil conditions. Legumes form aerenchymatous phellem on the surfaces of the root nodules (Loureiro et al. 1994; Parsons and Day 1990; Shimamura et al. 2002; Thomas et al. 2005; Walker et al. 1983), which connect with the root aerenchymatous phellem. In nodules that develop aerenchymatous phellem during waterlogging, the interiors acquire a pink color due to the presence of leghaemoglobin, which binds O_2 and acts as a carrier to enable respiration (James and Sprent 1999; Loureiro et al. 1994). Nitrogenase, which fixes nitrogen, is inactivated by O_2 (Gallon 1992; James et al. 1992). In rhizobium-infected plant cells, a low level of free O_2 is maintained by an O_2 diffusion barrier in the inner nodule cortex and by binding of O_2 by leghaemoglobin, which acts as carrier to enable respiration in the cytosol (Appleby 1984; Minchin 1997; Witty et al. 1987; Witty 1991). However, the nitrogenase activity of nodules is reduced under waterlogging or low-O_2 conditions (Bisseling et al. 1980; James et al. 1992). Nitrogenase activity was inhibited under waterlogged conditions when aerenchymatous phellem above the water was daubed with Vaseline to block the connection with air (Shimamura et al. 2002). This suggests that at least some O_2 and nitrogen (N_2) are needed for high levels of nitrogenase activity in the nodules. *Lotus uliginosus* is more tolerant of flooding than *Lotus corniculatus* (Justin and Armstrong 1987; James and Crawford 1998). This may be because *L. uliginosus*, but not *L. corniculatus*, has profuse aerenchymatous phellem on the nodules and nodulated roots, which help to supply O_2 and N_2 to the root nodules for nitrogen fixation under flooded conditions (James and Crawford 1998). Nitrogen gas transport through aerenchymatous phellem, and consumption in nitrogen fixation in nodules, was confirmed in ^{15}N labeling experiments (Saraswati et al. 1992; Walker et al. 1983). Taken together, these findings indicate that aerenchymatous phellem is important for diffusion of both O_2 and N_2 into submerged nodules on roots of some waterlogging-tolerant legumes.

4 Perspectives

Mechanisms of inducible lysigenous aerenchyma formation (e.g., regulation by ethylene, and signaling by Ca^{2+} and ROS) have been investigated mainly in roots of maize and rice. During lysigenous aerenchyma formation, cell lysis spreads out radially or tangentially, but some viable cortical cells remain. Therefore, ethylene alone cannot be the fine-control for the responses of individual cortical cells (i.e., death or viability), because ethylene gas would diffuse to all cells. Further study of the determination of cell fate (i.e., which cells eventually live or die) would provide valuable insights into cellular signaling that regulates this important adaptation to flooding.

Unlike the mechanisms of inducible lysigenous aerenchyma formation, the mechanisms of constitutive lysigenous aerenchyma formation are poorly understood. *Z. nicaraguensis* (a wild relative of maize) forms lysigenous aerenchyma constitutively even under well-aerated conditions (Mano et al. 2006). Quantitative trait loci associated with constitutive lysigenous aerenchyma formation have been mapped by the genetic analysis using a mapping population produced by crossing the maize inbred line (Mi29) and *Z. nicaraguensis* (Mano and Omori 2008, 2009). This genetic approach and further fine-mapping could identify the genes involved in constitutive lysigenous aerenchyma formation, which should help to elucidate the mechanisms involved in constitutive formation of aerenchyma.

Morphological, anatomical, and physiological studies of schizogenous aerenchyma and secondary aerenchyma have demonstrated that these aerenchyma types play key roles in flooding adaptation in several species. Soybean, one of the world's most important crops, forms secondary aerenchyma in response to waterlogged conditions (Shimamura et al. 2002, 2003, 2010; Thomas et al. 2005). Now that the whole genome sequence of soybean is available (Schmutz et al. 2010), it should be easier to identify genes involved in secondary aerenchyma formation and to elucidate the developmental processes involved in secondary aerenchyma formation. With the continual decline in the cost of transcriptome analyses, it should be possible to identify the genes responsible for the formation of schizogenous aerenchyma or secondary aerenchyma in other plants. Such knowledge might then be applied to future efforts to breed more waterlogging-tolerant crops.

Acknowledgments We thank Drs. W. Armstrong (University of Hull) and S. Nishiuchi (Nagoya University) for their critical readings and stimulating discussions.

References

Abiko T, Kotula L, Shiono K, Malik AI, Colmer TD, Nakazono M (2012) Enhanced formation of aerenchyma and induction of a barrier to radial oxygen loss in adventitious roots of *Zea nicaraguensis* contribute to its waterlogging tolerance as compared with maize (*Zea mays* ssp. *mays*). Plant Cell Environ 35:1618–1630

Angeles G (1992) The periderm of flooded and non-flooded *Ludwigia octovalvis* (Onagraceae). IAWA Bull 13:195–200

Appleby CA (1984) Leghemoglobin and *Rhizobium* respiration. Annu Rev Plant Physiol 35:443–478

Arber A (1920) Water plants: a study of aquatic angiosperms. Cambridge University Press, Cambridge

Arikado H, Adachi Y (1955) Anatomical and ecological responses of barley and some forage crops to the flooding treatment. Bull Fac Agric Mie Univ 11:1–29

Armstrong W (1968) Oxygen diffusion from the roots of woody species. Physiol Plant 21:539–543

Armstrong W (1971) Radial oxygen losses from intact rice roots as affected by distance from the apex, respiration and waterlogging. Physiol Plant 25:192–197

Armstrong W (1979) Aeration in higher plants. Adv Bot Res 7:236–332

Armstrong J, Armstrong W (1994) Chlorophyll development in mature lysigenous and schizogenous root aerenchyma provides evidence of continuing cortical cell viability. New Phytol 126:493–497

Armstrong J, Armstrong W, Beckett PM (1992) *Phragmites australis*: Venturi- and humidity-induced pressure flows enhance rhizome aeration and rhizosphere oxidation. New Phytol 120:197–207

Armstrong J, Armstrong W, Beckett PM, Halder JE, Lythe S, Holt R, Sinclair A (1996) Pathways of aeration and the mechanisms and beneficial effects of humidity- and Venturi-induced convections in *Phragmites australis* (Cav.) Trin ex Steud. Aquat Bot 54:177–197

Bailey-Serres J, Voesenek LACJ (2008) Flooding stress: acclimations and genetic diversity. Annu Rev Plant Biol 59:313–339

Bisseling T, Van Staveren W, Van Kammen A (1980) The effect of waterlogging on the synthesis of the nitrogenase components in bacteroids of *Rhizobium leguminosarum* in root nodules of *Pisum sativum*. Biochem Biophys Res Commun 93:687–693

Chae HZ, Uhm TB, Rhee SG (1994) Dimerization of thiol-specific antioxidant and the essential role of cysteine 47. Proc Natl Acad Sci U S A 91:7022–7026

Colmer TD (2003) Long-distance transport of gases in plants: a perspective on internal aeration and radial oxygen loss from roots. Plant Cell Environ 26:17–36

Colmer TD, Pedersen O (2008) Oxygen dynamics in submerged rice (*Oryza sativa*). New Phytol 178:326–334

Colmer TD, Voesenek LACJ (2009) Flooding tolerance: suites of plant traits in variable environments. Funct Plant Biol 36:665–681

Colmer TD, Peeters AJM, Wagemaker CAM, Vriezen WH, Ammerlaan A, Voesenek LACJ (2004) Expression of α-expansin genes during root acclimations to O_2 deficiency in *Rumex palustris*. Plant Mol Biol 56:423–437

Colmer TD, Cox MCH, Voesenek LACJ (2006) Root aeration in rice (*Oryza sativa*): evaluation of oxygen, carbon dioxide, and ethylene as possible regulators of root acclimatizations. New Phytol 170:767–778

Constable JVH, Longstreth DJ (1994) Aerenchyma carbon dioxide can be assimilated in *Typha latifolia* L. leaves. Plant Physiol 106:1065–1072

Coutts MP, Philipson JJ (1978) Tolerance of tree roots to waterlogging. II. Adaptation of Sitka spruce and Lodgepole pine to waterlogged soil. New Phytol 80:71–77

Coyle P, Philcox JC, Carey LC, Rofe AM (2002) Metallothionein: the multipurpose protein. Cell Mol Life Sci 59:627–647

Dittert K, Wätzel J, Sattelmacher B (2006) Responses of *Alnus glutinosa* to anaerobic conditions—mechanisms and rate of oxygen flux into the roots. Plant Biol 8:212–223

Drew MC, Fourcy A (1986) Radial movement of cations across aerenchymatous roots of *Zea mays* measured by electron probe X-ray microanalysis. J Exp Bot 37:823–831

Drew MC, Jackson MB, Giffard S (1979) Ethylene-promoted adventitious rooting and development of cortical air spaces (aerenchyma) in roots may be adaptive responses to flooding in *Zea mays* L. Planta 147:83–88

Drew MC, Jackson MB, Giffard SC, Campbell R (1981) Inhibition by silver ions of gas space (aerenchyma) formation in adventitious roots of *Zea mays* L. subjected to exogenous ethylene or to oxygen deficiency. Planta 153:217–224

Drew MC, He CJ, Morgan PW (2000) Programmed cell death and aerenchyma formation in roots. Trends Plant Sci 5:123–127

Evans DE (2003) Aerenchyma formation. New Phytol 161:35–49

Fraser L (1931) The reaction of *Viminaria denudata* to increased water content of the soil. Proc Linn Soc NSW 56:391–406

Gallon JR (1992) Reconciling the incompatible: N_2 fixation and O_2. New Phytol 122:571–609

Gibberd MR, Gray JD, Cocks PS, Colmer TD (2001) Waterlogging tolerance among a diverse range of *Trifolium* accessions is related to root porosity, lateral root formation and 'aerotropic rooting'. Ann Bot 88:579–589

Greenway H, Armstrong W, Colmer TD (2006) Conditions leading to high CO_2 (>5 kPa) in waterlogged-flooded soils and possible effects on root growth and metabolism. Ann Bot 98:9–32

Haque ME, Abe F, Kawaguchi K (2010) Formation and extension of lysigenous aerenchyma in seminal root cortex of spring wheat (*Triticum aestivum* cv. Bobwhite line SH 98 26) seedlings under different strengths of waterlogging. Plant Root 4:31–39

He CJ, Finlayson SA, Drew MC, Jordan WR, Morgan PW (1996a) Ethylene biosynthesis during aerenchyma formation in roots of maize subjected to mechanical impedance and hypoxia. Plant Physiol 112:1679–1685

He CJ, Morgan PW, Drew MC (1996b) Transduction of an ethylene signal is required for cell death and lysis in the root cortex of maize during aerenchyma formation induced by hypoxia. Plant Physiol 112:463–472

Hook DD, Brown CL, Kormanik PP (1971) Inductive flood tolerance in swamp tupelo (*Nyssa sylvatica* var. *biflora* (Walt.) Sarg.). J Exp Bot 22:78–89

Jackson MB, Armstrong W (1999) Formation of aerenchyma and the processes of plant ventilation in relation to soil flooding and submergence. Plant Biol 1:274–287

Jackson MB, Fenning TM, Jenkins W (1985) Aerenchyma (gas-space) formation in adventitious roots of rice (*Oryza sativa* L.) is not controlled by ethylene or small partial pressures of oxygen. J Exp Bot 36:1566–1572

James EK, Crawford RMM (1998) Effect of oxygen availability on nitrogen fixation by two *Lotus* species under flooded conditions. J Exp Bot 49:599–609

James EK, Sprent JI (1999) Development of N_2-fixing nodules on the wetland legume *Lotus uliginosus* exposed to conditions of flooding. New Phytol 142:219–231

James EK, Minchin FR, Sprent JI (1992) The physiology and nitrogen-fixing capability of aquatically and terrestrially grown *Neptunia plena*: the importance of nodule oxygen supply. Ann Bot 69:181–187

Jung J, Lee SC, Choi HK (2008) Anatomical patterns of aerenchyma in aquatic and wetland plants. J Plant Biol 51:428–439

Justin SHFW, Armstrong W (1987) The anatomical characteristics of roots and plant response to soil flooding. New Phytol 106:465–495

Justin SHFW, Armstrong W (1991) Evidence for the involvement of ethene in aerenchyma formation in adventitious roots of rice (*Oryza sativa* L.). New Phytol 118:49–62

Karahara I, Umemura K, Soga Y, Akai Y, Bando T, Ito Y, Tamaoki D, Uesugi K, Abe J, Yamauchi D, Mineyuki Y (2012) Demonstration of osmotically dependent promotion of aerenchyma formation at different levels in the primary roots of rice using a 'sandwich' method and X-ray computed tomography. Ann Bot 110:503–509

Kawai M, Samarajeewa PK, Barrero RA, Nishiguchi M, Uchimiya H (1998) Cellular dissection of the degradation pattern of cortical cell death during aerenchyma formation of rice roots. Planta 204:277–287

Konings H, Verschuren G (1980) Formation of aerenchyma in roots of *Zea mays* in aerated solutions, and its relation to nutrient supply. Physiol Plant 49:265–270

Laan P, Berrevoets MJ, Lythe S, Armstrong W, Blom CWPM (1989) Root morphology and aerenchyma formation as indicators of the flood-tolerance of *Rumex* species. J Ecol 77:693–703

Lempe J, Stevens KJ, Peterson RL (2001) Shoot responses of six *Lythraceae* species to flooding. Plant Biol 3:186–193

Liang F, Shen LZ, Chen M, Yang Q (2008) Formation of intercellular gas space in the diaphragm during the development of aerenchyma in the leaf petiole of *Sagittaria trifolia*. Aquat Bot 88:185–195

Loureiro MF, De Faria SM, James EK, Pott A, Franco AA (1994) Nitrogen-fixing stem nodules of the legume, *Discolobium pulchellum* Benth. New Phytol 128:283–295

Malik AI, Colmer TD, Lambers H, Schortemeyer M (2003) Aerenchyma formation and radial O_2 loss along adventitious roots of wheat with only the apical root portion exposed to O_2 deficiency. Plant Cell Environ 26:1713–1722

Malik AI, Islam AKMR, Colmer TD (2011) Transfer of the barrier to radial oxygen loss in roots of *Hordeum marinum* to wheat (*Triticum aestivum*): evaluation of four *H. marinum*-wheat amphiploids. New Phytol 190:499–508

Mano Y, Omori F (2008) Verification of QTL controlling root aerenchyma formation in a maize × teosinte "*Zea nicaraguensis*" advanced backcross population. Breed Sci 58:217–223

Mano Y, Omori F (2009) High-density linkage map around the root aerenchyma locus *Qaer1.06* in the backcross populations of maize Mi29 × teosinte "*Zea nicaraguensis*". Breed Sci 59:427–433

Mano Y, Omori F (2013) Relationship between constitutive root aerenchyma formation and flooding tolerance in *Zea nicaraguensis*. Plant Soil 370:447–460

Mano Y, Omori F, Takamizo T, Kindiger B, Bird RMK, Loaisiga CH (2006) Variation for root aerenchyma formation in flooded and non-flooded maize and teosinte seedlings. Plant Soil 281:269–279

Matsukura C, Kawai M, Toyofuku K, Barrero RA, Uchimiya H, Yamaguchi J (2000) Transverse vein differentiation associated with gas space formation—fate of the middle cell layer in leaf sheath development of rice. Ann Bot 85:19–27

McDonald MP, Galwey NW, Colmer TD (2001) Waterlogging tolerance in the tribe Triticeae: the adventitious roots of *Critesion marinum* have a relatively high porosity and a barrier to radial oxygen loss. Plant Cell Environ 24:585–596

McPherson DC (1939) Cortical air spaces in the roots of *Zea mays* L. New Phytol 38:190–202

Minchin FR (1997) Regulation of oxygen diffusion in legume nodules. Soil Biol Biochem 29:881–888

Nishiuchi S, Yamauchi T, Takahashi H, Kotula L, Nakazono M (2012) Mechanisms for coping with submergence and waterlogging in rice. Rice 5:2

Parlanti S, Kudahettige NP, Lombardi L, Mensuali-Sodi A, Alpi A, Perata P, Pucciariello C (2011) Distinct mechanisms for aerenchyma formation in leaf sheaths of rice genotypes displaying a quiescence or escape strategy for flooding tolerance. Ann Bot 107:1335–1343

Parsons R, Day DA (1990) Mechanism of soybean nodule adaptation to different oxygen pressures. Plant Cell Environ 13:501–512

Postma JA, Lynch JP (2011) Root cortical aerenchyma enhances the growth of maize on soils with suboptimal availability of nitrogen, phosphorus, and potassium. Plant Physiol 156:1190–1201

Rajhi I, Yamauchi T, Takahashi H, Nishiuchi S, Shiono K, Watanabe R, Mliki A, Nagamura Y, Tsutsumi N, Nishizawa NK, Nakazono M (2011) Identification of genes expressed in maize root cortical cells during lysigenous aerenchyma formation using laser microdissection and microarray analyses. New Phytol 190:351–368

Ranathunge K, Steudle E, Lafitte R (2003) Control of water uptake by rice (*Oryza sativa* L.): role of the outer part of the root. Planta 217:193–205

Saraswati R, Matoh T, Sekiya J (1992) Nitrogen fixation of *Sesbania rostrata*: contribution of stem nodules to nitrogen acquisition. Soil Sci Plant Nutr 38:775–780

Sato M, Bremner I (1993) Oxygen free radicals and metallothionein. Free Radic Biol Med 14:325–337

Schmutz J, Cannon SB, Schlueter J, Ma J, Mitros T, Nelson W, Hyten DL, Song Q, Thelen JJ, Cheng J, Xu D, Hellsten U, May GD, Yu Y, Sakurai T, Umezawa T, Bhattacharyya M, Sandhu D, Valliyodan B, Lindquist E, Peto M, Grant D, Shu S, Goodstein D, Barry K, Futrell-Griggs M, Abernathy B, Du J, Tian Z, Zhu L, Gill N, Joshi T, Libault M, Sethuraman A, Zhang X, Shinozaki K, Nguyen HT, Wing RA, Cregan P, Specht J, Grimwood J, Rokhsar D, Stacey G, Shoemaker RC, Jackson SA (2010) Genome sequence of the palaeopolyploid soybean. Nature 463:178–183

Schussler EE, Longstreth DJ (1996) Aerenchyma develops by cell lysis in roots and cell separation in leaf petioles in *Sagittaria lancifolia* (Alismataceae). Am J Bot 83:1266–1273

Scott DH, Wager H (1888) On the floating-root of *Sesbania aculeata*, Pers. Ann Bot 1:308–314

Seago JL Jr, Peterson CA, Enstone DE (1999) Cortical ontogeny in roots of the aquatic plant, *Hydrocharis morsus-ranae* L. Can J Bot 77:113–121

Seago JL Jr, Marsh LC, Stevens KJ, Soukup A, Votrubova O, Enstone DE (2005) A re-examination of the root cortex in wetland flowering plants with respect to aerenchyma. Ann Bot 96:565–579

Setter TL, Waters I (2003) Review of prospects for germplasm improvement for waterlogging tolerance in wheat, barley and oats. Plant Soil 253:1–34

Shiba H, Daimon H (2003) Histological observation of secondary aerenchyma formed immediately after flooding in *Sesbania cannabina* and *S. rostrata*. Plant Soil 255:209–215

Shimamura S, Mochizuki T, Nada Y, Fukuyama M (2002) Secondary aerenchyma formation and its relation to nitrogen fixation in root nodules of soybean plants (*Glycine max*) grown under flooded conditions. Plant Prod Sci 5:294–300

Shimamura S, Mochizuki T, Nada Y, Fukuyama M (2003) Formation and function of secondary aerenchyma in hypocotyl, roots and nodules of soybean (*Glycine max*) under flooded conditions. Plant Soil 251:351–359

Shimamura S, Yoshida S, Mochizuki T (2007) Cortical aerenchyma formation in hypocotyl and adventitious roots of *Luffa cylindrica* subjected to soil flooding. Ann Bot 100:1431–1439

Shimamura S, Yamamoto R, Nakamura T, Shimada S, Komatsu S (2010) Stem hypertrophic lenticels and secondary aerenchyma enable oxygen transport to roots of soybean in flooded soil. Ann Bot 106:277–284

Shiono K, Takahashi H, Colmer TD, Nakazono M (2008) Role of ethylene in acclimations to promote oxygen transport in roots of plants in waterlogged soils. Plant Sci 175:52–58

Shiono K, Ogawa S, Yamazaki S, Isoda H, Fujimura T, Nakazono M, Colmer TD (2011) Contrasting dynamics of radial O_2-loss barrier induction and aerenchyma formation in rice roots of two lengths. Ann Bot 107:89–99

Smirnoff N, Crawford RMM (1983) Variation in the structure and response to flooding of root aerenchyma in some wetland plants. Ann Bot 51:237–249

Steffens B, Sauter M (2009) Epidermal cell death in rice is confined to cells with a distinct molecular identity and is mediated by ethylene and H_2O_2 through an autoamplified signal pathway. Plant Cell 21:184–196

Steffens B, Geske T, Sauter M (2011) Aerenchyma formation in the rice stem and its promotion by H_2O_2. New Phytol 190:369–378

Stevens KJ, Peterson RL, Stephenson GR (1997) Morphological and anatomical responses of *Lythrum salicaria* L. (purple loosestrife) to an imposed water gradient. Int J Plant Sci 158:172–183

Stevens KJ, Peterson RL, Reader RJ (2002) The aerenchymatous phellem of *Lythrum salicaria* (L.): a pathway for gas transport and its role in flood tolerance. Ann Bot 89:621–625

Subbaiah CC, Sachs MM (2003) Molecular and cellular adaptations of maize to flooding stress. Ann Bot 90:119–127

Subbaiah CC, Bush DS, Sachs MM (1994) Elevation of cytosolic calcium precedes anoxic gene expression in maize suspension-cultured cells. Plant Cell 6:1747–1762

Teakle NL, Armstrong J, Barrett-Lennard EG, Colmer TD (2011) Aerenchymatous phellem in hypocotyl and roots enables O_2 transport in *Melilotus siculus*. New Phytol 190:340–350

Thomas AL, Guerreiro SMC, Sodek L (2005) Aerenchyma formation and recovery from hypoxia of the flooded root system of nodulated soybean. Ann Bot 96:1191–1198

Thornalley PJ, Vašák M (1985) Possible role for metallothionein in protection against radiation-induced oxidative stress. Kinetics and mechanism of its reaction with superoxide and hydroxyl radicals. Biochim Biophys Acta 827:36–44

Torres MA, Dangl JL (2005) Functions of the respiratory burst oxidase in biotic interactions, abiotic stress and development. Curr Opin Plant Biol 8:397–403

Trought MCT, Drew MC (1980) The development of waterlogging damage in young wheat plants in anaerobic solution cultures. J Exp Bot 31:1573–1585

Verboven P, Pedersen O, Herremans E, Ho QT, Nicolaï BM, Colmer TD, Teakle N (2012) Root aeration via aerenchymatous phellem: three-dimensional micro-imaging and radial O_2 profiles in *Melilotus siculus*. New Phytol 193:420–431

Visser EJW, Bögemann GM (2006) Aerenchyma formation in the wetland plant *Juncus effusus* is independent of ethylene. New Phytol 171:305–314

Visser EJW, Voesenek LACJ (2004) Acclimation to soil flooding—sensing and signal-transduction. Plant Soil 254:197–214

Visser EJW, Colmer TD, Blom CWPM, Voesenek LACJ (2000) Changes in growth, porosity, and radial oxygen loss from adventitious roots of selected mono- and dicotyledonous wetland species with contrasting types of aerenchyma. Plant Cell Environ 23:1237–1245

Walker BA, Pate JS, Kuo J (1983) Nitrogen fixation by nodulated roots of *Viminaria juncea* (Schrad. & Wendl.) Hoffmans. (Fabaceae) when submerged in water. Aust J Plant Physiol 10:409–421

Winkel A, Borum J (2009) Use of sediment CO_2 by submersed rooted plants. Ann Bot 103:1015–1023

Witty JF (1991) Microelectrode measurements of hydrogen concentrations and gradients in legume nodules. J Exp Bot 42:765–771

Witty JF, Skøt L, Revsbech NP (1987) Direct evidence for changes in the resistance of legume root nodules to O_2 diffusion. J Exp Bot 38:1129–1140

Wong HL, Sakamoto T, Kawasaki T, Umemura K, Shimamoto K (2004) Down-regulation of metallothionein, a reactive oxygen scavenger, by the small GTPase OsRac1 in rice. Plant Physiol 135:1447–1456

Xue T, Li X, Zhu W, Wu C, Yang G, Zheng C (2009) Cotton metallothionein GhMT3a, a reactive oxygen species scavenger, increased tolerance against abiotic stress in transgenic tobacco and yeast. J Exp Bot 60:339–349

Yamauchi T, Rajhi I, Nakazono M (2011) Lysigenous aerenchyma formation in maize root is confined to cortical cells by regulation of genes related to generation and scavenging of reactive oxygen species. Plant Signal Behav 6:759–761

Yamauchi T, Shimamura S, Nakazono M, Mochizuki T (2013) Aerenchyma formation in crop species: a review. Field Crops Res 152:8–16

Zhu J, Brown KM, Lynch JP (2010) Root cortical aerenchyma improves the drought tolerance of maize (Zea mays L.). Plant Cell Environ 33:740–749

Plant Internal Oxygen Transport (Diffusion and Convection) and Measuring and Modelling Oxygen Gradients

W. Armstrong and J. Armstrong

Abstract The parts played by oxygen diffusion (and in some species, convected 'air') in facilitating aerobic metabolism in plants subject to soil flooding and submergence are explored. Simple diffusion equations are used to illustrate how resistance and respiration interact to create oxygen gradients and experimental and modelling examples of gradients in roots and the limitations of diffusive transport are presented and discussed. Attention is drawn to the limiting effects of diffusion especially in non-wetland species such as *Arabidopsis thaliana*. Here, a paucity of root gas-space and a cortical cell configuration found also in crop species, such as pea, tomato and cotton, is particularly unsuited for long-distance oxygen transport. The contrast with other more flood-tolerant Brassicas is highlighted. The relative roles of root aerenchymas and barrier formation to radial oxygen loss in improving oxygen supply and supporting root extension and phytotoxin exclusion in flooded soils are considered. Methods for monitoring radial oxygen loss from roots and oxygen concentrations both within and external to the plant are discussed, as are the results of analogue and more complex mathematical models that predict and explain the role of both diffusive and convective transport in plant aeration. Finally, pressurized gas-flow mechanisms and their ability to overcome diffusion limitations are briefly described and discussed.

1 Introduction

Plants have no metabolically activated oxygen dispersal mechanisms other than cytoplasmic streaming which is relatively ineffective for long-distance transport (Armstrong 1979; Denison 1992). Some wetland plants have convective

W. Armstrong (✉) • J. Armstrong
Biological Sciences, University of Hull, Yorkshire HU6 7RX, UK

School of Plant Biology, University of Western Australia, Crawley, WA 6009, Australia
e-mail: w.armstrong@hull.ac.uk

J.T. van Dongen and F. Licausi (eds.), *Low-Oxygen Stress in Plants*, Plant Cell Monographs 21, DOI 10.1007/978-3-7091-1254-0_14, © Springer-Verlag Wien 2014

(pressurized) gas-flows, but overall, it is diffusion that dominates the oxygen supply process. However, diffusion is slow and through liquid is particularly slow. In this chapter we review the role of diffusion in terms of contributing to, and alleviating, respiratory oxygen stress, comment on the measurement and modelling of internal and external oxygen gradients and review convective oxygen transport.

2 Diffusion and Diffusive Resistance

Diffusion is the *net* movement of molecules by random motion from regions of high to lower concentration. It is revealed by concentration gradients. Diffusion rate, J, (mol s^{-1}) and flux (mol cm^{-2} s^{-1}) are proportional to gradient.

The principles of diffusion are perhaps best illustrated by the simplest case: *steady state planar* diffusion between an oxygen source, C_o (mol cm^{-3}), and sink ($C = 0$) at opposite ends of a tube, length L (cm) having an impermeable wall. Here,

$$J = D_o A \Delta C / L, \tag{1}$$

where ΔC is the concentration difference (mol cm^{-3}) between source and sink[1]; A is the cross-sectional area of source, sink and tube (cm^2) and D_o, the diffusion coefficient, is the proportionality constant equating to an *amount per unit time, area and gradient*, i.e. mol/[s $\times$ cm^2 $\times$ ($\Delta C/L$)] giving units of cm^2 s^{-1}.

The gradient, $\Delta C/L$, is linear (Fig. 1b) but although J is directed towards the sink, the directional movements of individual molecules are random and their individual velocities are not a function of the gradient. The velocity of the random walk process, and hence the value of D_o, depends upon molecular size, mass, temperature and the characteristics of the medium. In air, oxygen molecular velocities are ca. 0.44 km s^{-1} (990 mph), N_2 0.47 km s^{-1} and CO_2 0.38 km s^{-1}, but intermolecular collisions considerably slow *net* progress in any direction.[2] Water reduces velocity to 1/10,000 of that in air. In air at 20 °C, D_o is 0.201 cm^2 s^{-1}, while in water, 2.1 $\times$ 10^{-5} cm^2 s^{-1}. Water also reduces oxygen concentration to ca. 1/30 of that in air.[3] The net result is an *effective resistance* to O_2 diffusion in aqueous media at least 300,000 times greater than in air. The diffusive resistance of path L is L/D_oA (s cm^{-3}).

Plant organs rarely, if ever, fully approximate with the open-tube analogy and the *per cent* pore-space (porosity) can vary enormously with the number and

[1] Although often referred to as the 'driving force', ΔC does not affect molecular velocities.

[2] Mean free-path length = 67 nm.

[3] Expressing such sudden changes in oxygen concentration by volume can be avoided by expressing data in terms of oxygen partial pressure since this does not significantly alter at an air–water interface.

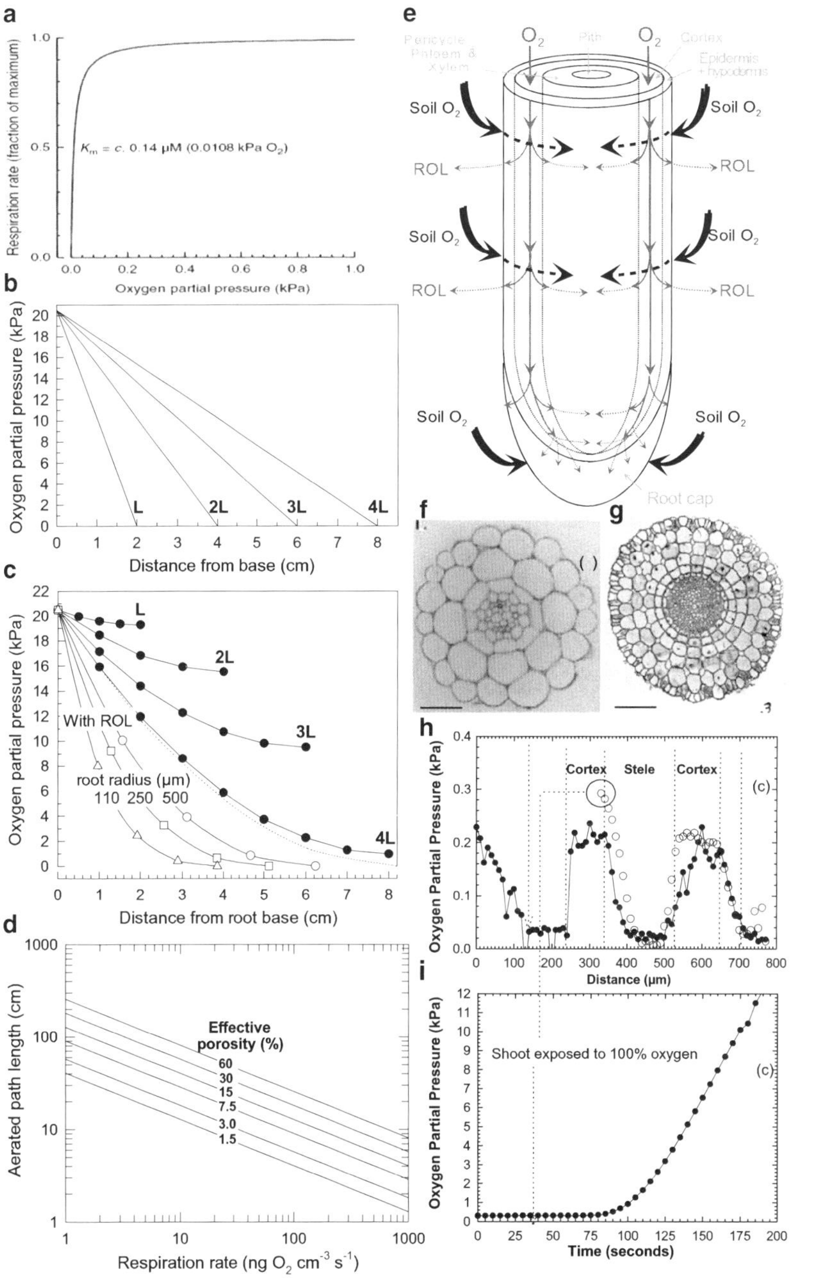

Fig. 1 (**a**) Michaelis–Menten O_2-dependency curve based on the respiration of isolated plant mitochondria (K_m 0.0108 kPa) and used for programming the relationship between the respiratory rate and O_2 concentration at any locus in respiring root segments (after Armstrong and Beckett 2011a). (**b**) The effect of increasing path length, *L*, on the O_2-concentration gradients in a simple tubular source–sink diffusion system of radius 0.05 cm and gas-filled porosity 3 %. (**c**) (i) *Closed symbols*: the effect of increasing path length, *L* to 2*L*, 3*L* and 4*L* on oxygen concentrations along a

distribution of cells: in stelar tissues and outer cortex/hypodermis there may be no gas-space or gas-space continuum. Cells (or solid objects in soils), by reducing gas-filled porosity, increase diffusive impedance; so does path tortuosity. Both can be accommodated by modifying D_o creating an effective diffusion coefficient, $D_e = D_o\varepsilon\tau$, and the resistance L/D_oA becomes L/D_eA; ε is the fractional porosity (<1) and τ (<1) expresses the tortuosity; A remains the overall cross-sectional area. Usually there is little axial tortuosity in cortical gas-space (Armstrong 1979); radial tortuosity is more likely.

In root aeration some oxygen transport is also radial, e.g. from inner cortex into the vascular cylinder, from inner cortex to outer cortex and soil (or hypoxic culture medium) or from soil (or aerated culture solution) to root (Fig. 1e). Radial diffusive resistance between cylinders can be expressed as $r_i \log_e r_o/r_i/D_eA_i$, where r_i is the radius of the inner cylinder, r_o, the radius of the outer cylinder and A_i, the surface area of the inner cylinder. The expression $r_i \log_e r_o/r_i$, the radial equivalent of L in (1), is an effective length and less than $r_o - r_i$. A relatively short radial path between cortex and root surface may offer a significantly greater resistance to oxygen diffusion than the cortical gas-phase path from the shoot. Diffusive resistance down 9 cm of root, radius 0.05 cm and porosity, 50 %, (1.12×10^4 s cm^{-3}) is less than 50 μm of radial liquid path through the epidermis and hypodermis ($\geq 2.5 \times 10^4$ s cm^{-3}) of a 1 cm long root segment (i.e. outer surface area of 0.314 cm^2).

Fig. 1 (continued) tubular diffusion system where tube wall is impermeable, porosity 3 % and respiratory demand (50 ng cm^3 s^{-1}) is uniformly distributed. Note that L is independent of radius. (ii) *Dotted line*: oxygen profile for maximum aerated length > 4L. (iii) *Open symbols*: As for (i) potential length 8.2 cm but permeable wall allowing ROL to an external flooded medium (porosity 40 % respiratory demand 50 ng cm^3 s^{-1}) and so shortening aerated length of root: (*open circle*) root radius 0.05 cm; (*open square*) root radius 0.025 cm; (*open triangle*) root radius 0.011 cm. (**d**) Data computed from (2) predicting the maximum distance to which oxygen will diffuse longitudinally through plant organs of different porosities and oxygen demand assuming uniformity of demand and porosity along the diffusion path and no radial leakage (redrawn from Armstrong 1979). (**e**) Diagram showing major paths for oxygen diffusion in roots. In anaerobic media oxygen diffuses from the shoot via the porous cortex and wherever the root wall is permeable; there will be radial oxygen loss (ROL). In aerobic soil or solution culture media there will be a radial influx of oxygen but with still the potential for transport from the shoot via cortical gas-space. (**f**) Transverse root sections of *Arabidopsis thaliana*; note lack of intercellular gas-space. Bar = 25 μm (after Dolan et al. 1993). (**g**) Transverse root sections of *Sinapis alba*: note concave quadrangular intercellular gas-spaces in the inner cortex. Bar = 135 μm (after Cormack 1947). (**h**) Oxygen transport in *Brassica napus*. Showing radial oxygen concentration profiles across a lateral root of an intact plant with gradients revealing oxygen diffusion from the cortex to the stele and to the external anaerobic agar medium (*closed symbols*: in-track; *open symbols*: out-track). (**i**) Oxygen transport in *Brassica napus*. Continuation of the out-track of (**h**) showing the rapid increase in cortical oxygen supply when the shoot was transferred from air to 100 % oxygen (after Voesenek et al. 1999)

3 Oxygen Concentration Requirements for Respiration

Mitochondria, the major sites of oxygen consumption by plant cells, and cytochrome oxidase, the principal terminal oxidase in plant respiration, located in the mitochondria, have a very strong affinity for oxygen that follows a Michaelis–Menten-type relationship. For the isolated enzyme, the oxygen concentration necessary for half-maximum consumption rate (the Michaelis–Menten constant, or K_m) is ca. 0.1 μM (1/2,700 of air saturation). K_ms for whole mitochondria in suspension range from 0.1 to 0.15 μM (Barzu and Satre 1970; Millar et al. 1994; Rawsthorne and LaRue 1986) which suggests that there is little of a diffusive barrier to oxygen entry to the mitochondria. For a K_m of 0.14 μM (0.0108 kPa), oxygen uptake should be close to maximum at a concentration (the critical oxygen pressure, COP) of ca. 14 μM (i.e. 1 kPa: 1/20 of air saturation) (Fig. 1a); 95 % of maximum respiration should be possible at ca. 1/100 of air saturation: 2.6 μM (0.2 kPa O_2) (Armstrong and Beckett 2011a).

Parenchymatous cell walls, cell membranes and cytoplasmic paths do not appear to greatly impede diffusive oxygen entry to cells (Nobel 1991)[4]: whole leaves with their labyrinth of gas-space can exhibit very low apparent K_m values: 0.33–1.1 μM (Laisk et al. 2007) inclusive of some uptake by the non-phosphorylating alternative oxidase (K_m 1.7 μM). Provided that cells are in close proximity to gas-spaces there may be few diffusive constraints on respiration. K_ms are usually found by monitoring consumption versus oxygen concentration in the medium bathing the material under investigation. Where uptake is apparently constrained at high oxygen concentrations, the effect can usually be traced to additional resistance and respiration between the site affected and the primary oxygen source (Armstrong and Beckett 2011a; Armstrong et al. 2009; Asplund and Curtis 2001; Curtis and Tuerk 2008; Shiao and Doran 2000). If the medium is a fluid, the diffusive boundary layer around the material may be a very significant additional diffusive resistance in series with any membrane, cytoplasmic or cell wall resistances between the monitored oxygen level in the medium and the mitochondria. The result can be a substantial O_2-concentration drop across the boundary layer (Fig. 2d) (Aguilar et al. 2003; Armstrong and Beckett 2011a; Armstrong et al. 2009; Asplund and Curtis 2001; Curtis and Tuerk 2008; Shiao and Doran 2000). Vigorous stirring reduces boundary layers (Asplund and Curtis 2001; Greenwood and Goodman 1971) but if the material travels with the flow, shearing forces will be less, and boundary layers less reduced. A consequence of both this and the use of bulky material, e.g. root segments, is that the K_m and COP values derived will no longer reflect mitochondrial oxygen concentrations. Instead they represent apparent K_ms

[4] The water-filled interstices through which oxygen diffuses in primary cell walls may occupy about 50 % of the total wall volume (fractional porosity, $\varepsilon = 0.5$) but with a tortuosity which doubles path length ($\tau = 0.5$) (Nobel 1991). For a planar cell wall of area 1 cm^2 and thickness of say 0.3 μm the diffusive resistance would be $0.3 \times 10^{-4}/2 \times 10^{-5} \times 0.5 \times 0.5 \times 1$, or 6 s cm^{-3}. This is a very small value (5 %) compared with the resistance across a non-occluded water path of say 25 μm which would be $25 \times 10^{-4}/2 \times 10^{-5}$, or 125 s cm^{-3}.

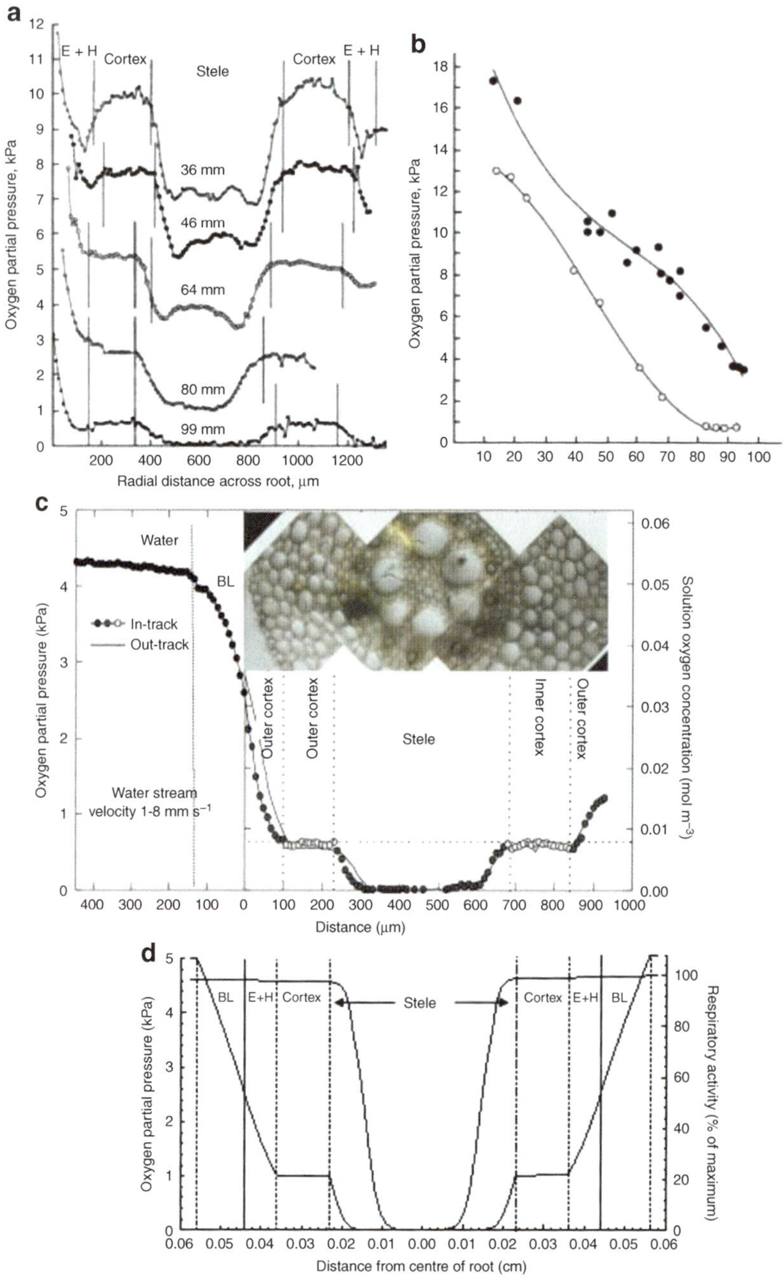

Fig. 2 (**a**) Five superimposed radial profiles taken sequentially down the length of a 6-day-old, 105-mm-long non-aerenchymatous maize primary root. *Vertical bars* indicate approximate boundaries of tissue cylinders. E + H = epidermal–hypodermal cell layers (after Darwent et al. 2003). (**b**) Longitudinal variation of cortical oxygen concentration in aerenchymatous and non-aerenchymatous maize primary roots (*Zea mays*, cv. LG11). Combined data from two aerenchymatous roots (*closed circle*), approx. lengths 95 and 96 mm (**d**), compared with a

and COPs, indicative of only that somewhere in the tissue (i.e. typically in the inner core, as this is furthest from the external source) is the mitochondrial COP being experienced (Armstrong and Beckett 2011a; Armstrong et al. 2009).

Without modifying influences, e.g. high internal CO_2 levels (Lammertyn et al. 2001) or some, as yet unspecified, down-regulating mechanism, aerobic respiration in plants should be maintained close to a maximum provided that internal oxygen concentrations do not fall below about 0.2 kPa anywhere in the tissue.

4 Gradients

Gradients are created by the consumption of oxygen at sinks (e.g. respiratory sites): since consumed oxygen cannot return to the system, a local drop in concentration (or partial pressure) occurs. This can only be redressed by oxygen moving from a more remote site, statistically more likely to come from the direction of higher concentration (the primary source). The ease and extent of replacement depends upon the resistance to molecular motion between source and sink; molecular velocity is not increased and more oxygen molecules per unit time do not contact respiratory sites. On the contrary, a gradient reflects only a failure of molecules to move back randomly towards the source.

In the simple steady state planar system with impermeable wall and a sink at the end of the path, the gradient is always linear and consumption by the sink, concentration-dependent (Fig. 1b). Lengthening the path to 2 L would halve the diffusion rate, but at all distances $< L$ concentrations would rise. Irrespective of how much the tube is lengthened, and the diffusion rate reduced, the resistance never completely prevents some oxygen reaching the sink; nowhere along the path will there be zero oxygen except at the surface of the sink.

If O_2 consumption had occurred not at the end of the tube, but uniformly along it, the situation is more akin to that in plants rooted in wet anaerobic soils where oxygen must diffuse through cortical intercellular spaces from shoot to root to support metabolism. If respiratory rate, M (mol cm^{-3} s^{-1}), was independent of oxygen concentration (a useful approximation at the mitochondrial level) the

Fig. 2 (continued) 95-mm-long non-aerenchymatous root (*open circle*) (after Darwent et al. 2003). (**c**) In-track (*symbols*) and out-track (*continuous line*) microelectrode radial O_2 profiles across an excised primary root of maize lying in a nutrient medium that was streamed lengthwise past it at a velocity of 1.8 mm s^{-1}. The dissolved O_2 concentration in the medium was 0.054 mol m^{-3} (modified from Gibbs et al. 1998). Transverse section of root inset—dark stainings in pericycle reveal track taken by the electrode (modified after Gibbs et al. 1998). (**d**) Mathematical modelling of oxygen distribution across a maize root similar to that in (**c**) but with dissolved oxygen in the medium at 5 kPa. Based on Michaelis–Menten kinetics and showing the predicted respiratory activities (*upper plot*) corresponding with the oxygen partial pressures (*lower plot*) across the root (Armstrong and Beckett, unpublished data)

oxygen distribution along the tube in the steady state is given by solving the equation:

$$C_x = C_o - Mx(2L - x)/2D_e, \quad (2)$$

where C_x is the O_2 concentration at any distance x along L and C_o is the source concentration at $x = 0$. The O_2 decline along L is now curvilinear (Fig. 1c); any increase in L lowers the concentrations throughout the system, rather than raising them. This strikingly illustrates how consumption along a diffusive path combines with diffusive resistance to reduce the oxygen supply to more remote points. Consequently, oxygen concentrations that impede respiration will tend to develop first in non-porous tissues where the diffusion path is entirely aqueous and the synergism between respiration and diffusive resistance greatest (Figs. 1g and 2a, d). Equation (2) can predict approximate limits to root growth if oxygen comes only from the shoot (Fig. 1d).

Equation (2) and others in spherical or cylindrical coordinates show how oxygen gradients may develop longitudinally and radially within, and to and from roots and within the gas-phase of soils and in soil crumbs (Armstrong 1979; Armstrong and Beckett 2011a; Armstrong et al. 1991b; Currie 1962, 1965; Greenwood and Goodman 1971; Sorrell et al. 2000). Measuring gradients could contribute to resolving whether, or at what point, respiratory down-regulation might occur in roots in response to lowered oxygen supply (Armstrong and Beckett 2011b). The consequences of limited oxygen supply within tissues (e.g. inhibited xylem loading) are discussed elsewhere (Gibbs and Greenway 2003a, b).

5 Shoot to Root Oxygen Transport

Oxygen supplied from shoot to root may come through stomata or lenticels, or be photosynthetic. Since photosynthetic oxygen is lost readily through stomata, root aeration normally should not vary greatly from day to night, except where photosynthetic tissues are submerged (Gaynard and Armstrong 1987; Pedersen et al. 2006; Waters et al. 1989; Winkel et al. 2013) or behind significant barriers to oxygen escape, e.g. phellogen of woody stems (Armstrong and Armstrong 2005b). Diurnal fluctuations in shoot to root transmission will then result and oxygen stresses may develop at night and disappear during the day (Armstrong and Armstrong 2005b; Pedersen et al. 2006; Waters et al. 1989; Winkel et al. 2013).

Diffusive resistances in shoots will obviously affect transport to roots. Where aerenchyma in emergent shoots (Jung et al. 2008; Yamasaki 1952) reaches the root–shoot junction, resistances will be relatively low; without aerenchyma, resistance might be substantial but such effects have been little explored. With woody species, lenticels (often hypertrophied) closest to the submergence level are those most involved in oxygen transport to the roots (Armstrong 1968; Denison 1992; Jackson and Armstrong 1999; Shimamura et al 2010). Gas-filled elements in the wood, and gas-spaces through the cambium (Hook and Brown 1972; Dittert et al. 2006), may play an important role for gas-transport in mature trees during partial submergence but the former remains to be proven. In submerged plants gas-films on the leaves can reduce resistances to oxygen and carbon dioxide exchange with the water body (Carvalho and Curtis 1998; Colmer and Pedersen 2008a, b; Waters et al. 1989; Winkel et al. 2011, 2013) as can a thin cuticle (Mommer et al. 2004; Sand-Jensen et al. 2005).

The root–shoot junction is another significant resistance although there is usually a gas-phase connexion through it (Gaynard and Armstrong 1987; Justin and Armstrong 1983). In *Phragmites australis* junction porosity is ca. 50 % (Armstrong and Armstrong 1990a) and resistance is very low; in other species such as rice the porosity can be much lower (<10 %) but the shortness of the path compensates for lower porosity. Claims of exceptionally high root–shoot junction resistance in rice (Groot et al. 2005) appear to have been miscalculated.

6 Oxygen Transport in Roots

Potentials for oxygen stress in roots are particularly high: oxygen from the shoot must navigate a series of resistances, within and from which oxygen will be consumed (Armstrong 1979; Colmer 2003a): in anaerobic media liquid-phase radial oxygen loss (ROL) occurs through non-porous tissues to the rooting medium (Figs. 1e, h and 3a). In aerobic media there is radial inward diffusion (Fig. 1e and 2a, c, d) again with consumption along the path (Armstrong and Drew 2002; Armstrong et al. 1991b) and often with an oxygen source of sub-atmospheric concentration.

The degree and distribution of cortical gas-space and the degree and distribution of respiratory demand determine to what extent roots will be sufficiently aerated (Armstrong 1979; Armstrong and Beckett 1987; Sorrell et al. 2000). Respiratory demand (per unit volume) is highest near the root tip (Armstrong et al. 1991b; Lemon and Wiegand 1962; Luxmoore et al. 1970) diminishing sub-apically as cortical cells expand and vacuolate, but pericycle, phloem, xylem parenchyma and lateral roots and initials remain as respiratory 'hot-spots' (Armstrong et al. 1991b; Sorrell et al. 2000).

Cortical gas-space is to some degree related to plant habitat preference: viewed in TS, small triquetrous spaces (cells in a hexagonal array and persisting as the root matures) are particularly prevalent in *non-wetland* species; diamond-shaped spaces (cells in cubic array and prone to later modification or collapse to form aerenchyma) characterize most *wetland plants* (Jung et al. 2008; Justin and Armstrong 1987; Smirnoff and Crawford 1983; Seago et al. 2005). Recent studies suggest that reactive oxygen species may be responsible for the cortical cell breakdown of lysigenous aerenchyma (Yamauchi et al. 2011).

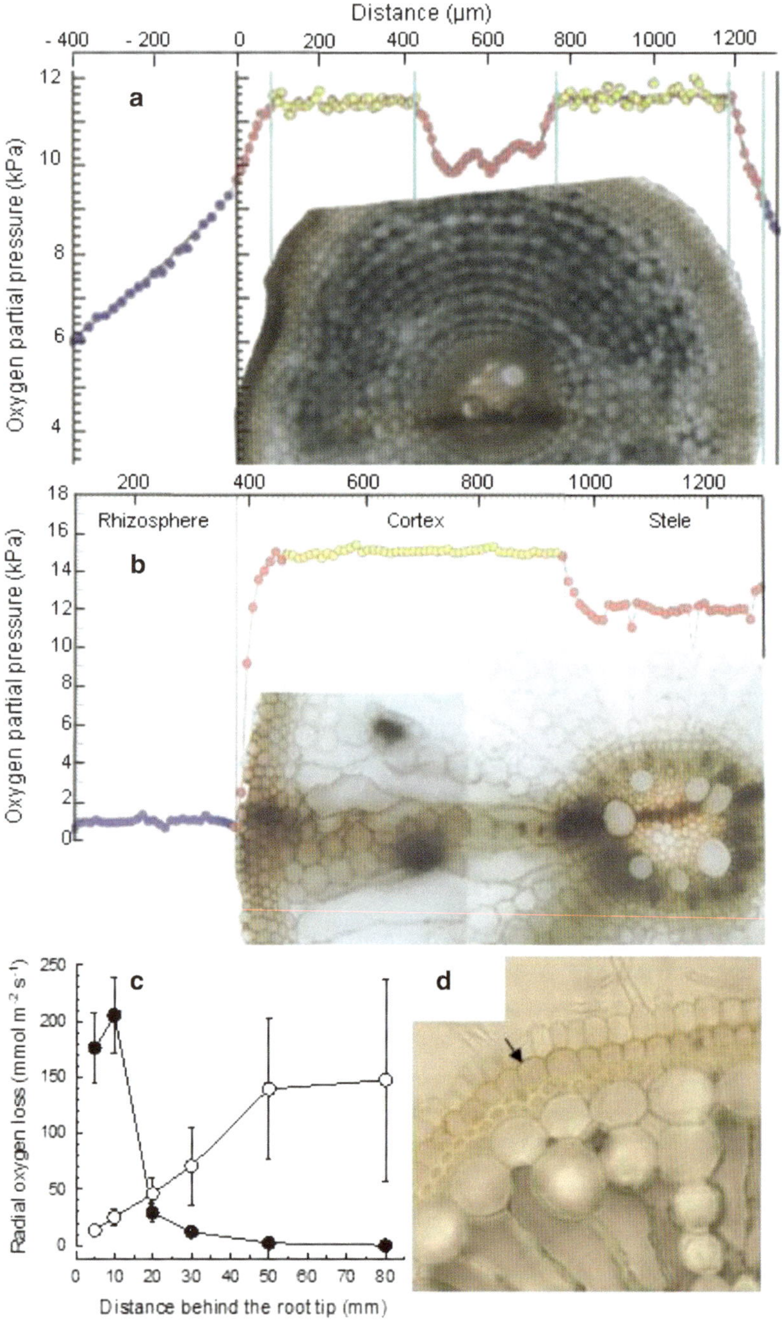

Fig. 3 (**a**) Radial oxygen profile through *Phragmites* adventitious root at 7 mm from the apex superimposed upon a photomicrograph at this distance showing the track (*brown-stained*) taken by the electrode (after Armstrong et al. 2000). (**b**) Radial oxygen profile through *Phragmites* adventitious root at 100 mm from the apex superimposed upon a photomicrograph at this distance showing the track (*brown-stained*) taken by the electrode. Note the steep gradient across

6.1 Non-wetland Plants

Hexagonal packing with cells in point contact limits *cortical* porosity to ≤9.3 % but *root porosities* in *non-wetland* species are generally <3 % because of tighter packing (Justin and Armstrong 1987). Although oxygen transport from the shoot is often measurable (Armstrong and Healy 1984; Armstrong et al. 1982, 1983; Greenwood and Goodman 1971), cortical concentrations can be extremely low even in quite short roots, e.g. *Brassica napus* in hypoxic agar (Fig 1h, i; Voesenek et al. 1999): here, the mitochondrial O_2-affinity curve (Fig. 1a) points to cortical respiration at 95 % of maximum, but at the centre of the stele (O_2 partial pressure ca. 0.01 kPa) respiration may have been <30 % of maximum. The potential for oxygen stress in Brassicas is exacerbated by the high surface area to volume of their narrow roots: ROL (and laterals: Armstrong et al. 1983) must significantly reduce the length to which they can grow before running short of oxygen (cf. Fig. 1c); the same applies to fine lateral roots on the adventitious roots of wetland species (Armstrong et al. 1996b). In Fig. 1g some ROL is evident as an O_2 gradient from root to agar. In wetland soils, microbial oxygen consumption plus chemical oxygen sinks may remove much more oxygen by ROL, reducing root extension still further. Roots of non-wetland plants rarely penetrate more than a few centimetres into flooded soil (Justin and Armstrong 1987) and in drained, well-aerated soil, can die back to their self-sustained aerated length if soil is temporarily flushed with nitrogen (Huck 1970) or flooded (*pers. observation of pea*). Among non-wetland species, the Brassica *Arabidopsis thaliana* is at an extreme with little if any root gas-space (Fig. 1f; Dolan et al. 1993): roots must depend almost entirely on oxygen diffusion from the rooting medium. Recent studies have confirmed its poor oxygen transport potential (Lee et al. 2011; Vashisht et al. 2011). Notable crop species with low root porosity and intolerance to soil flooding are pea (Armstrong et al. 1982), tomato (Cormack 1947) and cotton (Huck 1970). Transport in other Brassicas, *B. napus* and *Sinapis alba*, is traceable to cortical cells in cubic packing (Fig. 1g; Cormack 1947) while *Cardamine pratensis*, a wetland species, has both cubic packing and some aerenchyma developed from it (Justin and Armstrong 1987). *It is difficult to envisage any significant improvement in flood tolerance within a species without some gas-space provision and hence improved oxygen transport*

Fig. 3 (continued) epidermal–hypodermal cylinder due to barrier formation in the cell walls (after Armstrong et al. 2000). (**c**) Radial O_2 loss (ROL) along intact adventitious roots of rice (*Oryza sativa*) in an O_2-free root medium with shoots in air, at 30 °C ($n = 3 \pm$ SE). Plants had been grown in aerated (*open circles*) or stagnant deoxygenated 0.1 % agar (*closed circles*) in nutrient solution. Plants were 28- to 29-day-old; treatments imposed on day 14 (modified from Colmer et al. 2006). (**d**) Sector (iodine-stained) from near the base of a 100 mm long rice root that had been grown in 0.05 % stagnant deoxygenated agar and then exposed for 2 days to 0.174 mM sulphide in the same medium. Note the thickened out tangential wall of the exodermis (*arrow*) which presented a barrier to Fe^{2+} absorption and seems likely to be the principal barrier to radial oxygen loss (after Armstrong and Armstrong 2005a)

from shoot to root. Selecting for cortical cell organization (Maggio et al. 2001) may offer the best prospects for introducing flood tolerance to these non-wetland species.

Other features which will exacerbate aeration problems in some non-wetland species are (a) secondary growth where secondary tissues are low in porosity and (b) continuing meristematic activity of primary cortical tissues which must sustain demand on the already limited oxygen supply (Justin and Armstrong 1987).

6.2 Intermediate Species

This ill-defined grouping grows best in well-drained soils but they tolerate wetland conditions due to aerenchyma formation induced in flooded soil (Justin and Armstrong 1987). The benefits of some aerenchyma production in maize can be seen in Fig. 2b. Two other *intermediate* species highlighted recently, soybean and *Melilotus siculus*, improve root aeration by means of secondary growth producing aerenchymatous phellem in hypocotyl and root (Shimamura et al. 2003, 2010; Teakle et al. 2011; Thomas et al. 2005; Verboven et al. 2011), a feature also of the wetland species, *Lythrum salicaria* (Stevens et al. 2002).

Often it is not appreciated that even roots with low porosity may simultaneously receive oxygen from the shoot and the rooting environment (Fig. 1e). Oxygen gradients in Fig. 2a (and see Armstrong et al. 1994) reveal both pathways operating in a non-aerenchymatous maize root; each contributes differently at different positions along the root. At 36 and 46 mm from the base oxygen diffuses from rooting medium (a solid agar) to the outer cortex. Simultaneously, oxygen from the shoot diffuses radially from inner to outer cortex. The outer cortex thus receives oxygen from rooting medium and shoot; there is no net transfer from the medium to the inner cortex to supplement supplies to apex and stele. Radial inward transport has *buffered* the system; if it hadn't occurred, less oxygen would have reached the tip. At 64 and 80 mm, oxygen from the medium reaches the inner cortex, but not at 99 mm, perhaps due to greater respiratory demand in the outer cortex at this point. At all positions the shoot is probably supplying the greater part of this root's oxygen requirement. Buffering effects have also been modelled (Armstrong 1979).

Oxygen supply to roots from solution culture will be significantly impeded by boundary layer resistances unless there is forced aeration and vigorous mixing (Asplund and Curtis 2001; Carvalho and Curtis 1998; Greenwood and Goodman 1971; Shiao and Doran 2000). A partially aerated medium (with a partial oxygen pressure of 4.3 kPa) streaming past at 1.4 mm s^{-1} was insufficient to support fully the aerobic demands of this excised non-aerenchymatous maize root (Fig. 2c; Gibbs et al. 1998) creating near to zero [O_2] across much of the stele. Diffusion through 200–300 μm of boundary layer was responsible for an almost 2 kPa drop in partial pressure; a further 2 kPa drop is attributable to transport through, and consumption by, the non-porous outer tissues. The limited oxygen detection limits of oxygen microelectrodes (0.007–0.021 kPa) make it impossible to prove that the centre of the stele was anaerobic but modelling, using Michaelis–Menten kinetics, shows aerobic respiration still at ≤20 % of maximum where microelectrodes indicated tissue anoxia (cf. Fig. 2c, d). A second example is banana roots, where root hairs helped to create a boundary layer of ca. 1,000 μm with an oxygen deficit across it of 11 kPa (Aguilar et al. 2003).

6.3 Wetland Plants

Near the apex, roots are generally well-furnished with cortical gas-space from cubic cell-packing (potential porosity 21.4 %) in middle and/or inner, or throughout the cortex (Justin and Armstrong 1987), while aerenchyma usually forms sub-apically when in stagnant hypoxic conditions, e.g. agar or wet soils (Colmer 2003b; Jackson and Armstrong 1999; Justin and Armstrong 1987; Seago et al. 2005; Visser et al. 2000). Consequently, not only is diffusive resistance from shoot to root apex substantially reduced, so too is the respiratory rate per unit volume (Armstrong 1979; Armstrong et al. 1991b; Luxmoore et al. 1970). In some species extensive (constitutive) aerenchyma occurs whatever the soil conditions (Justin and Armstrong 1987). Also characteristic of *wetland plant* roots is a barrier to ROL (Fig. 3b, c) and phytotoxin entry caused by sub-apical suberization and lignification of exodermal cell walls (Fig. 3d) (Armstrong and Armstrong 2005a; Armstrong 1964, 1971, 1979; Colmer 2003a; Colmer et al. 1998; Garthwaite et al. 2008; Kotula and Steudle 2009; Shiono et al. 2011; Soukup et al. 2007); this may be much less strong in some dicotyledonous species (Laan et al. 1989; Visser et al. 2000) but can be extreme in some mangroves (Darwent 1997). Oxygen is freely lost to the rooting medium near to the apex of *Phragmites* roots (Fig. 3a) but barrier formation coupled to oxygen consumption is sufficient to prevent its loss sub-apically despite a cortical oxygen pressure of 15–16 kPa (Fig. 3b) (Armstrong et al. 2000). In a wetland soil the epidermis may not receive any oxygen sub-apically and this could contribute to its disintegration in wetland roots; here it received some oxygen from the bathing medium. It is interesting to note that the oxygen concentration only fell by 2–3 kPa across the high-oxygen-demand tissues, xylem parenchyma, phloem and pericycle. This is a consequence of their peripheral location, the relatively short diffusion path across them and probably low respiratory demand in the medulla. Wetland plants also tend to have unusually small proportional and absolute stelar areas (McDonald et al. 2002). These features, and suppressed or aerenchymatous secondary tissues (Justin and Armstrong 1987; Stevens et al. 2002), result in wetland plants usually rooting much deeper in wet soils (Justin and Armstrong 1987) than non-wetland species while probably still functioning aerobically throughout (Armstrong and Webb 1985).

The transfer of higher root porosity and a barrier to ROL from *H. marinum* into wheat through wide hybridization and production of *H. marinum*-wheat amphiploids to improve tolerance to flooded soil (Malik et al. 2011) is an exciting development as is the possibility for transferring constitutive aerenchyma and stronger barrier formation into maize from its wild (more flood-tolerant) relative, *Z. nicaraguensis* (Abiko et al. 2012; Mano and Omori 2007).

The requirement for barrier formation may be complex. It helps to increase aerated root lengths (Armstrong and Beckett 1987; Sorrell et al. 2000) but is possibly more concerned with the exclusion of phytotoxins than with oxygen conservation (Armstrong 1979; Garthwaite et al. 2008). From necessity its formation may be rapid and it is promoted by phytotoxins (Armstrong and Armstrong 2005a). If it did not form, continued ROL, by stimulating more external oxygen demand by proliferating aerobic microbes, could critically narrow any oxygenated rhizosphere and bring phytotoxin fronts closer to the root surface. The close proximity of lateral roots produces overlapping rhizospheres which probably afford better protection from phytotoxins without the need for strong barrier formation (Armstrong et al. 1996b; Sorrell et al. 2000). In some wetland species floating roots may generate their own oxygen by photosynthesis (Rich et al. 2011).

7 Measuring Oxygen Transport

The measurement of oxygen concentrations and diffusion within, and to and from, roots and shoots has relied so far mostly on various polarographic techniques involving root-sleeving cylindrical Pt electrodes (Armstrong 1967, 1979; Armstrong and Wright 1975) and naked and Clark-type microelectrodes (Armstrong 1994; Mancusco et al. 2000; Revsbech 1989). More recently there has been some use of fibre-optic (McLamore et al. 2010; Rolletschek et al. 2002) and planar optodes (optrodes) (Tschiersch et al. 2011).

The sleeving electrode (used in stagnant anaerobic agar medium) is unique in providing a cylindrical oxygen sink to measure ROL along roots and is separated from the root by a calculable diffusive resistance. The electrode is non-invasive, robust and highly sensitive, and enables continuous monitoring without loss of signal, helping to identify (Fig. 3c; Armstrong 1979; Colmer 2003a) and quantify barrier resistances (Garthwaite et al. 2008; Kotula and Steudle 2009), internal diffusive resistances (Armstrong et al. 1982), cortical oxygen concentrations and COPs for root respiration (Armstrong and Gaynard 1976; Armstrong et al. 2009) and for extension growth (Armstrong and Webb 1985). It has also helped to identify/quantify the effects on long-distance oxygen transport, of laterals (Armstrong and Healy 1984; Armstrong et al. 1983), respiration (Armstrong 1971; Armstrong and Gaynard 1976; Armstrong et al. 2009; Healy and Armstrong 1972; Waters et al. 1989) including the relative influences of cytochrome oxidase and the alternative oxidase (Webb and Armstrong 1983), and the effects of submergence level and photosynthesis on root oxygen levels and growth (Gaynard and Armstrong 1987; Waters et al. 1989). Changes in oxygen concentration attributable to real gas-pressure variation are also recordable (Armstrong and Armstrong 2005b; Waters et al. 1989).

Direct measurements of internal oxygen concentrations and gradients have been facilitated using polarographic microelectrodes (Aguilar et al. 2003; Armstrong et al. 1994, 2000; Colmer and Pedersen 2008a; Darwent et al. 2003; Gibbs et al. 1998; Revsbech et al. 1999; Verboven et al. 2011; Winkel et al. 2013) and fibre-optodes (Rolletschek et al. 2002, 2005). There are advantages and disadvantages to both: polarographic microelectrodes are simple to calibrate and use but need extreme care to avoid breakages; also they consume some oxygen and so potentially create their own gradients. However, with the least tissue-damaging sizes (tip diameters 1–12 μm) sink activity is minimal compared to respiratory demands. Fibre-optodes are perhaps more robust and don't consume oxygen but, to date, those used invasively have been of a size (30–50 μm) that will create significant damage. Also, their curvilinear linear response to oxygen concentration makes calibration less simple and it is not clear whether the calibration remains constant during invasive use. However, whichever sensor type is chosen, *if used invasively to record gradients, the track should be visibly recorded by subsequent sectioning. Such devices easily divert from their intended track*; *in roots strictly radial profiles can be difficult to obtain*. Examples of invasive measurements by microelectrodes have already been presented (Figs. 1h, i, 2a, b, c, and 3a, b); fibre-optodes have been used with great success to study the oxygen distribution within seeds both in the light and the dark and to link the findings to aerobic and hypoxic biochemistry (Rolletschek et al. 2002, 2005).

Vibrating (oscillating), self-referencing microelectrodes and fibre-optodes enable continuous recording of oxygen flux from solution culture to root (Mancusco et al. 2000; Mancuso and Boselli 2002; McLamore et al. 2010). Fixed near the root surface, the gradient and hence the flux are found using signals recorded at the extremes of the oscillations (e.g. 50 μm apart); amplitude and frequency do not distort the gradients. Respiratory uptake along roots has been determined and regular oscillations in demand revealed close to tips of *Glycine max*, *Phasaeolus vulgaris* and maize (McLamore et al. 2010). However, interpreting such results requires caution since no account was taken of any oxygen contribution from the shoots: respiratory rates deduced from the inward fluxes were lower than would be expected had all demand been met from the medium. This suggests either that the missing quantity was met from the shoot or that the roots were experiencing some oxygen stress; oscillations in ROL and internal oxygen pressures have been noted previously at cortical oxygen concentrations below the COP (Armstrong et al. 2009).

Finally, planar optodes (optically sensitive plate or foil) have revealed instantaneous two-dimensional records of oxygen distribution within sections of seeds and stems (Tschiersch et al. 2011) and O_2 distribution in the rhizosphere (Blossfeld et al. 2011; Jensen et al. 2005). The results appear to confirm data previously obtained by optode microsensors and are suitable for following changes in oxygen distribution with time.

8 Modelling

Electrical analogue- (Armstrong 1979; Armstrong and Wright 1976) and mathematical modelling using both analytical (Armstrong et al. 1991b; Beckett and Armstrong 1992; Greenwood and Goodman 1971) and numerical methods (Armstrong and Beckett 1985, 1987; Beckett et al. 1988; Bidel et al. 2000; Luxmoore et al. 1970) have been used successfully in interpretative, predictive, exploratory and teaching roles to study both diffusive oxygen transport and pressurized gas-flows in plants (Armstrong et al. 1996c, d; Beckett et al. 2001; Sorrell et al. 2000; Steinberg 1996). The more complex of early diffusion models (Luxmoore et al. 1970) accommodated changing diffusive resistance, respiratory demand and barrier permeability along roots, the presence or absence of an oxygen source external to the root and a Michaelis–Menten-type concentration dependence of respiration. However, no radial variation in structure and respiratory demand was modelled and the Michaelis–Menten response was based on that for whole root segments rather than being localized at the mitochondrial or cellular level. Later models have accommodated radial variation in root characteristics by treating them as multi-cylindrical (Armstrong and Beckett 1985, 1987; Sorrell et al. 2000) and, in the wetland case, by introducing a more realistic soil oxygen sink. Models for radial inward diffusion into roots have incorporated some dynamic capability and perhaps provided a more realistic Michaelis–Menten treatment (Armstrong and Beckett 2011a, b; Darwent et al. 2003). These could be usefully incorporated in a more comprehensive multi-cylindrical model covering root aeration in both the wetland and non-wetland conditions. Recent models on oxygen and CO_2 diffusion in soft fruits (Ho et al. 2006, 2010; Lammertyn et al. 2001, 2003) have also incorporated features that could be useful in root models. However, modelling always involves some compromises and assumptions and always there will be room for improvements.

Regarding roots, modelling has revealed that (1) lowering diffusive resistance by aerenchyma formation improves aeration more than the attendant loss of respiratory demand, (2) oxygen demands of root and rhizosphere act competitively for oxygen diffusing from the shoot, (3) if stelar or endodermal oxygen diffusivities decline sub-apically due to various secondary wall-deposits, anoxia/severe hypoxia could first arise sub-apically and ethanol production and reduced energy-charge could be a function of sub-apical stelar anoxia only, (4) declining stelar oxygen diffusivity sub-apically would raise the apparent COP for respiration but not extension, and any sub-apical stelar anoxia will raise oxygen levels elsewhere and may prolong root extension, (5) although barrier formation will increase aerated root length the effect will be greatest in non-aerenchymatous roots (Fig. 1c) and relatively small in aerenchymatous roots (Armstrong and Beckett 1987). Finally, mathematical modelling using the Michaelis–Menten response curve (Fig. 1a) indicates that total anoxia may never be reached in cases such as in Fig. 2c, d. Whether this applies in practice remains to be discovered, and what is meant by anoxia/anaerobic in this context needs to be defined or qualified. It raises the question of when oxygen supply may be inadequate and lead to the switch to anaerobic metabolism: we still do not know at what concentration the root cells will cease to utilize O_2. In Fig. 2c the point at which aerobic respiration and glycolysis would generate the same levels of ATP would be at ca. $\leq$0.01 kPa, viz. ca. the lower limit of oxygen detection by microelectrodes. Consequently, modelling suggests that the O_2 partial pressure at the centre of roots could be below the usual detection limits of O_2-microelectrodes while still supporting some aerobic respiration; perhaps aerobic and anaerobic metabolism might at some point occur simultaneously in the same cell?

9 Convective Gas-Flow

For some floating and emergent macrophytes, especially deep-water species and those whose rhizomes grow deep in anoxic waterlogged sediments resistances may be too great for the diffusion of atmospheric O_2 to keep underground systems adequately aerated. Such plants additionally generate pressurized (convective) through-flows of air (Dacey 1980; Schröder et al. 1986; Armstrong et al. 1988, 1991a; Brix et al. 1992; Sorrell and Boon 1994; White and Ganf 1998). These 'internal winds' (at speeds $\leq$ 10 cm s^{-1}) through the gas-space systems, supply O_2 at far higher rates than diffusion (Fig. 4a) (Armstrong et al. 1991a). Curvilinear dynamic pressure gradients down shoots have been demonstrated (Afreen et al. 2007; Armstrong and Armstrong 2009) (Fig. 4b) and dynamic pressures and convective flow rates also show curvilinear relationships to resistances in the flow pathway (Fig. 4b, c).

So far, three types of convective throughflow have been reported, induced by humidity gradients, thermal gradients or by the wind. Non-throughflow convections are considered relatively ineffective (Beckett et al. 1988).

9.1 *Humidity-Induced Convection*

Humidity-induced convection (HIC) has been found in species of *Phragmites*, *Equisetum*, *Typha*, *Eleocharis*, *Juncus*, *Baumea*, *Cyperus and Schoenoplectus* (Armstrong and Armstrong 1990a; Armstrong et al. 1991a, 1992; Bendix et al. 1994; Brix et al. 1992; Sorrell and Boon 1994; Sorrell et al. 1997; Tornbjerg et al. 1994) and almost certainly contributes to flows in waterlilies (Armstrong et al. 1991a; Dacey 1980; Grosse 1996; Mevi-Schutz and Grosse 1988a, b). HIC is induced in living shoots and depends upon the diffusion of gases through a stomatal or sub-stomatal partition whose narrow pores ($\leq$1–3 μm width, Fig. 4d) are more resistant to pressurized backflow than to diffusion (Armstrong and Armstrong 1994; Armstrong et al. 1996c, d; Bendix et al. 1994). Both physical (Armstrong and Armstrong 1994) and mathematical (Armstrong et al. 1996c, d) modelling have indicated an optimum pore diameter for HIC of ca. 0.2 μm, approximating that of

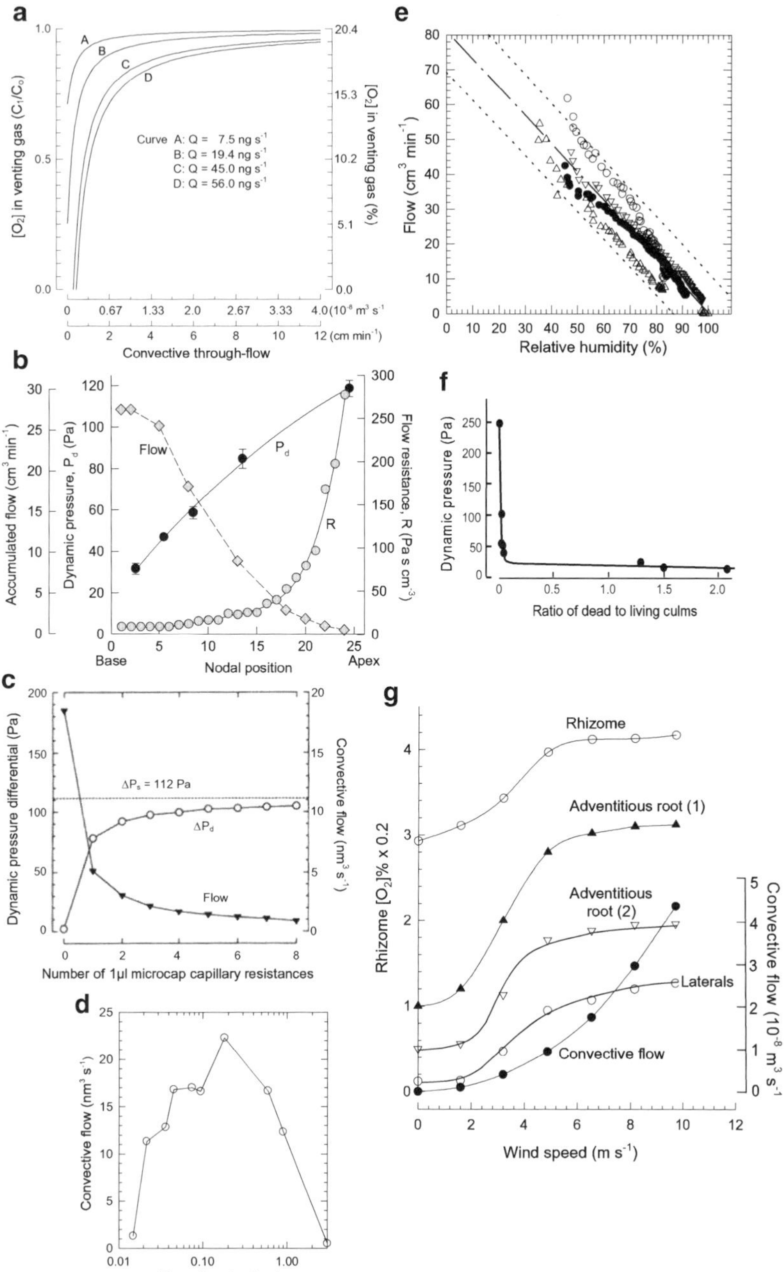

Fig. 4 (**a**) Results of modelling the influences of diffusion (from one end of the rhizome only), increasing convection rates and various respiratory demands on O_2 venting from the rhizome ($L = 50$ cm). A = rhizome respiration only; B = as A plus 10 adventitious roots and laterals (600 per root); C = as B plus O_2 consumption in rhizosphere of 50 ng cm^{-3} s^{-1}; D = as B plus O_2

stomatal pore width in *Phragmites* leaf sheaths. Pressurization, however, continues to increase as pore size diminishes.

During HIC, constant humidification of the plant's internal gas-spaces can reduce O_2 and N_2 concentrations to below atmospheric, inducing a gradient for the inward diffusion of drier atmospheric gases. This, together with water vapour continually produced from cells below the membrane, causes pressurization that drives gas-flows along the path of least resistance, through shoot/petiole and rhizome, with venting through more porous (often older) parts (Armstrong and Armstrong 1990a; Armstrong et al. 1996b; Bendix et al. 1994; Brix et al. 1992; Dacey 1980; Mevi-Schutz and Grosse 1988b). HIC is powered by the latent heat of evaporation to maintain humidity of the gas-spaces, and is most rapid under warm, dry (low relative humidity: RH), sunny conditions (Armstrong and Armstrong 1990b; Armstrong et al. 1996b; Bendix et al. 1994; Brix et al. 1992, 1996; Steinberg 1996). HIC may have aerated ancient Equisetales and possibly Calamites 350 million years ago (Armstrong and Armstrong 2009). Examples of flow pathways are:

Fig. 4 (continued) consumption in rhizosphere of 500 ng cm^{-3} s^{-1}. Note that for realistic O_2 demands (curves C, D) convection is necessary to maintain an O_2 supply throughout the rhizome. Also here, comparatively low convection rates predict close to atmospheric O_2 levels in the rhizome (after Armstrong, Armstrong and Beckett 1992). (**b**) *Equisetum telmateia* intact shoot: relationships between dynamic pressure gradient down shoot (*closed circles*), individual nodal resistances to flow (*diamonds*) and humidity-induced convective flow rates. Accumulated flow rates with distance down shoot (*grey circles*) were deduced from the pressure gradients and resistances. $T = 26$ °C; RH = 51 %; PPFR = 1,100 μmol m^{-2} s^{-1}; $n = 5$ readings ± SE; a typical example from five shoots investigated (after Armstrong and Armstrong 2009). (**c**) Effects of venting resistance on dynamic pressures and convective flows: experimental data from physical model using nuclepore membranes (area approx. 250 mm^2; thickness 10 μm; porosity approx. 10 %; pore diameter = 0.046 μm). Evaporative surface 4 mm below the membrane, speed of dry air across outer surface of membrane = 0.0532 m s^{-1}; isothermal conditions (after Armstrong and Armstrong 1994). (**d**) For a physical model, the observed relationship between humidity-induced convective flow rates and nuclepore membrane pore diameters. The water surface was 4 cm beneath the membrane; speed of dry air across outer surface of membrane = 0.056 ms^{-1}. $n = 5$ membranes per pore size; ±SE (after Armstrong and Armstrong 1994). (**e**) *Equisetum telmateia*: convective flow rates in relation to atmospheric humidity for four excised shoots. Shoot heights ca. 1 m; $T = 28$–32 °C; PPFR = 125–200 μmol m^{-2} s^{-1} (after Armstrong and Armstrong 2009). (**f**) *Phragmites australis*: relationship between dynamic pressure differentials and ratio of dead: living culms in the field (after Armstrong and Armstrong 1990b). (**g**) Venturi-induced convection in *Phragmites australis*. Effects of wind speed on convective flow rate and their influence on [O_2] venting from rhizome and on root oxygen efflux (arbitrary units). The experiment was performed in the dark to minimize humidity-induced convection. Lengths of adventitious roots 1 and 2 = 70 and 98 mm, respectively. Inflow culm: two-noded; length 240 mm; pith cavity diameter 5 mm. Outflow culm (exposed to controlled wind): six-noded; length = 790 mm; upper pith cavity diameter = 3.5 mm. $T = 21$–22.5 °C; RH = 42–48 % (after Armstrong et al. 1992)

Phragmites australis

stomata on leaf sheaths → intercellular spaces → leaf sheath aerenchyma → stem aerenchyma → stem pith cavity → rhizome pith cavity → venting via pith cavity of stubble

Equisetum. telmateia

stomata on branches → intercellular spaces → branch aerenchyma → stem aerenchyma → rhizome aerenchyma → venting via stubble or damaged shoots

Eleocharis sphacelata

stomata on emergent influx culm → intercellular spaces → cortical channels → submerged node → rhizome †cortex → node → venting via stomata or damage on older emergent culms

†there may be also some flow via pith.

10 Factors Affecting HIC

Rates of HIC are inversely proportional to external relative humidity (Fig. 4e) and to resistances to gas-flow in the plant's aeration system, e.g. lengths and diameters of channels and numbers of nodal diaphragms (Armstrong and Armstrong 2009; Sorrell et al. 1997).[5] Flow rates from excised shoots of *Equisetum telmateia* may be reduced by ca. 60 % when shoots are reconnected to the rhizome system because of resistance of the latter (Armstrong and Armstrong 2009). Also, blockages caused by insect damage and phytotoxins, and leakages caused by insect damage, can reduce or even prevent flows (Armstrong et al. 1996a). Some species can pressurize but have resistances too great to enable significant flows (Armstrong and Armstrong 2011; Brix et al. 1992). Conversely, large numbers of efflux culms lower resistance and increase flow rates (Sorrell et al. 1997; Sorrell and Boon 1994; Armstrong and Armstrong 1990b) (Fig. 4f). A shortage of efflux sites has been associated with increased alanine and *gamma*-aminobutyric acid in *Phragmites* rhizomes indicative of hypoxia (Rolletschek et al. 1998).

Flow rates can be directly related to the area of stomatal surface in some species, although upper parts of shoots tend to be less effective than lower parts (Afreen et al. 2007; Armstrong and Armstrong 2009). Also, the degrees to which stomata/other tissues are effective in inducing HIC and minimizing backflow are important. Moreover, wind reduces humid barrier layer effects around the plant and steepens diffusion gradients for HIC (Armstrong and Armstrong 2009; Steinberg 1996); also, RH may fall and stomata open as radiation increases (Armstrong and Armstrong 1990a). Rates of HIC are often highest around mid-day under dry warm conditions, fall sharply at sunset and approach zero during the night (Armstrong and Armstrong

[5] The degree to which the potential convection from an individual shoot is realized can be expressed as a delivery coefficient: $1 - (\Delta P_d/\Delta P_s)$ (Beckett et al. 2001) where ΔP_d = dynamic pressure at the base during convection and ΔP_s = static pressure differential (maximum pressure developed with outflow blocked and equal rates of incoming and outgoing molecules).

1990b, 2009; Armstrong et al. 1996b; Bendix et al. 1994; Brix et al. 1992, 1996; Steinberg 1996). HIC takes place via living shoots and only occurs to any great extent during the growing season; its potential is maximal when the shoots are fully grown.

10.1 Thermal Transpiration

Thermal transpiration (TT) per se is gaseous diffusion induced by a temperature gradient across a partition whose small pores resist pressurized backflow (similar to HIC) and can occur even where the gases on each side are dry. Net movement is from the cold side where the molecules are more concentrated, thus inducing pressurized flow on the warm side (Armstrong et al. 1996c, d; Grosse 1996; Leuning 1983; Steinberg 1996). TT, first described in waterlilies (Dacey 1980), is initiated in living leaves, where the 'porous membrane' is the boundary cell partition between the palisade and spongy mesophylls (Schröder et al. 1986).

Pathway of TT & HIC in *Nuphar*

Stomata on young leaf → intercellular spaces of palisade mesophyll → pores of boundary partition → lacunae of spongy mesophyll → lacunae of petiole → lacunae of rhizome → venting via gas-spaces of petiole and lamina of older leaves

In some waterlilies, e.g. *Nelumbo*, inflow and venting occur in the same leaf (Mevi-Schutz and Grosse 1988a, b; Matthews and Seymour 2013).

11 Factors Affecting TT

The cooler the ambient air and the warmer the interior of the leaf, the greater the temperature gradient across the boundary partition and the faster will be TT. The greater the areas of influx organs (stems/leaves) and the lower the internal resistances to gas-flow, including older efflux parts, the faster are the flows.

11.1 HIC and TT and PPFR

Gases in plants are humid and warm air holds more water vapour than cool air. If internal temperatures are above ambient, as may occur with high PPFR, HIC and TT will be additive. However, internal temperatures are often lower than ambient (Bendix et al. 1994; Sorrell and Boon 1994; Steinberg 1996) and TT will reverse direction and counter HIC. Humidity gradients are generally more effective than temperature differentials for creating flows (Armstrong et al. 1996c, d; Steinberg 1996). Maximum pressures for humidity gradients (external RH at zero and non-leaky partition, pore diameters ≤ 0.01 μm) could be 2,200–4,200 Pa at 20–30 °C, but for TT realistic temperature differentials of 1–4 °C would only induce 110–440 Pa. Hence, when HIC and TT operate together, HIC tends to be the main driving force for convection. Some pressurization can also result from entrapped photosynthetic O_2 (Armstrong and Armstrong 2005b; Waters et al. 1989).

11.2 Venturi-Induced Convection

Venturi-induced convection (VIC) found in *Phragmites australis* (Armstrong et al. 1992) is initiated by wind blowing across the tops of tall dead culms creating a suction which draws gases from these culms and the rhizome system; these gases are replaced by air entering the rhizomes via the stubble which is sheltered from wind-flow. The suction pressure ΔP (Pa) is developed: $\Delta P = -(\frac{1}{2}\rho V^2)$ where ρ is the viscosity of air (approx. 1.20–1.25 kg m^{-3}) and V the wind velocity (m s^{-1}).

11.2.1 Factors Affecting VIC

Venturi-induced flow is directly proportional to the wind-speed^2. Similarly to HIC, VIC is decreased by resistances and leakages in the plant. In general, flow rates are higher the greater the numbers of outflow culms and the greater their diameters, causing lower resistance (Armstrong et al. 1992). VIC occurs in any season and at any time of day if there is sufficient wind. A gusting wind may aerate the rhizomes more effectively than a steady wind speed at the same average value (Armstrong et al. 1992). Tall dead culms are most efficient as they catch the wind more; short broken culms tend to act as inflow routes. In fresh winds of 36–40 km h^{-1}, VIC rates >2.4 $cm^3\ s^{-1}$ have been recorded from single culms in situ in a reed bed, while wind-induced suction pressures for excised culms can be ca. −40 Pa.

Pathway of VIC in *Phragmites*

Inflow via pith cavity of stubble → rhizome pith cavity → pith cavities of tall dead culm → venting via tall dead culms due to wind-induced suction pressure

11.3 Beneficial Effects of Throughflow Convections

Comparatively slow rates of convection can be sufficient to raise O_2 concentrations in underground rhizomes to close to atmospheric (Fig. 4a, g) (Armstrong et al. 1992, 1996b; Sorrell et al. 1997). Roots and rhizome apices and buds are 'blind endings' and only directly aerated by diffusion. However, convections, by greatly enhancing rhizome aeration, can improve the aeration of rhizome apices and roots by increasing O_2 diffusion gradients near rhizome apices and at root–rhizome junctions. These effects can be detected as increased O_2 flux from phyllospheres around rhizome apices (Armstrong et al. 2006) and rhizosphere regions around adventitious and lateral roots compared to periods of non-convection, when aeration is by diffusion (Fig. 4g) (Armstrong and Armstrong 1990a; Armstrong et al. 1992). Convective through-flows in rhizomes also 'flush out' waste gases (Armstrong and Armstrong 1990a; Brix et al. 1996) and increase diffusion gradients for their removal from roots; similarly, methane can be removed (Arkebauer et al. 2001; Brix et al. 1996; Sorrell and Boon 1994).

Convections safeguard underground systems from periods of anoxia and hence should conserve food reserves. Furthermore, convections can increase potentials for oxidations of phytotoxins in rhizosphere and phyllosphere regions, helping root and underground bud survival in phytotoxic soils.

In *Phragmites*, with HIC as opposed to diffusion alone, greater numbers of roots can be supported and the penetration depth for rhizome and root growth can sometimes be doubled (Vretare-Strand 2002; Vretare-Strand and Weisner 2002; White et al. 2007). It is almost certain that convective ventilation increases survival and competitive ability in floating-leaved and emergent macrophytes (Sorrell and Tanner 2000; Sorrell and Hawes 2010).

12 Concluding Remarks

In terms of sustained tolerance of soil flooding or submergence an adequate gas-space provision for long-distance oxygen transport is essential and exciting developments are in progress. However, in order to achieve this it may be necessary in some cases to alter cortical cell configurations and this may present a greater challenge.

Much is now known of oxygen transport in plants and modelling has greatly helped our understanding. Models that can also accommodate the activities of individual cells in three-dimensions as well as tissues must be a goal.

In terms of oxygen monitoring, the most critical oxygen concentrations for metabolism, severely hypoxic (e.g. <0.2 kPa), are close to or below current sensitivity limits (0.007–0.02 kPa). Greater sensitivity with stability is required of microprobes to help quantify degrees and distribution of hypoxia in a more targeted way and reveal more of the dynamics of the switch from aerobic to anaerobic metabolism and vice versa. Such dynamics must be a daily, non-lethal occurrence in many partially submerged species such as rice but so far have attracted little attention.

References

Abiko T, Kotula L, Shiono K, Malik AI, Colmer TD, Nakazono M (2012) Enhanced formation of aerenchyma and induction of a barrier to radial oxygen loss in adventitious roots of *Zea nicaraguensis* contribute to its waterlogging tolerance as compared with maize (*Zea mays ssp mays*). Plant Cell Environ 35:1618–1630

Afreen F, Zobayed SMA, Armstrong J, Armstrong W (2007) Pressure gradients along whole culms and leaf sheaths, and other aspects of humidity-induced gas-transport in *Phragmites australis*. J Exp Bot 58:1651–1662

Aguilar EA, Turner DW, Gibbs DJ, Armstrong W, Sivasithamparam K (2003) Oxygen distribution and movement, respiration and nutrient loading in banana roots (*Musa* spp. L.) subjected to aerated and oxygen-depleted environments. Plant Soil 253:91–102

Arkebauer TJ, Chanton JP, Verma SB, Kim J (2001) Field measurements of internal pressurization in *Phragmites australis* (Poaceae) and implications for regulation of methane emissions in midlatitude prairie wetland. Am J Bot 88:653–658

Armstrong W (1964) Oxygen diffusion from the roots of some British bog plants. Nature 204:801–802

Armstrong W (1967) The use of polarography in the assay of oxygen diffusing from roots in anaerobic media. Physiol Plant 20:540–553

Armstrong W (1968) Oxygen diffusion from the roots of woody species. Physiol Plant 21:539–543

Armstrong W (1971) Radial oxygen losses from intact rice roots as affected by distance from the apex, respiration and waterlogging. Physiol Plant 25:192–197

Armstrong W (1979) Aeration in higher plants. In: Woolhouse HWW (ed) Advances in botanical research, vol 7. Academic, London, pp 225–332

Armstrong W (1994) Polarographic oxygen electrodes and their use in plant aeration studies. Proc R Soc Edinburgh B 102:511–528

Armstrong J, Armstrong W (1990a) Light-enhanced convective throughflow increases oxygenation in rhizomes and rhizosphere of *Phragmites australis* (Cav.) Trin. ex Steud. New Phytol 114:121–12

Armstrong J, Armstrong W (1990b) Pathways and mechanisms of oxygen transport in *Phragmites australis*. In: Cooper P, Findlater BC (eds) The use of constructed wetlands in water pollution control. Pergamon Press plc, Oxford, pp 529–533

Armstrong J, Armstrong W (1994) A physical model involving Nuclepore membranes to investigate the mechanism of humidity-induced convection in *Phragmites australis*. Proc R Soc Edinburgh B 102:529–540

Armstrong J, Armstrong W (2005a) Rice: sulphide-induced barriers to root radial oxygen loss, Fe^{2+} and water uptake, and lateral root emergence. Ann Bot 96:625–638

Armstrong W, Armstrong J (2005b) Stem photosynthesis not pressurized ventilation is responsible for light-enhanced oxygen supply to submerged roots of Alder (*Alnus glutinosa*). Ann Bot 96:591–612

Armstrong J, Armstrong W (2009) Record rates of pressurized gas-flow in the great horsetail, *Equisetum telmateia*. Were Carboniferous Calamites similarly aerated? New Phytol 184:202–215

Armstrong J, Armstrong W (2011) Reasons for the presence or absence of convective (pressurized) ventilation in the genus *Equisetum*. New Phytol 190:387–397

Armstrong W, Beckett PM (1985) Root aeration in unsaturated soil: a multi-shelled model of oxygen distribution and diffusion with and without sectoral blocking of the diffusion path. New Phytol 100:293–311
Armstrong W, Beckett PM (1987) Internal aeration and the development of stelar anoxia in submerged roots: a multishelled mathematical model combining axial diffusion of oxygen in the cortex with radial losses to the stele, the wall layers and the rhizosphere. New Phytol 105:221–245
Armstrong W, Beckett PM (2011a) Experimental and modelling data contradict the idea of respiratory down-regulation in plant tissues at an internal [O_2] substantially above the critical oxygen pressure for cytochrome oxidase. New Phytol 190:431–441
Armstrong W, Beckett PM (2011b) The respiratory down-regulation debate. New Phytol 190:276–278
Armstrong W, Drew MC (2002) Root growth and metabolism under oxygen deficiency. In: Waisel Y, Eshel A, Kafkafi U (eds) Roots the hidden half, 3rd edn. Marcel Dekker Inc, New York, pp 729–761
Armstrong W, Gaynard TJ (1976) The critical oxygen pressure for root respiration in intact plants. Physiol Plant 37:200–206
Armstrong W, Healy MT (1984) Oxygen diffusion in pea. III. Changes in the oxygen status of the primary pea root attributable to an ageing of the tissues. New Phytol 96:179–185
Armstrong W, Webb T (1985) A critical oxygen pressure for root extension in rice. J Exp Bot 36:1573–1582
Armstrong W, Wright EJ (1975) Radial oxygen loss from roots: the theoretical basis for the manipulation of flux data obtained by the cylindrical platinum electrode technique. Physiol Plant 35:21–26
Armstrong W, Wright EJ (1976) An electrical analogue to simulate the oxygen relations of roots in anaerobic media. Physiol Plant 36:383–387
Armstrong W, Healy MT, Webb T (1982) Oxygen diffusion in pea. I. Pore-space resistance in the primary root. New Phytol 91:647–659
Armstrong W, Healy MT, Lythe S (1983) Oxygen diffusion in pea. II. The oxygen status of the primary root as affected by growth, the production of laterals and radial oxygen loss. New Phytol 94:549–559
Armstrong J, Armstrong W, Beckett PM (1988) *Phragmites australis*—a critical appraisal of the ventilating pressure concept and an analysis of resistance to pressurised gas-flow and gaseous diffusion in horizontal rhizomes. New Phytol 110:383–390
Armstrong W, Armstrong J, Beckett PM, Justin SHFW (1991a) Convective gas-flows in wetland plant aeration. In: Jackson MB, Davies DD, Lambers H (eds) Plant life under oxygen stress. SPB Academic Publishing bv, The Hague, pp 283–302
Armstrong W, Beckett PM, Justin SHFW, Lythe S (1991b) Modelling and other aspects of root aeration. In: Jackson MB, Davies DD, Lambers H (eds) Plant life under oxygen stress. SPB Academic Publishing bv, The Hague, pp 267–282
Armstrong J, Armstrong W, Beckett PM (1992) *Phragmites australis*: Venturi- and humidity-induced pressure flows enhance rhizome aeration and rhizosphere oxidation. New Phytol 120:197–207
Armstrong W, Strange ME, Cringle S, Beckett PM (1994) Microelectrode and modelling study of oxygen distribution in roots. Ann Bot 74:287–299
Armstrong J, Armstrong W, Armstrong IB, Pittaway GR (1996a) Senescence, and phytotoxin, insect, fungal and mechanical damage: factors reducing convective gas-flows in *Phragmites australis*. Aquat Bot 54:211–216
Armstrong J, Armstrong W, Beckett PM, Halder JE, Lythe S, Holt R, Sinclair A (1996b) Pathways of aeration and the mechanisms and beneficial effects of humidity- and Venturi-induced convections in *Phragmites australis* (Cav.) Trin. Ex Steud. Aquat Bot 54:177–197

Armstrong W, Armstrong J, Beckett PM (1996c) Pressurised ventilation in emergent macrophytes: the mechanism and mathematical modelling of humidity-induced convection. Aquat Bot 54:121–136

Armstrong W, Armstrong J, Beckett PM (1996d) Pressurised aeration in wetland macrophytes: some theoretical aspects of humidity-induced convection and thermal transpiration. Folia Geobot 31:25–36

Armstrong W, Cousins D, Armstrong J, Turner DW, Beckett PM (2000) Oxygen distribution in wetland plant roots and permeability barriers to gas-exchange with the rhizosphere: a microelectrode and modelling study with *Phragmites australis*. Ann Bot 86:687–703

Armstrong J, Jones RE, Armstrong W (2006) Rhizome phyllosphere oxygenation in *Phragmites* and other species in relation to redox potential, convective gas-flow, submergence and aeration pathways. New Phytol 172:719–731

Armstrong W, Webb T, Darwent M, Beckett PM (2009) Measuring and interpreting respiratory critical oxygen pressures in roots. Ann Bot 103:281–294

Asplund PT, Curtis WR (2001) Intrinsic oxygen use kinetics of transformed plant root culture. Biotechnol Prog 17:481–489

Barzu O, Satre M (1970) Determination of oxygen affinity of respiratory systems using oxy-haemoglobin as oxygen donor. Ann Biochem 36:428–433

Beckett PM, Armstrong W (1992) The modelling of convection and diffusion-driven aeration in plants. In: Egginton S, Ross HF (eds) Oxygen transport in biological systems, SEB seminar series 51. Cambridge University Press, Cambridge, pp 253–293

Beckett PM, Armstrong W, Justin SHFW, Armstrong J (1988) On the relative importance of convective and diffusive gas-flows in plant aeration. New Phytol 110:463–468

Beckett PM, Armstrong W, Armstrong J (2001) A modelling approach to the analysis of pressure-flow in *Phragmites* stands. Aquat Bot 69:269–291

Bendix M, Tornbjerg T, Brix H (1994) Internal gas transport in *Typha latifolia* L. and *Typha angustifolia* L. 1. Humidity-induced pressurization and convective throughflow. Aquat Bot 49:75–89

Bidel LPR, Renault P, Pagès L, Rivière LM (2000) Mapping meristem respiration of *Prunus persica* (L) Batsch seedlings: potential respiration of the meristems, O_2 diffusional constraints and combined effects on root growth. J Exp Bot 51:755–768

Blossfeld S, Gansert D, Thiele B, Kuhn J, Lösch R (2011) The dynamics of oxygen concentration, pH value, and organic acids in the rhizosphere of Juncus spp. Soil Biol Biochem 43:1186–1197

Brix H, Sorrell BK, Orr PT (1992) Internal pressurization and convective gas flow in some emergent freshwater macrophytes. Limnol Oceanogr 37:1420–1433

Brix H, Sorrell BK, Schierup H-H (1996) Gas fluxes achieved by in situ convective flow in *Phragmites australis*. Aquat Bot 54:151–163

Carvalho EB, Curtis WR (1998) Characterization of fluid-flow resistance in root cultures with a convective flow tubular bioreactor. Biotechnol Bioeng 60:375–384

Colmer TD (2003a) Aerenchyma and an inducible barrier to radial oxygen loss facilitate root aeration in upland, paddy and deep water rice (*Oryza sativa* L.). Ann Bot 91:301–309

Colmer TD (2003b) Long-distance transport of gases in plants: a perspective on internal aeration and radial oxygen loss from roots. Plant Cell Environ 26:17–36

Colmer TD, Pedersen O (2008a) Oxygen dynamics in submerged rice (*Oryza sativa*). New Phytol 178:326–334

Colmer TD, Pedersen O (2008b) Underwater photosynthesis and respiration in leaves of submerged wetland plants: gas films improve CO_2 and O_2 exchange. New Phytol 177:918–926

Colmer TD, Cox MCH, Voesenek LACJ (2006) Root aeration in rice (Oryza sativa): evaluation of oxygen, carbon dioxide and ethylene as possible regulators of root acclimatizations. New Phytol 170:767–778

Colmer TD, Gibberd MR, Wiengweera A, Tinh TK (1998) The barrier to radial oxygen loss from roots of rice (Oryza sativa L.) is induced by growth in stagnant solution. J Exp Bot 49:1431–1436

Cormack RGH (1947) A comparative study of developing epidermal cells in white mustard and tomato roots. Am J Bot 34:310–314
Currie JA (1962) The importance of aeration in providing the right conditions for plant growth. J Sci Food Agric 13:380–385
Currie JA (1965) Diffusion within the soil microstructure: a structural parameter for soil. J Soil Sci 16:279–289
Curtis WR, Tuerk AL (2008) Oxygen transport in plant tissue culture systems. In: Gupta SD, Ibaraki Y (eds) Plant tissue culture engineering. Springer, New York, pp 173–186
Dacey JWH (1980) Internal winds in water lilies: an adaptation for life in anaerobic sediments. Science 210:1017–1019
Darwent MJ (1997) The development and use of microelectrodes for the study of oxygen transport and distribution in roots. PhD thesis, University of Hull, Yorkshire
Darwent MJ, Armstrong W, Armstrong J, Beckett PM (2003) Exploring the radial and longitudinal aeration of primary maize roots by means of Clark-type oxygen microelectrodes. Russ J Plant Physiol 50:723–732
Denison RF (1992) Mathematical modeling of oxygen diffusion and respiration in legume root nodules. Plant Physiol 98:901–907
Dittert K, Wötzel J, Sattelmacher B (2006) Responses of *Alnus glutinosa* to anaerobic conditions—mechanisms and rate of oxygen flux into the roots. Plant Biol 8:212–223
Dolan L, Janmaat K, Willemsen V, Linstead P, Poethig S, Roberts K, Scheres B (1993) Cellular organisation of the *Arabidopsis thaliana* root. Development 119:71–84
Garthwaite AJ, Armstrong W, Colmer TD (2008) Assessment of the O_2 diffusivity across the barrier to radial O_2 loss in adventitious roots of *Hordeum marinum*. New Phytol 179:405–416
Gaynard TJ, Armstrong W (1987) Some aspects of internal plant aeration in amphibious habitats. In: Crawford RMM (ed) Amphibious and intertidal plants, British Ecological Society special symposium 5. Blackwells, Oxford, pp 303-320
Gibbs J, Greenway H (2003a) Mechanisms of anoxia tolerance in plants. I. Growth, survival and anaerobic catabolism. Funct Plant Biol 30:1–47
Gibbs J, Greenway H (2003b) Mechanisms of anoxia tolerance in plants. II. Energy requirements for maintenance and energy distribution to essential processes. Funct Plant Biol 30:999–1036
Gibbs J, Turner DW, Armstrong W, Darwent MJ, Greenway H (1998) Response to oxygen deficiency in maize roots I. Development of O_2-deficiency in the stele reduces radial solute transport to the xylem. Funct Plant Biol 25:745–758
Greenwood DJ, Goodman D (1971) Studies on the oxygen supply to the roots of mustard seedlings (*Sinapsis alba* L.). New Phytol 70:85–96
Groot TT, van Bodegom PM, Meijer HAJ, Harren FJM (2005) Gas transport through the root-shoot transition zone of rice tillers. Plant Soil 277:107–116
Grosse W (1996) Pressurised ventilation in floating-leaved aquatic macrophytes. Aquat Bot 54:137–150
Healy MT, Armstrong W (1972) The effectiveness of internal oxygen transport in a mesophyte (*Pisum sativum* L.). Planta 103:302–309
Ho QT, Verlinden BE, Verboven P, Vandewalle S, Nicolaï BM (2006) A permeation-diffusion-reaction model of gas transport in cellular tissue of plant materials. J Exp Bot 57:4215–4224
Ho QT, Verboven P, Verlinden BE, Nicolaï B (2010) A model for gas transport in pear fruit at multiple scales. J Exp Bot 61:2071–2081
Hook DD, Brown CL (1972) Permeability of the cambium to air in trees adapted to wet habitats. Bot Gaz 133:304–310
Huck MG (1970) Variation in taproot elongation rate as influenced by composition of the soil air. Agron J 62:815–818
Jackson MB, Armstrong W (1999) Formation of aerenchyma and the processes of plant ventilation in relation to soil flooding and submergence. Plant Biol 1:274–287
Jensen SI, Kühl M, Glud RN, Jørgensen LB, Priemé A (2005) Oxic microzones and radial oxygen loss from roots of *Zostera marina*. Mar Ecol Prog Ser 293:49–58

Jung J, Lee SC, Choi H-K (2008) Anatomical patterns of aerenchyma in aquatic and wetland plants. J Plant Biol 51:428–439

Justin SHFW, Armstrong W (1983) Oxygen transport in the salt-marsh genus *Puccinellia* with particular reference to the diffusive resistance of the root-shoot junction and the use of paraffin oil as a diffusive barrier in plant studies. J Exp Bot 34:980–986

Justin SHFW, Armstrong W (1987) The anatomical characteristics of roots and plant response to soil flooding. New Phytol 106:465–495

Kirk GHD, Greenway H, Atwell BJ, Ismail AM, Colmer TD (2014) Adaptation of rice to flooded soils. Prog Bot 75:215–253

Kotula L, Steudle E (2009) Measurements of oxygen permeability coefficient of rice (Oryza sativa L.) roots using a new perfusion technique. J Exp Bot 60:567–580

Laan P, Smolders A, Blom CWPM, Armstrong W (1989) The relative roles of internal aeration, radial oxygen losses, iron exclusion and nutrient balance in flood-tolerance in *Rumex* species. Acta Bot Neerl 38:131–145

Laisk A, Oja V, Eichelmann H (2007) Kinetics of leaf oxygen uptake represent in planta activities of respiratory electron transport and terminal oxidases. Physiol Plant 131:1–9

Lammertyn J, Franck C, Verlinden BE, Nicolaï BM (2001) Comparative study of the O_2, CO_2 and temperature effect on respiration between 'Conference' pear cell protoplasts in suspension and intact pears. J Exp Bot 52:1769–1777

Lammertyn J, Scheerlinck N, Jancso' KP, Verlinden BE, Nicolaï BM (2003) A respiration-–diffusion model for 'Conference' pears I: model development and validation. Postharvest Biol Technol 30:29–42

Lee SC, Mustroph A, Sasidharan R, Vashisht D, Pedersen O, Oosumi T, Voesenek LACJ, Bailey-Serres J (2011) Molecular characterization of the submergence response of the *Arabidopsis thaliana* ecotype Columbia. New Phytol 190:457–471

Lemon ER, Wiegand CL (1962) Soil aeration and plant root relations. II. Root respiration. Agron J 54:171–175

Leuning R (1983) Transport of gases into leaves. Plant Cell Environ 6:181–194

Luxmoore RJ, Stolzy LH, Letey J (1970) Oxygen diffusion in the soil-plant system. I. A model; II. Respiration, permeability and porosity of consecutive excised segments of maize and rice roots; III. Oxygen concentration profiles predicted for maize roots; IV. Oxygen concentration profiles predicted for rice roots. Agron J 62:322–332

Maggio A, Hasegawa PM, Bressan RA, Consiglio MF, Joly J (2001) Unravelling the functional relationship between root anatomy and stress tolerance. Aust J Plant Physiol 28:999–1004

Malik AL, Islam AKMR, Colmer TD (2011) Transfer of the barrier to radial oxygen loss in roots of *Hordeum marinum* to wheat (*Triticum aestivum*): evaluation of four H. marinum-wheat amphiploids. New Phytol 190:499–508

Mancusco S, Papeschi G, Marras AM (2000) A polarographic, oxygen-selective, vibrating micro-electrode system for the spatial and temporal characterization of transmembrane fluxes in plants. Planta 211:384–389

Mancuso S, Boselli ÆM (2002) Characterisation of the oxygen fluxes in the division, elongation and mature zones of *Vitis* roots: influence of oxygen availability. Planta 214:767–774

Mano Y, Omori F (2007) Breeding for flooding tolerant maize using 'teosinte' as a germplasm resource. Plant Root 1:17–21

Matthews PGD, Seymour RS (2013) Stomata actively regulate internal aeration of the sacred lotus Nelumbo nucifera. Plant Cell Environ doi: 10.111/pce.12163

McDonald MP, Galwey NW, Colmer TD (2002) Similarity and diversity in adventitious root anatomy as related to root aeration among a range of wetland and dryland grass species. Plant Cell Environ 25:441–451

McLamore ES, Jaroch D, Chatni MR, Porterfield DM (2010) Self-referencing optrodes for measuring spatially resolved, real-time metabolic oxygen flux in plant systems. Planta 232:1087–1099

Mevi-Schutz J, Grosse W (1988a) The importance of water vapour for the circulating air-flow through *Nelumbo nucifera*. J Exp Bot 39:1231–1236
Mevi-Schutz J, Grosse W (1988b) A two-way gas-transport system in *Nelumbo nucifera*. Plant Cell Environ 11:27–34
Millar AH, Bergersen FJ, Day DA (1994) Oxygen affinity of terminal oxidases in soybean mitochondria. Plant Physiol Biochem 32:847–852
Mommer L, Pedersen O, Visser EJW (2004) Acclimation of a terrestrial plant to submergence facilitates gas exchange under water. Plant Cell Environ 27:1281–1287
Nobel PS (1991) Physicochemical and environmental plant physiology. Academic, San Diego
Pedersen O, Vos H, Colmer TD (2006) Oxygen dynamics during submergence in the halophytic stem succulent *Halosarcia pergranulata*. Plant Cell Environ 29:1388–1399
Rawsthorne S, LaRue TA (1986) Metabolism under microaerobic conditions of mitochondria from Cowpea nodules. Plant Physiol 81:1097–1102
Revsbech NP (1989) An oxygen microsensor with a guard cathode. Limnol Oceanogr 34:474–478
Revsbech NP, Pedersen O, Reichardt W, Briones A (1999) Microsensor analysis of oxygen and pH in the rice rhizosphere under field and laboratory conditions. Biol Fertil Soils 29:379–385
Rich SM, Ludwig M, Pedersen O, Colmer TD (2011) Aquatic adventitious roots of the wetland plant Meionectes brownie can photosynthesize: implications for root function during flooding. New Phytol 190:311–319
Rolletschek H, Bumiller A, Henze R, Kohl J-G (1998) Implications of missing efflux sites on convective ventilation and amino acid metabolism in *Phragmites australis*. New Phytol 140:211–217
Rolletschek H, Borisjuk L, Koschorrek M, Wobus U, Weber H (2002) Legume embryos develop in a hypoxic environment. J Exp Bot 53:1099–1107
Rolletschek H, Radchuk R, Klukas C, Schreiber F, Wobus U, Borisjuk L (2005) Evidence of a key role for photosynthetic oxygen release in oil storage in developing soybean seeds. New Phytol 167:777–786
Sand-Jensen K, Pedersen O, Binzer T, Borum J (2005) Contrasting oxygen dynamics in the freshwater isoetid *Lobelia dortmanna* and the marine seagrass *Zostera marina*. Ann Bot 96:613–623
Schröder P, Grosse W, Woermann D (1986) Localisation of thermo-osmotically active partitions in young leaves of *Nuphar lutea*. J Exp Bot 37:1450–1461
Seago JL Jr, Marsh LC, Stevens KJ, Soukup A, Votrubova O, Enstone DE (2005) A re-examination of the root cortex in wetland flowering plants with respect to aerenchyma. Ann Bot 96:565–579
Shiao T-L, Doran PM (2000) Root hairiness: effect on fluid flow and oxygen transfer in hairy root cultures. J Biotechnol 83:199–210
Shimamura S, Mochizuki T, Nada Y, Fukuyama M (2003) Formation and function of secondary aerenchyma in hypocotyl, roots and nodules of soybean (*Glycine max*) under flooded conditions. Plant Soil 251:351–359
Shimamura S, Yamamoto R, Nakamura T, Shimada S, Komatsu S (2010) Stem hypertrophic lenticels and secondary aerenchyma enable oxygen transport to roots of soybean in flooded soil. Ann Bot 106:277–284
Shiono K, Ogawa S, Yamazaki S, Isoda H, Fujimura T, Nakazono M, Colmer TD (2011) Contrasting dynamics of radial O_2-loss barrier induction and aerenchyma formation in rice roots of two lengths. Ann Bot 107:89–99
Smirnoff N, Crawford RMM (1983) Variation in the structure and response to flooding of root aerenchyma in some wetland plants. Ann Bot 51:237–249
Sorrell BK, Boon PI (1994) Convective gas flow in *Eleocharis sphacelata R.* Br.: methane transport and release from wetlands. Aquat Bot 47:197–212
Sorrell BK, Hawes I (2010) Convective gas flow development and the maximum depths achieved by helophyte vegetation in lakes. Ann Bot 105:165–174

Sorrell BK, Tanner CC (2000) Convective gas flow and internal aeration in *Eleocharis sphacelata* in relation to water depth. J Ecol 88:778–789

Sorrell BK, Brix H, Orr PT (1997) *Eleocharis sphacelata*: internal gas transport pathways and modelling of aeration by pressurized flow and diffusion. New Phytol 136:433–442

Sorrell BK, Mendelssohn IA, McKee KL, Woods R (2000) Ecophysiology of wetland plant roots: a modelling comparison of aeration in relation to species distribution. Ann Bot 86:675–685

Soukup A, Armstrong W, Schreiber L, Franke R, Votrubová O (2007) Apoplastic barriers to radial oxygen loss (ROL) and solute penetration: a chemical and functional comparison of the exodermis of two wetland species—*Phragmites australis* and *Glyceria maxima*. New Phytol 173:264–278

Steinberg SL (1996) Mass and energy exchange between the atmosphere and leaf influence gas pressurization in aquatic plants. New Phytol 134:587–600

Stevens KJ, Peterson RL, Reader RJ (2002) The aerenchymatous phellem of *Lythrum salicaria* (L.): a pathway for gas transport and its role in flood tolerance. Ann Bot 89:621–625

Teakle NL, Armstrong J, Barrett-Lennard EG, Colmer TD (2011) Aerenchymatous phellem in hypocotyl and roots enables O_2 transport in *Melilotus siculus*. New Phytol 190:340–350

Thomas AL, Guerreiro SMC, Sodek L (2005) Aerenchyma formation and recovery from hypoxia of the flooded root system of nodulated soybean. Ann Bot 96:1191–1198

Tornbjerg T, Bendix M, Brix H (1994) Internal gas transport in *Typha latifolia* L. and *Typha angustifolia* L. 2. Convective throughflow pathways and ecological significance. Aquat Bot 49:91–105

Tschiersch H, Liebsch G, Stangelmayer A, Borisjuk L, Rolletschek H (2011) Planar oxygen sensors for non-invasive imaging in experimental biology. In: Minin I (ed) Microsensors. InTech, New York, pp 281–294. ISBN 978-953-307-170-1. doi: 10.5772/17893

Vashisht D, Hesselink A, Pierik R, Ammerlaan JMH, Bailey-Serres J, Visser EJW, Pedersen O, van Zanten M, Vreugdenhill D, Jamar DCL, Voesenek LACJ, Sasidharan R (2011) Natural variation of submergence tolerance among *Arabidopsis thaliana* accessions. New Phytol 190:299–310

Verboven P, Pedersen O, Herremans E, Ho QT, Nicolaı BM, Colmer TD, Teakle N (2011) Root aeration via aerenchymatous phellem: three-dimensional micro-imaging and radial O_2 profiles in *Melilotus siculus*. New Phytol 193:420–431

Visser EJW, Colmer TD, Blom CWPM, Voesenek LACJ (2000) Changes in growth, porosity, and radial oxygen loss from adventitious roots of selected mono- and dicotyledonous wetland species with contrasting types of aerenchyma. Plant Cell Environ 23:1237–1245

Voesenek LACJ, Armstrong W, Bogemann GM, Colmer TD (1999) A lack of aerenchyma and high rates of radial oxygen loss from the root base contribute to waterlogging intolerance in *Brassica napus*. Funct Plant Biol 26:87–93

Vretare-Strand V (2002) The influence of ventilation systems on water depth penetration of emergent macrophytes. Freshwater Biol 47:1097–1105

Vretare-Strand V, Weisner SEB (2002) Interactive effects of pressurized ventilation, water-depth and substrate conditions on *Phragmites australis*. Oecologia 131:490–497

Waters I, Armstrong W, Thomson CJ, Setter TL, Adkins S, Gibbs J, Greenway H (1989) Diurnal changes in radial oxygen loss and ethanol metabolism in roots of submerged and non-submerged rice seedlings. New Phytol 113:439–451

Webb T, Armstrong W (1983) The effects of CN and salicylhydroxamic acid on the root respiration of pea seedlings. Plant Physiol 72:280–286

White SD, Ganf GG (1998) The influence of convective flow on rhizome length in *Typha domingensis* over a water depth gradient. Aquat Bot 62:57–70

White SD, Deegan BM, Ganf GG (2007) The influence of water level fluctuations on the potential for convective flow in the emergent macrophytes *Typha domingensis* and *Phragmites australis*. Aquat Bot 86:369–376

Winkel A, Colmer TD, Pedersen O (2011) Leaf gas films of *Spartina anglica* enhance rhizome and root oxygen during tidal submergence. Plant Cell Environ 34:2083–2092

Winkel A, Colmer TD, Ismail AM, Pedersen O (2013) Internal aeration of paddy field rice (*Oryza sativa* L.) during complete submergence: importance of light and floodwater O_2. New Phytol 197(4):1193–1203
Yamasaki T (1952) Studies on the 'excess moisture injury' of upland crops in overmoist soil from the viewpoint of soil chemistry and plant physiology (Japanese with English summary). Bull Nat Inst Agric Sci (Japan) B 1:1–92
Yamauchi T, Rajhi I, Nakazono M (2011) Lysigenous aerenchyma formation in maize root is confined to cortical cells by regulation of genes related to generation and scavenging of reactive oxygen species. Plant Signal Behav 65:759–761

Biogenesis of Adventitious Roots and Their Involvement in the Adaptation to Oxygen Limitations

Margret Sauter and Bianka Steffens

Abstract Adventitious root formation is an adaptive response that contributes to flooding tolerance. Formation of adventitious roots is controlled by auxin and was shown to depend on auxin synthesis, transport, and signaling. Preexisting and newly formed adventitious root primordia emerge upon soil waterlogging or submergence to replace the primary root system which becomes easily dysfunctional in oxygen-poor soil or flood waters. Adventitious roots reduce the distance for oxygen and nutrient supply from the shoot. Formation of aerenchyma and of a barrier to radial oxygen loss support root oxygenation of adventitious roots. In semiaquatic species such as rice, adventitious root growth is induced by ethylene, which rapidly accumulates in submerged tissues. Gibberellic acid (GA) promotes ethylene-induced root growth whereas abscisic acid is strongly inhibitory such that an altered hormone homeostasis with elevated ethylene and GA and lowered ABA levels favors root growth.

1 Introduction

Plants develop throughout their whole life cycle during which they have to cope with changing environmental conditions. Plant development and growth depend on a functional root system. Roots are essential for uptake of water and nutrients, and anchorage to the soil but, at the same time, they impose a stationary way of life. This means that plants have to endure stresses with little possibility of escape in most cases. Among the abiotic stresses that plants encounter are soil waterlogging and flooding. These can rapidly result in O_2 shortage in flooded organs. Underwater, ethylene accumulates because of reduced gas diffusion and enhanced biosynthesis (Jackson 1985; Kende 1993; Zarembinski and Theologis 1997; Van der Straeten et al. 2001).

M. Sauter • B. Steffens (✉)
Plant Developmental Biology and Plant Physiology, Institute of Botany, University of Kiel, 24118 Kiel, Germany
e-mail: msauter@bot.uni-kiel.de; bsteffens@bot.uni-kiel.de

J.T. van Dongen and F. Licausi (eds.), *Low-Oxygen Stress in Plants*, Plant Cell Monographs 21, DOI 10.1007/978-3-7091-1254-0_15, © Springer-Verlag Wien 2014

Ethylene regulates many adaptive reactions during flooding including shoot elongation, aerenchyma formation, and adventitious root growth.

2 Adventitious Roots Develop in Response to Soil Waterlogging or Flooding

Flooding causes oxygen shortage and consequently energy shortage in roots which affects carbohydrate metabolism, water relation, and gas exchange among others. When exposed to flooding, roots can easily become completely dysfuntional. Therefore, plants which are permanently or frequently flooded have adapted their root system to survive these conditions. Mangroves such as *Sonneratia alba* are frequently exposed to flooding and their roots encounter hypoxic or anoxic stress on a regular basis. These plants develop respiratory roots called pneumatophores to improve oxygen supply (Purnobasuki and Suzuki 2004). When the primary root system suffers from anoxic stress, secondary roots can take over for the primary root system. By definition, secondary roots are lateral or adventitious roots. Lateral roots develop from within another root while adventitious roots develop at the shoot, e.g., from cotyledons of *Magnifera indica* (Li et al. 2012), branches in *Fraxinus americana* (Paolillo and Bassuk 2005), or at stem nodes of *Oryza sativa* (rice; Bleecker et al. 1986). Shoot-borne roots that are formed in cereals during normal development are occasionally termed crown roots (Hochholdinger et al. 2004). Secondary roots develop from quiescent or differentiated cells and this involves activation or dedifferentiation and establishment of a root apical meristem. Controlled cell proliferation and cell patterning from root founder cells result in the formation of an organized root primordium (Malamy and Benfey 1997; Petricka et al. 2012).

Adventitious root formation was observed in pteridophytes (Ridge and Sack 1992), monocots, and dicots. In some grasses such as *Zea mays* (maize) crown roots and in some conifers adventitious roots can make up the largest part of the root system. Adventitious root formation is triggered by exogenous factors such as wounding, light, and flooding while this process is mediated endogenously by hormones such as auxin and ethylene. Adventitious root formation as a result of waterlogging or flooding was reported for *Solanum lycopersicum* (tomato; Vidoz et al. 2010), *Larix laricina* (Calvo-Polanco et al. 2012), *Luffa cylindrica* (Shimamura et al. 2007), *Triticum aestivum* (wheat; Trought and Drew 1980; Barrett-Lennard et al. 1988), *Hordeum vulgare* (barley; Pang et al. 2007), rice (Bleecker et al. 1986), and maize (Abiko et al. 2012). In deepwater rice cv Pin Gaew 56 (PG56), adventitious root primordia emerge from the nodes within 8 to 10 h of partial submergence (Lorbiecke and Sauter 1999; Bleecker et al. 1986, Fig. 1).

In semiaquatic or wetland species including rice, *Sagittaria lancifolia* (Raskin and Kende 1983; Webb and Armstrong 1983; Schussler and Longstreth 1996) and

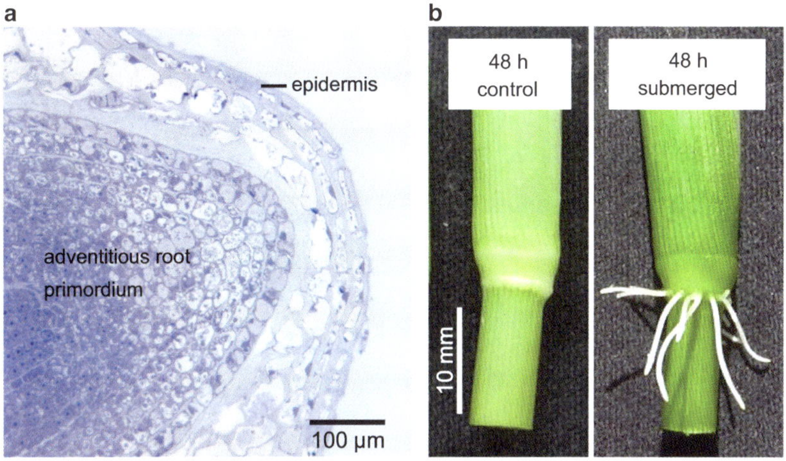

Fig. 1 Submergence induces adventitious root growth in deepwater rice cv Pin Gaew 56 (PG56). (**a**) Light microscopy picture of a cross-section through a node showing adventitious root primordia and overlying cell layers. (**b**) Adventitious root growth at the third youngest node of 12-week-old rice plants partially submerged for 48 h or left in air as a control

Zea nicaraguensis, a wild relative of maize, aerenchyma are constitutively formed in roots (Abiko et al. 2012). Aerenchyma are gas-filled spaces that provide a low-resistance gas transport within the plant. These gas spaces are formed as a result of lysigeny (cell death) or schizogeny (cell separation). In rice roots but also in maize or wheat roots, lysigenous aerenchyma formation is enhanced during soil waterlogging thus improving gas exchange including oxygen supply (Drew et al. 1981; Justin and Armstrong 1991; He et al. 1996; Steffens et al. 2011). During waterlogging, O_2 supply to the root depends on internal gas diffusion. Aerenchyma formation in adventitious roots kept in stagnant water occurs in wheat after 1–2 days while in rice aerenchyma formation is induced within 12 h (Webb and Jackson 1986). A comparison of the porosity of adventitious roots of wheat and rice plants grown in stagnant solution revealed that adventitious roots of wheat possess about 15 % porosity in comparison to about 40 % porosity of rice roots (Wiengweera et al. 1997; Colmer 2003). In addition, rice roots possess an inducible barrier against radial oxygen loss (ROL) at the basal region which is missing in wheat roots (Armstrong 1971; Thomson et al. 1992; Colmer 2003). The wetland species *Phragmites australis* and *Glyceria maxima* also possess a gas diffusion barrier in the outer cell layers of the root (Armstrong et al. 2000; Colmer 2003; Soukup et al. 2007; Shiono et al. 2011). The microstructure of this barrier is unknown but suberization and lignification of the exodermis and/or sclerenchyma in roots of rice and waterlogging-tolerant *Zea nicaraguensis* reduce loss of oxygen (Ranathunge et al. 2011; Shiono et al. 2011; Abiko et al. 2012). Reduced ROL increases oxygen diffusion to the root apex. The dryland species wheat, maize, and *Hordeum* form no or only a minor barrier for ROL (Thomson et al. 1992; Garthwaite et al. 2003; Abiko et al. 2012). In summary, adventitious roots develop as an adaptation to soil waterlogging or flooding in semiaquatic and wetland plants. Adventitious roots can form aerenchyma and a diffusion barrier to prevent ROL. As a consequence, semiaquatic species survive longer periods of flooding than less well adapted dryland species.

3 A Complex Signaling Network Regulates Secondary Root Formation

Most of the plants functional root system develops postembryonically. The use of *Arabidopsis thaliana* has contributed substantially to understand the development of primary and lateral roots and has provided insight into the molecular basis of root growth (Casimiro et al. 2003; Péret et al. 2009; Petricka et al. 2012). In contrast to the root system of Arabidopsis with a primary root and lateral roots, cereals possess a more complex, fibrous root system with a primary root, lateral roots, and additional pole-borne (seminal) and shoot-borne adventitious roots. Due to its more complex nature, our understanding of root development in cereals lags behind what we know for Arabidopsis despite the importance of crops such as rice, maize, or barley in agriculture. The primary root system of rice can be supported or replaced by adventitious roots during periods of flooding when the primary root system succumbs to oxygen shortage. Root development is a reiterative process. Hence, the developmental program that results in the formation of new root primordia and in root growth is repeated with each lateral root and is presumably also repeated in adventitious root development.

3.1 Auxin is the Major Hormone Initiating Adventitious Root Formation

Auxin is known to induce dedifferentiation of cells in the process of root apical meristem formation from quiescent or differentiated cells and auxin also promotes adventitious root biogenesis. In rice, adventitious root primordia are formed at each newly developing stem node as part of normal development (Bleecker et al. 1986). More adventitious roots are formed in plants overexpression *OsYUCCA1* which encodes a flavin monooxygenase-like enzyme involved in auxin biosynthesis (Yamamoto et al. 2007, Fig. 2). Mutant analysis revealed that adventitious root formation was the result of elevated auxin content. Mutants affected in auxin synthesis or signaling are frequently affected in secondary root development. Auxin-overproducing mutants such as *YUCCA1*, *SUPERROOT1*, or *SUPERROOT2* from Arabidopsis form adventitious roots at their hypocotyl (Boerjan et al. 1995; Delarue et al. 1988; Zhao et al. 2001).

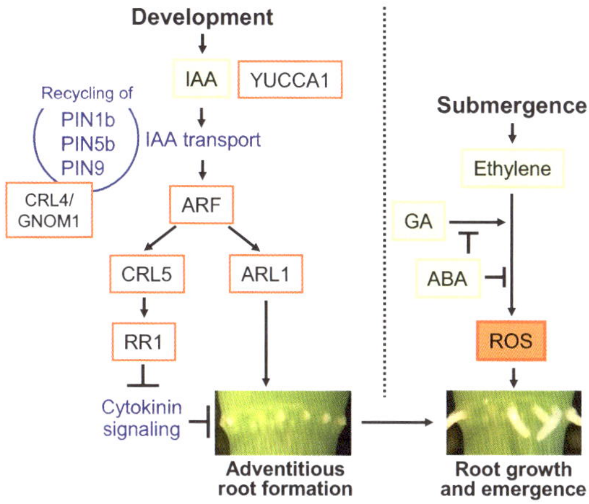

Fig. 2 Adventitious root biogenesis and growth regulation in rice. Adventitious roots develop at each new node through enhanced auxin biosynthesis via flavin monooxygenase-like enzymes such as YUCCA1. Auxin induces dedifferentiation of cells to initiate adventitious root formation. The ARF-GEF CRL4/GNOM1 mediates polar auxin transport likely through relocalization and recycling of PIN proteins. Two LBD genes, *CRL5* and *ARL1*, are direct targets of transcription factors of the ARF family. Upregulation of the negative regulator of cytokinin signaling RR1 through CRL5 results in inhibition of cytokinin signaling and enhanced adventitious root formation. In addition, ARL1 mediates auxin-induced root formation. Ethylene which accumulates in submerged tissues promotes adventitious root growth. GA enhances ethylene-induced root growth while ABA acts as a root growth inhibitor. ROS act downstream of ethylene to mediate the root growth response. Detailed explanation in the text

Adventitious root formation depends on directed auxin transport. Influx carriers of the AUX/LAX family and members of the PINFORMED family which act as efflux carriers are responsible for polar auxin transport. In the shoot, efflux carriers are localized at the basal cell pole thus promoting basipetal auxin flux. GNOM, an ADP ribosylation factor-GTP exchange factor (ARF-GEF) is important for vesicle trafficking and polar auxin transport. The Arabidopsis GNOM protein maintains the recycling of PINFORMED1 (PIN1) from endosomes to the plasma membrane which is required for auxin-mediated gravitropism as well as lateral root formation (Geldner et al. 2004). In rice, CROWN ROOTLESS4 (CRL4)/OsGNOM1 is similarly involved in lateral and adventitious root formation by regulating polar auxin transport likely through OsPIN1b, OsPIN5b, and OsPIN9 (Liu et al. 2009, Fig. 2).

In waterlogging-intolerant tomato flooding- and ethylene-induced adventitious root formation is likewise dependent on polar auxin transport (Vidoz et al. 2010). Tomato plants with a defect in the *DIAGEOTROPICA* (*DGT*) gene do not form adventitious roots in response to flooding or following auxin treatment (Vidoz et al. 2010). *DGT* encodes a type A cyclophilin, a member of a conserved family of peptidyl-prolyl cis-trans isomerase proteins that functions in auxin responses such as gravitropism. *dgt* mutants are less sensitive to auxin (Oh et al. 2006) and auxin-treated *dgt* mutants produce less ethylene.

3.2 Transcription Factors of the GRAS and the AP2/ERF Families Regulate Auxin-Induced Adventitious Root Biogenesis

GRAS (GIBBERELLIN INSENSITIVE, GAI; REPRESSOR OF GAI-3, RGA; SCARECROW, SCR) proteins are transcription factors encoded by a large gene family some of which are involved in the regulation of plant growth and development in response to light, hormone, and hypoxia signaling. Over 30 and 50 GRAS genes were identified in Arabidopsis and rice, respectively (Tian et al. 2004). Nine members of the GRAS gene family were induced in Arabidopsis root cells within 2 h of hypoxia (Mustroph et al. 2009). GRAS proteins share characteristic leucine-rich repeats involved in protein–protein interaction, a VHIID motif for DNA–protein interaction, a basic region responsible for nuclear localization, PFYRE and SAW domains, and an LXXLL motif. Some members of the GRAS protein family such as SHORT-ROOT (SHR), SCR and SCARECROW-LIKE (SCL) are involved in the establishment of the embryonic root meristem (Di Laurenzio et al. 1986; Sabatini et al. 1999; Helariutta et al. 2000; Sassa et al. 2001; Kamiya et al. 2003; Lim et al. 2005). In Arabidopsis, *SHR* is transcribed in the stele and the SHR protein moves to activate *SCR* expression in the quiescent center and endodermal and cortex initials (Malamy and Benfey 1997; Konishi and Sugiyama 2006; Lucas et al. 2011). The SHR/SCR pathway activates cell division and generates cortical and endodermal cell layers. In addition, SHR is essential for the establishment of lateral and adventitious root primordia in Arabidopsis (Benfey et al. 1993; Lucas et al. 2011, Table 1). In pine and *Castanea sativa* (chestnut), adventitious root formation is also induced by auxin and is dependent on polar auxin transport (Sánchez and Vieitez 1991; Díaz-Sala et al. 1996; Ballester et al. 1999). Transcript abundance of the pine SCARECROW-LIKE gene *PrSCL1*, of the chestnut *CsSCL1* and of genes encoding expansin proteins increased within 24 h of adventitious root induction by exogenous application of auxin suggesting a role in auxin-mediated induction of adventitious root formation (Hutchison et al. 1999; Sánchez et al. 2007). The expression of *PrSHR*, a putative ortholog of the Arabidopsis *SHR* gene, increased during adventitious root formation in pine (Solé et al. 2008, Table 1).

Eight ASYMMETRIC LEAVES2 (AS2)/LATERAL ORGAN BOUNDARIES (LOB) domain (LBD) genes were also induced in Arabidopsis root cells after 2 h hypoxia (Mustroph et al. 2009). Some members of the AS2/LOB gene family are early auxin-induced genes that were shown to be involved in secondary root formation (Table 1). The plant-specific AS2/LOB family is characterized by an AS2/LOB domain that spans approximately 100 conserved amino acids with a cysteine repeat, the so-called C block, a GAS block and a coiled coil

Table 1 Genes with a function in lateral and/or adventitious root initiation

Protein family	Adventitious root formation	Lateral root formation	Plant species	References
GH3	*GH3.3*	–	*Arabidopsis thaliana*	Gutierrez et al. (2012)
	GH3.5	–	*Arabidopsis thaliana*	Gutierrez et al. (2012)
	GH3.6	–	*Arabidopsis thaliana*	Gutierrez et al. (2012)
ARF-GEF	*CRL4/OsGNOM1*	*CRL4/ OsGNOM1*	Rice	Liu et al. (2009)
AS2/LOB	*CRL1/ARL1*	–	Rice	Inukai et al. (2005), Liu et al. (2005)
	RTCS	–	Maize	Taramino et al. (2007)
	–	*LBD16*	*Arabidopsis thaliana*	Shuai et al. (2002)
	–	*LBD29*	*Arabidopsis thaliana*	Shuai et al. (2002)
AP2/ERF	*ANT*	–	*Arabidopsis thaliana*	Elliott et al. (1996), Mizukami and Fischer (2000), Kitomi et al. (2011)
	CRL5	–	Rice	
GRAS	*SCR*	*SCR*	*Arabidopsis thaliana*	Malamy and Benfey (1997), Lucas et al. (2011)
	CsSCL1	–	Chestnut	Sánchez et al. (2007)
	PrSCL1	–	Pine	Sánchez et al. (2007)
	SHR	–	*Arabidopsis thaliana*	Lucas et al. (2011)
	PrSHR	–	Pine	Solé et al. (2008)
Others	*PtRR13*	–	Populus	Ramírez-Carvajal et al. (2009)
	DGT	–	Tomato	Vidoz et al. (2010)

Plant species and references were included in the table

(Shuai et al. 2002). The AS2/LOB domain is sufficient for DNA-binding activity and capable of mediating interactions with other proteins (Husbands et al. 2007). The *LOB* gene of Arabidopsis was the first member identified. It is expressed at the boundary of lateral organs throughout plant development (Shuai et al. 2002). In Arabidopsis, lateral root initiation is regulated by *LBD16* and *LBD29* (Shuai et al. 2002). In rice, the initiation of adventitious root primordia is controlled by CROWN ROOTLESS1 (CRL1; Inukai et al. 2005) also referred to as ADVENTITIOUS ROOTLESS1 (ARL1; Liu et al. 2005). *CRL1/ARL1* (short: *ARL1*) is an auxin- and ethylene-responsive AS2/LOB domain gene. ARL1 shares high sequence similarity with LBD16 and LBD29 from Arabidopsis. Expression of all three LBD genes is auxin-induced mediated by a transcription factor of the AUXIN RESPONSE FACTOR (ARF) family that binds to a specific *cis*-element in their promoter. Microarray analysis comparing wild-type rice and *arl1* mutant revealed that *ARL1* regulates genes encoding for proteins with functions in cell cycle progression, chromatin remodeling and a microtubule-associated protein suggesting a role for these proteins in auxin-induced ARL1-mediated adventitious root growth (Coudert et al. 2011). ROOTLESS CONCERNING CROWN AND SEMINAL ROOTS (RTCS), a putative ortholog of ARL1 is involved in seminal and crown root formation in maize (Taramino et al. 2007). The AS2/LOB domains of ARL1 and RTCS are nearly 100 % identical whereas outside of the AS2/LOB domain sequences are about 50 % identical (Taramino et al. 2007) indicating that the AS2/LOB domain might by crucial for root initiation at the shoot.

3.3 Role of Other Hormones in Adventitious Root Biogenesis and Growth

Cytokinins, jasmonic acid (JA), and abscisic acid (ABA) antagonize adventitious root formation and growth. Transcriptional regulators of the APETALA2/ETHYLENE RESPONSE FACTOR (AP2/ERF) gene family, e.g., AINTEGUMENTA (ANT) in Arabidopsis or CRL5 in rice regulate shoot organ and floral growth (Elliott et al. 1996; Mizukami and Fischer 2000), and adventitious root formation through repression of cytokinin signaling (Kitomi et al. 2011, Table 1). CRL5 is an auxin-responsive transcriptional regulator with two AP2 domains that shares 50 % similarity with ANT from Arabidopsis. CRL5 upregulates type-A response regulators such as *OsRR1*, a negative regulator of cytokinin signaling resulting in adventitious root formation (Fig. 2). Similar to mutants with reduced *ARL1* expression *CRL5* mutants have fewer adventitious roots. The *crl5/arl1* double mutant does not possess any adventitious roots (Liu et al. 2005; Kitomi et al. 2011). In *Populus* cuttings, the cytokinin type-B response regulator PtRR13 negatively regulates adventitious root formation (Ramírez-Carvajal et al. 2009).

JA-deficient *dde1* and *dde2-2* mutants with a defect in the *DELAYED DEHISCENCE* (*DDE*) gene that encodes for a protein involved in JA biosynthesis have more adventitious roots than wild-type plants indicating that JA inhibits adventitious root initiation (Gutierrez et al. 2012). According to a recently proposed model for adventitious root formation in Arabidopsis, transcription factors of the ARF family, *ARF6* and *ARF8* positively and *ARF17* negatively regulate three members of the GRETCHEN HAGEN3 (GH3) family (Gutierrez et al. 2012, Table 1). *GH3* genes encode amino acid-conjugating enzymes. GH3.3, GH3.5, and GH3.6 control JA-homeostasis through conjugation of JA to different amino acids. JA-Ile inhibits adventitious root formation. GH3.3, GH3.5, and GH3.6 are involved in reducing free JA content by conjugating JA to JA-Trp, JA-Met, or JA-Asp. This way JA-Ile content is lowered and JA signaling through the JA receptor CORONATE INSENSITIVE1 (COI1) is prevented resulting in adventitious root formation.

In rice, adventitious root primordia form but do not emerge until an appropriate signal is perceived. In the presence of ethylene, gibberellic acid (GA) accelerates adventitious root growth. ABA inhibits both, ethylene-induced and ethylene-induced and GA-promoted adventitious root growth (Suge 1985; Lorbiecke and Sauter 1999; Steffens et al. 2006). Elevated ethylene levels, elevated GA activity, and reduced ABA levels within 3 h of submergence favor adventitious root growth in rice (Hoffmann-Benning and Kende 1992; Azuma et al. 1995).

In conclusion, adventitious root formation is induced by auxin and is regulated at the level of auxin biosynthesis and polar auxin transport. Cytokinin and JA are antagonists of auxin-induced adventitious root formation. A network of transcriptional regulators of the GRAS and the AP2/ERF protein families controls adventitious root formation but only little information is available on how these regulators mediate adventitious root formation in response to submergence. In rice, ethylene is mainly responsible for adventitious root emergence and growth during submergence. GA promotes and ABA inhibits both, root emergence and growth (Fig. 2).

3.4 Role of Reactive Oxygen Species in Adventitious Root Growth

The homeostasis between and localization of the reactive oxygen species (ROS) hydrogen peroxide (H_2O_2) and superoxide anion radicals (O_2^-) are necessary for primary and secondary root growth regulation. Distinct localization of O_2^- and H_2O_2 in growing primary Arabidopsis roots is regulated by UPBEAT1, a bHLH transcriptional regulator (Tsukagoshi et al. 2010). Studies using diphenylene iodonium (DPI), an inhibitor of the plasma membrane localized NADPH oxidase, revealed that O_2^- is required for primary root growth in Arabidopsis (Dunand et al. 2007). In rice, H_2O_2 mediates ethylene-induced adventitious root growth (Steffens et al. 2012).

4 Breakthrough News from Adventitious Roots

Adventitious roots emerge from within the shoot and therefore have to traverse the maternal tissue without damaging the apical meristem. In Arabidopsis, cell wall loosening occurs in cortex cells in front of a growing lateral root primordium to facilitate its emergence (Swarup et al. 2008; Takahashi et al. 2010). In rice, adventitious roots emerge from the stem nodes. To facilitate secondary root emergence, epidermal cells overlying adventitious root primordia undergo cell death (Lorbiecke and Sauter 1999; Mergemann and Sauter 2000). Root growth and epidermal cell death are coordinately regulated by ethylene. At the third node of deepwater rice cv PG56, ethylene promotes epidermal cell death within 1.5 h and adventitious root emergence within 7–10 h (Steffens and Sauter 2005, 2009; Steffens et al. 2012). Epidermal cell death is a crucial event that can render the plant susceptible to invasion by pathogens if uncontrolled. It is therefore restricted to sites overlying adventitious root primordia. In wild-type and *arl1* with fewer adventitious roots epidermal cell death is strictly linked to the presence of a root primordium implicating that the root emits a positional signal. Recently, the mechanical force that is excreted by a growing root primordium was identified as the positional signal that triggers and locally restricts epidermal cell death (Steffens et al. 2012). In addition to a mechanical trigger, epidermal cells must perceive ethylene to undergo cell death. Combined application of ethylene and force are in fact sufficient to ectopically induce local epidermal cell death. ROS mediate ethylene signaling of both adventitious root elongation and programmed cell death. METALLOTHIONEIN2b (MT2b) is an ROS scavenger in rice. In epidermal cells overlying adventitious root primordia *MT2b* is downregulated by ethylene and ROS (Steffens and Sauter 2009). In epidermal cells, reduced ROS scavenging through the MT2b is crucial for ROS accumulation and cell death induction while ROS homeostasis appears to be controlled differently in adventitious roots. Identifying the mechanical receptor that perceives an emerging adventitious root and the signaling pathway that alters ROS levels has yet to be achieved.

References

Abiko T, Kotula L, Shiono K, Malik AI, Colmer TD, Nakazono M (2012) Enhanced formation of aerenchyma and induction of a barrier to radial oxygen loss in adventitious roots of *Zea nicaraguensis* contribute to its waterlogging tolerance as compared with maize (*Zea mays* ssp. mays). Plant Cell Environ. doi:10.1111/j.1365-3040.2012.02513.x

Armstrong W (1971) Radial oxygen losses from intact rice roots as affected by distance from the apex, respiration, and waterlogging. Physiol Plant 25:192–197

Armstrong W, Cousins D, Armstrong J, Turner DW, Beckett PM (2000) Oxygen distribution in wetland plant roots and permeability barriers to gas-exchange with the rhizosphere: a micro-electrode and modelling study with *Phragmites australis*. Ann Bot 86:687–703

Azuma T, Hirano T, Deki Y, Uchida N, Yasuda T, Yamaguchi T (1995) Involvement of the decrease in levels of abscisic acid in the internodal elongation of submerged floating rice. J Plant Physiol 146:323–328

Ballester A, San-José MC, Vidal N, Fernández-Lorenzo JL, Vieitez AM (1999) Anatomical and biochemical events during in vitro rooting of microcuttings from juvenile and mature plants of chestnut. Ann Bot 83:619–629

Barrett-Lennard EG, Leighton PD, Buwalda F, Gibbs J, Armstrong W, Thomson CJ, Greenway H (1988) Effects of growing wheat in hypoxic nutrient solution and of subsequent transfer to aerated solutions. I. Growth and carbohydrate status of shoots and roots. Aust J Plant Physiol 15:585–598

Benfey PN, Linstead PJ, Roberts K, Schiefelbein JW, Hauser MT, Aeschbacher RA (1993) Root development in Arabidopsis: four mutants with dramatically altered root morphogenesis. Development 119:57–70

Bleecker AB, Schuette JL, Kende H (1986) Anatomical analysis of growth and development patterns in the internode of deepwater rice. Planta 169:490–497

Boerjan W, Cervera MT, Delarue M, Beeckman T, Dewitte W, Bellini C, Caboche M, Van Onckelen H, Van Montagu M, Inzé D (1995) Superroot, a recessive mutation in Arabidopsis, confers auxin overproduction. Plant Cell 7:1405–1419

Calvo-Polanco M, Señorans J, Zwiazek JJ (2012) Role of adventitious roots in water relations of tamarack (*Larix laricina*) seedlings exposed to flooding. BMC Plant Biol. doi:10.1186/1471-2229-12-99

Casimiro I, Beeckman T, Graham N, Bhalerao R, Zhang H, Casero P, Sandberg G, Bennett MJ (2003) Dissecting Arabidopsis lateral root development. Trends Plant Sci 8:165–171

Colmer TD (2003) Aerenchyma and an inducible barrier to radial oxygen loss facilitate root aeration in upland, paddy and deep-water rice (*Oryza sativa* L.). Ann Bot 91:301–309

Coudert Y, Bès M, Le TV, Pré M, Guiderdoni E, Gantet P (2011) Transcript profiling of crown rootless1 mutant stem base reveals new elements associated with crown root development in rice. BMC Genomics 12:387

Delarue M, Prinsen E, Onckelen HV, Caboche M, Bellini C (1988) Sur2 mutations of Arabidopsis thaliana define a new locus involved in the control of auxin homeostasis. Plant J 14:603–611

Di Laurenzio L, Wysocka-Diller J, Malamy JE, Pysh L, Helariutta Y, Freshour G, Hahn MG, Feldman KA, Benfey PN (1986) The *SCARECROW* gene regulates an asymmetric cell division that is essential for generating the radial organization of Arabidopsis root. Cell 86:423–433

Díaz-Sala C, Hutchison KW, Goldfarb B, Greenwood MS (1996) Maturation-related loss in rooting competence by stem cuttings: the role of auxin transport, metabolism and tissue sensitivity. Physiol Plant 97:481–490

Drew MC, Jackson MB, Giffard SC, Campbell R (1981) Inhibition by silver ions of gas space (aerenchyma) formation in adventitious roots of *Zea mays* L. subjected to exogenous ethylene or to oxygen deficiency. Planta 153:217–224

Dunand C, Crèvecoeur M, Penel C (2007) Distribution of superoxide and hydrogen peroxide in Arabidopsis root and their influence on root development: possible interaction with peroxidases. New Phytol 174:332–341

Elliott RC, Betzner AS, Huttner E, Oakes MP, Tucker WQ, Gerentes D, Perez P, Smyth DR (1996) AINTEGUMENTA, an APETALA2-like gene of Arabidopsis with pleiotropic roles in ovule development and floral organ growth. Plant Cell 8:155–168

Garthwaite AJ, von Bothmer R, Colmer TD (2003) Diversity in root aeration traits associated with waterlogging tolerance in the genus *Hordeum*. Funct Plant Biol 30:875–889

Geldner N, Richter S, Vieten A, Marquardt S, Torres-Ruiz RA, Mayer U, Jürgens G (2004) Partial loss-of-function alleles reveal a role for GNOM in auxin transport-related, post-embryonic development of Arabidopsis. Development 131:389–400

Gutierrez L, Mongelard G, Floková K, Pacurar DI, Novák O, Staswick P, Kowalczyk M, Pacurar M, Demailly H, Geiss G, Bellini C (2012) Auxin controls Arabidopsis adventitious root initiation by regulating jasmonic acid homeostasis. Plant Cell 24:2515–2527

He CJ, Morgan PW, Drew MC (1996) Transduction of an ethylene signal is required for cell death and lysis in the root cortex of maize during aerenchyma formation induced by hypoxia. Plant Physiol 112:463–472

Helariutta Y, Fukaki H, Wysocka-Diller J, Nakajima K, Jung J, Sena G, Hauser MT, Benfey PN (2000) The *SHORT-ROOT* gene controls radial patterning of the Arabidopsis root through radial signalling. Cell 101:555–567

Hochholdinger F, Park WJ, Sauer M, Woll K (2004) From weeds to crops: genetic analysis of root development in cereals. Trends Plant Sci 9:42–48

Hoffmann-Benning S, Kende H (1992) On the role of abscisic acid and gibberellin in the regulation of growth in rice. Plant Physiol 99:1156–1161

Husbands A, Bell EM, Shuai B, Smith HM, Springer PS (2007) LATERAL ORGAN BOUNDARIES defines a new family of DNA-binding transcription factors and can interact with specific bHLH proteins. Nucleic Acids Res 35:6663–6671

Hutchison KW, Singer PB, McInnis S, Diaz-Sala C, Greenwood MS (1999) Expansins are conserved in conifers and expressed in hypocotyls in response to exogenous auxin. Plant Physiol 120:827–832

Inukai Y, Sakamoto T, Ueguchi-Tanaka M, Shibata Y, Gomi K, Umemura I, Hasegawa Y, Ashikari M, Kitano H, Matsuoka M (2005) Crown rootless1, which is essential for crown

root formation in rice, is a target of an AUXIN RESPONSE FACTOR in auxin signaling. Plant Cell 17:1387–1396
Jackson MB (1985) Ethylene and responses of plants to soil waterlogging and submergence. Annu Rev Plant Physiol 36:145–174
Justin SHFW, Armstrong W (1991) Evidence for the involvement of ethylene in aerenchyma formation in adventitious roots of rice (*Oryza sativa* L.). New Phytol 118:49–62
Kamiya N, Itoh JI, Morikami A, Nagato Y, Matsuoka M (2003) The *SCARECROW* gene's role in asymmetric cell divisions in rice plants. Plant J 36:45–54
Kende H (1993) Ethylene biosynthesis. Annu Rev Plant Physiol Plant Mol Biol 44:283–307
Kitomi Y, Ito H, Hobo T, Aya K, Kitano H, Inukai Y (2011) The auxin responsive AP2/ERF transcription factor CROWN ROOTLESS5 is involved in crown root initiation in rice through the induction of OsRR1, a type-A response regulator of cytokinin signaling. Plant J 67: 472–484
Konishi M, Sugiyama M (2006) A novel plant-specific family gene, ROOT PRIMORDIUM DEFECTIVE 1, is required for the maintenance of active cell proliferation. Plant Physiol 140:591–602
Li YH, Zou MH, Feng BH, Huang X, Zhang Z, Sun GM (2012) Molecular cloning and characterization of the genes encoding an auxin efflux carrier and the auxin influx carriers associated with the adventitious root formation in mango (*Mangifera indica* L.) cotyledon segments. Plant Physiol Biochem 55:33–42
Lim J, Jung JW, Lim C, Lee M, Kim B, Kim M, Bruce WB, Benfey PN (2005) Conservation and diversification of *SCARECROW* in maize. Plant Mol Biol 59:619–630
Liu H, Wang S, Yu X, Yu J, He X, Zhang S, Shou H, Wu P (2005) ARL1, a LOB-domain protein required for adventitious root formation in rice. Plant J 43:47–56
Liu S, Wang J, Wang L, Wang X, Xue Y, Wu P, Shou H (2009) Adventitious root formation in rice requires OsGNOM1 and is mediated by the OsPINs family. Cell Res 19:1110–1119
Lorbiecke R, Sauter M (1999) Adventitious root growth and cell-cycle induction in deepwater rice. Plant Physiol 119:21–29
Lucas M, Swarup R, Paponov IA, Swarup K, Casimiro I, Lake D, Peret B, Zappala S, Mairhofer S, Whitworth M, Wang J, Ljung K, Marchant A, Sandberg G, Holdsworth MJ, Palme K, Pridmore T, Mooney S, Bennett MJ (2011) Short-root regulates primary, lateral, and adventitious root development in Arabidopsis. Plant Physiol 155:384–398
Malamy JE, Benfey PN (1997) Organization and cell differentiation in lateral roots of *Arabidopsis thaliana*. Development 124:33–44
Mergemann H, Sauter M (2000) Ethylene induces epidermal cell death at the site of adventitious root emergence in rice. Plant Physiol 124:609–614
Mizukami Y, Fischer RL (2000) Plant organ size control: AINTEGUMENTA regulates growth and cell numbers during organogenesis. Proc Natl Acad Sci U S A 97:942–947
Mustroph A, Zanetti ME, Jang CJ, Holtan HE, Repetti PP, Galbraith DW, Girke T, Bailey-Serres J (2009) Profiling translatomes of discrete cell populations resolves altered cellular priorities during hypoxia in Arabidopsis. Proc Natl Acad Sci U S A 106:18843–18848
Oh K, Ivanchenko MG, White TJ, Lomax TL (2006) The *diageotropica* gene of tomato encodes a cyclophilin: a novel player in auxin signalling. Planta 224:133–144
Pang J, Cuin T, Shabala L, Zhou M, Mendham N, Shabala S (2007) Effect of secondary metabolites associated with anaerobic soil conditions on ion fluxes and electrophysiology in barley roots. Plant Physiol 145:266–276
Paolillo DJ Jr, Bassuk NL (2005) On the occurrence of adventitious branch roots on root axes of trees. Am J Bot 92:802–809
Péret B, De Rybel B, Casimiro I, Benková E, Swarup R, Laplaze L, Beeckman T, Bennett MJ (2009) Arabidopsis lateral root development: an emerging story. Trends Plant Sci 14:399–408
Petricka JJ, Winter CM, Benfey PN (2012) Control of Arabidopsis root development. Annu Rev Plant Biol 63:563–590

Purnobasuki H, Suzuki M (2004) Aerenchyma formation and porosity in root of a mangrove plant, *Sonneratia alba* (Lythraceae). J Plant Res 117:465–472
Ramírez-Carvajal GA, Morse AM, Dervinis C, Davis JM (2009) The cytokinin type-B response regulator PtRR13 is a negative regulator of adventitious root development in Populus. Plant Physiol 150:759–771
Ranathunge K, Lin J, Steudle E, Schreiber L (2011) Stagnant deoxygenated growth enhances root suberization and lignifications, but differentially affects water and NaCl permeabilities in rice (*Oryza sativa* L.) roots. Plant Cell Environ 34:1223–1240
Raskin I, Kende H (1983) How does deep water rice solve its aeration problem. Plant Physiol 72:447–454
Ridge RW, Sack FD (1992) Cortical and cap sedimentation in gravitropic Equisetum roots. Am J Bot 79:328–334
Sabatini S, Beis D, Wolkenfelt H, Murfett J, Guilfoyle T, Malamy J, Benfey P, Leyser O, Bechtold N, Weisbeek P, Scheres B (1999) An auxin-dependent distal organizer of pattern and polarity in the *Arabidopsis* root. Cell 99:463–472
Sánchez C, Vielba JM, Ferro E, Covelo G, Solé A, Abarca D, de Mier BS, Díaz-Sala C (2007) Two SCARECROW-LIKE genes are induced in response to exogenous auxin in rooting-competent cuttings of distantly related forest species. Tree Physiol 27:1459–1470
Sánchez MC, Vieitez AM (1991) In vitro morphogenetic competence of basal sprouts and crown branches of mature trees. Tree Physiol 8:59–70
Sassa N, Matsushita Y, Nakamura T, Nyunoya H (2001) The molecular characterization and in situ expression pattern of pea *SCARECROW* gene. Plant Cell Physiol 42:385–394
Schussler EE, Longstreth DJ (1996) Aerenchyma develops by cell lysis in roots and cell separation in leaf petioles *in Sagittaria lancifolia* (Alismataceae). Am J Bot 83:1266–1273
Shimamura S, Yoshida S, Mochizuki T (2007) Cortical aerenchyma formation in hypocotyl and adventitious roots of *Luffa cylindrica* subjected to soil flooding. Ann Bot 100:1431–1439
Shiono K, Ogawa S, Yamazaki S, Isoda H, Fujimura T, Nakazono M, Colmer TD (2011) Contrasting dynamics of radial O_2-loss barrier induction and aerenchyma formation in rice roots of two lengths. Ann Bot 107:89–99
Shuai B, Reynaga-Peña CG, Springer PS (2002) The *lateral organ boundaries* gene defines a novel, plant-specific gene family. Plant Physiol 129:747–761
Solé A, Sánchez C, Vielba JM, Valladares S, Abarca D, Díaz-Sala C (2008) Characterization and expression of a *Pinus* radiata putative ortholog to the Arabidopsis SHORT-ROOT gene. Tree Physiol 28:1629–1639
Soukup A, Armstrong W, Schreiber L, Franke R, Votrubová O (2007) Apoplastic barriers to radial oxygen loss and solute penetration: a chemical and functional comparison of the exodermis of two wetland species, *Phragmites australis* and *Glyceria maxima*. New Phytol 173:264–278
Steffens B, Geske T, Sauter M (2011) Aerenchyma formation in the rice stem and its promotion by H_2O_2. New Phytol 190:369–378
Steffens B, Kovalev A, Gorb SN, Sauter M (2012) Emerging roots alter epidermal cell fate through mechanical and reactive oxygen species signaling. Plant Cell. http://www.plantcell.org/cgi/doi/10.1105/tpc.112.101790
Steffens B, Sauter M (2005) Epidermal cell death in rice (*Oryza sativa* L.) is regulated by ethylene, gibberellin and abscisic acid. Plant Physiol 139:713–721
Steffens B, Sauter M (2009) Epidermal cell death in rice is confined to cells with a distinct molecular identity and is mediated by ethylene and H_2O_2 through an autoamplified signal pathway. Plant Cell 21:184–196
Steffens B, Wang J, Sauter M (2006) Interactions between ethylene, gibberellin and abscisic acid regulate emergence and growth rate of adventitious roots in deepwater rice. Planta 223:604–612
Suge H (1985) Ethylene and gibberellin: regulation of internodal elongation and nodal root development in floating rice. Plant Cell Physiol 26:607–614
Swarup K, Benková E, Swarup R, Casimiro I, Péret B, Yang Y, Parry G, Nielsen E, De Smet I, Vanneste S, Levesque MP, Carrier D, James N, Calvo V, Ljung K, Kramer E, Roberts R,

Graham N, Marillonnet S, Patel K, Jones JD, Taylor CG, Schachtman DP, May S, Sandberg G, Benfey P, Friml J, Kerr I, Beeckman T, Laplaze L, Bennett MJ (2008) The auxin influx carrier LAX3 promotes lateral root emergence. Nat Cell Biol 10:946–954

Takahashi K, Shimada T, Kondo M, Tamai A, Mori M, Nishimura M, Hara-Nishimura I (2010) Ectopic expression of an esterase, which is a candidate for the unidentified plant cutinase, causes cuticular defects in *Arabidopsis thaliana*. Plant Cell Physiol 51:123–131

Taramino G, Sauer M, Stauffer JL Jr, Multani D, Niu X, Sakai H, Hochholdinger F (2007) The maize (*Zea mays* L.) RTCS gene encodes a LOB domain protein that is a key regulator of embryonic seminal and post-embryonic shoot-borne root initiation. Plant J 50:649–659

Thomson CJ, Colmer TD, Watkins ELJ, Greenway H (1992) Tolerance of wheat (*Triticum aestivum* cvs Gamenya and Kite) and triticale (*Triticosecale* cv. Muir) to waterlogging. New Phytol 120:335–344

Tian C, Wan P, Sun S, Li J, Chen M (2004) Genome-wide analysis of the GRAS gene family in rice and Arabidopsis. Plant Mol Biol 54:519–532

Trought MCT, Drew MC (1980) The development of waterlogging damage in young wheat plants in anaerobic solution cultures. J Exp Bot 31:1573–1585

Tsukagoshi H, Busch W, Benfey PN (2010) Transcriptional regulation of ROS controls transition from proliferation to differentiation in the root. Cell 143:606–616

Van Der Straeten D, Zhou Z, Prinsen E, Van Onckelen HA, Van Montagu MC (2001) A comparative molecular-physiological study of submergence response in lowland and deepwater rice. Plant Physiol 125:955–968

Vidoz ML, Loreti E, Mensuali A, Alpi A, Perata P (2010) Hormonal interplay during adventitious root formation in flooded tomato plants. Plant J 63:551–562

Webb T, Armstrong W (1983) The effects of anoxia and carbohydrates on the growth and viability of rice, pea and pumpkin roots. J Exp Bot 34:579–603

Webb J, Jackson MB (1986) A transmission and cryo-scanning electron microscopy study of the formation of aerenchyma (cortical gas-filled space) in adventitious roots of rice (*Oryza sativa*). J Exp Bot 37:832–841

Wiengweera A, Greenway H, Thomson CJ (1997) The use of agar nutrient solution to simulate lack of convection in waterlogged soils. Ann Bot 80:115–123

Yamamoto Y, Kamiya N, Morinaka Y, Matsuoka M, Sazuka T (2007) Auxin biosynthesis by the YUCCA genes in rice. Plant Physiol 143:1362–1371

Zarembinski TI, Theologis A (1997) Expression characteristics of *Os-ACS1* and *Os-ACS2*, two members of the 1-aminocyclopropane-1-carboxylate synthase gene family in rice (*Oryza sativa* L. cv. Habiganj Aman II) during partial submergence. Plant Mol Biol 33:71–77

Zhao Y, Christensen SK, Fankhauser C, Cashman JR, Cohen JD, Weigel D, Chory J (2001) A role for flavin monooxygenase-like enzymes in auxin biosynthesis. Science 291:306–309

Part V
Ecophysiological Adaptations

Underwater Photosynthesis and Internal Aeration of Submerged Terrestrial Wetland Plants

Ole Pedersen and Timothy D. Colmer

Abstract Submergence impedes plant gas exchange with the environment. Survival depends upon internal aeration to provide O_2 throughout the plant body, although short-term anoxia can be tolerated. During nights, plants rely on O_2 entry from the floodwater and pO_2 in roots declines so that some tissues become severely hypoxic or even anoxic. Underwater photosynthesis is the main daytime O_2 source and also provides sugars. Capacity for photosynthesis under water, like in air, is determined by available CO_2 and light; however, slow diffusion in water often limits CO_2 supply. Underwater photosynthesis in some wetland species is enhanced by gas films on superhydrophobic leaf surfaces. Leaf gas films also increase night-time O_2 uptake by submerged plants. Flooding events are forecast to increase and understanding of plant submergence tolerance should enable predictions of possible impacts on vegetation communities and also aid breeding of improved submergence tolerance in rice.

1 The Submergence Environment

The slow diffusion of gases in water compared with in air presents a challenge to submerged terrestrial plants (Armstrong 1979) as oxygen (O_2) and carbon dioxide (CO_2) uptake are greatly impeded. Diffusion of gases in water is approximately 10,000-fold slower than in air, so that since the diffusive boundary layers (DBLs)

O. Pedersen (✉)
Freshwater Biological Laboratory, University of Copenhagen, Universitetsparken 4, 3rd floor, 2100 Copenhagen, Denmark

Institute of Advanced Studies, The University of Western Australia, Crawley, WA 6009, Australia

School of Plant Biology, The University of Western Australia, Crawley, WA 6009, Australia
e-mail: opedersen@bio.ku.dk

T.D. Colmer
School of Plant Biology, The University of Western Australia, Crawley, WA 6009, Australia

J.T. van Dongen and F. Licausi (eds.), *Low-Oxygen Stress in Plants*, Plant Cell Monographs 21, DOI 10.1007/978-3-7091-1254-0_16, © Springer-Verlag Wien 2014

adjacent to surfaces are of similar thickness in both environments, the resulting apparent resistance to gas exchange is 10,000-fold higher when under water (Vogel 1994). Submerged aquatic plants have thus developed adaptive features of their leaves to reduce the DBL, and also to reduce other resistances (e.g. thin cuticle and thin leaves), to facilitate gas exchange in the aqueous environment (Sculthorpe 1967; Colmer et al. 2011).

In addition to the slow diffusion of gases, the solubility of O_2 in water is relatively low; 1 L of air contains 33-fold more O_2 than 1 L of water at 20 °C at sea level (Stumm and Morgan 1996). The amount of dissolved O_2 in a particular water body is determined by a combination of the water temperature and salinity and the surrounding O_2 partial pressure (pO_2). The pO_2 in the atmosphere decreases with elevation (at atmospheric equilibrium, less O_2 is dissolved in the water of a mountain lake than at sea level) and increases with depth (the absolute pressure increases with 101 kPa per 10 m depth). Temperature and salinity can differ substantially in the various habitats of terrestrial wetland plants (e.g. freshwater to hypersaline wetlands; near-freezing to warm waters at, e.g. 40 °C). Like for O_2, the solubility of CO_2 also decreases with increasing temperature and salinity and increases with pressure. The chemistry of CO_2 in water is more complicated than for O_2, as CO_2 reacts with water itself when it dissolves and forms a pH-dependent equilibrium where CO_2 dominates below pH 6.3, HCO_3^- between pH 6.3 and 10.2 and above pH 10.2 CO_3^{2-} dominates and this form cannot be utilised by photosynthetic organisms.

In addition to severe CO_2 limitation, the photosynthesis of inundated plants can also be limited by light. Light is attenuated in an exponential fashion due to water itself, but more importantly due to suspended particles, pigments in planktonic algae and dissolved humic substances (Kirk 1994). In turbid floodwaters, light absorption can be as much as 90 % in the upper 0.1 m, but more typical values for a 90 % reduction of light penetration in floodwaters is 0.5–2 m (Vervuren et al. 2003; Winkel et al. 2013).

2 Leaf Adaptations for Underwater Gas Exchange

Leaf morphology determines DBL resistances to exchange of dissolved gases and ions (Madsen and Sand-Jensen 1991). The DBL resistance to CO_2 uptake reduces underwater photosynthesis in submerged plants and is a large component of the overall apparent resistance to gas exchange between chloroplasts and the surrounding floodwater (Black et al. 1981). Morphological traits that reduce the effective DBL resistance, by decreasing the distance to the 'leading-edge' (Vogel 1994) and thus reducing the path-length across the DBL, include leaf shapes of small, dissected/lobed and/or strap-like leaves (Sculthorpe 1967). In addition, aquatic leaves lack trichomes thus facilitating water movement adjacent to the surfaces and so avoiding development of thicker DBLs. Leaves of aquatic species also tend to be thin, in extreme cases being only two cell layers thick, shortening internal diffusion-path lengths and thus reducing the overall resistance to CO_2 diffusion to chloroplasts (Madsen and Sand-Jensen 1991; Maberly and Madsen 2002).

In addition to these morphological traits, leaves of aquatic species also have very reduced cuticles, or this layer can even be absent. Diffusion into and across the epidermis is the pathway for dissolved gas-exchange as aquatic leaves lack stomata (Sculthorpe 1967), or if present, the stomata are non-functional (Pedersen and Sand-Jensen 1992). Diffusion path-length to chloroplasts is also minimised by having these organelles in all epidermal cells and in sub-epidermal cells the chloroplasts are positioned towards the exterior (Sculthorpe 1967).

Submerged aquatic plants also display physiological adaptations to increase the CO_2 concentration at Rubisco, the site of carboxylation; these are referred to as carbon concentrating mechanisms (CCMs) (Maberly and Madsen 2002; Raven et al. 2008). In submerged aquatic plants, CCMs include HCO_3^- use (Prins and Elzenga 1989), C_4 (Magnin et al. 1997), C_3–C_4 intermediates (Keeley 1999) and CAM photosynthesis (Keeley 1998). CCMs increase underwater net photosynthesis (P_N) in CO_2 limited aquatic environments. In addition, CAM has been shown to diminish photorespiration in the aquatic species, *Isoetes australis* (Pedersen et al. 2011b).

Leaves of terrestrial wetland plants lack most of the features described above for aquatic species and so suffer from larger diffusion resistances that limit CO_2 uptake for photosynthesis when under water. Some terrestrial species, however, can produce submergence acclimated leaves (Mommer and Visser 2005) and some possess leaf gas films (Raskin and Kende 1983; Colmer and Pedersen 2008b), Fig. 1; features which can also reduce the apparent resistance to CO_2 uptake by these species when submerged. Below, we evaluate in more detail underwater P_N by leaves of terrestrial wetland plants.

3 Underwater Photosynthesis in Leaves of Terrestrial Plants

The overall beneficial effects of aquatic leaf traits for underwater P_N, as well as the generally poor performance of leaves of terrestrial plants, were clearly demonstrated in Sand-Jensen et al. (1992). These authors highlighted that: (1) underwater P_N on a mass basis increased from terrestrial, then amphibious, to aquatic leaf types; and (2) that Danish low-land stream waters commonly contain CO_2 above air-equilibrium values, allowing even some terrestrial species to have adequate P_N for growth when submerged in these habitats. This pioneering study is considered in detail in Colmer et al. (2011).

Species of many terrestrial wetland plants produce new leaves when submerged and these can display some acclimation to enhance underwater gas exchange (Mommer et al. 2005, 2007). The best example is the 69-fold higher underwater

Fig. 1 Microelectrode set up in an experimental field pond at the International Rice Research Institute (The Philippines) to measure root pO_2 during complete submergence (**a**), bubble formation due to extensive underwater photosynthesis (**b**), and gas film on the superhydrophobic leaf surface of submerged rice, *Oryza sativa* (**c**). The system was set up with O_2 microelectrodes positioned into adventitious roots in the soil and then the field was flooded to completely submerge plants. Data from the experiment are shown in Figs. 3 and 4. Photos by Ole Pedersen

P_N due to a reduction in cuticle resistance in *Rumex palustris* (Mommer et al. 2006b). Mommer et al. (2007) found that seven terrestrial wetland species formed a thinner cuticle as a response to submergence leading to enhanced underwater gas exchange, but the degree of this response was not correlated with submergence tolerance. These acclimations in submerged leaves of terrestrial species are much more subtle than the altered leaf development displayed by some amphibious heterophyllous species, which produce true aquatic leaf types when under water (Nielsen 1993).

Here, we consider for terrestrial wetland plants how rates of underwater P_N compare with those in air. The few data available show that P_N under water is substantially lower than in air (Colmer et al. 2011). Rates of underwater P_N vary not only with species, but also with environmental conditions. Rice leaves at the ambient CO_2 of 70 mmol m^{-3} in a submergence field pond at the International Rice Research Institute (The Philippines) resulted in P_N under water at only 5 % of the rate in air (Winkel et al. 2013). However, CO_2 enrichment several-fold above the level in these ponds has been recorded in flooded rice fields in Thailand; 20–180-fold air-equilibrium (Setter et al. 1987) and in India; 31–217-fold

(Ram et al. 1999). At 200 mmol CO_2 m^{-3}, underwater P_N by rice was 27 % of that in air (Pedersen et al. 2009). Similar to rice, at 200 mmol CO_2 m^{-3} *Hordeum marinum* grown at high mineral nutrition also had underwater P_N at 28 % of that in air (Pedersen et al. 2010). By contrast, field-collected leaves of three other wetland species (*Phalaris arundinacea*, *Typha latifolia* and *Phragmites australis*) had underwater P_N at 200 mmol CO_2 m^{-3} (Colmer and Pedersen 2008b) estimated to be approximately 15 % of that in air (Colmer et al. 2011).

In the cases where underwater respiration of the lamina has also been measured, a crude 24 h C-balance of this tissue can be estimated. At 200 mmol CO_2 m^{-3} and assuming 12 h light and 12 h darkness, a positive C-balance of about 100 mmol C m^{-2} d^{-1} is estimated for lamina of rice (Pedersen et al. 2009). The C-balances of submerged lamina of four other wetland species can be estimated also at 200 mmol CO_2 m^{-3} and ranged from approximately 35 to 70 mmol C m^{-2} d^{-1}; calculated from Colmer and Pedersen (2008b). Thus, although P_N under water is substantially less than in air, the lamina C-balance appears to be positive.

The importance of photosynthesis during submergence is further highlighted by enhanced plant survival when light is provided, e.g. rice (Adkins et al. 1990; Das et al. 2009). Light also enhances survival during submergence of other wetland species (Vervuren et al. 1999, 2003; Mommer et al. 2006a). Similarly, survival of *Arabidopsis thaliana* was improved two- to threefold by light (16 h d^{-1}) as compared with in continuous darkness (Vashisht et al. 2011). Some shading studies have found that survival of submerged rice was highest under moderate levels of light (Adkins et al. 1990; Das et al. 2009). Adkins et al. (1990) highlighted that at high light extensive algal growth results in competition with rice for light and CO_2 during the day and for O_2 during the night.

In some situations, underwater photosynthesis will be CO_2 limited, rather than limited by light (Mommer and Visser 2005). For submerged terrestrial wetland plants, CO_2 limitation is severe when near air-equilibrium (approximately 10–15 mmol CO_2 m^{-3}). Underwater photosynthesis only becomes CO_2 saturated at 100-fold concentrations higher than air-equilibrium in *Phragmites australis* (Colmer and Pedersen 2008b) and at 20-fold higher concentrations for rice (Pedersen et al. 2009). As discussed above, CO_2 enrichment above air-equilibrium is common in many water bodies, including in flooded rice fields in Thailand at 20–180-fold air-equilibrium (Setter et al. 1987) and in India at 31–217-fold (Ram et al. 1999). The beneficial effects of higher dissolved CO_2 on submergence tolerance (growth and/or survival) have been documented for rice (Setter et al. 1989) and for *Hordeum marinum* (Pedersen et al. 2010). In the case of submerged rice, CO_2 enrichment to approximately 290 mmol m^{-3} enhanced by twofold the growth of two cultivars, compared with water at air-equilibrium which would have contained approximately 10 mmol CO_2 m^{-3}, at 30 °C (Setter et al. 1989). Thus, CO_2 levels in the floodwaters determine rates of underwater P_N with consequences for tissue sugar levels, e.g. *Hordeum marinum* (Pedersen et al. 2010), O_2 supply to roots, e.g. *Eriophorum angustifolium* (Gaynard and Armstrong 1987), growth and ultimately survival, e.g. rice and *Hordeum marinum* (Setter et al. 1989; Pedersen et al. 2010). Hence, Pedersen et al. (2010) suggested that future assessments of submergence tolerance in plants should be conducted at defined CO_2 levels and controlled environment treatments might require CO_2 enrichment of the water to better reflect many field environments.

4 Leaf Gas Films and Underwater Photosynthesis

Leaf surface hydrophobicity (i.e. water repellence) is a feature that sheds off water in wet aerial environments (Smith and McClean 1989; Brewer and Smith 1997) and promotes 'self cleansing', enhancing leaf performance and reputably lowering susceptibility to pathogens (Neinhuis and Barthlott 1997). Several terrestrial wetland plants possess superhydrophobic leaves that retain a thin gas film when submerged (Fig. 1b, c) (Raskin and Kende 1983; Colmer and Pedersen 2008b). The leaf gas films have been shown to enhance underwater gas exchange (CO_2 and O_2 in light and O_2 in darkness), functioning as a 'physical gill' similar to those known for aquatic insects and spiders (Thorpe and Crisp 1947; Raven 2008; Pedersen and Colmer 2012). CO_2 that enters the gas film can rapidly diffuse to stomata. By contrast, for leaves without gas films, a major proportion of the CO_2 and O_2 entry might transverse the cuticle (Mommer et al. 2004). O_2 (Frost-Christensen et al. 2003) and CO_2 (Frost-Christensen and Floto 2007) both can, albeit relatively slowly, permeate cuticles of amphibious plants.

The beneficial effect of leaf gas films on underwater P_N was not only demonstrated by the marked decreases of P_N when the gas films were removed (Fig. 2), but also leaves with gas films had higher rates of underwater P_N than leaves of species without gas films (Colmer and Pedersen 2008b; Colmer et al. 2011). At dissolved CO_2 concentrations of relevance to field conditions, underwater P_N was enhanced four- to fivefold by gas films on leaves of rice (Fig. 2). When gas films were removed artificially from leaves of completely submerged rice, tissue sugar levels declined (Pedersen et al. 2009; Winkel et al. 2013). Thus, leaf gas films enhance underwater P_N and submergence tolerance.

5 Sources of O_2 in the Light

Several studies have documented that terrestrial plants survive complete submergence better in natural light–dark-cycles compared with complete darkness (Vervuren et al. 2003; Mommer et al. 2007; Vashisht et al. 2011). The enhanced survival presumably results from a combination of better internal aeration (O_2 produced in photosynthesis) as well as the sugars produced.

For submerged rice (Fig. 3), root aeration is greatly enhanced already soon after sunrise where adventitious root pO_2 increases from <1 to >10 kPa in less than 2 h. The data also show that root pO_2 continues to be a function of incoming light throughout the day (transient reductions in light are followed by transient steep

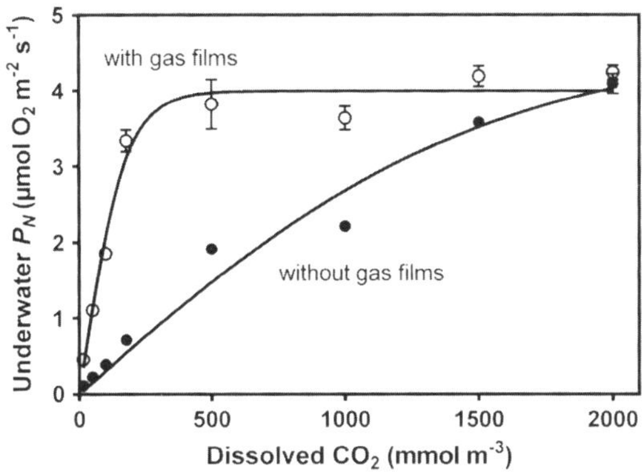

Fig. 2 Underwater net photosynthesis (P_N) *versus* dissolved CO_2 for rice leaf segments with or without gas films. Underwater P_N was measured as net O_2 evolution from leaf segments of approximately 2 cm^2 in closed glass vials with a range of dissolved CO_2 concentrations and PAR of 350 µmol photons m^{-2} s^{-1} (for methods, see Pedersen et al. 2013). Leaf gas films were either intact or removed experimentally by brushing with a dilute detergent. Reproduced from Pedersen et al. (2009)

declines in root pO_2) and at dusk, root pO_2 declines rapidly to predawn levels (Fig. 3). Similar relationships between light and internal aeration have been shown in several other in situ field studies both of aquatic plants (e.g. Greve et al. 2003; Borum et al. 2005; Sand-Jensen et al. 2005; Holmer et al. 2009; Pedersen et al. 2011a; Rich et al. 2013) and completely submerged terrestrial wetland plants (Pedersen et al. 2006; Winkel et al. 2011). The relationship between incoming light (and thus underwater P_N) is reinforced by the data analyses in Fig. 4a where root pO_2 is graphed against light. This example taken from the field study by Winkel et al. (2013) demonstrates the strong dependence on light for root aeration of completely submerged rice.

Transient peaks in tissue pO_2 at dawn can occur in submerged terrestrial wetland plants. Such peaks in tissue pO_2 are likely to arise from enhanced initial P_N fuelled by accumulated respiratory CO_2 following a dark period with net respiration (Waters et al. 1989; Colmer and Pedersen 2008a). In the case of rice, these peaks occurred in artificial dark-light switches in laboratory experiments (Waters et al. 1989; Colmer and Pedersen 2008a) but were not observed in field recordings of submerged rice in which O_2 increased markedly but without a transient peak (Winkel et al. 2013). Interestingly, a large transient peak was observed in situ for a stem-succulent halophyte (*Tecticornia pergranulata*) submerged in a salt lake even though the increase in morning light was more gradual than the sudden switches in laboratory experiments. The succulent tissues with impeded gas exchange with the surrounding floodwater might have contributed to this dawn peak in *T. pergranulata* (Pedersen et al. 2006). Regardless of species and environmental conditions, any peaks in pO_2 are only transient as the underwater P_N soon becomes CO_2 limited as determined by the rate of CO_2 entry from the floodwater, so that tissue pO_2 decreases to a new *quasi* steady-state.

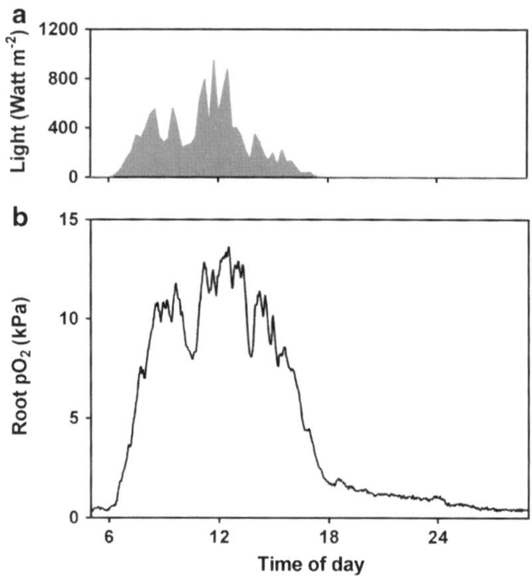

Fig. 3 Incident light (**a**) and root pO_2 of completely submerged rice (**b**) measured in situ in a field experiment (Fig. 1). An O_2 microelectrode was inserted into the adventitious root approximately 10 mm below the root–shoot junction which was about 50 mm below the soil surface. Data extracted from Winkel et al. (2013) for the first full day of submergence

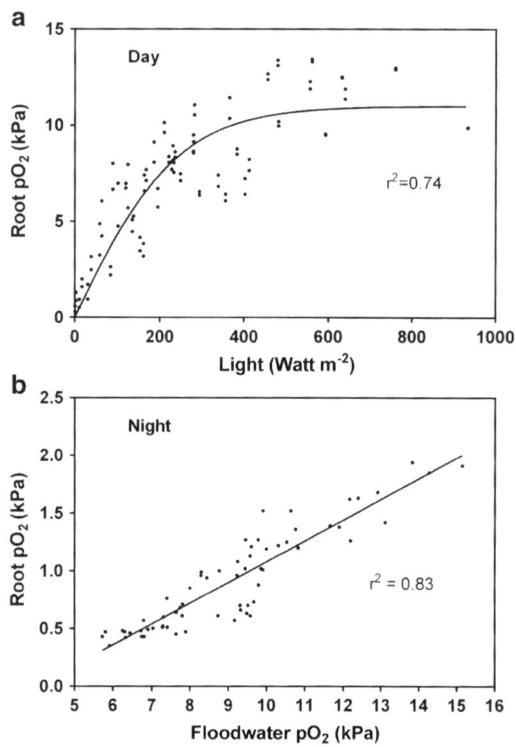

Fig. 4 Root pO_2 as a function of incident light during the day (**a**) and of floodwater pO_2 during the night (**b**) for rice when completely submerged in a field situation (Fig. 1). Data shown are extracted from Fig. 3. Data reproduced from Winkel et al. (2013)

The implications of dynamics in tissue pO_2 are of interest to consider further. Day-night dynamics have been shown to have consequences for energy metabolism. During light periods, roots of completely submerged rice had adequate energy for root extension when photosynthetically derived O_2 reached root tips, whereas during dark periods O_2 declined, root extension ceased and ethanolic fermentation occurred (Waters et al. 1989). Increased night-time activity of pyruvate decarboxylase in roots of submerged rice (Mohanty and Ong 2003) supports the earlier observations of ethanol production during dark periods (Waters et al. 1989) and presumably contributes to survival of anoxia (Gibbs and Greenway 2003). An area requiring study is whether the rapid morning increases in tissue pO_2 result in any oxidative stress, this being an intriguing possibility since emphasis has been placed on increased reactive oxygen species (ROS) following O_2 re-entry upon de-submergence, e.g. for rice (Ushimaru et al. 1994; Santosa et al. 2007). In addition, damage from ROS in hypoxic tissues of submerged rice might also occur (Santosa et al. 2007), and the possible aggravation of such processes by fluctuating pO_2 in submerged rice should be evaluated. The dynamics in pO_2 within tissues of submerged plants would also presumably influence how putative O_2-sensing (Bailey-Serres et al. 2012) might contribute to acclimation of plants during submergence.

6 Sources of O_2 in the Dark

During night-time, completely submerged plants rely on an influx of O_2 from the surrounding floodwater to sustain aerobic respiration in shoots and roots. The O_2 initially present in the aerenchyma at nightfall has been shown to be insufficient to sustain respiration; in *Zostera marina*, the O_2 in the aerenchyma can only sustain respiration for 8–13 min (Sand-Jensen et al. 2005). Figure 4b shows that internal root aeration of submerged rice during the night relies on floodwater pO_2 diffusing into the shoot and further down to the roots. Extrapolation of the regression line in Fig. 4b to the intercept on the x-axis suggests that even at this position measured only 10 mm from the root–shoot junction, anoxia would occur if floodwater pO_2 had decreased below approximately 4.5 kPa. However, root pO_2 will depend on, in addition to floodwater pO_2, other environmental and plant factors. Firstly, the second key environmental parameter after floodwater pO_2 is mixing of the water; flow/turbulence results in erosion of the DBLs around the leaves and thereby a higher flux of O_2 into the shoot at the same external bulk floodwater pO_2 (Binzer et al. 2005). Important plant factors include (1) internal resistance to longitudinal O_2 diffusion, determined by tissue porosity and the diffusion path-length (Armstrong 1979), (2) respiration (also influenced by temperature) (Armstrong 1979), (3) radial O_2 loss along the diffusion pathway (from roots but potentially also the buried sheath bases (both dependent upon sediment demand for O_2 and whether roots possess a barrier to ROL (Armstrong 1979; Colmer 2003; Pedersen et al. 2011a) and (4) the shoot-to-root ratio, i.e. capacity of shoot uptake to satisfy the sink demand in roots (and rhizomes if present) (Borum et al. 2006).

Leaf gas films have been shown to enhance O_2 uptake from floodwater when in darkness. Laboratory experiments with completely submerged rice showed that *quasi* steady-state root pO_2 was approximately 3.4 kPa (10–15 mm behind the root tip) and declined essentially to zero upon removal of the leaf gas films (Pedersen et al. 2009). Similarly, in situ measurements of rhizome pO_2 in the tidal halophyte, *Spartina anglica*, showed that during tidal inundation at night-time pO_2 remained higher in plants with gas films (5–7 kPa) than in those where the gas films had been removed (approximately 1 kPa) (Winkel et al. 2011).

7 Outlook

Underwater photosynthesis and internal aeration are crucial to plant survival of submergence. Studies are few of underwater photosynthesis and O_2 dynamics in completely submerged terrestrial plants, whereas mechanisms of internal aeration of below-ground organs are well understood. Underwater photosynthesis leads to dynamic changes in pO_2 within submerged plants as light availability changes during the daytime and photosynthesis ceases each night. Plants must cope with these large internal fluctuations in O_2 and especially root tissues will have low pO_2 (or even be anoxic) during nights and a resupply of O_2 in the daytime. Finally, improved knowledge on C budgets of submerged terrestrial plants is needed to provide a more complete understanding of the contribution of underwater photosynthesis beyond the benefits to submerged plants of improved O_2 status. Flooding events are predicted to increase in the future, so understanding of submergence tolerance should aid efforts aimed at breeding of rice to better withstand floods.

References

Adkins SW, Shiraishi T, McComb JA (1990) Submergence tolerance of rice—a new glasshouse method for the experimental submergence of plants. Physiol Plant 80:642–646

Armstrong W (1979) Aeration in higher plants. Adv Bot Res 7:225–332

Bailey-Serres J, Fukao T, Gibbs DJ, Holdsworth MJ, Lee SC, Licausi F, Perata P, Voesenek LACJ, van Dongen JT (2012) Making sense of low oxygen sensing. Trends Plant Sci 17:129–138

Binzer T, Borum J, Pedersen O (2005) Flow velocity affects internal oxygen conditions in the seagrass *Cymodocea nodosa*. Aquat Bot 83:239–247

Black MA, Maberly SC, Spence DHN (1981) Resistances to carbon dioxide fixation in four submerged freshwater macrophytes. New Phytol 89:557–568

Borum J, Pedersen O, Greve TM, Frankovich TA, Zieman JC, Fourqurean JW, Madden CJ (2005) The potential role of plant oxygen and sulphide dynamics in die-off events of the tropical seagrass, *Thalassia testudinum*. J Ecol 93:148–158

Borum J, Sand-Jensen K, Binzer T, Pedersen O, Greve TM (2006) Oxygen movement in seagrasses. In: Larkum AWD, Duarte CM (eds) Seagrasses: biology, ecology and conservation. Springer, Dordrecht, pp 255–270

Brewer CA, Smith WK (1997) Patterns of leaf surface wetness for montane and subalpine plants. Plant Cell Environ 20:1–11
Colmer TD (2003) Long-distance transport of gases in plants: a perspective on internal aeration and radial oxygen loss from roots. Plant Cell Environ 26:17–36
Colmer TD, Pedersen O (2008a) Oxygen dynamics in submerged rice (*Oryza sativa*). New Phytol 178:326–334
Colmer TD, Pedersen O (2008b) Underwater photosynthesis and respiration in leaves of submerged wetland plants: gas films improve CO_2 and O_2 exchange. New Phytol 177:918–926
Colmer TD, Winkel A, Pedersen O (2011) A perspective on underwater photosynthesis in submerged terrestrial wetland plants. AoB Plants 2011:plr030
Das KK, Panda D, Sarkar RK, Reddy JN, Ismail AM (2009) Submergence tolerance in relation to variable floodwater conditions in rice. Environ Exp Bot 66:425–434
Frost-Christensen H, Floto F (2007) Resistance to CO_2 diffusion in cuticular membranes of amphibious plants and the implication for CO_2 acquisition. Plant Cell Environ 30:12–18
Frost-Christensen H, Jørgensen LB, Floto F (2003) Species specificity of resistance to oxygen diffusion in thin cuticular membranes from amphibious plants. Plant Cell Environ 26:561–569
Gaynard TJ, Armstrong W (1987) Some aspects of internal plant aeration in amphibious habitats. In: Crawford RMM (ed) Plant life in aquatic and amphibious habitats. Blackwell Scientific Publishers, Oxford, p 452
Gibbs J, Greenway H (2003) Mechanisms of anoxia tolerance in plants. I. Growth, survival and anaerobic catabolism. Funct Plant Biol 30:1–47
Greve TM, Borum J, Pedersen O (2003) Meristematic oxygen variability in eelgrass (*Zostera marina*). Limnol Oceanogr 48:210–216
Holmer M, Pedersen O, Krause-Jensen D, Olesen B, Petersen MH, Schopmeyer S, Koch M, Lomstein BA, Jensen HS (2009) Sulfide intrusion in the tropical seagrasses *Thalassia testudinum* and *Syringodium filiforme*. Estuar Coast Shelf Sci 85:319–326
Keeley JE (1998) CAM photosynthesis in submerged aquatic plants. Bot Rev 64:121–175
Keeley JE (1999) Photosynthetic pathway diversity in a seasonal pool community. Funct Ecol 13:106–118
Kirk JTO (1994) Light and photosynthesis in aquatic ecosystems. Cambridge University Press, New York
Maberly SC, Madsen TV (2002) Freshwater angiosperm carbon concentrating mechanisms: processes and patterns. Funct Plant Biol 29:393–405
Madsen TV, Sand-Jensen K (1991) Photosynthetic carbon assimilation in aquatic macrophytes. Aquat Bot 41:5–40
Magnin NC, Cooley BA, Reiskind JB, Bowes G (1997) Regulation and localization of key enzymes during the induction of Kranz-less, C_4-type photosynthesis in *Hydrilla verticillata*. Plant Physiol 115:1681–1689
Mohanty B, Ong BL (2003) Contrasting effects of submergence in light and dark on pyruvate decarboxylase activity in roots of rice lines differing in submergence tolerance. Ann Bot 91: 291–300
Mommer L, Visser EJW (2005) Underwater photosynthesis in flooded terrestrial plants: a matter of leaf plasticity. Ann Bot 96:581–589
Mommer L, Pedersen O, Visser EJW (2004) Acclimation of a terrestrial plant to submergence facilitates gas exchange under water. Plant Cell Environ 27:1281–1287
Mommer L, Pons TL, Wolters-Arts M, Venema JH, Visser EJW (2005) Submergence-induced morphological, anatomical, and biochemical responses in a terrestrial species affect gas diffusion resistance and photosynthetic performance. Plant Physiol 139:497–508
Mommer L, Lenssen JPM, Huber H, Visser EJW, De Kroon H (2006a) Ecophysiological determinants of plant performance under flooding: a comparative study of seven plant families. J Ecol 94:1117–1129
Mommer L, Pons TL, Visser EJW (2006b) Photosynthetic consequences of phenotypic plasticity in response to submergence: *Rumex palustris* as a case study. J Exp Bot 57:283–290

Mommer L, Wolters-Arts M, Andersen C, Visser EJW, Pedersen O (2007) Submergence-induced leaf acclimation in terrestrial species varying in flooding tolerance. New Phytol 176:337–345
Neinhuis C, Barthlott W (1997) Characterization and distribution of water-repelent, self-cleaning plant surfaces. Ann Bot 79:667–677
Nielsen SL (1993) A comparison of aerial and submerged photosynthesis in some Danish amphibious plants. Aquat Bot 45:27–40
Pedersen O, Colmer TD (2012) Physical gills prevent drowning of many wetland insects, spiders and plants. J Exp Biol 215:705–709
Pedersen O, Sand-Jensen K (1992) Adaptations of submerged *Lobelia dortmanna* to aerial life form—morphology, carbon sources and oxygen dynamics. Oikos 65:89–96
Pedersen O, Vos H, Colmer TD (2006) Oxygen dynamics during submergence in the halophytic stem succulent *Halosarcia pergranulata*. Plant Cell Environ 29:1388–1399
Pedersen O, Rich SM, Colmer TD (2009) Surviving floods: leaf gas films improve O_2 and CO_2 exchange, root aeration, and growth of completely submerged rice. Plant J 58:147–156
Pedersen O, Malik AI, Colmer TD (2010) Submergence tolerance in *Hordeum marinum*: dissolved CO_2 determines underwater photosynthesis and growth. Funct Plant Biol 37:524–531
Pedersen O, Pulido C, Rich SM, Colmer TD (2011a) *In situ* O_2 dynamics in submerged *Isoetes australis*: varied leaf gas permeability influences underwater photosynthesis and internal O_2. J Exp Bot 62:4691–4700
Pedersen O, Rich SM, Pulido C, Cawthray GR, Colmer TD (2011b) Crassulacean acid metabolism enhances underwater photosynthesis and diminishes photorespiration in the aquatic plant *Isoetes australis*. New Phytol 190:332–339
Pedersen O, Colmer TD, Sand-Jensen K (2013) Underwater photosynthesis of submerged plants—recent advances and methods. Front Plant Sci 4:140
Prins HBA, Elzenga JTM (1989) Bicarbonate utilization: function and mechanism. Aquat Bot 34: 59–83
Ram PC, Singh AK, Singh BB, Singh VK, Singh HP, Setter TL, Singh VP, Singh RK (1999) Environmental characterization of floodwater in eastern India: relevance to submergence tolerance of lowland rice. Exp Agric 35:141–152
Raskin I, Kende H (1983) How does deep water rice solve its aeration problem? Plant Physiol 72: 447–454
Raven JA (2008) Not drowning but photosynthesizing: probing plant plastrons. New Phytol 177: 841–845
Raven JA, Giordano M, Beardall J (2008) Insights into the evolution of CCMs from comparisons with other resource acquisition and assimilation. Physiol Plant 133:4–14
Rich SM, Pedersen O, Ludwig M, Colmer TD (2013) Shoot atmospheric contact is of little importance to aeration of deeper portions of the wetland plant *Meionectes brownii*; submerged organs mainly acquire O_2 from the water column or produce it endogenously in underwater photosynthesis. Plant Cell Environ 36:213–223
Sand-Jensen K, Pedersen MF, Nielsen SL (1992) Photosynthetic use of inorganic carbon among primary and secondary water plants in streams. Freshw Biol 27:283–293
Sand-Jensen K, Pedersen O, Binzer T, Borum J (2005) Contrasting oxygen dynamics in the freshwater isoetid *Lobelia dortmanna* and the marine seagrass *Zostera marina*. Ann Bot 96: 613–623
Santosa IE, Ram PC, Boamfa EI, Laarhoven LJJ, Reuss J, Jackson MB, Harren FJM (2007) Patterns of peroxidative ethane emission from submerged rice seedlings indicate that damage from reactive oxygen species takes place during submergence and is not necessarily a post-anoxic phenomenon. Planta 226:193–202
Sculthorpe CD (1967) The biology of aquatic vascular plants. Edward Arnold Ltd., London
Setter TL, Kupkanchanakul T, Kupkanchanakul K, Bhekasut P, Wiengweera A, Greenway H (1987) Concentrations of CO_2 and O_2 in floodwater and in internodal lacunae of floating rice growing at 1–2 metre water depths. Plant Cell Environ 10:767–776

Setter TL, Waters I, Wallace I, Bekhasut P, Greenway H (1989) Submergence of rice. I. Growth and photosynthetic response to CO_2 enrichment of floodwater. Aust J Plant Physiol 16:251–263
Smith WK, McClean TM (1989) Adaptive relationship between leaf water repellency, stomatal distribution, and gas exchange. Am J Bot 76:465–469
Stumm W, Morgan JJ (1996) Aquatic chemistry. Wiley, New York
Thorpe WH, Crisp DJ (1947) Studies on plastron respiration; the respiratory efficiency of the plastron in *Aphelocheirus*. J Exp Biol 24:270–303
Ushimaru T, Shibasaka M, Tsuji H (1994) Resistance to oxidative injury in submerged rice seedlings after exposure to air. Plant Cell Physiol 35:211–218
Vashisht D, Hesselink A, Pierik R, Ammerlaan JMH, Bailey-Serres J, Visser EJW, Pedersen O, van Zanten M, Vreugdenhil D, Jamar DCL, Voesenek LACJ, Sasidharan R (2011) Natural variation of submergence tolerance among *Arabidopsis thaliana* accessions. New Phytol 190: 299–310
Vervuren PJA, Beurskens SMJH, Blom CWPM (1999) Light acclimation, CO_2 response and long-term capacity of underwater photosynthesis in three terrestrial plant species. Plant Cell Environ 22:959–968
Vervuren PJA, Blom CWPM, De Kroon H (2003) Extreme flooding events on the Rhine and the survival and distribution of riparian plant species. J Ecol 91:135–146
Vogel S (1994) Life in moving fluids: the physical biology of flow. Princeton University Press, Princeton
Waters I, Armstrong W, Thompson C, Setter T, Adkins S, Gibbs J, Greenway H (1989) Diurnal changes in radial oxygen loss and ethanol metabolims in roots of submerged and non-submerged rice seedlings. New Phytol 113:439–451
Winkel A, Colmer TD, Pedersen O (2011) Leaf gas films of *Spartina anglica* enhance rhizome and root oxygen during tidal submergence. Plant Cell Environ 34:2083–2092
Winkel A, Colmer TD, Ismail AM, Pedersen O (2013) Internal aeration of paddy field rice (*Oryza sativa* L.) during complete submergence—importance of light and floodwater O_2. New Phytol 197:1193–1203

Different Survival Strategies Amongst Plants to Cope with Underwater Conditions

Hans van Veen, Divya Vashisht, Laurentius A.C.J. Voesenek, and Rashmi Sasidharan

Abstract Many plants experience flooding at some point during their life cycle. The underwater environment creates a carbon and energy crisis for the plant, for which two successful strategies have been identified, quiescence and escape. During quiescence, growth is actively reduced until the water levels recede, whereas escape encompasses rapid upward shoot elongation to establish air contact. An inherent cost is associated with flood-induced elongation, which is also reflected by the difference in managing energy production and expenditure compared to plants adopting a quiescence strategy.

The underwater elongation, via a combination of cell elongation and division, is mainly driven by changes in the internal gaseous composition of ethylene, carbon dioxide, and oxygen. Interestingly, the same internal and environmental cues induce contrasting growth responses, depending on the species. The underlying hormonal network and molecular components constituting these differences amongst wetland species are further discussed.

1 Growth Strategies Determine Plant Survival and the Occupied Niche

Terrestrial plant species face many challenges in an underwater environment. Primarily, the severely reduced gas exchange in the aqueous surroundings, ~10.000 times slower compared to air, can lead to a dramatic oxygen (O_2) shortage in the belowground tissue. In the above ground tissue photosynthesis is severely reduced due to the poor exchange of carbon dioxide (CO_2) with the environment (Mommer and Visser 2005). In the absence of O_2 mitochondrial respiration is

H. van Veen (✉) • D. Vashisht • L.A.C.J. Voesenek • R. Sasidharan
Plant Ecophysiology, Institute of Environmental Biology, Utrecht University, Padualaan 8, 3584 CH Utrecht, The Netherlands
e-mail: H.vanVeen@uu.nl

J.T. van Dongen and F. Licausi (eds.), *Low-Oxygen Stress in Plants*, Plant Cell Monographs 21, DOI 10.1007/978-3-7091-1254-0_17, © Springer-Verlag Wien 2014

arrested. Therefore, plants have to solely rely on glycolysis for energy production, leading to a mere two to four ATP molecules produced per molecule hexose. It is essential that fermentation pathways are activated to regenerate NAD^+ to sustain the energy production through glycolysis.

Interestingly, a large range of species experience partial to complete submergence during some stage of their life cycle. Indeed, a shift from an exclusive terrestrial to an amphibious lifestyle is estimated to have occurred more than 200 times throughout the diversification of angiosperms (Cook 1999). A common response to submergence is the formation of new "aquatic" leaves, which have an increased specific leaf area (SLA), thinner epidermal cell walls, and increased chloroplast orientation towards the leaf surface. These traits minimize the diffusion distance and thus resistance between the chloroplast and environment and thereby improve gas exchange underwater. This leaf plasticity is a widespread response that somewhat alleviates adverse underwater conditions (Mommer et al. 2005, 2007; Sand-Jensen and Frost-Christensen 1999). Apart from this general response, two successful strategies to cope with complete submergence are quiescence and escape (Bailey-Serres and Voesenek 2008).

The quiescence strategy aims to reduce energy expenditure by ceasing non-essential processes. Furthermore, growth and development of new tissues are reduced. In this way quiescent plants minimize oxygen consumption and preserve valuable carbohydrates for regrowth once the water levels recede. Indeed, species that successfully achieve a quiescent state can survive prolonged periods of flooding and associated hypoxia, like for instance *Phalaris arundinacea* which has been reported to survive complete submergence for more than 100 days (Mommer et al. 2006) and *Acorus calamus* where both the rhizomes and leaves are remarkably anoxia-tolerant (Schlüter and Crawford 2001). Furthermore, rice cultivars lacking underwater growth have significantly improved survival during complete submergence compared to varieties that continue elongation growth (Setter and Laureles 1996; Ram et al. 2002).

Plants employing an escape strategy upon submergence show a rapid upward elongation of shoot tissue towards the water surface. The subsequently established air contact allows for resumption of aerial photosynthesis and re-aeration of underwater organs via aerenchyma tissue. Rhizomes of some wetland species also utilize an elongation response to escape from anoxic sediments (Summers and Jackson 1994).

Major costs are associated with shoot elongation. This was neatly demonstrated in quiescent and escaping rice varieties by pharmaceutical manipulation of hormone levels leading to the artificial activation or prevention of elongation growth (Setter and Laureles 1996). Indeed, quiescent plants in which elongation was stimulated had an increased mortality under complete submergence. Similarly, the escaping plants in which elongation was inhibited had an increased survival.

However, when elongation leads to successful aerial contact it greatly improves plant survival and reduces biomass loss, as has been demonstrated for *Rorippa amphibia* (Akman et al. 2012) and *Rumex palustris* (Pierik et al. 2009). Interestingly, establishment of air contact is by itself not always enough to improve plant performance. When *Rumex palustris*, which employs an escape strategy, regains aerial contact there is a significant increase in biomass. In contrast, when the close relative *Rumex acetosa*, which lacks the escape response, was artificially allowed to regain air contact, no such increase in biomass was observed. A key difference between both species is the petiole porosity, which is more than double in *R. palustris*, indicating that improved aeration is indeed a crucial aspect of the escape strategy (Pierik et al. 2009).

The nature of flooding events can differ dramatically, depending on the water supply, site elevation, water runoff and the soil drainage capabilities. This variation in flooding characteristics is the basis for a large amount of plant diversity in wetland areas (Silvertown et al. 1999). Indeed, a strong relationship exists between the flooding tolerance of a species and the frequency of flooding events (Van Eck et al. 2004). Given that the escape strategy is only beneficial if aerial contact can be made, one would expect this trait to be most prevalent on sites experiencing prolonged and relatively shallow floods. On the other hand, the quiescent strategy would be expected on sites subjected to deep or very short floods. Indeed, a study of the species distribution and growth responses on the river Rhine floodplains demonstrated this relationship (Voesenek et al. 2004). Thus fine-tuning of the adaptation strategy to the environment is crucial for plant survival.

2 Growth Management: Balancing Energy Generation and Consumption

A consequence of adopting the quiescence strategy is that the plant still suffers from reduced photosynthesis and reduced oxygen levels. The belowground tissue rapidly experiences hypoxia upon flooding, with high levels of fermentation. The aboveground tissue usually retains oxygen levels ranging from 3 to 22 % depending on plant morphology, photosynthetic capacities, and light conditions (Mommer et al. 2007). It is therefore essential for plants that adopt the quiescence strategy to keep energy production and consumption to a minimum to conserve valuable carbohydrates. In contrast, escaping plants require rapid mobilization of reserves to fuel expensive elongation growth.

2.1 Catabolism to Support Growth and Maintenance

The reduced photosynthesis observed underwater means that most species will depend on reserve carbohydrates to supply energy for cellular maintenance. Indeed, amylases, glycolytic, and other catabolic enzymes are invariably activated in various plant species upon exposure to flooding and low oxygen stress (Arpagaus and Braendle 2000; Harada and Ishizawa 2003; Mustroph and Albrecht 2003; Umeda and Uchimiya 1994). Interestingly, rice cultivars varying in the extent of underwater elongation showed a strong correlation between carbohydrate consumption and elongation growth (Das et al. 2005; Fukao et al. 2006). Furthermore, a mutant rice line that is unable to activate alpha-amylase failed to elongate underwater (Lee et al. 2009), Similarly, *Rumex palustris* lost its ability to elongate underwater when depleted of starch (Groeneveld and Voesenek 2003).

These studies experimentally demonstrate the importance of reserve mobilization in fuelling elongation. Indeed, species that show an escape response under anoxic conditions display a characteristic Pasteur effect. The anoxia leads to a halt of respiratory energy production and the plants are thus dependent on the glycolysis coupled with fermentation to fuel elongation growth. The lower ATP yield per molecule hexose is compensated for by increasing the rate of glycolysis, the so-called Pasteur effect. The Pasteur effect has proven surprisingly efficient in generating energy for elongation in rhizomes rich in starch substrate. A study on pondweed, arrowhead, and rice shows that due to the Pasteur effect the adenylate energy status remains unchanged for rice and is even improved in pondweed and arrowhead (Ishizawa et al. 1999; Summers et al. 2000). Furthermore, a strong correlation between ethanol production and elongation is observed during the anoxic germination of coleoptiles from various cultivars (Magneschi et al. 2009a).

Mobilizing carbohydrate resources is widespread in both flooded terrestrial plants and anoxic rhizomes. For escaping plants, the adverse conditions are only of temporary nature until aerial contact is made. However, during a quiescent strategy adverse conditions are not ameliorated. In this case, accessing reserve carbohydrates and energy production should preferably be kept to a minimum. The extent of reduced energy production in quiescent plants has not been studied in great detail. Nevertheless, long-term anoxia in rice coleoptiles is found to decrease fermentation rate despite sufficient substrate availability without adverse effects on survival (Colmer et al. 2001; Zhang and Greenway 1994). This does suggest a possibility for delicate control of metabolism and some potential mechanisms have been identified.

It has been proposed that plants can lower their respiration upon reduced oxygen availability, despite oxygen levels not dropping below the critical oxygen pressure for mitochondrial cytochrome C oxidase (Geigenberger 2003; Gupta et al. 2009), discussed elsewhere in this book by Paepke et al. (2013). The advantages of respiratory downregulation are potentially twofold. Firstly, decreased consumption means that tissue oxygen content is subsequently maintained at a higher level and possibly preventing ineffective fermentation-dependent ATP production. Secondly, the energy production and carbon consumption are reduced allowing reserves to last longer, subsequently prolonging survival. The exact mechanisms of the respiratory regulation are still unclear. Nevertheless, a recent study suggests that pyruvate and TCA cycle intermediates are important determinants of respiration (Zabalza et al. 2008). The notion of oxygen-dependent regulation of respiration has been opposed, as a mathematical model using Michaelis–Menten respiration kinetics and laws of gas diffusion correlates with experimental data (Armstrong and

Beckett 2011; Armstrong et al. 2009). However, so far experimental and modeling evidence has remained inconclusive, and further work is required to assess whether species adopting a quiescence strategy actively reduce respiration rates.

2.2 Conserving Reserves via Efficient Energy Usage

The severe carbon and energy limitation upon submergences forces a prioritization of the most essential processes. A study of the energy allocation patterns in germinating rice coleoptiles revealed different processes that consume ATP upon high- or low oxygen conditions (Edwards et al. 2012). While the anoxic coleoptiles spent all their available ATP on the synthesis of the four main polymers (proteins, cell wall components, lipids, and nucleic acids) and potassium uptake, under normoxic conditions only 35 % of the available ATP was allocated to these essential processes.

In absolute terms every aspect of biosynthesis was decreased under anoxia. However, relative to the total amount spent on essential processes, ATP was preferentially allocated towards lipid and protein synthesis. Interestingly, a comparison of rice cultivars varying in elongation showed a strong correlation between anoxia-induced escape, ATP production, and rate of protein synthesis (Edwards et al. 2012). Similarly, increased protein biosynthesis rates were also observed during the anoxia-induced escape response in *Sagittaria pygmaea*, arrowhead (Ishizawa et al. 1999). However, in seedling maize root tips, with reduced elongation, new protein biosynthetic rates were very low (Chang et al. 2000). Furthermore, in the model plant *Arabidopsis*, which lacks hypoxia-induced escape responses, there was a selective and significant repression (70 %) of mRNA translation (Branco-Price et al. 2008; Sachs et al. 1980). Thus the repression of protein synthesis observed in maize and *Arabidopsis* suggests a major energy saving feature during the quiescence strategy, while the continued protein synthesis in rice and arrowhead, despite being energetically expensive, is essential for sustaining the escape response.

Studies on *Arabidopsis* have identified a protein kinase complex, termed SnRK1, that plays a key role in regulating the transcriptome of more than a 1,000 genes involved in metabolism (Baena-González et al. 2007). SnRK1 is activated upon low energy conditions, extended darkness, and hypoxia and alters the cellular priorities by reducing biosynthesis and increasing catabolism. Recently it was shown that Trehalose-6-Phosphate (T6P), an intermediate in the pathway of trehalose metabolism, could inhibit SnRK1 activity in *Arabidopsis* (Zhang et al. 2009). An across kingdom study on hypoxic trancriptomic responses revealed that the activation of T6P phosphatase, breaking down T6P, is common amongst the plant kingdom. This ultimately leads to reduced T6P and increased SnRK1 activity (Mustroph et al. 2010). With respect to the growth strategies, quiescent rice was also found to induce T6P phosphatase, whereas in the elongating variety this regulation was absent (Jung et al. 2010). Metabolic regulation via T6P could be an essential mechanism to match the energy production with the requirements of either quiescence or escape.

Ion transport and homeostasis is one of the largest energy sinks in plant cells and tissues (Pang and Shabala 2010). Not surprisingly, several studies report a dramatic drop in potassium uptake or even a potassium loss upon hypoxia, such as *Vitis* roots, barley roots, and rice coleoptiles (Colmer et al. 2001; Mugnai et al. 2011; Pang et al. 2007). Furthermore, a loss of membrane permeability was observed upon anoxia for rice coleoptiles and *Vitis* roots (Colmer et al. 2001; Mugnai et al. 2011), with a reduction of energy expenditure on potassium transport in anoxic rice coleoptiles (Colmer et al. 2001; Edwards et al. 2012).

Additionally, a cytoplasmic acidification is observed in tissues facing anoxia (Menegus et al. 1991; Roberts et al. 1984; Roberts et al. 1985), which could potentially act as an energy-conserving mechanism for transport processes. However, this cytoplasmic pH drop was initially identified as negative aspect of flooding stress, as several intolerant crop species were unable to limit or revert the pH drop, whereas rice was able to limit the extent of acidification. Additionally, a loss of membrane potential is also observed upon hypoxia (Zhang and Greenway 1995). Interestingly, hypoxia-associated soil toxins are able to induce similar changes in membrane potential in aerated roots (Pang et al. 2007). Currently, hypoxia-associated changes in pH and membrane potentials are seen as a representative of a new metabolic steady state that optimizes the cytoplasmic conditions for fermentative enzymes. Furthermore, the lower proton motive force and reduced transporter activity, that are characteristic of the new steady state, leads to reduced cellular energy costs (Felle 2005, 2010).

Another way to cope with hypoxia-induced ATP deficiency is the induction of alternative metabolic pathways, catalyzed by enzymes using pyrophosphate (PPi) as an energy donor instead of ATP (Huang et al. 2008). A detailed description by Mustroph et al. (2013) of these PPi utilizing enzymes in the context of hypoxia responses in plants can be found elsewhere in this book.

Several mechanisms for energy and carbon management have been identified in the plant kingdom. It is likely that many of these represent conserved mechanisms, like those involving SnRK1. These adaptations are likely to be most pronounced during the quiescence strategy, where there is no elongation that requires a lot of resources.

3 Cellular Mechanisms Determining Growth

In *Rumex palustris* underwater shoot elongation occurs primarily via rapid elongation of the leaf petiole. In existing petioles it is exclusively mediated by cell elongation and occurs throughout the entire length of this organ (Voesenek et al. 1990). On the other hand, the underwater shoot elongation in rice occurs in the stem internodal regions which can be divided into three zones (Bleecker et al. 1986). In the basal zone of the internode both cell division and elongation occur. Submergence mainly induces growth in this segment (Kutschera and Kende 1988) where division is stimulated independent of the increase in cell size (Lorbiecke and Sauter 1998). In the second zone cells show an increased elongation, whereas in the next region the cells differentiate. Here submergence also delays the rate of differentiation shifting the balance more towards elongation (Bleecker et al. 1986).

In *Nymphoides peltata* (fringed waterlily) the balance between division and elongation during underwater escape is age-dependent (Ridge and Amarasinghe 1984). The mature leaves show predominantly cell elongation of the petiole, whereas in young leaves the balance is shifted more towards cell division. The two main driving forces for growth, cell division and cell elongation, are further discussed below.

3.1 Cellular Expansion

Cell elongation is a biophysical process, where through osmotic-driven water uptake an outward-driven turgor pressure is applied on the cell wall. Depending on the extensibility of the cell wall it yields to the turgor pressure facilitating cell growth. Therefore, osmoregulation and cell wall modifications are the main targets of growth control.

The osmolarity of deepwater rice only marginally increases upon submergence, whereas big changes in cell wall extensibility of internodes are observed (Kutschera and Kende 1988). Indeed, several studies support the notion of cell enlargement via the regulation of cell wall extensibility (Sasidharan et al. 2011), whereas changes in turgor pressure rarely drive cell expansion (Cosgrove 2005). Indeed, expression of osmoregulatory aquaporins in deepwater rice was independent of hormonal growth cues and instead water status of the tissue was the dominant determining factor (Malz and Sauter 1999; Ridge and Osborne 1989).

The plant primary cell wall is composed of cellulose microfibrils that are interconnected by hemicellulose, a xyloglucan polymer, and surrounded by pectin (Sasidharan et al. 2011). Several enzymes can facilitate loosening of the cell wall by targeting the three main components, cellulose, hemicellulose, and pectin. Expansins loosen the connections between cellulose microfibrils and hemicellulose, whereas Xyloglucan Endotransglucosylase/Hydrolases (XTHs) can cleave or rejoin xyloglucan polymers (hemicellulose). Pectin methylesterases (PMEs) catalyze the esterification of the pectin polysaccharides and can thereby alter the fluidity of pectin polysaccharide matrix. As most of these enzymes have a low pH optimum, cell wall acidification is an essential aspect of cell wall loosening.

Submerged plants adopting an escape strategy strongly upregulate cell wall modifying activity in the rapidly elongating shoots. Indeed, upon submergence expansin expression and activity are increased in the elongating petioles of *R. palustris* (Vreeburg et al. 2005) and internodes of deepwater rice (Cho and Kende 1997). The non-elongating *Rumex acetosa* and rice varieties did not show a similar induction upon submergence (Fukao et al. 2006; Vriezen et al. 2000). Interestingly, such a correlation between expansin expression and elongation was absent in coleoptile elongation of anoxic rice (Magneschi et al. 2009b), suggesting alternative modes of regulating cell wall loosening. For instance, a role of XTHs was identified in the underwater elongation of *Sagittaria pygmaea* (Arrowhead) (Ookawara et al. 2005). PMEs have not yet been associated with the underwater escape response. However, these enzymes are involved in growth regulatory processes upon different environmental cues, and a role in the escape strategy therefore seems likely.

During underwater elongation cell wall acidification, mediated by the proton ATPase, occurs in a wide range of escaping species such *Rumex palustris*, *Nymphoides peltata*, and *Potamogeton distinctus* (Koizumi et al. 2011; Malone and Ridge 1983; Vreeburg et al. 2005). Acidification of the apoplastic space is a potent mechanism to activate already present cell wall modifying enzymes allowing a rapid response upon submergence. Interestingly, apoplastic alkalinization inhibits growth in root tissue (Staal et al. 2011). This inhibition can be relieved by the application of the fungal toxin fusicoccin that activates the proton-ATPase thereby acidifying the cell wall. Regulation of ATPase activity could therefore be instrumental in the regulation of cellular expansion and growth mediating the escape or quiescence strategy.

3.2 Cell Division

The increased cell division observed in some species requires proper control of cell cycle progression. However, the role of these processes in mediating underwater elongation has so far, received little attention. Nevertheless, some work on deepwater rice has identified a set of genes that are regulated at various cell cycle stages: namely the mitotic (M), gap1 (G1), DNA synthesis (S), and gap2 (G2) phase. Cell cycle progression is regulated by the sequential expression of cyclin-dependent kinases (CDKs). CDK activity can be modulated by binding of cyclins and requires phosphorylation by CDK-activating kinases (CAKs).

At the basal zone of deepwater rice internodes, where cell division takes place, a strong induction of two cyclins, which mediate G2 to M phase transition, was observed upon gibberellin (GA) treatment (Sauter et al. 1995). Because of its dominant role in mediating escape, GA is often used to mimic flooding as it also induces internode elongation. However, it is not clear if the cyclin induction was the result of GA or an indirect result of an increasing number of cells in the G2 phase compared to other mitotic states.

DNA synthesis (S phase) is clearly stimulated by submergence and GA, as was demonstrated by the cell cycle-independent induction of a *CDK* gene, *cdcOs-2* (Lorbiecke and Sauter 1998). Additionally, a cyclin involved with G1 to S transition, *cycA1;1*, was transcriptionally activated (Fabian et al. 2000). Also a CAK (R2), another cell cycle regulatory component was transcriptionally induced in a

GA-dependent manner (Sauter 1997). R2 accelerates the progression through the S phase, and enhanced growth was observed when R2 was overexpressed in rice cell suspensions (Fabian-Marwedel et al. 2002). To complement the changes in cell cycle progression, factors involved in DNA replication, such as histone H3 and RPA1, show concomitant regulation (van der Knaap et al. 1997; van der Knaap and Kende 1995).

In conclusion, a strong regulation of cell cycle progression is apparent during elongation of deepwater rice. The increased growth observed upon overexpression of R2 demonstrates their potential as potent growth regulators. Negative regulation of cell cycle control could be an alternative way of growth inhibition in species adopting a quiescence strategy, although there have so far been no investigations to support this hypothesis.

4 Regulation of Underwater Growth Strategies

Upon complete submergence quiescent and escaping species experience the same initial environmental and physiological cues. However, distinct signaling cascades are observed for different species and strategies that mediate either growth or energy homeostasis.

4.1 Passive Ethylene Accumulation Leads to a Hormonal Cascade

A range of amphibious plant species show a strong elongation response when supplied with exogenous ethylene. Indeed, this gaseous plant hormone has been suggested to be the driving force of underwater elongation (Ridge 1987). The balance between biosynthesis and outward diffusion determines the internal ethylene concentration, and both are affected by submergence. Methionine is the precursor of ethylene biosynthesis. After the formation of *S*-adenosyl-methionine (SAM), it is converted to 1-aminocyclopropane-1-carboxylic acid (ACC) by ACC synthetase (ACS). Finally, ACC is converted to ethylene by ACC oxidase (ACO), in an oxygen-requiring reaction.

The activity and transcription of both ACS and ACO are upregulated upon submergence in the escaping *Rumex palustris* and rice (Rieu et al. 2005; Van Der Straeten et al. 2001; Vriezen et al. 1999). Despite increases in gene expression, ethylene production was not increased in *R. palustris*, and was found to be due to reduced ACO activity within the plant during submergence (Banga et al. 1996; Vriezen et al. 1999). Not the biosynthesis, but the severely reduced gas diffusion results in the accumulation of extremely high internal levels of ethylene, ranging from 1 to 12 μL L^{-1} (Banga et al. 1996; Jackson et al. 1987; Métraux and Kende

1983; Samarakoon and Horton 1984; Voesenek et al. 1993; Voesenek and Sasidharan 2013).

Ethylene signaling is further modulated by changes in receptor abundance. Both in rice and *R. palustris* transcription of the ethylene receptor is activated upon submergence (Vriezen et al. 1997; Watanabe et al. 2004). Without bound ethylene, the signal transduction is a strong negative regulator of ethylene signaling. With saturating levels of ethylene during submergence, the additional receptors have no effect. However, when plants emerge out of the water, the establishment of aerial contact leads to a rapid release of ethylene (Voesenek et al. 1993) and a consequent dramatic suppression of ethylene signal transduction, providing a potential mechanism to shut down elongation when aerial contact is made.

Interestingly, both escaping and quiescent species accumulate high levels of ethylene, but respond to it in contrasting ways (Banga et al. 1996; Keith et al. 1986; Pierik et al. 2006; Voesenek et al. 1993). In both deepwater rice and *R. palustris*, ethylene accumulation led to an increase in GA levels, and inhibition of GA biosynthesis hampered elongation (Benschop et al. 2006; Ilya and Kende 1984). Indeed, cell expansion and cell division in deepwater rice internodes is strongly regulated by GA. Application of GA to quiescent cultivars elicits an elongation response (Setter and Laureles 1996), which means that quiescent plants are not resource-limited and retain the capacity to elongate.

Interestingly, it was shown that underwater elongation in deepwater rice is highly dependent on ABA (Hoffmann-Benning and Kende 1992). Plants treated with ABA were unable to elongate and only upon the addition of extra exogenous GA could an elongation response be restored. Indeed, a rapid drop of ABA was observed upon ethylene treatment. Not surprisingly, the ABA catabolic enzyme, *OsABA8ox1*, was strongly induced upon submergence in deepwater rice (Saika et al. 2006). Despite the importance of ABA catabolism for underwater elongation, a similar drop of ABA levels was also observed in non-elongating rice varieties (Fukao and Bailey-Serres 2008; Hattori et al. 2009; Ram et al. 2002). This suggests that interactions between ethylene and ABA with GA are of primary importance for differences in the elongation responses of different rice varieties.

Also in *R. palustris* the accumulation of GA is dependent on an initial breakdown of ABA (Benschop et al. 2006). However, upon submergence or ethylene exposure the non-elongating relative *R. acetosa* does not downregulate ABA levels (Benschop et al. 2005). The importance of internal ABA levels in regulating elongation responses in *Rumex* was further highlighted by a study on *R. palustris* ecotypes with different elongation rates. Ecotypes with the fastest elongation response showed the strongest reduction in ABA levels and concomitant expression of an ABA biosynthesis gene (*RpNCED1*, *9-cis-epoxycarotenoid dioxygenases*), whereas the slow elongating ecotypes had the smallest reduction in ABA levels and *RpNCED1* expression (Chen et al. 2010). In contrast to rice, the interaction of ethylene with ABA is most important in explaining variation between different species and genotypes. Interestingly, in the elongating monocot *Scirpus mucronatus* there was also a clear correlation between the degree of reduction in ABA levels and the extent of underwater elongation (Lee et al. 1996).

Apart from ABA and GA, a role for auxin has been found to play an important role in the underwater elongation in some species. The most convincing work is on *Ranunculus scleratus* where application of a polar transport inhibitor prevented petiole extension, and where ethylene was found to induce polar auxin transport (Musgrave and Walters 1973; Rijnders et al. 1996). Interestingly, a role of auxin in *R. palustris* has also been identified. Inhibiting polar auxin transport, chemically or by lamina (auxin source) removal, transiently reduced the underwater elongation response (Cox et al. 2006). Correspondingly, petiole auxin levels were highly dependent on the leaf blade. Furthermore, a rapid accumulation of auxin in the ab- and adaxial fragments of the elongating petioles was observed (Cox et al. 2004).

4.2 Role of Low O_2 and CO_2 in Regulating Growth Strategies

Despite the wealth of evidence for ethylene-induced elongation growth, some species respond primarily to different cues. Furthermore, it remains unclear whether ethylene responses can explain the full underwater elongation or quiescent response. For instance, in rice coleoptiles, internodes, and arrowhead tubers (*Sagittaria pygmaea*), optimal elongation was achieved by a combination of reduced O_2, and increased CO_2 and ethylene availability rather than an ethylene stimulus by itself (Ishizawa and Esashi 1984; Métraux and Kende 1984). Also in *R. palustris* a lowering of the oxygen levels increased ethylene-induced elongation, whereas it had the contrasting effect on the non-elongating *R. acetosa* (Voesenek et al. 1997). Furthermore, inhibition of ethylene signaling does not completely abolish underwater elongation in *R. palustris* (Van Veen et al. unpublished). Under anoxic conditions, rice germinates and elongates independently from ethylene (Alpi and Beevers 1983). Additionally, *Potamogeton pectinatus* rhizomes showed a strong elongation under pure anoxic conditions and reduced oxygen levels of 5–8 % (Summers and Jackson 1994).

Clearly, a specific blend of O_2, CO_2 and ethylene is most effective in stimulating elongation, and maybe also in suppressing growth. An increase in ethylene levels to usually saturating levels upon submergence has been clearly established. However, the effects of submergence on the shoot O_2 and CO_2 levels are less robust. The internal levels of these gases are dependent on the balance between photosynthesis and respiration, and diffusive resistances. These vary greatly between flooding conditions such as turbidity, depth and light penetration, and anatomy of tissue; all factors known to affect underwater plant survival (Mommer et al. 2006; Ram et al. 2002).

Hypoxia has dramatic effects on metabolism and recently a low oxygen-dependent transcription factor was identified (Gibbs et al. 2011; Licausi et al. 2011). Low oxygen-dependent transcription could be important to activate the cell division or expansion mediating elongation under anoxic conditions. Alternatively, metabolic changes such as altered adenylate charge can act as a secondary messenger of low oxygen conditions (Geigenberger et al. 2010).

How plants translate high CO_2 and reduced non-limiting O_2 conditions into altered elongation behavior is still unclear, but crucial to the underwater growth responses.

4.3 Calcium and ROS as Essential Components

In addition to the multitude of gaseous signals that regulate escape and quiescent strategies, processes downstream of calcium (Ca^{2+}) signaling have been shown to be essential for elongation in deepwater rice. The mutant *cipk15* was unable to elongate upon flooding because of the inability to activate starch reserves (Lee et al. 2009). CIPK15 is a CBL-interacting protein kinase (CIPKs), where CBLs (calcineuring B-like proteins) bind free Ca^{2+} and so are the main transducers of calcium signaling (Batistic and Kudla 2004).

The role of Ca^{2+} as a key signaling component during changes in O_2 availability has been supported by elevation of Ca^{2+} levels in maize cells as a result of anoxia-induced Ca^{2+} mobilization. Subsequent manipulations of the calcium dynamics resulted in altered hypoxic responses (Subbaiah et al. 1994a, b, 1998). Other submergence stress-related aspects regulated by elevated Ca^{2+} are ionic homeostasis, aerenchyma formation and cell death, and expression of anoxia-inducible genes (Subbaiah and Sachs 2003). Interestingly, in *Sagittaria pygmaea* (arrowhead) anoxic elongation was dependent specifically on Ca^{2+} uptake from the growth media. However, differences with regard to uptake from the growth media, the effect of Ca^{2+} transport inhibitors and temporal dynamics suggests the involvement of alternative physiological processes between maize and arrowhead (Subbaiah et al. 1994a, b; Tamura et al. 2001).

A large degree of cross talk exists between reactive oxygen species (ROS) and calcium in plant stress signaling, and also upon hypoxia (Bowler and Fluhr 2000). Furthermore, a large overlap exists between ROS and hypoxia-induced transcriptomic changes (Pucciariello et al. 2012). Interestingly, a monomeric G protein of plants, ROPGAP4, was shown to be essential in regulating alcohol dehydrogenase (ADH) induction in *Arabidopsis*. ROPGAP4 activity increased upon hypoxia, which in turn activated the production of hydrogen peroxide (H_2O_2) via a Ca^{2+}-dependent NADPH oxidase (Baxter-Burrell et al. 2002). In rice another heterotrimeric G protein (RGA1), which plays a role in aerenchyma formation, was identified directly downstream of ethylene and H_2O_2 signaling (Steffens and Sauter 2009, 2010).

Although ROS have been identified as potential cell wall modifying agents that could stimulate growth (Liszkay et al. 2003; Schopfer and Liszkay 2006), so far only a role for calcium has been identified in mediating underwater elongation (Lee et al. 2009; Tamura et al. 2001). The strong interactions observed between the two in other studies (Baxter-Burrell et al. 2002; Bowler and Fluhr 2000) suggest a complex signaling network on which the regulation of either quiescence or escape is highly reliant.

5 One Signal Two Responses

Interestingly all species are expected to have a similar internal gaseous composition as the result of complete submergence. The same signals result in elongation for some species, whereas others adopt a quiescence strategy. Hormonal reprogramming by an interaction between ethylene and other hormones such as ABA, GA, and auxin is instrumental in the difference between the two main growth strategies. However, there is an array of underlying molecular processes mediating these different hormonal interactions that vary between floodplain species giving rise to opposite growth responses.

5.1 Group VII ERFs

Rice genetics has received much attention because of its agronomical importance. Interestingly, two mapping studies both identified candidates of the same transcription factor gene-family that play a role in mediating escape or quiescence in rice, the group VII Ethylene Response Factors (group VII ERFs) (Voesenk and Bailey-Serres 2009).

A QTL was identified on chromosome 9 that explained 69 % of the variation in flooding tolerance in rice (Xu and Mackill 1996). The region responsible for this variation is a cluster of three genes, *SUB1A*, *SUB1B*, and *SUB1C*, all encoding group VII ERFs that are induced upon submergence. Where *SUB1B* and *SUB1C* are present in all rice cultivars, *SUB1A* is present only in a select number of genotypes. Those with the *SUB1A-1* allele invariably show a high degree of tolerance to submergence (Xu et al. 2006).

The introgression of the tolerant *SUB1* gene cluster into an agronomically interesting rice variety revealed many novel functions. The *SUB1A* genotype showed a reduced elongation response underwater, with concomitant reduction of carbohydrate reserve consumption and cell wall loosening enzymes (Fukao et al. 2006). Furthermore, the *SUB1* genotypes induce two suppressors of GA signaling upon submergence, Slender Rice-1 (SLR1) and SLR1 Like-1 (SLRL1), thereby decreasing the sensitivity to GA (Fukao and Bailey-Serres 2008). Indeed, SUB1A-1 confers tolerance to flooding by activating a quiescence strategy upon submergence.

Another study looking specifically at elongation responses of rice identified three other major QTLs, whose combined effect could close to completely restore elongation in the quiescent parental line (Hattori et al. 2009). The QTL on chromosome 12 encoded two group VII ERFs, *Snorkel1* and *-2* (*SK1*, *SK2*). Interestingly, the *SK* genes were found to be present only in genotypes and wild relatives that showed a deepwater elongation response. SK1 and SK2 may act via GA biosynthesis to confer an escape strategy, as this was the main difference between the parental lines.

Thus, members of the same gene family (group VII ERFs) affect the hormone signaling responsible for the growth strategies in opposite ways. Furthermore, the induction of *SK1* transcription is regulated directly downstream of ethylene signaling, demonstrated by binding of an ethylene signaling component (EIL1B) to the *SK* promotor and a range of hormone treatments. Similarly, *SUB1A* transcriptional activation is regulated by ethylene, and in addition by ROS and sucrose (Fukao et al. 2011; Kudahettige et al. 2010). However, *SUB1C* is activated by GA, and thus regulated further downstream of the flooding induced hormonal signaling network.

Clearly, there is a lot of variation in group VII ERF behavior amongst different rice genotypes which has proven to be essential in adapting to a range of different environments. Interestingly, exploration of SUB1 loci in a range of wild rice species confirms the flooding tolerance mediated by SUB1A-1, in close relatives of *O. sativa*. However, very tolerant wild rice species lack the *SUB1A* locus, but do have *SUB1C* homologs, and therefore have a *SUB1A*-independent mechanism conferring tolerance (Niroula et al. 2012).

Naturally, this specific family of transcription factors has sparked interest and detailed studies have also been done on related members in *Arabidopsis*. Here they function primarily in regulating hypoxia-induced expression of genes such *ADH* and *PDC* (Hinz et al. 2010; Licausi et al. 2010). Most interestingly, in *Arabidopsis* these transcription factors are subject to post-translational regulation depending on oxygen availability. Normal oxygen levels lead to specific N-terminal modifications that target the protein for proteolysis. However, under hypoxia N-terminal modifications do not occur, thereby stabilizing these transcription factors and permitting the activation of downstream target genes (Gibbs et al. 2011; Licausi et al. 2011). Interestingly, in rice no group VII ERFs have been detected yet with similar oxygen-dependent proteolysis. It would be interesting to see if resistance to oxygen-dependent proteolysis of group VII ERFs is a more common response amongst wetland species and if these transcription factors also mediate the growth strategies in wetland species other than rice.

6 Summary and Future Perspectives

The persistence of plants at flood prone locations is dependent on successfully employing either the escape or quiescent strategy. The fitness consequence of these growth strategies is now well established and their biology is not only of academic interest. Understanding the regulation of these growth strategies will also be highly relevant for the development of flood-tolerant crops.

The two essential mechanisms of the growth strategies are balancing the energy production and consumption, and modifying extensibility of the cell wall or cell cycle progression. Both are highly regulated processes during flooding and the molecular mechanisms are slowly being elucidated. Regardless of the growth strategy, submerged plants accumulate saturating levels of ethylene and specific changes in O_2 and CO_2. In rice the differences in growth strategy is predominantly due to the variation of the interactions of GA with ABA and ethylene. In other species, the ethylene and ABA interaction has been shown to be crucial. At the same time, several plants also use auxin as a primary growth hormone for underwater elongation.

An especially interesting case is *Lotus tenuis*, which can either escape or go into quiescence depending on the extent of flooding (Manzur et al. 2009). The regulatory mechanisms mediating the two strategies in a single plant is still unknown. Different flooding conditions have a signature make up of the major regulatory components, ethylene, O_2, CO_2, calcium, and ROS. This specific blend can be exploited to differentiate between activating or repressing the signaling pathways leading to the suitable growth response. Despite that plants respond differently to a variety of gas treatments, there are no detailed investigations regarding responses to specific blend of signals mimicking different flooding regimes. This presents one of the future challenges in flooding research that will provide a significant insight into the regulation of flooding responses.

In contrast to *Lotus tenuis*, most plant species still only have the capacity to employ one strategy upon flooding. So far only the group VII ERFs in rice have been identified as mediating different growth responses. In the plant kingdom a great variation exists in this transcription factor family and the generality amongst wetland plants is uncertain. Future challenges will be to unravel more of these switch points via which environmental cues can mediate a quiescent or escape response. Knowledge of the underlying signaling network will reveal the main targets of selection pressures exerted by different flooding regimes and help with the development of flooding tolerance in crop species.

References

Akman M, Bhikharie AV, McLean EH et al (2012) Wait or escape? Contrasting submergence tolerance strategies of *Rorippa amphibia*, *Rorippa sylvestris* and their hybrid. Ann Bot. doi:10.1093/aob/mcs059

Alpi A, Beevers H (1983) Effects of O_2 concentration on rice seedlings. Plant Physiol 71:30–34

Armstrong W, Beckett PM (2011) Experimental and modelling data contradict the idea of respiratory down-regulation in plant tissues at an internal [O] substantially above the critical oxygen pressure for cytochrome oxidase. New Phytol 190:431–441. doi:10.1111/j.1469-8137.2010.03537.x

Armstrong W, Webb T, Darwent M, Beckett PM (2009) Measuring and interpreting respiratory critical oxygen pressures in roots. Ann Bot 103:281–293. doi:10.1093/aob/mcn177

Arpagaus S, Braendle R (2000) The significance of alpha-amylase under anoxia stress in tolerant rhizomes (*Acorus calamus* L.) and non-tolerant tubers (*Solanum tuberosum* l., var. Désirée). J Exp Bot 51:1475–1477

Baena-González E, Rolland F, Thevelein JM, Sheen J (2007) A central integrator of transcription networks in plant stress and energy signalling. Nature 448:938–942. doi:10.1038/nature06069

Bailey-Serres J, Voesenek LACJ (2008) Flooding stress: acclimations and genetic diversity. Annu Rev Plant Biol 59:313–339. doi:10.1146/annurev.arplant.59.032607.092752

Banga M, Slaa EJ, Blom C, Voesenek LACJ (1996) Ethylene biosynthesis and accumulation under drained and submerged conditions (a comparative study of two *Rumex* species). Plant Physiol 112:229–237

Batistic O, Kudla J (2004) Integration and channeling of calcium signaling through the CBL calcium sensor/CIPK protein kinase network. Planta 219:915–924. doi:10.1007/s00425-004-1333-3

Baxter-Burrell A, Yang Z, Springer PS, Bailey-Serres J (2002) RopGAP4-dependent Rop GTPase rheostat control of *Arabidopsis* oxygen deprivation tolerance. Science 296:2026–2028. doi:10.1126/science.1071505

Benschop JJ, Jackson MB, Gühl K et al (2005) Contrasting interactions between ethylene and abscisic acid in *Rumex* species differing in submergence tolerance. Plant J 44:756–768. doi:10.1111/j.1365-313X.2005.02563.x

Benschop JJ, Bou J, Peeters AJM et al (2006) Long-term submergence-induced elongation in *Rumex palustris* requires abscisic acid-dependent biosynthesis of gibberellin1. Plant Physiol 141:1644–1652. doi:10.1104/pp. 106.082636

Bleecker AB, Schuette JL, Kende H (1986) Anatomical analysis of growth and developmental patterns in the internode of deepwater rice. Planta 169:490–497. doi:10.1007/BF00392097

Bowler C, Fluhr R (2000) The role of calcium and activated oxygens as signals for controlling cross-tolerance. Trends Plant Sci 5:241–246

Branco-Price C, Kaiser KA, Jang CJH et al (2008) Selective mRNA translation coordinates energetic and metabolic adjustments to cellular oxygen deprivation and reoxygenation in *Arabidopsis thaliana*. Plant J 56:743–755. doi:10.1111/j.1365-313X.2008.03642.x

Chang WWP, Huang L, Shen M et al (2000) Patterns of protein synthesis and tolerance of anoxia in root tips of maize seedlings acclimated to a low-oxygen environment, and identification of proteins by mass spectrometry. Plant Physiol 122:295–317

Chen X, Pierik R, Peeters AJM et al (2010) Endogenous abscisic acid as a key switch for natural variation in flooding-induced shoot elongation. Plant Physiol 154:969–977. doi:10.1104/pp. 110.162792

Cho HT, Kende H (1997) Expansins and internodal growth of deepwater rice. Plant Physiol 113:1145–1151

Colmer TD, Huang S, Greenway H (2001) Evidence for down-regulation of ethanolic fermentation and K^+ effluxes in the coleoptile of rice seedlings during prolonged anoxia. J Exp Bot 52:1507–1517

Cook CDK (1999) The number and kinds of embryo-bearing plants which have become aquatic: a survey. Perspect Plant Ecol Evol Syst 2:79–102. doi:10.1078/1433-8319-00066

Cosgrove DJ (2005) Growth of the plant cell wall. Nat Rev Mol Cell Biol 6:850–861. doi:10.1038/nrm1746

Cox MCH, Benschop JJ, Vreeburg RAM et al (2004) The roles of ethylene, auxin, abscisic acid, and gibberellin in the hyponastic growth of submerged *Rumex palustris* petioles. Plant Physiol 136:2948–2960. doi:10.1104/pp. 104.049197

Cox MCH, Peeters AJM, Voesenek LACJ (2006) The stimulating effects of ethylene and auxin on petiole elongation and on hyponastic curvature are independent processes in submerged *Rumex palustris*. Plant Cell Environ 29:282–290

Das KK, Sarkar RK, Ismail AM (2005) Elongation ability and non-structural carbohydrate levels in relation to submergence tolerance in rice. Plant Sci 168:131–136. doi:10.1016/j.plantsci. 2004.07.023

Edwards JM, Roberts TH, Atwell BJ (2012) Quantifying ATP turnover in anoxic coleoptiles of rice (*Oryza sativa*) demonstrates preferential allocation of energy to protein synthesis. J Exp Bot. doi:10.1093/jxb/ers114

Fabian T, Lorbiecke R, Umeda M, Sauter M (2000) The cell cycle genes *cycA1;1* and *cdc2Os-3* are coordinately regulated by gibberellin in planta. Planta 211:376–383

Fabian-Marwedel T, Umeda M, Sauter M (2002) The rice cyclin-dependent kinase-activating kinase R2 regulates S-phase progression. Plant Cell 14:197–210

Felle H (2005) pH regulation in anoxic plants. Ann Bot 96:519–532. doi:10.1093/aob/mci207
Felle H (2010) pH signaling during anoxia. In: Mancuso S, Shabala S (eds) Waterlogging signalling and tolerance in plants. Springer, Berlin, Heidelberg
Fukao T, Bailey-Serres J (2008) Submergence tolerance conferred by *Sub1A* is mediated by SLR1 and SLRL1 restriction of gibberellin responses in rice. Proc Natl Acad Sci U S A 105:16814
Fukao T, Xu K, Ronald PC, Bailey-Serres J (2006) A variable cluster of ethylene response factor-like genes regulates metabolic and developmental acclimation responses to submergence in rice. Plant Cell 18:2021–2034. doi:10.1105/tpc.106.043000
Fukao T, Yeung E, Bailey-Serres J (2011) The submergence tolerance regulator SUB1A mediates crosstalk between submergence and drought tolerance in rice. Plant Cell 23:412–427. doi:10.1105/tpc.110.080325
Geigenberger P (2003) Response of plant metabolism to too little oxygen. Curr Opin Plant Biol 6:247–256. doi:10.1016/S1369-5266(03)00038-4
Geigenberger P, Riewe D, Fernie AR (2010) The central regulation of plant physiology by adenylates. Trends Plant Sci 15:98–105. doi:10.1016/j.tplants.2009.11.004
Gibbs DJ, Lee SC, Isa NM et al (2011) Homeostatic response to hypoxia is regulated by the N-end rule pathway in plants. Nature 479:415–418. doi:10.1038/nature10534
Groeneveld HW, Voesenek LACJ (2003) Submergence-induced petiole elongation in *Rumex palustris* is controlled by developmental stage and storage compounds. Plant Soil 253:115–123
Gupta KJ, Zabalza A, van Dongen JT (2009) Regulation of respiration when the oxygen availability changes. Physiol Plant 137:383–391. doi:10.1111/j.1399-3054.2009.01253.x
Harada T, Ishizawa K (2003) Starch degradation and sucrose metabolism during anaerobic growth of pondweed (*Potamogeton distinctus* A. Benn.) turions. Plant Soil 253:125–135. doi:10.1023/A:1024585015697
Hattori Y, Nagai K, Furukawa S et al (2009) The ethylene response factors *SNORKEL1* and *SNORKEL2* allow rice to adapt to deep water. Nature 460:1026–1030. doi:10.1038/nature08258
Hinz M, Wilson IW, Yang J et al (2010) Arabidopsis RAP2.2: an ethylene response transcription factor that is important for hypoxia survival. Plant Physiol 153:757–772. doi:10.1104/pp. 110.155077
Hoffmann-Benning S, Kende H (1992) On the role of abscisic acid and gibberellin in the regulation of growth in rice. Plant Physiol 99:1156–1161
Huang S, Colmer TD, Millar AH (2008) Does anoxia tolerance involve altering the energy currency towards PPi? Trends Plant Sci 13:221–227. doi:10.1016/j.tplants.2008.02.007
Ilya R, Kende H (1984) Role of gibberellin in the growth response of submerged deep water rice. Plant Physiol 76:947
Ishizawa K, Esashi Y (1984) Gaseous factors involved in the enhanced elongation of rice coleoptiles under water. Plant Cell Environ 7:239–245. doi:10.1111/1365-3040.ep11589438
Ishizawa K, Murakami S, Kawakami Y, Kuramochi H (1999) Growth and energy status of arrowhead tubers, pondweed turions and rice seedlings under anoxic conditions. Plant Cell Environ 22:505–514
Jackson MB, Waters I, Setter T, Greenway H (1987) Injury to rice plants caused by complete submergence; a contribution by ethylerie (ethene). J Exp Bot 38:1826–1838. doi:10.1093/jxb/38.11.1826
Jung KH, Seo YS, Walia H et al (2010) The submergence tolerance regulator *Sub1A* mediates stress-responsive expression of *AP2/ERF* transcription factors. Plant Physiol 152:1674–1692. doi:10.1104/pp. 109.152157
Keith KA, Raskin I, Kende H (1986) A comparison of the submergence response of deepwater and non-deepwater rice. Plant Physiol 80:479
Koizumi Y, Hara Y, Yazaki Y et al (2011) Involvement of plasma membrane H^+-ATPase in anoxic elongation of stems in pondweed (*Potamogeton distinctus*) turions. New Phytol 190:421–430. doi:10.1111/j.1469-8137.2010.03605.x
Kudahettige NP, Pucciariello C, Parlanti S et al (2010) Regulatory interplay of the Sub1A and CIPK15 pathways in the regulation of α-amylase production in flooded rice plants. Plant Biol 13:611–619. doi:10.1111/j.1438-8677.2010.00415.x

Kutschera U, Kende H (1988) The biophysical basis of elongation growth in internodes of deepwater rice. Plant Physiol 88:361

Lee T-M, Shieh Y-J, Chou C-H (1996) Abscisic acid inhibits shoot elongation of *Scirpus mucronatus*. Physiol Plant 97:1–4. doi:10.1111/j.1399-3054.1996.tb00470.x

Lee KW, Chen PW, Lu CA et al (2009) Coordinated responses to oxygen and sugar deficiency allow rice seedlings to tolerate flooding. Sci Signal 2:ra61. doi:10.1126/scisignal.2000333

Licausi F, van Dongen JT, Giuntoli B et al (2010) *HRE1* and *HRE2*, two hypoxia-inducible ethylene response factors, affect anaerobic responses in *Arabidopsis thaliana*. Plant J 62:302–315. doi:10.1111/j.1365-313X.2010.04149.x

Licausi F, Kosmacz M, Weits DA et al (2011) Oxygen sensing in plants is mediated by an N-end rule pathway for protein destabilization. Nature 479:419–422. doi:10.1038/nature10536

Liszkay A, Kenk B, Schopfer P (2003) Evidence for the involvement of cell wall peroxidase in the generation of hydroxyl radicals mediating extension growth. Planta 217:658–667. doi:10.1007/s00425-003-1028-1

Lorbiecke R, Sauter M (1998) Induction of cell growth and cell division in the intercalary meristem of submerged deepwater rice (*Oryza sativa* L.). Planta 204:140–145

Magneschi L, Kudahettige RL, Alpi A, Perata P (2009a) Comparative analysis of anoxic coleoptile elongation in rice varieties: relationship between coleoptile length and carbohydrate levels, fermentative metabolism and anaerobic gene expression. Plant Biol 11:561–573. doi:10.1111/j.1438-8677.2008.00150.x

Magneschi L, Kudahettige RL, Alpi A, Perata P (2009b) Expansin gene expression and anoxic coleoptile elongation in rice cultivars. J Plant Physiol 166:1576–1580. doi:10.1016/j.jplph.2009.03.008

Malone M, Ridge I (1983) Ethylene-induced growth and proton excretion in the aquatic plant Nymphoides peltata. Planta 157:71–73. doi:10.1007/BF00394542

Malz S, Sauter M (1999) Expression of two PIP genes in rapidly growing internodes of rice is not primarily controlled by meristem activity or cell expansion. Plant Mol Biol 40:985–995

Manzur ME, Grimoldi AA, Insausti P, Striker GG (2009) Escape from water or remain quiescent? *Lotus tenuis* changes its strategy depending on depth of submergence. Ann Bot 104:1163–1169. doi:10.1093/aob/mcp203

Menegus F, Cattaruzza L, Mattana M et al (1991) Response to anoxia in rice and wheat seedlings: changes in the pH of intracellular compartments, glucose-6-phosphate level, and metabolic rate. Plant Physiol 95:760

Métraux JP, Kende H (1983) The role of ethylene in the growth response of submerged deep water rice. Plant Physiol 72:441–446

Métraux J-P, Kende H (1984) The cellular basis of the elongation response in submerged deep-water rice. Planta 160:73–77. doi:10.1007/BF00392468

Mommer L, Visser EJW (2005) Underwater photosynthesis in flooded terrestrial plants: a matter of leaf plasticity. Ann Bot 96:581–589. doi:10.1093/aob/mci212

Mommer L, Pons TL, Wolters-Arts M et al (2005) Submergence-induced morphological, anatomical, and biochemical responses in a terrestrial species affect gas diffusion resistance and photosynthetic performance. Plant Physiol 139:497–508. doi:10.1104/pp. 105.064725

Mommer L, Lenssen JPM, Huber H et al (2006) Ecophysiological determinants of plant performance under flooding: a comparative study of seven plant families. J Ecol 94:1117–1129. doi:10.1111/j.1365-2745.2006.01175.x

Mommer L, Wolters-Arts M, Andersen C et al (2007) Submergence-induced leaf acclimation in terrestrial species varying in flooding tolerance. New Phytol 176:337–345. doi:10.1111/j.1469-8137.2007.02166.x

Mugnai S, Marras AM, Mancuso S (2011) Effect of hypoxic acclimation on anoxia tolerance in *Vitis* roots: response of metabolic activity and K^+ fluxes. Plant Cell Physiol 52:1107–1116. doi:10.1093/pcp/pcr061

Musgrave A, Walters J (1973) Ethylene-stimulated growth and auxin transport in *Ranunculus sceleratus* petioles. New Phytol 72:783–789

Mustroph A, Albrecht G (2003) Tolerance of crop plants to oxygen deficiency stress: fermentative activity and photosynthetic capacity of entire seedlings under hypoxia and anoxia. Physiol Plant 117:508–520. doi:10.1034/j.1399-3054.2003.00051.x
Mustroph A, Lee SC, Oosumi T et al (2010) Cross-kingdom comparison of transcriptomic adjustments to low-oxygen stress highlights conserved and plant-specific responses. Plant Physiol 152:1484–1500. doi:10.1104/pp. 109.151845
Mustroph A, Hess N, Sasidharan R (2013) Hypoxic energy metabolism and PPi as an alternative energy currency. In: van Dongen J, Licausi F (eds) Low-oxygen stress in plants. Springer, Vienna
Niroula RK, Pucciariello C, Ho VT et al (2012) *SUB1A*-dependent and -independent mechanisms are involved in the flooding tolerance of wild rice species. Plant J. doi:10.1111/j.1365-313X.2012.05078.x
Ookawara R, Satoh S, Yoshioka T, Ishizawa K (2005) Expression of α-expansin and xyloglucan endotransglucosylase/hydrolase genes associated with shoot elongation enhanced by anoxia, ethylene and carbon dioxide in arrowhead (*Sagittaria pygmaea* Miq.) tubers. Ann Bot 96(4): 693–702
Paepke S, Ramírez-Aguilar S, Antonio C (2013) Oxygen consumption under hypoxic conditions. In: van Dongen J, Licausi F (eds) Low-oxygen stress in plants. Springer, Vienna
Pang J, Shabala SN (2010) Membrane transporters and waterlogging. In: Mancuso S, Shabala S (eds) Waterlogging signalling and tolerance in plants. Springer, Berlin, Heidelberg
Pang J, Cuin T, Shabala L et al (2007) Effect of secondary metabolites associated with anaerobic soil conditions on ion fluxes and electrophysiology in barley roots. Plant Physiol 145:266–276. doi:10.1104/pp. 107.102624
Pierik R, Tholen D, Poorter H et al (2006) The Janus face of ethylene: growth inhibition and stimulation. Trends Plant Sci 11:176–183. doi:10.1016/j.tplants.2006.02.006
Pierik R, van Aken JM, Voesenek LACJ (2009) Is elongation-induced leaf emergence beneficial for submerged *Rumex* species? Ann Bot 103:353–357. doi:10.1093/aob/mcn143
Pucciariello C, Parlanti S, Banti V et al (2012) Reactive oxygen species-driven transcription in *Arabidopsis* under oxygen deprivation. Plant Physiol 159:184–196. doi:10.1104/pp. 111.191122
Ram PC, Singh BB, Singh AK et al (2002) Submergence tolerance in rainfed lowland rice: physiological basis and prospects for cultivar improvement through marker-aided breeding. Field Crop Res 76:131–152. doi:10.1016/S0378-4290(02)00035-7
Ridge I (1987) Ethylene and growth control in amphibious plants. In: Crawford RMM, Spence DHN (eds) Plant life in aquatic and amphibious habitats. Blackwell Scientific Publications, Oxford
Ridge I, Amarasinghe I (1984) Ethylene and growth control in the fringed waterlily (*Nymphoides peltata*): stimulation of cell division and interaction with buoyant tension in petioles. Plant Growth Regul 2:235–249. doi:10.1007/BF00124772
Ridge I, Osborne DJ (1989) Wall extensibility, wall pH and tissue osmolality: significance for auxin and ethylene-enhanced petiole growth in semi-aquatic plants. Plant Cell Environ 12: 383–393. doi:10.1111/j.1365-3040.1989.tb01954.x
Rieu I, Cristescu SM, Harren FJM et al (2005) *RP-ACS1*, a flooding-induced 1-aminocyclopropane-1-carboxylate synthase gene of *Rumex palustris*, is involved in rhythmic ethylene production. J Exp Bot 56:841–849. doi:10.1093/jxb/eri078
Rijnders JGHM, Barendse GWM, Blom CWPM, Voesenek LACJ (1996) The contrasting role of auxin in submergence-induced petiole elongation in two species from frequently flooded wetlands. Physiol Plant 96:467–473. doi:10.1111/j.1399-3054.1996.tb00460.x
Roberts JK, Callis J, Jardetzky O et al (1984) Cytoplasmic acidosis as a determinant of flooding intolerance in plants. Proc Natl Acad Sci U S A 81:6029–6033
Roberts JKM, Andrade FH, Anderson IC (1985) Further evidence that cytoplasmic acidosis is a determinant of flooding intolerance in plants. Plant Physiol 77:492
Sachs MM, Freeling M, Okimoto R (1980) The anaerobic proteins of maize. Cell 20:761–767

Saika H, Okamoto M, Miyoshi K et al (2006) Ethylene promotes submergence-induced expression of *OsABA8ox1*, a gene that encodes ABA 8′-hydroxylase in rice. Plant Cell Physiol 48: 287–298. doi:10.1093/pcp/pcm003

Samarakoon AB, Horton RF (1984) Petiole growth in *Ranunculus sceleratus* L.: ethylene synthesis and submergence. Ann Bot 54:263–270

Sand-Jensen K, Frost-Christensen H (1999) Plant growth and photosynthesis in the transition zone between land and stream. Aquat Bot 63:23–35

Sasidharan R, Voesenek LA, Pierik R (2011) Cell wall modifying proteins mediate plant acclimatization to biotic and abiotic stresses. Crit Rev Plant Sci 30:548–562. doi:10.1080/07352689.2011.615706

Sauter M (1997) Differential expression of a CAK (cdc2-activating kinase)-like protein kinase, cyclins and *cdc2* genes from rice during the cell cycle and in response to gibberellin. Plant J 11:181–190

Sauter M, Mekhedov SL, Kende H (1995) Gibberellin promotes histone H1 kinase activity and the expression of *cdc2* and cyclin genes during the induction of rapid growth in deepwater rice internodes. Plant J 7:623–632

Schlüter U, Crawford RMM (2001) Long-term anoxia tolerance in leaves of *Acorus calamus* L. and *Iris pseudacorus* L. J Exp Bot 52:2213–2225

Schopfer P, Liszkay A (2006) Plasma membrane-generated reactive oxygen intermediates and their role in cell growth of plants. Biofactors 28:73–81

Setter TL, Laureles EV (1996) The beneficial effect of reduced elongation growth on submergence tolerance in rice. J Exp Bot 47:1551–1559

Silvertown J, Dodd ME, Gowing DJG, Mountford JO (1999) Hydrologically defined niches reveal a basis for species richness in plant communities. Nature 400:61–63

Staal M, De Cnodder T, Simon D et al (2011) Apoplastic alkalinization is instrumental for the inhibition of cell elongation in the *Arabidopsis* root by the ethylene precursor 1-aminocyclopropane-1-carboxylic acid. Plant Physiol 155:2049–2055. doi:10.1104/pp. 110.168476

Steffens B, Sauter M (2009) Heterotrimeric G protein signaling is required for epidermal cell death in rice. Plant Physiol 151:732–740. doi:10.1104/pp. 109.142133

Steffens B, Sauter M (2010) G proteins as regulators in ethylene-mediated hypoxia signaling. Plant Signal Behav 5:375–378

Subbaiah CC, Sachs MM (2003) Calcium-mediated responses of maize to oxygen deprivation. Russ J Plant Physiol 50:752–761. doi:10.1023/B:RUPP.0000003273.44823.cd

Subbaiah CC, Bush DS, Sachs MM (1994a) Elevation of cytosolic calcium precedes anoxic gene expression in maize suspension-cultured cells. Plant Cell 6:1747–1762. doi:10.1105/tpc.6.12.1747

Subbaiah CC, Zhang J, Sachs MM (1994b) Involvement of intracellular calcium in anaerobic gene expression and survival of maize seedlings. Plant Physiol 105:369–376

Subbaiah C, Bush D, Sachs M (1998) Mitochondrial contribution to the anoxic Ca^{2+} signal in maize suspension-cultured cells. Plant Physiol 118:759–771

Summers JE, Jackson MB (1994) Anaerobic conditions strongly promote extension by stems of overwintering tubers of *Potamogeton pectinatus* L. J Exp Bot 45:1309–1318. doi:10.1093/jxb/45.9.1309

Summers JE, Ratcliffe RG, Jackson MB (2000) Anoxia tolerance in the aquatic monocot *Potamogeton pectinatus* absence of oxygen stimulates elongation in association with an unusually large pasteur effect. J Exp Bot 51:1413–1422

Tamura S, Kuramochi H, Ishizawa K (2001) Involvement of calcium ion in the stimulated shoot elongation of arrowhead tubers under anaerobic conditions. Plant Cell Physiol 42:717–722

Umeda M, Uchimiya H (1994) Differential transcript levels of genes associated with glycolysis and alcohol fermentation in rice plants (*Oryza sativa* L.) under submergence stress. Plant Physiol 106:1015–1022

van der Knaap E, Kende H (1995) Identification of a gibberellin-induced gene in deepwater rice using differential display of mRNA. Plant Mol Biol 28:589–592 van der Knaap E, Jagoueix S, Kende H (1997) Expression of an ortholog of replication protein A1 (RPA1) is induced by gibberellin in deepwater rice. Proc Natl Acad Sci U S A 94:9979–9983

Van Der Straeten D, Zhou Z, Prinsen E et al (2001) A comparative molecular-physiological study of submergence response in lowland and deepwater rice. Plant Physiol 125:955–968

Van Eck WHJM, van de Steeg HM, Blom CWPM, de Kroon H (2004) Is tolerance to summer flooding correlated with distribution patterns in river floodplains? A comparative study of 20 terrestrial grassland species. Oikos 107:393–405. doi:10.1111/j.0030-1299.2004.13083.x

Voesenek LACJ, Perik PJM, Blom CWPM, Sassen MMA (1990) Petiole elongation in *Rumex* species during submergence and ethylene exposure: the relative contributions of cell division and cell expansion. J Plant Growth Regul 9:13–17. doi:10.1007/BF02041936

Voesenek LACJ, Banga M, Thier RH et al (1993) Submergence-induced ethylene synthesis, entrapment, and growth in two plant species with contrasting flooding resistances. Plant Physiol 103:783–791

Voesenek LACJ, Vriezen WH, Smekens M et al (1997) Ethylene sensitivity and response sensor expression in petioles of *Rumex* species at low O_2 and high CO_2 concentrations. Plant Physiol 114:1501–1509

Voesenek LACJ, Rijnders JHGM, Peeters A et al (2004) Plant hormones regulate fast shoot elongation under water: from genes to communities. Ecology 85:16–27

Voesenk LACJ, Bailey-Serres J (2009) Plant biology: genetics of high-rise rice. Nature 460(7258): 959–960. doi:10.1038/460959a

Voesenek LACJ, Sasidharan R (2013) Ethylene - and oxygen signalling - drive plant survival during flooding. Plant Biol 15:426–435.

Vreeburg RAM, Benschop JJ, Peeters AJM et al (2005) Ethylene regulates fast apoplastic acidification and expansin A transcription during submergence-induced petiole elongation in Rumex palustris. Plant J 43:597–610. doi:10.1111/j.1365-313X.2005.02477.x

Vriezen WH, van Rijn CP, Voesenek LACJ, Mariani C (1997) A homolog of the *Arabidopsis thaliana ERS* gene is actively regulated in *Rumex palustris* upon flooding. Plant J 11:1265–1271

Vriezen WH, Hulzink R, Mariani C, Voesenek LACJ (1999) 1-aminocyclopropane-1-carboxylate oxidase activity limits ethylene biosynthesis in *Rumex palustris* during submergence. Plant Physiol 121:189–196

Vriezen WH, De Graaf B, Mariani C, Voesenek LACJ (2000) Submergence induces expansin gene expression in flooding-tolerant *Rumex palustris* and not in flooding-intolerant *R. acetosa*. Planta 210:956–963

Watanabe H, Saigusa M, Hase S et al (2004) Cloning of a cDNA encoding an ETR2-like protein (Os-ERL1) from deep water rice (*Oryza sativa* L.) and increase in its mRNA level by submergence, ethylene, and gibberellin treatments. J Exp Bot 55:1145–1148. doi:10.1093/jxb/erh110

Xu K, Mackill DJ (1996) A major locus for submergence tolerance mapped on rice chromosome 9. Mol Breed 2:219–224

Xu K, Xu X, Fukao T et al (2006) *Sub1A* is an ethylene-response-factor-like gene that confers submergence tolerance to rice. Nature 442:705–708. doi:10.1038/nature04920

Zabalza A, van Dongen JT, Froehlich A et al (2008) Regulation of respiration and fermentation to control the plant internal oxygen concentration. Plant Physiol 149:1087–1098. doi:10.1104/pp.108.129288

Zhang Q, Greenway H (1994) Anoxia tolerance and anaerobic catabolism of aged beetroot storage tissues. J Exp Bot 45:567–575. doi:10.1093/jxb/45.5.567

Zhang Q, Greenway H (1995) Membrane transport in anoxic rice coleoptiles and storage tissues of beetroot. Aust J Plant Physiol 22:965. doi:10.1071/PP9950965

Zhang Y, Primavesi LF, Jhurreea D et al (2009) Inhibition of SNF1-related protein kinase1 activity and regulation of metabolic pathways by trehalose-6-phosphate. Plant Physiol 149:1860–1871. doi:10.1104/pp. 108.133934

Part VI
Agronomical and Horticultural Aspects of Low-Oxygen Stress

Hypoxic Storage of Fruit

Quang Tri Ho, Kim Buts, Els Herremans, Maarten L.A.T.M. Hertog, Pieter Verboven, and Bart M. Nicolaï

Abstract Controlled and modified atmosphere storage are popular technologies to extend the storage life of fruit and vegetables, and pome fruit in particular. They aim at lowering the respiratory activity of the product by reducing and increasing the O_2 and CO_2 partial pressures, respectively, in addition to applying a low temperature. The optimal gas composition is critical, as a too low O_2 partial pressure in combination with a too high CO_2 partial pressure may induce off-flavours and storage disorders in the fruit. In this chapter, we will investigate the physiological processes in fruit exposed to hypoxic conditions during controlled or modified atmosphere storage. We will pay particular attention to storage disorders that may develop as a consequence of severe hypoxia.

1 Introduction

Fruit are perishable products and need to be stored at appropriate conditions in order to extend their storage and shelf life. As post-harvest changes of many quality attributes such as sugar content, firmness, acidity and aroma are directly or indirectly related to the respiratory activity of the fruit, modern storage technologies aim at reducing respiration. This is often accomplished by reducing the temperature as much as possible but above the critical temperature beyond which chilling or freezing injury starts to develop. To further reduce respiration, the O_2 partial pressure in the storage atmosphere may be reduced and the CO_2 partial pressure increased. When the atmosphere is actively manipulated through scrubbers in combination with a control system this is called *Controlled Atmosphere* (CA) *storage*; when the atmosphere is passively changed through respiratory O_2 consumption and CO_2 production this is called *Modified Atmosphere* (MA) *storage*.

Q.T. Ho • K. Buts • E. Herremans • M.L.A.T.M. Hertog • P. Verboven • B.M. Nicolaï (✉)
Flanders Center of Postharvest Technology, BIOSYST-MeBioS, Katholieke Universiteit Leuven, Box 2428, Willem de Croylaan 42, B-3001, Leuven (Heverlee), Belgium
e-mail: bart.nicolai@biw.kuleuven.be

J.T. van Dongen and F. Licausi (eds.), *Low-Oxygen Stress in Plants*, Plant Cell Monographs 21, DOI 10.1007/978-3-7091-1254-0_18, © Springer-Verlag Wien 2014

The latter is often implemented in combination with packaging in plastic foils with well-defined gas transport properties (*Modified Atmosphere Packaging* or MAP). In addition to reducing respiration and, thus, delaying fruit ripening and senescence, CA and MA also control physiological and pathological diseases and/or insect infestation (Yahia 2006).

The optimal storage atmosphere composition is critical, as a too low O_2 partial pressure in combination with a too high CO_2 partial pressure may induce fermentation in the fruit. This causes off-flavours (e.g. ethanol) and storage disorders (e.g. flesh browning) (Veltman et al. 1999a; Franck et al. 2003a, 2007). Therefore, the O_2 and CO_2 partial pressure in commercial cool stores should be kept at an optimal value that needs to be determined in advance. This requires a large experimental effort, and several attempts have been done to model gas exchange of fruit and vegetables in order to reduce experimental work and increase our understanding of the physiology of fruit stored in hypoxic conditions.

In this chapter, we will first discuss the respiration behaviour of fruit and how it changes during its different developmental stages. We will then describe storage technologies such as controlled and modified atmosphere storage that exploit the response of fruit to hypoxic conditions and that are now commercially used on a wide scale. Subsequently, we will address physiological storage disorders related to hypoxia. We will focus on browning disorders and address their physiological causes.

2 Controlled and Modified Atmosphere Storage

2.1 Fruit Respiration

Living produce such as fruit needs to respire in order to provide the energy and metabolic building blocks it needs. Plant respiration is the oxidative breakdown of complex substrates normally present in the cell, such as starch, sugars and organic acids to CO_2, water and heat. In hypoxia or anoxia, the aerobic metabolism switches to fermentative metabolism with, amongst others, ethanol as an end product, which is far less efficient from the energetic point of view. During respiration, energy is created in the form of adenosine triphosphate (ATP), which can be further used for anabolic reactions, growth, regulation and maintenance of the cellular organisation and membrane integrity of living cells. Respiration plays an important role during storage of fruits since it implies the loss of respiration substrate, and hence, the loss of quality (Kader 1988).

Fruit development can be subdivided into three, not always clearly distinguishable, physiological stages including growth, maturation and senescence (Wills et al. 1998), and respiration is affected by the development stage of fruit. The growth stage is the period of cell division and enlargement. During this stage, the respiration decreases and reaches to minimum. The maturation stage involves biochemical processes that transform the fruit from an inedible organ to a tasteful product. Following maturation, senescence is the period when the processes of catabolic reactions surpass the synthesis reactions, leading to the deterioration of fruit. Fruit ripening encompasses the changes taking place in the later stages of maturation and the beginning of senescence. In so-called climacteric fruit such as apple and pear, the ripening process is normally associated with a rise in respiratory activity. Respiration can, therefore, be considered as an indicator of the natural end of a period of active synthesis and maintenance, and the beginning of the actual senescence of climacteric fruit. Hence, the suggested optimal harvest time for long-term controlled atmosphere storage of climacteric fruit is at the pre-climacteric stage where the respiration rate is still low before ripening starts (Lammertyn 2001).

The temperature, O_2 concentration and CO_2 concentration of the storage atmosphere are considered to be three main environmental factors to control respiration during storage. Temperature is the most significant factor, since all biochemical reactions taking place during ripening are retarded at low temperature resulting in a low respiration rate (Hertog et al. 1998; Lammertyn et al. 2001; Helena Gomes et al. 2010; Caleb et al. 2012; Gwanpua et al. 2012). Moreover, low temperature storage reduces the growth of fungi, bacteria and insects (Thompson 1998; Yahia 2006).

The relation between O_2 consumption and respiration has been described extensively in the literature over the past decades. A decrease of the O_2 concentration in the storage atmosphere results in a retarded metabolism in many fruit and vegetable species such as carrots, tomatoes, mandarin (Hertog et al. 1998; Luengwilai et al. 2007; Iqbal et al. 2009; Martínez-Sánchez et al. 2011), apples, pears and chicory (Peppelenbos and van't Leven 1996; Hertog et al. 1998; Lammertyn et al. 2001; Helena Gomes et al. 2010; Ho et al. 2010a, 2011; Gwanpua et al. 2012).

The role of CO_2 on the respiration metabolism of fruit is more ambiguous (Conesa et al. 2007). Increased CO_2 levels have been shown to retard respiration rate of broccoli, Belgian endives and sprouts (Peppelenbos et al. 1996; Hertog et al. 1998; East et al. 2009), whereas no effect was found on the respiration rate of onions, lettuce and spinach (Kubo et al. 1990; Conesa et al. 2007). A high CO_2 concentration might inhibit growth of spoilage microorganisms (Conesa et al. 2007) but may cause brown discoloration and softening of flesh tissue (Ke et al. 1990; Conesa et al. 2007; Pedreschi et al. 2008). Storage of fruit at high CO_2 (>5 kPa) level caused reduction in ascorbic acid level in pear and guava (Franck et al. 2003a; Singh and Pal 2008) or a decrease of acidity in potato (Angós et al. 2008).

2.2 CA and MA Storage

Post-harvest storage techniques to reduce the respiratory activity by limiting the O_2 concentration, increasing the CO_2 level, and in combination with a low temperature are widely used to extend the shelf life of fruits and vegetables. In MA packaging,

Table 1 Advised storage conditions for some commercial pome fruit cultivars in Belgium (Schenk 2007)

Cultivars	Cool storage (°C)	CA
Apple		
– Golden	0.5	2–3 % CO_2
– Gloster		2–3 % O_2
– Mutsu		
– Gala		
– Elstar	1	1–2 % CO_2
– Idared		2–3 % O_2
– Kanzi		
– Jonagold	0.8	2–3 % CO_2
– Jonagold mutants		1–2 % O_2
– Jonagored		
– Pinova		
– Greenstar		
– Belgica		
– Coxs Orange Pippin	3–3.5	<1 % CO_2
– Boskoop		2–3 % O_2
Pear		
– Conference	−1/−0.5	<0.7 % CO_2
		2.5–3 % O_2
– Doyenne du Comice	−0.5/0	<0.7 % CO_2

the fruit are kept fresh for a longer period by using a gas permeable foil allowing, to some extent, gas exchange between the internal atmosphere in the package and the external atmosphere (Lee et al. 1991). The internal atmosphere of the package can be generated by the O_2 consumption and CO_2 production of product itself, or injected with a gas mixture into the package to obtain fast equilibrium, or a combination of both. In CA, fruit and vegetables are stored at a reduced O_2 and/or elevated CO_2 concentration. Deviations with respect to the set point are corrected by the automated measurement and control system. O_2 is removed by injecting N_2 gas that is produced from normal air by a membrane gas separator or active coal scrubber, while CO_2 is removed with a calcium hydroxide scrubber (irreversible formation of calcium carbonate) or absorbed by an active coal scrubber. Storage under controlled atmosphere is widely used commercially to store all kinds of horticulture products. Some typical storage conditions are listed in Table 1.

3 Characterisation of Hypoxia-Related Physiological Storage Disorders

The optimal storage conditions in CA or MA storage are always a compromise: the temperature must be decreased to minimise metabolic activity while avoiding chilling or freezing damage; the O_2 partial pressure must be decreased to minimise respiration yet avoiding fermentation and an elevated CO_2 partial pressure helps to maintain colour but may induce storage disorders. External factors can thus interrupt, restrict or accelerate normal metabolic processes and potentially cause physiological storage disorders. Disorders occurring during long-term storage of pome fruit can cause economic loss, especially when disordered fruit cannot be distinguished externally from sound fruit.

Browning is an important category of disorders in pome fruit that appear in hypoxic storage conditions. It is characterised by internal browning of the flesh and the presence of cavities (see Fig. 1). Besides post-harvest environmental factors, adverse pre-harvest conditions during growth are important (Kays 1991). Browning disorders can cause economic losses due to the fact that damaged fruit cannot be distinguished externally from sound fruit (Franck et al. 2007). So far, no reliable prediction method exists for brown disorders (Neuwald et al. 2008). An accurate non-destructive test method for sorting and removing fruit with internal browning would be readily accepted by the apple industry (McGlone et al. 2005).

Several techniques have been investigated for non-destructively detecting internal browning. McGlone et al. (2005) reported the use of near infrared reflectance spectroscopy for browning measurements in "Braeburn" apples at realistic grading speeds (approximately 5 fruit per second). Because of the limited penetration depth of NIR radiation in fruit tissue, transmission measurements are necessary. The advantage of this technology is that it is cheap, fast and can be combined with existing grading lines. However, the transmitted NIR spectrum needs to be related to the variable of interest via multivariate statistics. As this model varies with cultivar, orchard, season, temperature, used spectrophotometer, etc., there is a need for extensive calibration measurements (Nicolaï et al. 2007).

X-ray computer tomography (CT) imaging (Lammertyn et al. 2003a) and magnetic resonance imaging (MRI) (Wang and Wang 1989; Lammertyn et al. 2003a; Hernández-Sánchez et al. 2007) have been successfully used to detect brown heart in "Conference" pears. X-ray CT imaging is based on differences in mass density and absorption of the material, and contrasting greyscales indicate the local redistribution of water due to membrane damage. MR, on the other hand, employs static fields and radio frequencies in order to obtain information on proton mobility (of the water fraction) in biological systems. Lammertyn et al. (2003a) concluded that MRI was the most appropriate technique to study the development of core breakdown disorder during post-harvest storage, compared with X-ray CT, since its sensitivity is higher, especially in the case of incipient brown discoloration. They also discovered that incipient flesh browning was already present after 2 months of storage under browning inducing conditions (no cooling period, 10 % CO_2, 1 % O_2, −1 °C) and that the brown zone did not grow spatially during storage but only the intensity of brown discoloration increased (Lammertyn et al. 2003a). The disadvantages of MRI and X-ray CT are the high capital cost of the equipment and the low speed of measurement. Also, special safety measures are required because of the ionising radiation in the case of X-ray CT, and the need for strong magnetic fields for performing MRI measurements.

Fig. 1 Browning disorders in "Conference" pear (*left*) and "Braeburn" apple (*right*) in CA storage

4 Physiological Background of Browning Disorders in Pome Fruit

4.1 Browning is Related to Respiration and Gas Exchange

There is a considerable body of evidence that the development of browning in several pome fruit depends on the storage conditions: low O_2 levels and high CO_2 levels increase the susceptibility for browning considerably (Lammertyn et al. 2000; Gong et al. 2001; Veltman et al. 2003; Franck et al. 2003a, 2007; Pedreschi et al. 2008; de Castro et al. 2008; Lee et al. 2012). Interestingly, late picked fruit are far much more susceptible than early picked fruit (Lammertyn et al. 2000; Arzani et al. 2009). Lee et al. (2012) observed that flesh browning was associated with increased fermentative metabolic products such as acetaldehyde and ethanol. This suggests that browning is somehow related to the respiratory activity that greatly increases during climacteric ripening.

Further, large fruit were found to be more susceptible than small fruit (Lammertyn et al. 2000; Ho et al. 2008). While there might be confounding between the effect of size and picking time, this and the often concentric boundary of the brown area in affected fruit indicates that gas diffusion might play a role. Indeed, respiration and gas diffusion in bulky plant organs are tightly linked (Ho et al. 2010b). Cell wall and membrane degradation by senescence may cause the leaking of the cellular liquid into the intra-cellular spaces, reduce the diffusion of gases through the fruit (Schotsmans et al. 2004; James and Jobling 2009) and lead to O_2 depletion and the build-up of toxic quantities of CO_2.

4.2 Considerable Gas Gradients May Occur in Fruit During Hypoxic Storage

Respiration consumes O_2 and produces CO_2. As a consequence, the O_2 and CO_2 concentration in the centre of the fruit are below and above that of the storage atmosphere, respectively. This results in a driving force for gas transport—inward for O_2 and outward for CO_2. Gas transport through the skin occurs mainly through the pores in the fruit skin; nevertheless, the cuticular route does contribute to both CO_2 and O_2 transport (Cheng et al. 1998). In young apples, the pores in the fruit skin are stomata with guard cells regulating their opening and closure. In mature apples, the stomata are no longer functional, most of them being closed, completely covered with wax (Veraverbeke et al. 2001) or transformed into lenticels. These lenticels do not function actively like stomata but close progressively owing to suberisation of substomatal cells (Park 1991). A high gas resistance of skin tissue including epidermis and subepidermis is believed to be caused by a very dense structure of skin tissue (Schotsmans et al. 2004; Verboven et al. 2008). In cortex tissue of fleshy fruits, an internal ventilation system consisting of continuous gaseous channels is presented (Mendoza et al. 2007; Verboven et al. 2008) (Fig. 2). This system is formed by intercellular spaces interconnected with narrow capillary tubes. These intercellular spaces are supposed to be formed during cell division and growth (Lapsley et al. 1992). Using synchrotron X-ray tomography, Verboven et al. (2008) found that the gas exchange mechanisms in plant organs critically depend on insights in the three-dimensional (3D) structural arrangement of cells and voids.

As it is difficult to measure the O_2 and CO_2 concentration in vivo, models have been developed to describe gas exchange of fruit with their environment (Mannapperuma et al. 1991; Lammertyn et al. 2003b; Ho et al. 2008, 2010a, 2011). Ho et al. (2008) developed a permeation–diffusion reaction model to study gas exchange of intact pear at the macroscale level that. It accounts for both diffusion and pressure driven exchange of these gasses and incorporates anisotropic transport properties. O_2 depletion and CO_2 production by respiration and fermentation were modelled by means of Michaelis–Menten kinetics that was modified to account for O_2 and CO_2 inhibition effects. A computer vision system was used to reconstruct the actual geometry of the pear and the model equations were discretised over this geometry. The model was validated successfully under steady and transient conditions at 1 °C. Based on an in silico study, the authors found that considerable gradients of metabolic gases may exist in fruit, hereby invalidating earlier models (Cameron and Yang 1982; Amarante et al. 2001) in which it was assumed that gas transport could be lumped. Higher values of the Michaelis–Menten constants of the respiration of intact fruit compared to those of cortex tissue could be attributed to the gas exchange barrier function of fruit tissue (Ho et al. 2013). The larger the fruit, the lower and higher the O_2 and CO_2 partial pressures in the fruit centre, respectively, indicating a larger susceptibility to fermentation and storage disorders, consistent with the results of Lammertyn et al. (2000). The model predicted considerable gradients in Conference pear and

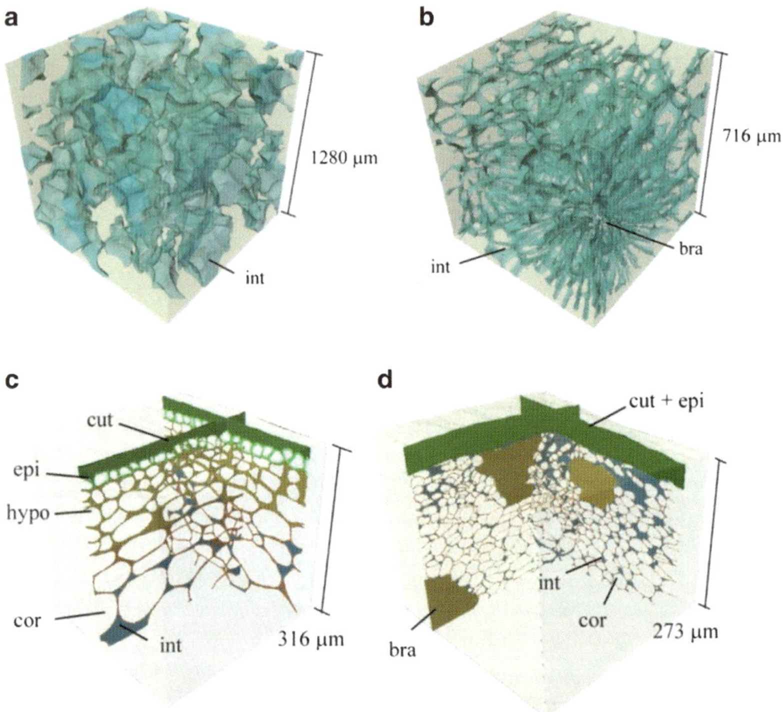

Fig. 2 3D rendering of the void network of apple (**a**) and pear (**b**) fruit cortex. The marked components are the intercellular void spaces (int) and the brachysclereids (bra). While the voids between apple parenchyma are large and form an incompletely connected network, those of pear are very small and form a complete network throughout the cortex sample without preferential direction. (**c**) and (**d**) Structural 3D models of epidermis and outer cortex of apple and pear, respectively (Verboven et al., Plant Physiol. 2008, 147:518–527)

different apple cultivars (Braeburn, Jonagold, Kanzi) (Ho et al. 2010b). In Braeburn—a cultivar susceptible to browning—stored at ultralow O_2 conditions, the predicted O_2 concentration was as low as 0.014 kPa (0.29 μM) at 1 % O_2, 2.5 % CO_2 and 1 °C, while for Jonagold it was equal to 0.102 kPa (2.1 μM). This may explain why, in contrast to Braeburn, Jonagold apple fruit can be stored for up to 10 months without risk of developing of storage disorders due to fermentation (Saquet et al. 2000).

In a further gas exchange modelling study, Ho et al. (2011) constructed a multiscale model to study the gas transport in apple fruit in more detail. In a first step, the previously developed macroscale model (Ho et al. 2008) was used to calculate the macroscopic gas concentration profile in a Jonagold apple that was stored in commercial storage conditions of 1 % O_2, 2.5 % CO_2 and 1 °C. Then, in a second step, microscale simulations were carried out with a model that incorporated the actual 3D microstructure of tissue at the position near the core of the fruit where the O_2 concentration was found to be the lowest (Fig. 3). This region was

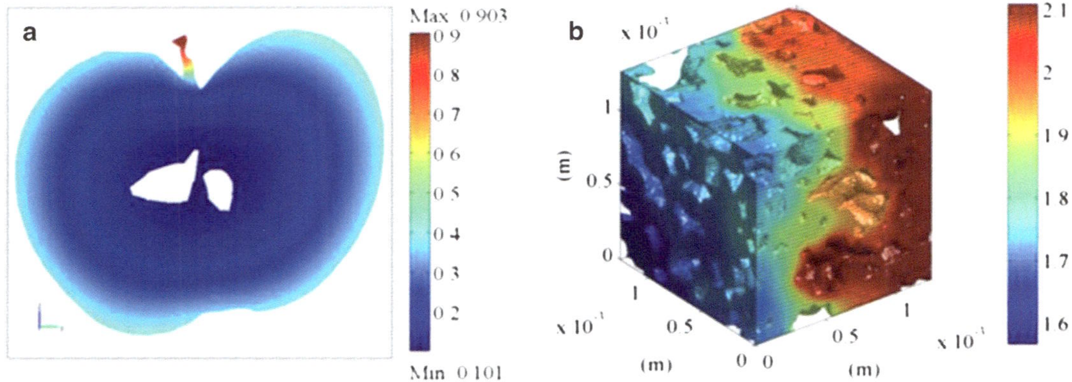

Fig. 3 (**a**) O_2 distribution of the apple intact fruit at typical commercial storage conditions (1 kPa O_2, 2.5 kPa CO_2 and 1 °C). The *colour bar* indicates the O_2 partial pressure (kPa). (**b**) Simulated intra-cellular O_2 concentration of microscopic tissue computed from the lowest O_2 concentration near the core of the macroscale simulation. The *colour bar* indicates O_2 concentration in cytoplasm (μM) (Ho QT et al. Plant Physiol. 2011, 155: 1158–1168)

considered to be the most susceptible to anoxia. The corresponding local intracellular O_2 concentration near the core was (1.45 ± 0.12) μM, well above the threshold of the Michaelis–Menten constant K_{m,O_2} of cytochrome *c* oxidase, the rate limiting enzyme in the oxidative phosphorylation, which is 0.14 μM (Millar et al. 1994). In such conditions, fermentation would be unlikely. The multiscale analysis was repeated for an O_2 concentration of 0.2 % O_2, which is known to cause fermentation symptoms inside the fruit. The macroscale model predicted that the lowest O_2 concentration was 0.15 μM, still higher than the K_m of cytochrome *c* oxidase. However, the local O_2 concentration calculated at the microscale at this position decreased to as low as 0.088 μM, well below the K_m of cytochrome *c* oxidase. The analysis clearly demonstrated that the onset of fermentation depends on the respiratory activity as well as on the microscopic structure of the tissue. Further, the multiscale model clearly showed that the local O_2 concentration in the cells may be much lower than what is estimated by simpler macroscale models.

4.3 Browning Development Is a Consequence of Major Changes in the Respiratory Metabolism

The effect of O_2 and CO_2 on the central metabolism of pear has been investigated by proteomics techniques. A problem is that the latter are destructive: even severe storage conditions cause browning in only a subset of the exposed fruit, and it is impossible to predict whether a fruit that is sampled would later on have developed browning. On the other hand, browning most probably leads to considerable alterations in the metabolism and massive proteolysis, so post hoc analysis of brown tissue does not necessarily provide information on the chain of events that leads to browning. Most authors have, therefore, focused on tissue samples of fruit that were stored in severe storage conditions but did not develop the disorder yet at the time of sampling ("sound" fruit, in contrast to unaffected, "healthy" fruit and "brown" fruit). The hypothesis was then that at least the metabolomic and proteomic changes in the tissue due to these conditions may provide clues on the underlying biochemistry of browning.

Pedreschi et al. (2007) found major changes in the central metabolism of sound pear fruit compared to both healthy and brown fruit. While triosephosphate isomerase was almost absent in healthy tissue and totally absent in brown tissue, this enzyme was upregulated in sound tissue, indicating an accelerated glycolytic flux consistent with an attempt of the cells to increase their ATP production when oxidative phosphorylation is stalled. Also, malate dehydrogenase, fumarase and malic enzyme were upregulated in sound tissue as compared to healthy and brown tissue. The latter catalyses the oxidative decarboxylation of L-malate to pyruvate and provides the hypoxic cells with NADPH and pyruvate for biosynthesis of defence related compounds and fatty acids needed for membrane repair. These results were also confirmed in experiments with slices of pear fruit exposed to severe gas conditions (1 kPa O_2, 10 kPa CO_2 and −1 °C, Pedreschi et al. 2009). Additionally, an upregulation of transketolase was found, indicating activation of the pentose phosphate pathway in hypoxic conditions. Also, NAD-dependent malate dehydrogenase appeared to be upregulated. In line with these proteomics results, Lee et al. (2012) observed that browning was related to elevated acetaldehyde, ethanol and ethyl ester levels in Braeburn apples.

These results indicate that browning development is a consequence of an imbalance between oxidative and reductive processes caused by too low oxygen or too high carbon dioxide conditions in hypoxic zones of the fruit which lead to a deficiency of reducing equivalents for defensive mechanisms, cell damage repair processes and biosynthesis reactions. Quantitative techniques, such as metabolic flux analysis, are essential to further elucidate the biochemical chain of events, which lead to core breakdown.

4.4 Browning of Tissue Is a Consequence of Membrane Damage

Because of the changes in the central metabolism, notably in the malic acid concentration, the intra-cellular pH at high CO_2 concentration levels may change. Moreover, it is possible that a change of the pH may result in loss of the membrane integrity and formation of reactive oxygen species such as super oxide species during the respiration (Veltman et al. 1999a, 2003; Saquet et al. 2003; Franck et al. 2003a, 2007).

The browning symptoms are due to the enzymatic oxidation of phenolic compounds by polyphenol oxidase (PPO) to *o*-quinones, which are very reactive and form brown coloured polymers (Mathew and Parpia 1971; Mayer 1987). The initial reaction, catalysed by PPO, uses O_2 as co-substrate. The important factors involved in enzymatic browning are (1) the phenolics concentration, (2) the PPO activity and (3) other factors such as L-ascorbic acid (L-AA) (L-AA is able to convert *o*-quinones back to diphenols) and peroxidases (which react also with phenolics using H_2O_2 as co-substrate) (Amiot et al. 1992; Nicolas et al. 1994; Franck et al. 2007). Since PPO and its substrate are located in different cell compartments (cytoplasm/plastids and vacuole, respectively) (Nicolas et al. 1994; Dixon and Paiva 1995), PPO activity was found not to be a limiting factor in the enzymatic browning (Amiot et al. 1992; Veltman et al. 1999b) and enzymatic browning is a direct consequence of membrane disintegration.

The importance of the antioxidant system was confirmed by several authors who investigated the effect of diphenylamine (DPA, a known inhibitor of a superficial scald in apple caused by oxidation of certain components in the apple skin); treatment immediately after harvest was found to prevent internal browning in different apple cultivars (Fernández-Trujillo et al. 2001; Argenta et al. 2002; de Castro et al. 2008; Mattheis and Rudell 2008; Felicetti et al. 2011). Further,

Pink Lady apples under CO_2 stress showed increased flesh browning (de Castro et al. 2008). In this study, a high H_2O_2 level, indicating tissue stress, and low ascorbic acid concentration were related to the disorder. Ascorbic acid may prevent tissues against oxidative damage by scavenge superoxide and reduce H_2O_2 (Blokhina and Fagerstedt 2010). In Braeburn, no clear association between ascorbic acid level and browning was found (Felicetti et al. 2011); however, high levels of SOD, catalase and peroxidase—enzymes of the ascorbic acid pathway—were correlated with low levels of Braeburn browning disorder (Toivonen et al. 2003). Franck et al. (2003a, b) investigated the changes of L-AA in pear during post-harvest storage in normal air and hypoxic conditions. The rate of decline of L-AA was more than fourfold faster under browning-inducing conditions than under optimal post-harvest conditions. The ascorbic acid concentration profile was found asymmetrical indicating that the sampling position is quite crucial and that analysing tissue cut randomly can give misleading results. It was found that all the brown tissue lay in between the contour line of 0.4 mg 100 g^{-1} FW, which corresponded very well with the threshold value of 0.37 mg 100 g^{-1} FW, derived from a logistic regression analysis of the ascorbic acid concentration of the inner core of 20 sound and 20 disordered pears. Again, since all analyses were unavoidably done post hoc, no direct causal relationship could be established, and, hence, it was not clear whether a low ascorbic acid concentration in brown tissue would be the cause rather than the consequence of browning. Interestingly, Pedreschi et al. (2007) found that ascorbate peroxidase was upregulated in sound tissue.

4.5 *Biomarkers for Browning*

Several attempts have been made to identify biomarkers that may be used to predict the susceptibility to browning in an early stage.

Based on a proteomics experiment on both intact pear and pear tissue slices, Pedreschi et al. (2007) found a pronounced decrease in expression of PR proteins due to hypoxia compared with air conditions. In brown pear tissue (sub-optimally stored under limiting levels of oxygen or high levels of carbon dioxide), the action of the enzyme PPO probably contributed to the total downregulation of these allergenic protein compared to the sound tissue. According to the authors, the total downregulation of these allergenic proteins during CA storage of pears might be related to browning.

Pedreschi et al. (2009) investigated metabolic markers for browning. Gluconic acid was found to accumulate in brown tissue, indicating impairment of the pentose phosphate pathway most probably with an insufficient production of NADPH for membrane repair processes and to maintain the cellular redox state. The accumulation of gluconic and threonic acids might also be the result of ascorbic acid catabolism due to lack of reducing equivalents. The imbalance between GABA and putrescine compared to a reduced glutamate concentration in brown tissue may be considered as an indicator of hypoxic stress. The concentration of other compounds, which are believed to be related to an oxidative stress response such as trehalose and putrescine, was also considerably higher in brown tissue than in sound tissue. Finally, the concentration of some sugars, which are typically found in xyloglucans, also increased during the core browning process, possibly indicating primary cell wall breakdown due to enzymatic processes or by action of hydroxyl radicals. Again, as of now, it is unclear whether these biomarkers can be used to predict the development of browning.

5 Conclusions and Outlook

Controlled and modified atmosphere storage are popular techniques to extend the storage life of fruit and vegetables, and pome fruit in particular. Their success is due to a reduction of the respiratory activity of the product and associated quality degrading reactions. This reduction is mainly realised by reducing and increasing the O_2 and CO_2 partial pressures, respectively, usually in combination with a reduction of the temperature. However, the optimal storage conditions are always a compromise that may lead to storage disorders. Browning is such a disorder that is characterised by brown spot extending from the core towards the boundary, and in a later phase the formation of cavities. The physiological cause of browning is an imbalance between oxidative and reductive processes that is a consequence of hypoxic zones in the fruit. The latter are due to the steep gas gradients that have been observed and modelled in fruit stored under CA conditions. Because of the reduced respiration rate, deteriorating membranes can no longer be repaired and decompartmentalisation starts. Eventually, phenolic compounds mix with polyphenol oxidase and brown polymers develop. While the broad scheme of the development of the disorder has been unravelled, many details still need to be revealed. An improved insight in the involved biochemical pathways requires the use of metabolic flux analysis.

As the presence of browning disorders is not visible externally, more research is required in developing non-destructive techniques for measuring browning. Both NIR transmittance spectroscopy and X-ray tomography are promising techniques, but many problems remain to be solved. Also, for practical applications it would be most helpful to have biomarkers that can be used to predict the susceptibility of fruit with respect to browning. Both metabolic and proteomic biomarkers have been identified, but their predictive value still needs to be validated, and mRNA biomarkers still have to be identified.

Acknowledgements The authors thank the Research Council of the K.U. Leuven (OT 08/023), the Flanders Fund for Scientific Research (project G.0603.08) and the Institute for the Promotion of Innovation by Science and Technology in Flanders (project IWT-050633, IWT scholarship SB/0991469) for financial support. Quang Tri Ho is postdoctoral fellow of the Flanders Fund for Scientific Research (FWO Vlaanderen).

References

Amarante C, Banks NH, Ganesh S (2001) Relationship between character of skin cover of coated pears and permeance to water vapour and gases. Postharvest Biol Technol 21:291–301

Amiot MJ, Tacchini M, Aubert S, Nicolas J (1992) Phenolic composition and browning susceptibility of various apple cultivars at maturity. J Food Sci 4:958–962

Angós I, Vírseda P, Fernández T (2008) Control of respiration and color modification on minimally processed potatoes by means of low and high O_2/CO_2 atmospheres. Postharvest Biol Technol 48:422–430

Argenta LC, Fan X, Mattheis JP (2002) Responses of 'Fuji' apples to short and long duration exposure to elevated CO_2 concentration. Postharvest Biol Technol 24:13–24

Arzani K, Khoshghalb H, Malakouti MJ, Barzegar M (2009) Polyphenoloxidase activity, polyphenol and ascorbic acid concentrations and internal browning in Asian pear (*Pyrus serotina* Rehd.) fruit during storage in relation to time of harvest. Eur J Hortic Sci 74(2):61–65

Blokhina O, Fagerstedt KV (2010) Oxidative metabolism, ROS and NO under oxygen deprivation. Plant Physiol Biochem 48:359–373

Caleb OJ, Mahajan PV, Opara UL, Witthuhn CR (2012) Modelling the respiration rates of pomegranate fruit and arils. Postharvest Biol Technol 64:49–54

Cameron AC, Yang S (1982) A simple method for the determination of resistance to gas diffusion in plant organs. Plant Physiol 70:21–23

Cheng Q, Banks NH, Nicholson SE, Kingsley AM, Mackay BR (1998) Effect of temperature on gas exchange of 'Braeburn' apples. NZ J Crop Hortic Sci 26:299–306

Conesa A, Verlinden BE, Artés-Hernández F, Nicolaï BM, Artés F (2007) Respiration rates of fresh-cut bell peppers under superatmospheric and low oxygen with or without high carbon dioxide. Postharvest Biol Technol 45:81–88

de Castro E, Barrett DM, Jobling J, Mitcham EJ (2008) Biochemical factors associated with a CO_2-induced flesh browning disorder of Pink Lady apples. Postharvest Biol Technol 48:182–191

Dixon RA, Paiva NL (1995) Stress-induced phenylpropanoid metabolism. Plant Cell 7:1085–1097

East AR, Trejo Araya XI, Hertog MLATM, Nicholson SE, Mawson AJ (2009) The effect of controlled atmospheres on respiration and rate of quality change in 'Unique' feijoa fruit. Postharvest Biol Technol 53:66–71

Felicetti E, Mattheis JP, Zhu Y, Fellman JK (2011) Dynamics of ascorbic acid in 'Braeburn' and 'Gala' apples during on-tree development and storage in atmospheres conducive to internal browning development. Postharvest Biol Technol 61:95–102

Fernández-Trujillo JP, Nock JF, Watkins CB (2001) Superficial scald, carbon dioxide injury and changes of fermentation products and organic acids in 'Cortland' and 'Law Rome' apples after high carbon dioxide stress treatment. J Am Soc Hortic Sci 126:235–241

Franck C, Baetens M, Lammertyn J, Verboven P, Nicolaï BM (2003a) Ascorbic acid content during fruit development and postharvest storage of conference pears. J Agric Food Chem 51:4757–4763

Franck C, Baetens M, Lammertyn J, Scheerlinck N, Nicolaï BM (2003b) Ascorbic acid mapping to study core breakdown. Postharvest Biol Technol 30:133–142

Franck C, Lammertyn J, Ho QT, Verboven P, Verlinden B, Nicolaï BM (2007) Browning disorders in pear: a review. Postharvest Biol Technol 43:1–13

Gong Y, Toivonen PMA, Lau OL, Wiersma PA (2001) Antioxidant system level in 'Braeburn' apple is related to its browning disorder. Bot Bull Acad Sin 42:259–264

Gwanpua SG, Verlinden BE, Hertog MLATM, Bulens I, Van de Poel B, Van Impe J, Nicolaï BM, Geeraerd AH (2012) Kinetic modeling of firmness breakdown in 'Braeburn' apples stored under different controlled atmosphere conditions. Postharvest Biol Technol 67:68–74

Helena Gomes M, Beaudry RM, Almeida DPF, Xavier Malcata F (2010) Modelling respiration of packaged fresh-cut 'Rocha' pear as affected by oxygen concentration and temperature. J Food Eng 96:74–79

Hernández-Sánchez N, Hills BP, Barreiro P, Marigheto N (2007) An NMR study on internal browning in pears. Postharvest Biol Technol 44:260–270
Hertog MLATM, Peppelenbos HW, Evelo RG, Tijskens LM (1998) A dynamic and generic model on the gas exchange of respiring produce: the effects of oxygen, carbon dioxide and temperature. Postharvest Biol Technol 14:335–349
Ho QT, Verboven P, Verlinden BE, Lammertyn J, Vandewalle S, Nicolaï BM (2008) A continuum model for metabolic gas exchange in pear fruit. PLoS Comput Biol 4(3):e1000023
Ho QT, Verboven P, Verlinden BE, Nicolaï BM (2010a) A model for gas transport in pear fruit at multiple scales. J Exp Bot 61:2071–2081
Ho QT, Verboven P, Verlinden B, Schenk A, Delele M, Rolletschek H, Vercammen J, Nicolaï BM (2010b) Genotype effects on internal gas gradients in apple fruit. J Exp Bot 61:2745–2755
Ho QT, Verboven P, Verlinden BE, Herremans E, Wevers M, Carmeliet J, Nicolaï BM (2011) A 3-D multiscale model for gas exchange in fruit. Plant Physiol 155(3):1158–1168
Ho QT, Verboven P, Verlinden BE, Schenk A, Nicolaï BM (2013) Controlled atmosphere storage may lead to local ATP deficiency in apple. Postharvest Biol Technol 78:103–112
Iqbal T, Rodrigues FAS, Mahajan PV, Kerry JP (2009) Mathematical modeling of the influence of temperature and gas composition on the respiration rate of shredded carrots. J Food Eng 91:325–332
James HJ, Jobling JJ (2009) Contrasting the structure and morphology of the radial and diffuse flesh browning disorders and CO_2 injury of 'Cripps Pink' apples. Postharvest Biol Technol 53:36–42
Kader AA (1988) Respiration and gas exchange of vegetables. In: Weichmann J (ed) Postharvest physiology of vegetables. Marcel Dekker, New York, pp 25–43
Kays SJ (1991) Postharvest physiology of perishable plant products. Van Nostrand Reinhold, New York
Ke D, van Gorsel H, Kader AA (1990) Physiological and quality responses of 'Bartlett' pears to reduced O_2 and enhanced CO_2 levels and storage temperature. J Am Soc Hortic Sci 115:435–439
Kubo Y, Inaba A, Nakamura R (1990) Respiration and ethylene production in various harvested crops held in CO_2-enriches atmospheres. J Am Soc Hortic Sci 115:975–978
Lammertyn J (2001) A respiration-diffusion model to study core breakdown in conference pears. PhD dissertation, KU Leuven
Lammertyn J, Aerts M, Verlinden BE, Schotsmans W, Nicolaï BM (2000) Logistic regression analysis of factors influencing core breakdown in 'Conference' pears. Postharvest Biol Technol 20:25–37
Lammertyn J, Franck C, Verlinden BE, Nicolaï BM (2001) Comparative study of the O_2, CO_2 and temperature effect on respiration between 'Conference' pear cells in suspension and intact pears. J Exp Bot 52:1769–1777
Lammertyn J, Dresselaers T, Van Hecke P, Jancsók P, Wevers M, Nicolaï BM (2003a) MRI and X-ray CT study of spatial distribution of core breakdown in 'Conference pears'. Magn Reson Imaging 21:805–815
Lammertyn J, Scheerlinck N, Jancsók P, Verlinden BE, Nicolaï BM (2003b) A respiration–diffusion model for Conference pears I: model development and validation. Postharvest Biol Technol 30:29–42
Lapsley KG, Escher FE, Hoehn E (1992) The cellular structure of selected apple varieties. Food Struct 11:339–349
Lee DS, Haggar PE, Lee J, Yam KL (1991) Model of fresh produce respiration in modified atmospheres based on principles of enzyme kinetics. J Food Sci 56:1580–1585
Lee J, Mattheis JP, Rudell DR (2012) Antioxidant treatment alters metabolism associated with internal browning in 'Braeburn' apples during controlled atmosphere storage. Postharvest Biol Technol 68:32–42
Luengwilai K, Sukjamsai K, Kader AA (2007) Responses of 'Clemenules Clementine' and 'W. Murcott' mandarins to low oxygen atmospheres. Postharvest Biol Technol 44:48–54

Mannapperuma JD, Singh RP, Montero ME (1991) Simultaneous gas diffusion and chemical reaction in foods stored in modified atmospheres. J Food Eng 14:167–183

Martínez-Sánchez A, Tudela JA, Luna C, Allende A, Gil MI (2011) Low oxygen levels and light exposure affect quality of fresh-cut Romaine lettuce. Postharvest Biol Technol 59:34–42

Mathew AG, Parpia HAB (1971) Food browning as a polyphenol reaction. Adv Food Nutr Res 19:75–145

Mattheis JP, Rudell DR (2008) Diphenylamine metabolism in 'Braeburn' apples stored under conditions conducive to the development of internal browning. J Agric Food Chem 56:3381–3385

Mayer AM (1987) Polyphenoloxidase in plants-recent progress. Phytochemistry 26:11–20

McGlone VA, Martinsen PJ, Clark CJ, Jordan RB (2005) On-line detection of brown heart in Braeburn apples using near infrared transmission measurements. Postharvest Biol Technol 37:142–151

Mendoza F, Verboven P, Mebatsion HK, Kerckhofs G, Wevers M, Nicolaï BM (2007) Three-dimensional pore space quantification of apple tissue using X-ray computed microtomography. Planta 226:559–570

Millar AH, Bergersen FJ, Day DA (1994) Oxygen affinity of terminal oxidases in soybean mitochondria. Plant Physiol Biochem 32:847–852

Neuwald DA, Kittemann D, Streif J (2008) Possible Prediction of physiological storage disorders in 'Braeburn' apples comparing fruit of different orchards. In Streif J, McCormick R (eds) Proceedings of the international conference on ripening regulation and postharvest fruit quality. Acta horticultute, vol 796. ISHS, Belgium

Nicolaï BM, Beullens K, Bobelyn E, Peirs A, Saeys W, Theron KI, Lammertyn J (2007) Nondestructive measurement of fruit and vegetable quality by means of NIR spectroscopy: a review. Postharvest Biol Technol 46:99–118

Nicolas J, Richard-Forget FC, Goupy P, Amiot M, Aubert S (1994) Enzymatic browning reactions in apple and apple products. Crit Rev Food Sci Nutr 34:109–157

Park YM (1991) Seasonal changes in resistance to gas diffusion of McIntosh apples in relation to development of lenticel structure. J Korean Soc Hortic Sci 32:329–334

Pedreschi R, Vanstreels E, Carpentier S, Hertog M, Lammertyn J, Robben J, Noben J, Swennen R, Vanderleyden J, Nicolaï BM (2007) Proteomic analysis of core breakdown disorder in Conference pears (*Pyrus communis* L.). Proteomics 7(12):2083–2099

Pedreschi R, Hertog M, Robben J, Noben JP, Nicolaï BM (2008) Physiological implications of controlled atmosphere storage of 'Conference' pears (Pyrus communis L.): a proteomic approach. Postharvest Biol Technol 50:110–116

Pedreschi R, Franck C, Lammertyn J, Erban A, Kopka J, Hertog M, Verlinden BE, Nicolaï BM (2009) Metabolic profiling of 'Conference' pears under low oxygen stress. Postharvest Biol Technol 51(2):123–130

Peppelenbos HW, van't Leven J (1996) Evaluation of four types of inhibition for modeling the influence of carbon dioxide on oxygen consumption of fruit and vegetables. Postharvest Biol Technol 7:27–40

Peppelenbos HW, Tijskens LMM, van't Leven J, Wikinson EC (1996) Modeling oxidative and fermentative carbon dioxide production of fruits and vegetables. Postharvest Biol Technol 9:283–295

Saquet AA, Streif J, Bangerth F (2000) Changes in ATP, ADP and pyridine nucleotide levels related to the incidence of physiological disorders in 'Conference' pears and 'Jonagold' apples during controlled atmosphere storage. J Hortic Sci Biotechnol 75:243–249

Saquet AA, Streif J, Bangerth F (2003) Energy metabolism and membrane lipid alterations in relation to brown heart development in Conference pears during delayed controlled atmosphere storage. Postharvest Biol Technol 30:123–132

Schenk A (2007) Advies bewaaromstandigheden 2007. Fruitteeltnieuws 20:12–16

Schotsmans W, Verlinden BE, Lammertyn J, Nicolaï BM (2004) The relationship between gas transport properties and the histology of apple. J Sci Food Agric 84:1131–1140

Singh SP, Pal RK (2008) Controlled atmosphere storage of guava (*Psidium guajava* L.) fruit. Postharvest Biol Technol 47:296–306

Thompson AK (1998) Controlled atmosphere storage of fruits and vegetables. CAB International, London

Toivonen PMA, Wiersma PA, Gong YP, Lau OL (2003) Levels of antioxidant enzymes and lipid soluble antioxidants are associated with susceptibility to internal browning in 'Braeburn' apples. Acta Hortic 600:57–61

Veltman RH, Sanders MG, Persijn ST, Peppelenbos HW, Oosterhaven J (1999a) Decreased ascorbic acid levels and brown core development in pears (*Pyrus communis* cv. communis). Physiol Plant 107:39–45

Veltman RH, Larrigaudiere C, Wichers HJ, Van Schaik ACR, Van der Plas LHW, Oosterhaven J (1999b) PPO activity and polyphenol content are not limiting factors during brown cove development in pears (*Pyrus communis* L. cv. Conference). J Plant Physiol 154:697–702

Veltman RH, Lentheric I, Van der Plas LHW, Peppelenbos HW (2003) Browning in pear fruit (*Pyrus communis* L. cv Conference) may be a result of a limited availability of energy and antioxidants. Postharvest Biol Technol 28:295–302

Veraverbeke EA, Lammertyn J, Saevels S, Nicolaï BM (2001) Changes in chemical wax composition of three different apple (*Malus domestica* Borkh.) cultivars during storage. Postharvest Biol Technol 23:197–208

Verboven P, Kerckhofs G, Mebatsion HK, Ho QT, Temst K, Wevers M, Cloetens P, Nicolaï BM (2008) 3-D gas exchange pathways in pome fruit characterised by synchrotron X-ray computed tomography. Plant Physiol 147:518–527

Wang CY, Wang PC (1989) Nondestructive detection of core breakdown in 'Bartlett' pears with nuclear magnetic resonance imaging. HortScience 24:106–109

Wills R, McGlasson B, Graham D, Joyce D (1998) Postharvest, an introduction to the physiology and handling of fruit, vegetables and ornamentals. CAB International, Wallingford

Yahia EM (2006) Modified and controlled atmospheres for tropical fruits. Stewart Postharvest Rev 5:1–10

Low Oxygen Stress in Horticultural Practice

Wessel L. Holtman, Berry J. Oppedijk, Marco Vennik, and Bert van Duijn

Abstract Gasses (especially oxygen and carbon dioxide) in the root environment are an important factor for plant performance. Based on experimental data, we will explain the most important factors determining oxygen concentration in the root environment and describe effects of "too" low oxygen concentrations on the plant. Importantly, a main conclusion is that growers can control oxygen concentrations in the root zone by applying the right irrigation strategy.

1 Impact of Oxygen Stress on Plants

The direct effect of different oxygen concentrations in the root environment on the development of cucumber plants has been measured (Grodan et al. 2004; Fig. 1). These experiments showed that growth was retarded already at oxygen levels lower than 3.5 mg/L (i.e., about 35 % of the oxygen that can be maximally present in water under these conditions, this equals to an equilibrium with about 8 % oxygen in air). Also for other horticultural crops, such as rose and tomato, similar critical levels for the oxygen availability have been determined.

In the remainder of this chapter, we will express oxygen concentrations in solution as the percentage of oxygen in air with which the measured solution would be in equilibrium. So no matter in what medium oxygen was monitored, 0 % always means that no oxygen is present at all, whereas 21 % means that a solution was air saturated. It should be emphasized that experimental data on oxygen as described in this chapter have been obtained using chemico-optical oxygen sensors (see also chapter "Methods and Techniques to Measure Molecular Oxygen in Plants"; Ast and Draaijer 2014), these are considered as the most reliable oxygen measurement tool for continuous measurements in the root zone.

W.L. Holtman (✉) • B.J. Oppedijk • M. Vennik • B. van Duijn
Fytagoras BV, Sylviusweg 72, 2333 BE Leiden, The Netherlands
e-mail: wessel.holtman@fytagoras.nl

J.T. van Dongen and F. Licausi (eds.), *Low-Oxygen Stress in Plants*, Plant Cell Monographs 21, DOI 10.1007/978-3-7091-1254-0_19, © Springer-Verlag Wien 2014

Fig. 1 Cucumber plants, after 3 weeks on nutrient solution, containing 10 (*left plant*), 3.5 (*middle plant*), or 0.5 mg oxygen per liter (*right plant*)

Below the so-called critical oxygen pressure [COP; see also chapter "Plant Internal Oxygen Transport (Diffusion and Convection) and Measuring and Modelling Oxygen Gradients"; Armstrong and Armstrong 2014], both the roots and shoot will be affected due to the lack of oxygen for respiratory energy metabolism (see chapter "Oxygen Consumption Under Hypoxic Conditions"; Paepke et al. 2014). In the root environment, the development of roots will slow down and roots will eventually die. In a recent study (Vennik et al. 2011), using pot gerbera as a model plant, it was shown that when plants were grown at high water content in the substrate (i.e., 80 % V/V in the lower part of the pot) for a period of 45 days, oxygen concentrations dropped dramatically (lower than 5 %) (Fig. 2a). As a result, roots showed a dark color, they had a slimy structure, and gradually died (Fig. 2a). In conditions where water content was set to 50 %, the oxygen concentrations were about 21 %. Under these circumstances, roots showed a firm structure and they had a white color (Fig. 2b). During the experimental period, the development of aboveground parts of the plant (leaves and flowers) was not visibly affected under both conditions. Apparently, waterlogging did not visibly affect shoot growth as long as some parts of the root remained well aerated. However, it may be expected that on the long term, in particular when further unfavorable growing conditions come together simultaneously, growth may well be affected and also the plants might become more susceptible to pathogens.

In these experiments, it was also shown that by decreasing water contents to 50 % for 20 days, after an initial period of 45 days at 80 %, roots recovered (data not shown). Apparently, the plant has a flexible character, and the ability to develop roots when circumstances are favorable.

At low oxygen concentrations, also ethylene may be produced (see chapter "Role of Ethylene and Other Plant Hormones in Orchestrating the Responses to Low Oxygen Conditions"; Steffens and Sauter 2014); moreover, hypoxia leads to reduced root pressure such that the transport of water and nutrients to the upper parts of the plant will decrease. Low oxygen concentration also may result in the formation of adventitious roots (Fig. 3a; see also chapter "Biogenesis of Adventitious Roots and Their Involvement in the Adaptation to Oxygen Limitations"; Sauter and Steffens 2014), and aerenchyma (Fig. 3b), i.e., air canals in the roots of plants that enable exchange of gasses between the stem and the roots (Morard and Silvestre 1996; see also chapter "Aerenchyma Formation in Plants"; Takahashi et al. 2014).

Some plant species, particularly those that commonly grow under waterlogged, and thus low oxygen conditions, such as rice, are capable of forming aerenchyma. Formation of aerenchyma can be considered as a regular mechanism to deal with these conditions. Many plant species grown in horticulture however are incapable of producing aerenchyma. Formation of adventitious roots is another strategy to overcome the lack of oxygen. However, the formation of thcsc roots requires additional energy, and therefore goes on account of yield. It is clear that this should be avoided in horticultural practice.

For the upper parts of plants, low oxygen concentrations in the root zone will result in closure of the stomata and wilting of leaves (Else et al. 1995; Bradford and

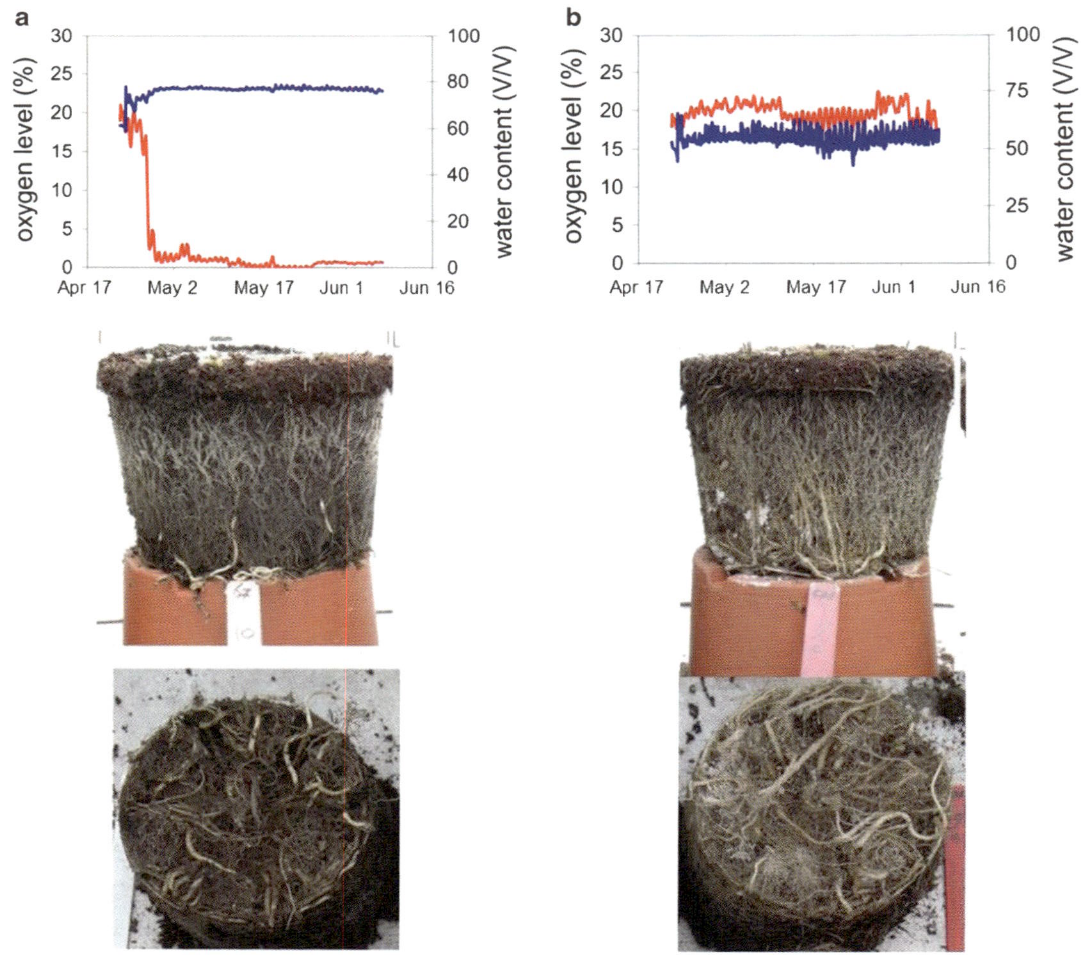

Fig. 2 Oxygen and water content profiles during the growth of gerbera. *Left* (**a**) a situation where water content in the containers was kept continuously high at

Fig. 3 (**a**) Formation of adventitious roots in maize; (**b**) formation of aerenchyma

Hsiao 1982). Low oxygen will also result in reduced photosynthesis (Ashraf and Mehmood 1990; Zakrzhevskii and Ladygina 1989; Morard and Silvestre 1996), reduced growth and development of the plants (see Fig. 1). The effects of low oxygen concentration on yield have not been validated in those studies. Actually no professional grower is keen to take the risk of growing plants under suboptimal conditions. However, in the case of cucumber, it is likely that flowering will start later and less fruits will be produced during the growing period due to the retarded plant development and growth.

2 Controlling of the Oxygen Concentration Around the Roots

Generally, it is known that the oxygen concentration in the root zone depends on several parameters, such as the number of roots; crop activity; the occurrence and density of microorganisms; temperature; the water content of the substrate; water distribution throughout the substrate; the water flow; and the type and age of the substrate.

For growers, oxygen can only be considered as a factor important for plant growth when it can be measured and controlled. In this paragraph, we focus on research aiming to determine those parameters that affect oxygen levels in the root environment most strongly. Consequently, it is explored how growers can supply the roots with sufficient oxygen.

From extensive measurements in stone wool slabs during the growth of roses in a greenhouse, it was elucidated that important factors determining oxygen concentrations in the root zone are the age and the type of substrate. Figure 4 shows changes of the oxygen concentrations through time as measured at 12 different positions in the lower part of the two kinds of stone wool slabs (1.5 years old and 8 years old, respectively). During the cultivation of roses, plants live for many years on the same slab, so both the slabs and plants differ in age. Each line represents the

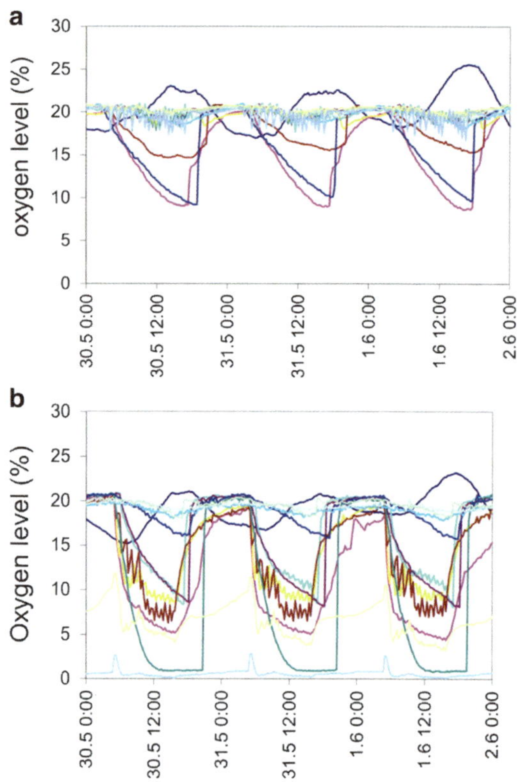

Fig. 4 Relation between oxygen and water contents in stone wool slabs of different age as recorded during 3 successive days. (**a**) 1.5 years old and (**b**) 8 years during growth of *rose*

oxygen level at an individual sensor. Remarkably, in the younger slabs, oxygen levels always remained above 10 %, and never became critical. In contrast, in slabs from 8 years old, located in another section of the same green house, very low oxygen levels were measured at several positions within the slab.

Water samples from both slabs were taken to determine the oxygen consumption by aerobic microorganisms present in the water. In the laboratory, samples were saturated with oxygen and only just before measuring the oxygen consumption rate within the samples, the vials were tightly closed. Noninvasive optical oxygen measurements showed that in samples from the older slabs oxygen was depleted within a few hours, whereas in samples taken from younger slabs this could take a day. Hence, microbiological oxygen consumption was much higher in the older slabs. The increase of microbial activity during time is probably caused by the accumulation of degradable carbon sources in the slabs (organic plant material and root exudates).

Further observations of both slabs showed that the physical structure of the older slab had changed significantly. The fibers which form the physical structure of a new stone wool slab were broken in the old slabs, and at the bottom of the old slabs most of the stone wool was replaced by plant roots. Many of these roots had a woody appearance and were probably not functional any longer. This change in slab structure resulted in a lower total water content in the older slab compared to the 1-year old slab. Also, the vertical distribution of water in the slab was changed. Due to the loss of capillary function, the older slab was relatively dryer in the upper part of the slab, but wetter in the lower part. Because of this we speculate that inside the older slabs, the volume with optimal conditions for root development and functioning has decreased as compared to younger slabs. Due to its heterogeneous character, the application of optimal irrigation in older slabs is more difficult.

Measurements during rose cultivation showed that under some circumstances, oxygen stress may occur in the root environment. Many other measurements carried out during the last 10 years, both in vegetable crops (e.g., cucumber, pepper, and tomato) and ornamentals (e.g., gerbera and rose) showed similar fluctuations of oxygen levels during the day. Apart from substrates like perlite, with a very open structure, low oxygen levels were detected in several types of substrates (like stone wool, potting soils, and coir). Oxygen levels were found particularly low in the lower parts of the substrates where the water content is relatively high (data not shown).

For a long time, the root zone has been regarded as a black box, with little possibilities for steering, and only little is known about how oxygen stress can be avoided in practice. To investigate this, oxygen consumption of some horticultural crops was estimated, and next it was determined how sufficient oxygen could be supplied to the roots. To measure oxygen consumption of the root systems of cucumber and tomato in a production greenhouse, sections of a stone wool slab carrying one plant were positioned in plastic boxes, which were totally filled with oxygen saturated water. Through time, the oxygen concentration was measured in these boxes. Remarkably, the initial decrease in oxygen levels at positions close to the roots was very fast, i.e., within 15 min concentration dropped from 21 % to less than 5 % (data not shown). From these oxygen profiles, we could calculate that those plants could consume up to 20–40 mg of oxygen per plant per hour. This means that in a green house situation at least 4 L fresh water per hour is needed to supply those plants with enough oxygen. For the plant density in a professional green house (several plants per meter square), with irrigation volumes, e.g., for pepper maximally 7 L/m^2/day, and draining percentages of 20–50 % (depending per crop), it can be concluded that the *supply of oxygen to roots can only be partly (10–20 %) realized by the oxygen which is dissolved in the irrigation water.*

From later studies, it became clear that by far the most important factor determining oxygen concentrations in the root environment is the *irrigation strategy of the farmer*. Figure 5 shows typical patterns of oxygen profiles and water content in a stone wool slab during growth of roses. Clearly, in blue, the subsequent irrigations during the morning can be detected by the resulting increases in water contents. It can been seen that during the night oxygen concentration are high, i.e., 21 %, but after several irrigations in the morning, resulting in a water content of over 80 %, the oxygen concentrations rapidly drop and stay low during the rest of the day. Only at the end of the afternoon, when irrigation stops, and the water content decreases both due to draining of the slab and because of the water consumption by the plant, oxygen concentrations increases again. In Fig. 5, 3 successive days are depicted and

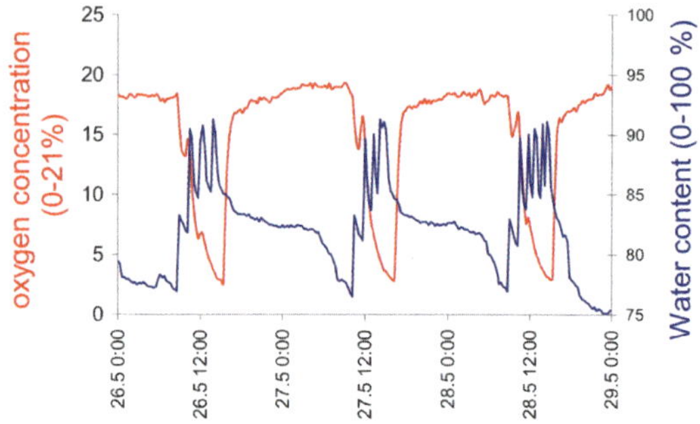

Fig. 5 Diurnal relationship between oxygen level (%, *red line*) and water content (%, *blue line*) in stone wool substrate during growth of *roses*

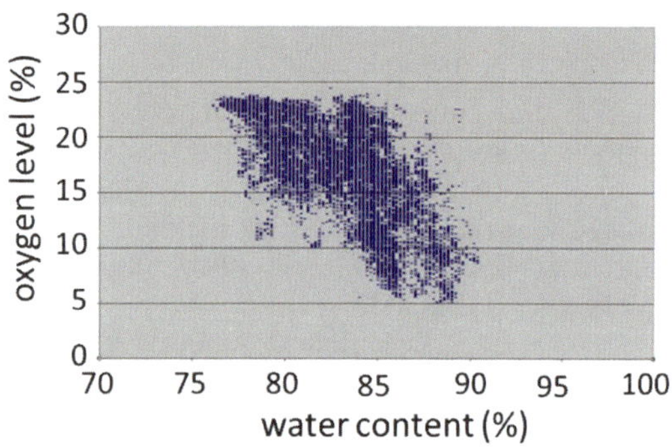

Fig. 6 Relationship between water content and oxygen levels during the growth of Rhododendron, in 3 L containers

the relationship between water content (irrigation) and oxygen levels was repeatedly observed. We carefully zoomed into the profiles and since oxygen levels in the afternoon only increase after water contents drop, it can be concluded that the supply of oxygen to lower parts of the slabs is caused by the diffusion of oxygen through pores in the substrate, which are waterlogged after several irrigations, but open up again after draining of the substrate. *In other words, draining of a substrate is very much related to water content and so are oxygen levels.*

These typical patterns, showing the relationship between water content and oxygen concentration were found for several crops and for several substrates. So, in general, it can be concluded that above certain moisture concentrations in the substrate, there is a serious risk of oxygen stress. In earlier studies, oxygen measurements during growth of Rhododendron in 3 L containers filled with potting soil showed a severe drop of oxygen levels at water contents above 75 % (Fig. 6). As a rule of thumb, based on several measurements which have been done during the last years, we consider that water content over 75 % are likely to result in critical oxygen levels. Depending on the type of substrate, and the plant species, threshold values of water content may be different.

3 Conclusion

In this chapter, experimental data show effects of oxygen stress on plant development. Certainly, data from young cucumber plants show that growth and development is retarded when oxygen levels are below 8 % (Fig. 1). Data from our studies on the gerbera root zone show that roots may die when water levels are too high and subsequently, oxygen levels drop too low for too long. Remarkably, for a certain period growers may not even notice that roots are harmed due to oxygen stress, as the shoot of the plants may show no visible effects immediately. Under certain conditions however, when plant growth demands optimal root activity, a weakened root system may lead to retarded plant development and consequently less yield. Also, a weakened root system may be more susceptible to certain pathogens.

The data shown in this chapter also show that it is relatively easy to avoid oxygen stress. Together with an extension service, a farmer should consider root oxygenation while optimizing the water management strategy. It should be noted that the optimal strategy might be slightly different for every single situation. An optimal water volume of each irrigation cycle should be defined and the duration of irrigation during the day, the intervals between irrigation cycles, and also the draining percentages of the rooting substrate should be determined. It should be kept in mind that for horticultural crops, oxygen supply to the root mainly occurs through open, aereous pores in a substrate, which may be logged at water contents above approximately 75 %. So, using oxygen sensors and water content sensors, irrigation strategies can be optimized and risk for oxygen stress can be reduced dramatically.

References

Armstrong W, Armstrong J (2014) Plant internal oxygen transport (diffusion and convection) and measuring and modelling oxygen gradients. In: van Dongen JT, Licausi F (eds) Low oxygen stress in plants. Springer, Heidelberg

Ashraf M, Mehmood S (1990) Effects of waterlogging on growth and some physiological parameters of four Brassica species. Plant Soil 121:203–209

Ast C, Draaijer A (2014) Methods and techniques to measure molecular oxygen in plants. In: van Dongen JT, Licausi F (eds) Low oxygen stress in plants. Springer, Heidelberg

Bradford KJ, Hsiao TC (1982) Stomatal behavior and water relations of waterlogged tomato plants. Plant Physiol 70:1508–1513

Else MA, Davies WJ, Malone M, Jackson MB (1995) A negative hydraulic message from oxygen-deficient roots of tomato plants? Plant Physiol 109:1017–1024

Grodan, TNO, PPO (2004) The gasfactor in cultivation management. Report of EET Kiem project 02002. This project was funded by the Ministry of Economic Affairs

Morard P, Silvestre J (1996) Plant injury due to oxygen deficiency in the root environment of soilless culture: a review. Plant Soil 184:243–254

Paepke C, Ramírez-Aguilar S, Antonio C (2014) Oxygen consumption during hypoxic conditions. In: van Dongen JT, Licausi F (eds) Low oxygen stress in plants. Springer, Heidelberg

Sauter J, Steffens B (2014) Biogenesis of adventitious roots and their involvement in the adaptation to oxygen limitations. In: van Dongen JT, Licausi F (eds) Low oxygen stress in plants. Springer, Heidelberg

Steffens B, Sauter M (2014) Role of ethylene and other plant hormones in orchestrating the responses to low oxygen conditions. In: van Dongen JT, Licausi F (eds) Low oxygen stress in plants. Springer, Heidelberg

Takahashi H, Yamauchi T, Colmer TD, Nakazono M (2014) Aerenchyma formation in plants. In: van Dongen JT, Licausi F (eds) Low oxygen stress in plants. Springer, Heidelberg

Vennik M, van Duijn B, Verhagen H, Baas R, Berkers P (2011) Onderzoek naar beworteling met Gerbera als modelgewas. Research financed by Productschap Tuinbouw, PT Project number 13785

Zakrzhevskii DA, Ladygina ON (1989) Effect of hypoxia on functional activity of pea and soybean leaves. Plant Physiol 36:465–471

Inducing Hypoxic Stress Responses by Herbicides That Inhibit Amino Acid Biosynthesis

Ana Zabalza and Mercedes Royuela

Abstract Nowadays, there are two main types of herbicides that inhibit amino acid biosynthesis: glyphosate, which inhibits aromatic amino acid synthesis via specific inhibition of 5-*enol*pyruvyl-shikimate-3-phosphate synthase (EPSPS), and inhibitors of branched-chain amino acid synthesis that act via specific inhibition of acetolactate synthase (ALS). Both types of inhibitors share different aspects of their mode of action, such as the induction of fermentation. Although this physiological effect resembles the hypoxic stress response, it was detected under aerobic conditions and it was not related to a change in respiratory rates or to a decrease in the energy charge. Fermentative induction has also been detected after treatment with other compounds inhibiting amino acid biosynthesis, such as glufosinate or ketol-acid reductoisomerase inhibitors suggesting that it can be considered as part of a general plant response to the stress conditions caused by this kind of herbicide treatment. Fermentation seems to be part of the impaired carbon metabolism detected in these treated plants that prevent the optimal utilization of available carbohydrates and energy.

Abbreviations

ADH	Alcohol dehydrogenase
AEC	Adenylate energy charge
ALS	Acetolactate synthase
CPCA	1,1-Cyclopropanedicarboxylic acid
EPSPS	5-Enolpyruvyl-shikimate-3-phosphate synthase

A. Zabalza (✉) • M. Royuela
Dpto. Ciencias Medio Natural, Universidad Pública de Navarra, 31006 Pamplona, Spain
e-mail: ana.zabalza@unavarra.es

J.T. van Dongen and F. Licausi (eds.), *Low-Oxygen Stress in Plants*, Plant Cell Monographs 21, DOI 10.1007/978-3-7091-1254-0_20, © Springer-Verlag Wien 2014

KARI	Ketol-acid reductoisomerase
PDC	Pyruvate decarboxylase
RC	Redox charge

1 Herbicides and the Importance of the Mode of Action Studies

In modern agricultural systems, crop plants have to compete with weeds that are wild plants adapted to grow competitively in open field sites. Herbicides are compounds that selectively kill or arrest growth of these weeds. They have played a significant role in the large increase of agricultural productivity achieved over the past 50 years.

Herbicidal compounds inhibit specific molecular target sites within critical plant biochemical and/or physiological pathways. The terms *mechanism of action* and *primary or target site* are commonly used to describe the location at which the herbicide potently inhibits an important process. This site is expected to show the fastest response to the herbicide. The term *mode of action* may be defined as the complete sequence of events leading to plant injury, from the first response to the death of the treated plant. The mode of action includes all areas of interaction between an herbicide and a crop or test species.

Studying the mechanism and mode of action of herbicides provides an important tool to learn more about plant physiology and biochemistry. In this context, the specific nature of the interaction between herbicides and their respective target sites makes it interesting to use these molecules for studying and dissecting the intricacy of plant biochemical and physiological processes (Dayan et al. 2010).

Not only might the study of the mode of action of commercial herbicides help to discovery novel processes in plant physiology, it might also support the development of new herbicidal molecules. Over the last years, social and environmental pressure is leading to increasing restrictions on the use of existing formulations and many herbicides have already been removed from the market in many countries. In this context, it is important to continue to develop new agrochemicals that are effective and safe and alleviate the selection pressure of the few mechanism of action of present herbicides. The study of the mode of action would help for the evaluation of chemicals for their suitability as lead structures in herbicide discovery. Knowing exactly which plant processes are directly related to plant death would help in the discovery of new products able to induce the same kind of plant responses from different mechanism of action.

2 Overview of Herbicides Inhibiting Amino Acid Biosynthesis

Herbicides are classified based on their mechanism of action, which includes at least 16 main characterized mechanisms that encompass all major plant functions, including amino acid synthesis. Herbicides, specifically inhibiting the biosynthesis of amino acids, are developed since the early 1970s and are still of great agronomic and commercial importance. Inhibition of plant amino acid biosynthesis has become a major target of herbicide development as only plants and microorganisms can synthesize all required amino acids themselves. Animals cannot synthesize histidine, isoleucine, leucine, lysine, methionine, threonine, tryptophan, phenylalanine, and valine, which are therefore called essential amino acids, which have considerable dietary significance. Consequently, herbicides that act upon the synthesis pathway of these essential amino acids are very likely not toxic for animals.

Three sites for amino acid biosynthesis that include three different enzyme systems are important sites of herbicide action (Duke 1990).

1. Inhibition of glutamine synthetase.
2. Inhibition of branched-chain amino acid synthesis, specifically inhibition of acetolactate synthase (ALS) = acetohydroxyacid synthase (AHAS).
3. Inhibition of aromatic amino acid synthesis, specifically inhibition of 5-enolpyruvyl-shikimate-3-phosphate synthase (EPSPS).

In the following paragraphs, these sites of action are briefly described.

2.1 Inhibition of Glutamine Synthetase

The ammonium salt of glufosinate 2-amino-4-(hydroxy(methyl)phosphonoyl) butanoic acid is a major nonselective, postemergence herbicide introduced in 1981. It is the only commercial herbicide that inhibits glutamine synthetase (Leason et al. 1982), which catalyzes the conversion of L-glutamate to L-glutamine in the presence of ATP and ammonia. The discovery and development of glufosinate are of particular interest since this herbicide may be regarded as having a natural origin (Cobb and Reade 2010): two species of *Streptomyces* were reported to produce bialaphos, a tripeptide containing glufosinate bound to two molecules of L-alanine. Biolaphos itself is inactive but is rapidly hydrolyzed in plants to form the phytotoxic glufosinate.

Phytotoxicity after glufosinate treatment is certainly rapid, since leaf chlorosis, desiccation, and necrosis can be observed within 2 days and subsequent plant death results 3 days later. Since glutamine synthetase is potently inhibited, the rapid accumulation of ammonia affects uncoupling of the electron flow from proton transport. This indirect effect on membrane polarization affects a number of metabolic pathways, such as increased protein degradation and inhibition of photosynthetic carbon assimilation (Cobb and Reade 2010).

2.2 Inhibition of ALS

There are five different chemical classes of herbicides that share the same site of action: inhibition of acetolactate synthase (ALS), a key enzyme in the biosynthesis of the branched-chain amino acids. The sulfonylureas, imidazolinones, triazolopyrimidines, sulfonylaminocarbnonyl-triazolines, and pyrimidinyl-oxy-benzoates have emerged since the 1980s and have proved to be potent, selective broad-spectrum inhibitors of plant growth. Due to their high efficacy, they are used at very low field rates, in the range of grams per hectare while other kinds of herbicides are used in the range of kilograms per hectare. The efficacy and potency of ALS inhibitors has ensured the continued success of these herbicides, which have rapidly challenged, and in some instances replaces traditional products, especially in cereals and soybeans (Cobb and Reade 2010). Currently, ALS inhibitors represent the second biggest class of herbicidal active ingredients. A major research effort has progressed to understand their mode of action to develop new products to inhibit the synthesis of branched-chain amino acids and will be discussed in detail in the next chapter.

2.3 Inhibition of EPSPS

Glyphosate (*N*-(phosphonomethyl)glycine) is a wide-spectrum, nonselective post-emergence herbicide that is currently the most popular herbicide; since its commercial introduction in 1974, glyphosate has become the dominant herbicide worldwide. Glyphosate inhibits the biosynthesis of aromatic amino acids, which takes place in chloroplasts via the shikimate pathway. Glyphosate specifically inhibits the activity of the enzyme 5-enolpyruvylshikimate-3-phosphate synthase (EPSPS; EC 2.5.1.19) (Steinrücken and Amrhein 1980) by acting as an analog of the second substrate, phospho*enol*pyruvate.

Biotechnology has increased the importance of glyphosate as glyphosate tolerant maize, soybean, cotton, and canola were commercialized. Most of them depend on the target-based tolerance and have been developed with the insertion of a glyphosate-resistant EPSPS gene from *Agrobacterium* sp. (Tan et al. 2006). Since transgenic glyphosate-resistant crops were introduced in 1996, this technology was widely adopted in the USA and South America.

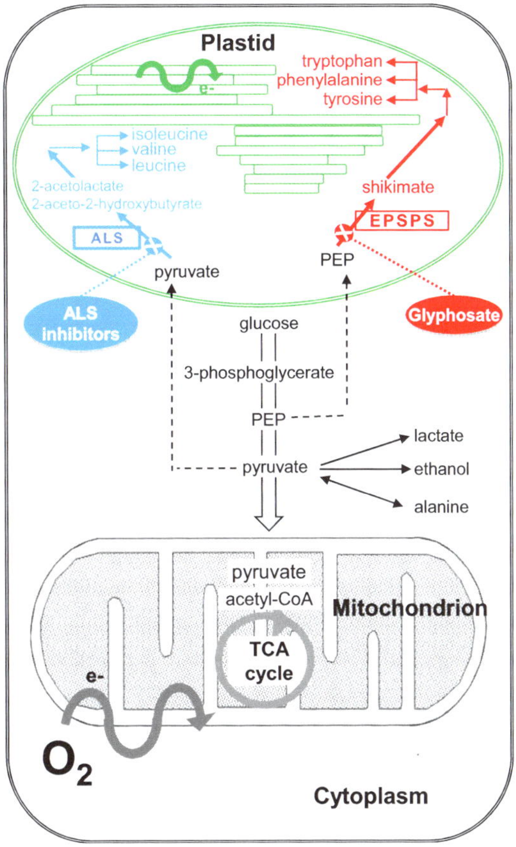

Fig. 1 Overview of the cellular localization of the target sites of herbicides inhibiting amino acid biosynthesis. Simplified view of pathways with pyruvate playing a central role in linking fermentative metabolism and amino acid biosynthesis. *ALS* acetolactate synthase, *EPSPS* 5-enolpyruvyl-shikimate-3-phosphate synthase, *e* electrons, *PEP* phosphoenolpyruvate

3 Common Physiological Effects of EPSPS and ALS Inhibitors

Although all three types of herbicides inhibiting amino acid biosynthesis were introduced here, this book chapter will focus mainly on ALS and EPSPS inhibitors. In particular, different aspect of their mode of action will be discussed. The main point will focus on their effect to induce the hypoxic response and fermentative metabolism.

Both ALS and EPSPS are nuclear-encoded enzymes, but their catalytic site is located in the plastid (Fig. 1) (Singh 1999; Schmid and Amrhein 1999). Both activities are well-established target sites of herbicides that inhibit branched-chain amino acid biosynthesis and glyphosate, respectively.

Even though the first primary action mechanism of these herbicides is widely known, it is not fully understood how plants actually die after the inhibition of EPSPS or ALS. For both types of herbicides that inhibit amino acid biosynthesis, the sequence of events leading from herbicide application to plant death is still being debated.

Although it is not clearly understood how plants die following ALS or EPSPS inhibition, several physiological effects have been well described. Interestingly, both types of herbicides have been shown to share some aspects in their mode of action. Both types of herbicides, the EPSPS inhibitor and ALS inhibitors, have been reported to arrest growth of treated plants and this growth arrest is followed by a slow death (Wittenbach and Abell 1999; Gruys and Sikorski 1999). Plants treated with glyphosate may not show symptoms of the treatment for 7–10 days; after that glyphosate symptoms include growth inhibition, chlorosis, necrosis, and subsequent plant death. In the case of ALS inhibitors, plant death can even take longer. Although nanomolar concentrations of the herbicide can inhibit in vitro ALS activity within minutes, physical symptoms may take days to develop; firstly, appearing as chlorosis and necrosis in young meristematic regions of both shoots and roots. Under optimal growth conditions, plant death may follow within 10 days, although up to 2 months may elapse when weed growth is slow.

Both types of herbicides have been described to affect free amino acid content (Wang 2001; Shaner and Reider 1986). Recent studies have shown that the proportion of amino acids whose pathway were specifically inhibited (branched-chain or aromatic) experiences a transient (the first 24 or 48 h) decrease followed by a general increase in the free amino acid pool (Orcaray et al. 2010). It was suggested (Rhodes et al. 1987) that this increased pool size might be derived from an increase in protein turnover with higher degradation and lower synthesis rates. Despite a particular amino acid biosynthetic pathway is inhibited, plants treated with ALS or EPSPS inhibitors will maintain nonlimiting amino acid levels due to increasing proteolytic activities. This would explain the retarded appearance of phytotoxic symptoms.

Moreover, both types of herbicides have been described to share more aspects in their mode of action at the level of primary and secondary metabolism. These herbicides stimulate secondary metabolism derived from the shikimate pathway (Orcaray et al. 2010, 2011). Glyphosate- or ALS inhibitors-treated plants showed impaired carbon metabolism with an accumulation of carbohydrates in the leaves and roots (Zabalza et al. 2004; Orcaray et al. 2012).

4 Hypoxic Response After Inhibiting the Biosynthesis of Amino Acids

ALS catalyzes the condensation of either two molecules of pyruvate to form acetolactate or one molecule of pyruvate with one molecule of 2-ketobutyrate to form 2-aceto-2-hydroxybutyrate (Singh 1999). Blockage of ALS-catalyzed reactions would result in higher availability of the ALS substrates, such as pyruvate. Considering that pyruvate is one of the major intermediates of many biosynthetic pathways other than branched-chain amino acid biosynthesis (Fig. 1), pyruvate is expected to be readily diverted into other metabolic pathways, such as ethanolic and lactic fermentative pathways. Indeed, the activity of various pyruvate-consuming enzymes was induced such as pyruvate decarboxylase (PDC), alcohol dehydrogenase (ADH), lactate dehydrogenase, and alanine aminotransferase in pea roots of pea plants treated with ALS inhibitors but grown under *aerobic* conditions (Gaston et al. 2002). Furthermore, a significant increase in gene expression of PDC in *Arabidopsis* plants treated with ALS inhibitors for 24 h was described (Das et al. 2010).

Although induction of fermentation after ALS inhibition correlates well with the increased availability of pyruvate, no pyruvate accumulation was detected in ALS-treated plants. Apparently, the regulation of pyruvate metabolism was efficient enough to balance the endogenous pyruvate levels of the cell. While ALS inhibition did not elicit pyruvate accumulation, it induced internal anoxia in roots. This depletion of the oxygen availability inside the root was related to a short-term increase in the respiratory rate. After several days, however, the oxygen consumption of roots from herbicide-treated plants decreased again to similar values as determined for nontreated roots (Zabalza et al. 2011). Similarly, previous studies had demonstrated that feeding pyruvate to the roots led to an increase of the oxygen consumption rate, which ultimately led to anoxia (Zabalza et al. 2009). Thus, it seems that, following ALS inhibition, the pyruvate is efficiently channeled to PDC and the TCA-cycle, as both fermentation and respiration increase in the short term. Alternative oxidase (AOX) is also induced after ALS inhibition (Gaston et al. 2003). Again, pyruvate could be mediating this response as pyruvate, like other α-keto acids, has a significant stimulating effect on the activity of AOX (Millar et al. 1993; Vanlerbergue et al. 1995).

Interestingly, both the EPSPS inhibitor and ALS inhibitors were reported to share another feature in their mode of action: the induction of alcoholic fermentation (Zabalza et al. 2005; Orcaray et al. 2012). Considering this, similarities at the level of carbon metabolism are expected and therefore effects on key parameters of carbon metabolism were evaluated and compared (Fig. 2).

Carbohydrate accumulation in leaves and roots has been reported after ALS and EPSPS inhibition (Zabalza et al. 2004; Orcaray et al. 2012). In both cases, carbohydrate accumulation can only be due to a lower translocation to sink tissues since soluble carbohydrate accumulated in roots before accumulation in leaves occurred. Moreover, both accumulation in roots and leaves happened before photosynthetic activity declined (Zabalza et al. 2004; Orcaray et al. 2010). Figure 2 shows the sucrose accumulation reported in roots after ALS and EPSPS inhibition. Under these conditions, the gradient of sugars required for long-distance transport is abolished; phloem transport is inhibited by a decrease in sink strength.

In general, no increase in the total respiratory activity was detected after application of either of both types of herbicides, even though carbohydrates were sufficiently available as substrate. Having said this, it should be noted however that after ALS treatments a short and transient increase of the respiration was observed (Zabalza et al. 2011, Fig. 2). Nevertheless, these data show that the increased

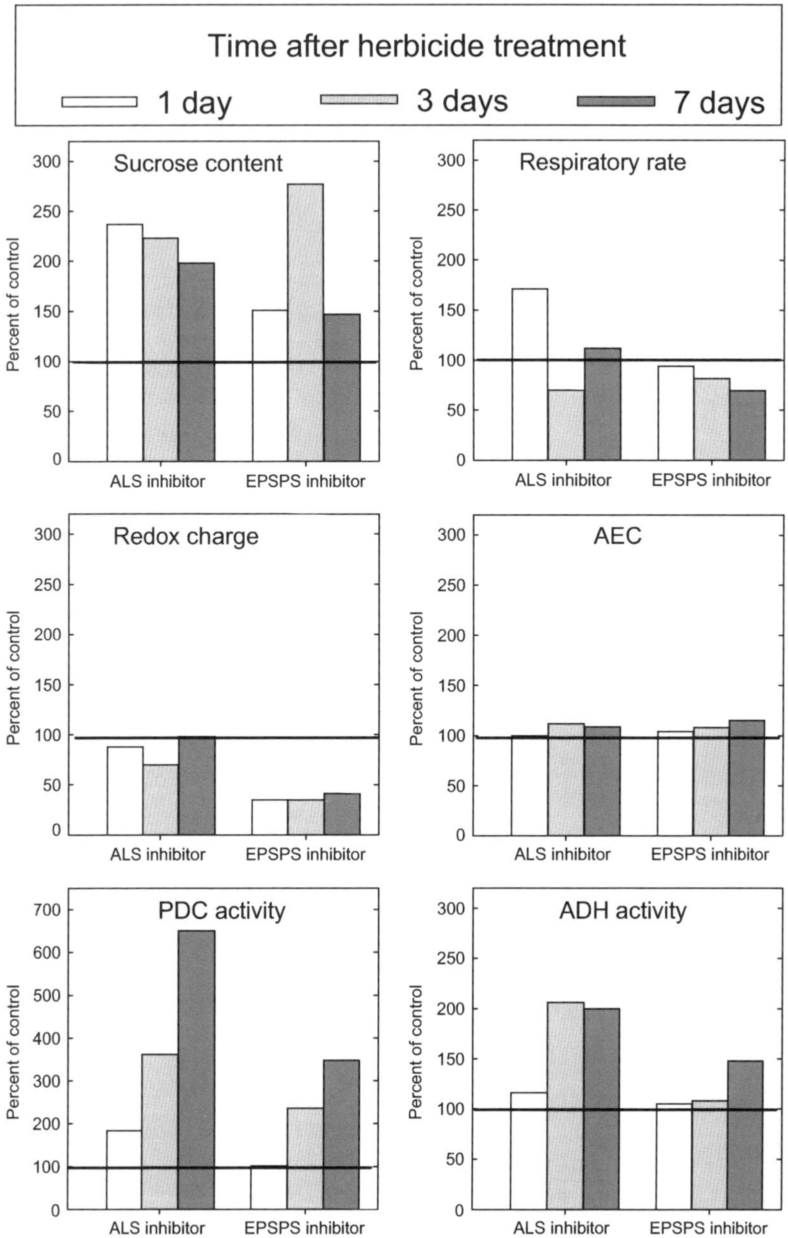

Fig. 2 Comparison of the main effects of imazethapyr (ALS inhibitor) and glyphosate (EPSPS inhibitor) on pea roots, expressed as the percentage of the nontreated control. Original data are in Zabalza et al. (2011) and Orcaray et al. (2012). Redox charge = ($NADH^+$ + NADPH)/(NAD^+ + NADH + $NADP^+$ + NADPH); AEC: adenylate energy charge = (ATP + 0.5 ADP)/(ATP + ADP + AMP)

fermentative activity cannot simply be explained by a drop of the plant internal oxygen concentration due to an increase in the rate of oxygen consumption. To find an alternative explanation for the induction of fermentation, the effect of the herbicide treatments on the general redox charge and energy state in roots of the treated plants was assessed (Fig. 2). These ratios are important measurements of the cellular energy status, and deviations from the standard values are considered to be sensitive indicators of metabolic disturbances. Redox charge (RC) was defined as (NADH + NADPH)/(NAD^+ + NADH + $NADP^+$ + NADPH) and energy state was evaluated by the adenylate energy charge (AEC) calculated as (ATP + 0.5 ADP)/(ATP + ADP + AMP).

The AEC ratio represents the amount of metabolically available energy stored in the adenine nucleotide pool. Usually, under low oxygen conditions, changes in the AEC or RC correlate well with the induction of PDC and ADH. A decrease in ATP availability has been proposed as an indirect measure of the oxygen status in plants (Bailey-Serres and Chang 2005). Usually, the hypoxic response of cells goes along with a low AEC and high RC conditions, as oxygen limitation affects ATP synthesis and oxidation of NADH. Nevertheless, in herbicide-treated roots, the induction of fermentation did not correlate with an increase in the RC, nor with a decrease in the AEC of the tissue (Fig. 2). On the contrary, although fermentation was induced, the AEC of these plants was higher and the RC was lower than that of the control plants.

Despite some different patterns, ALS and EPSPS inhibitors share more aspects in their mode of action. Fermentation is one of the least ATP-producing metabolic pathways that are activated after ALS and EPSPS inhibition, but not the only one. Also, the less efficient respiratory alternative pathway is induced after ALS inhibition as shown by oxygen isotope discrimination analyses using mass spectrometry (Gaston et al. 2003) and after glyphosate treatment as was shown by gene expression profiling (Zhu et al. 2008).

Growth arrest provoked by the ALS or EPSPS inhibition is due to a metabolic alteration that impairs carbohydrate utilization, leading to carbohydrate accumulation first in the roots and then in the shoots. This metabolic alteration induces changes in the specific amino acid content as ratio of the total free amino acid pool (Zabalza et al. 2011; Orcaray et al. 2010) and provokes an imbalance in the carbon/nitrogen metabolism. Concomitantly, this will result in metabolic imbalances that prevent the optimal utilization of available carbohydrates and energy. Since growth is hindered, the available ATP will not be consumed at the rate it would be expected to under noninhibited growth conditions. And thus, the energy charge will never become limiting in cells of plants treated with ALS or EPSPS inhibitors. Indeed, an increase in the AEC was detected (Fig. 2).

Although the induction of fermentation was observed after both ALS and EPSPS inhibition, one should be careful to conclude that this is the immediate consequence of higher substrate availability for the fermentation pathways only. Fermentation can also be regarded as a general physiological response after a stress situation as has been reported for other abiotic stresses, such as low temperature and osmotic stress (Dolferus et al. 1994; Kürsteiner et al. 2003). These two different explanations are, however, not mutually exclusive and they may even act in concert. Indeed, the analysis of fermentative activity following treatment of herbicides other than ALS inhibitors that do not directly affect a pyruvate-consuming enzyme showed that root of pea plants treated with glyphosate also showed induction of aerobic fermentation (Orcaray et al. 2012).

The inhibition of branched-chain amino acid biosynthesis at the level of ALS enzyme as the target of herbicides raised the issue of whether another enzyme of the same pathway might be the target for inhibitors that, therefore, may behave as herbicides. In this context, glufosinate and ketol-acid reductoisomerase (KARI) inhibitors were evaluated. KARI is the second enzyme in the branched-chain amino acid biosynthesis, whereas glufosinate inhibits amino acid biosynthesis by specifically inhibiting glutamine synthetase in the biosynthetic pathway of glutamine (Leason et al. 1982). Compounds, such as 2-dimethylphosphinoyl-2-hydroxyacetic acid (Hoe 704) and 1,1-cyclopropanedicarboxylic acid (CPCA), are very potent and selective inhibitors of KARI (Schulz et al. 1988; Gerwick et al. 1993) although they have not been commercialized. The effects of the application of glufosinate, Hoe 704, or CPCA to pea plants resulted in ethanolic fermentation in pea roots, similar to what was observed after the application of glyphosate or ALS inhibitor; the specific activities of both PDC and ADH were increased after 3 and 7 days from the onset of the inhibitors treatment (Fig. 3).

Fermentative metabolism is induced after inhibiting the branched-chain amino acid biosynthetic pathway at the level of the first and the second enzymes, after inhibiting aromatic amino acid biosynthesis and after inhibiting glutamine synthetase. As indicated earlier, two characteristics have been proposed to be related to the induction of fermentation under aerobic conditions after ALS inhibition. First, ALS inhibition would involve an increased availability of pyruvate for use by other enzymes, such as PDC; and second, some other abiotic stresses not related to hypoxia can be responsible for the induction of the expression of ADH and PDC. Fermentative induction after treatment with glyphosate, glufosinate, or KARI inhibitors cannot be easily or directly related to substrate availability because neither of these compounds inhibits a pyruvate-consuming enzyme. Thus, it seems that the fermentative response can be considered a physiological effect induced under stress although the induction after ALS inhibition might be due to both characteristics. This point of view might explain the faster response after ALS inhibiting herbicides than other inhibitors, because the two explanations may act in concert.

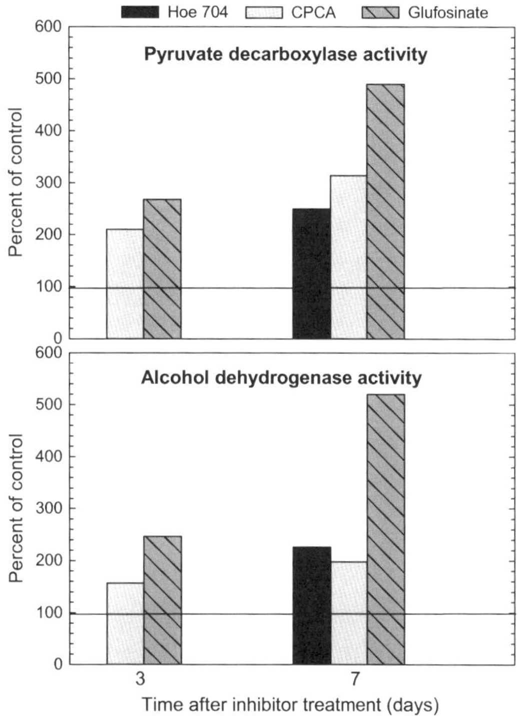

Fig. 3 Comparison of the effects of glufosinate, Hoe 704, and CPCA (KARI inhibitors) on pyruvate decarboxylase and alcohol dehydrogenase enzymatic activities of pea roots, expressed as the percentage of the nontreated control. Original data are in Zabalza et al. (2005) and Orcaray et al. (2012)

5 Role of Fermentation in the Mode of Action of Inhibitors of Amino Acid Biosynthesis

As indicated before, the increased free amino pool detected after amino acid biosynthesis inhibition has been suggested to be derived from an increase in protein turnover (Rhodes et al. 1987). Indeed, it was shown that although protein synthesis occurs after ALS inhibitor treatment, the amino acids that comprise these proteins do not contain newly incorporated nitrogen; they contain nitrogen that is mainly scavenged from protein degradation (Zabalza et al. 2006). Plants treated with amino acid biosynthesis inhibitors not only had increased specific enzymatic activities of fermentative enzymes but also showed increased levels of PDC and ADH enzymes as detected by western blotting (Zabalza et al. 2005; Orcaray et al. 2012). It is noticeable that plants treated with these classes of herbicides had increased levels of PDC and ADH protein synthesis in a situation in which amino acids have to be scavenged from existing proteins. This reuse of existing proteins in the synthesis of fermentative enzymes suggests that fermentation plays an important role in the plant's response to herbicide treatment.

The induction of PDC and ADH by inhibitors of amino acid biosynthesis seems to reflect a situation similar to hypoxia, but under aerobic conditions. Nevertheless, the role of fermentation in the mode of action of inhibitors remains to be elucidated.

On the one hand, ethanol production can be toxic to the plant and enhances the toxicity of the herbicides. On the other hand, the induction of fermentation can be considered to be part of a general plant response to the stress conditions caused by the herbicide treatment. Indeed, it was suggested that fermentation has a general function in aerobic metabolism under various stress conditions and that it might be an important switch in regulating carbohydrate metabolism (Tadege et al. 1999). Correlating the induction of fermentation to plant death is not easy due to the multiple effects fermentation has on a number of metabolic processes. In this context, experiments evaluating the effects of amino acid biosynthesis inhibitors on transgenic plants lacking fermentative enzymes would be very helpful.

Independent of what exactly the role of the hypoxic response is in the mode of action of the herbicides, the induction of fermentation has been detected after the application of different types of herbicides. This effect can thus be considered to represent physiological markers of herbicidal activity. Such markers can help in the search for new herbicidal active ingredients that are based on natural products to decrease the use of synthetic compounds. Thus, the use of physiological markers to evaluate the potential herbicidal activity of natural compounds can be very useful.

6 Concluding Remarks

The herbicides inhibiting branched-chain or aromatic amino acid biosynthesis share different aspects of their mode of action, such as free amino acid and carbohydrate accumulation. Both types of herbicides induce fermentative metabolism in plants grown under aerobic conditions. Although this physiological effect might resemble the hypoxic response of plants, the fermentative response was not related to a change in respiratory activity or to a decrease in the energy charge. These metabolic disturbances might be due to changes in relative ratios of free amino acid contents provoking an imbalance in the carbon/nitrogen metabolism.

Fermentative induction has been detected after treatment with other compounds inhibiting amino acid biosynthesis, suggesting that it can be considered part of a general plant response to the stress conditions caused by the herbicide treatment. Indeed, it has been proposed that fermentation can have a more general function in aerobic metabolism under various stress conditions and that it might be an important switch in regulating carbohydrate metabolism.

References

Bailey-Serres J, Chang R (2005) Sensing and signalling in response to oxygen deprivation in plants and other organisms. Ann Bot 96:507–518

Cobb AH, Reade JPH (2010) Herbicides and plant physiology. Wiley-Blackwell, Oxford, UK

Das M, Reichman JR, Haberer G, Welzl G, Aceituno FF, Mader MT, Watrud LS, Pfleeger TG, Gutiérrez RA, Schäffner AR, Olszyk DM (2010) A composite transcriptional signature differentiates responses towards closely related herbicides in *Arabidopsis thaliana* and *Brassica napus*. Plant Mol Biol 72:545–556

Dayan FE, Duke SO, Grossmann K (2010) Herbicides as probes in plant biology. Weed Sci 58:340–350

Dolferus R, Jacobs M, Peacock WJ, Dennis ES (1994) Differential interactions of promoter elements in stress responses of the Arabidopsis adh gene. Plant Physiol 105:1075–1087

Duke SO (1990) Overview of herbicide mechanisms of action. Environ Health Perspect 87:263–271

Gaston S, Zabalza A, González EM, Arrese-Igor C, Aparicio-Tejo PM, Royuela M (2002) Imazethapyr, an inhibitor of the branched-chain amino acid biosynthesis, induces aerobic fermentation in pea plants. Physiol Plant 114:524–532

Gaston S, Ribas-Carbo M, Busquets S, Berry JA, Zabalza A, Royuela M (2003) Changes in mitochondrial electron partitioning in response to herbicides inhibiting branched-chain amino acid biosynthesis in soybean. Plant Physiol 133:1351–1359

Gerwick BC, Mireless LC, Eilers RJ (1993) Rapid diagnosis of ALS/AHAS-resistant weeds. Weed Technol 7:519–524

Gruys KJ, Sikorski JA (1999) Inhibitors of tryptophan, phenylalanine, and tyrosine biosynthesis as herbicides. In: Singh BK (ed) Plant amino acids: biochemistry and biotechnology. Marcel Dekker, New York

Kürsteiner O, Dupuis I, Kuhlemeier C (2003) The pyruvate decarboxylase1 gene of Arabidopsis is required during anoxia but not other environmental stresses. Plant Physiol 132:968–978

Leason M, Cunliffe D, Parkin D, Lea PJ, Miflin BJ (1982) Inhibition of pea leaf glutamine synthetase by methionine sulphoximine, phosphinothricin and other glutamate analogues. Phytochemistry 21:855–857

Millar AH, Wiskich JT, Whelan J, Day DA (1993) Organic acid activation of the alternative oxidase of plant mitochondria. FEBS Lett 329:259–262

Orcaray L, Igal M, Marino D, Zabalza A, Royuela M (2010) The possible role of quinate in the mode of action of glyphosate and acetolactate synthase inhibitors. Pest Manag Sci 66:262–269

Orcaray L, Igal M, Zabalza A, Royuela M (2011) Role of exogenously supplied ferulic and p-coumaric acids in mimicking the mode of action of acetolactate synthase inhibiting herbicides. J Agric Food Chem 59:10162–10168

Orcaray L, Zulet A, Zabalza A, Royuela M (2012) Impairment of carbon metabolism induced by the herbicide glyphosate. J Plant Physiol 169:27–33

Rhodes D, Hogan AL, Deal L, Jamieson GC, Haworth P (1987) Amino-acid metabolism of Lemna minor L. 2. Responses to chlorsulfuron. Plant Physiol 84:775–780

Schmid J, Amrhein N (1999) The shikimate pathway. In: Singh BK (ed) Plant amino acids: biochemistry and biotechnology. Marcel Dekker, New York

Schulz A, Sponemann P, Kocher H, Wegenmayer F (1988) The herbicidally active experimental compound Hoe 704 is a potent inhibitor of the enzyme acetolactate reductoisomerase. FEBS Lett 238:375–378

Shaner DL, Reider ML (1986) Physiological responses of corn (Zea mays) to Ac 243,997 in combination with valine, leucine, and isoleucine. Pestic Biochem Physiol 25:248–257

Singh BK (1999) Biosynthesis of valine, leucine and isoleucine. In: Singh BK (ed) Plant amino acids: biochemistry and biotechnology. Marcel Dekker, New York

Steinrücken HC, Amrhein N (1980) The herbicide glyphosate is a potent inhibitor of 5-enolpyruvylshikimic acid-3-phosphate synthase. Biochem Biophys Res Commun 94:1207–1212

Tadege M, Dupuis I, Kuhlemeier C (1999) Ethanolic fermentation: new functions for an old pathway. Trends Plant Sci 4:320–325

Tan S, Evans R, Singh B (2006) Herbicidal inhibitors of amino acid biosynthesis and herbicide-tolerant crops. Amino Acids 30:195–204

Vanlerbergue GC, Day DA, Wiskich JT, Vanlerbergue AE, McIntosh L (1995) Alternative oxidase activity in tobacco leaf mitochondria: dependence on tricarboxylic acid cycle-mediated redox regulation and pyruvate activation. Plant Physiol 109:353–361

Wang CY (2001) Effect of glyphosate on aromatic amino acid metabolism in purple nutsedge (Cyperus rotundus). Weed Technol 15:628–635

Wittenbach VA, Abell LM (1999) Inhibition of valine, leucine and isoleucine biosynthesis. In: Singh BK (ed) Plant amino acids: biochemistry and biotechnology. Marcel Dekker, New York

Zabalza A, Orcaray L, Gastón S, Royuela M (2004) Carbohydrate accumulation in leaves of plants treated with the herbicide chlorsulfuron or imazethapyr is due to a decrease in sink strength. J Agric Food Chem 52:7601–7606

Zabalza A, González EM, Arrese-Igor C, Royuela M (2005) Fermentative metabolism is induced by inhibiting different enzymes of the branched-chain amino acid biosynthesis pathway in pea plants. J Agric Food Chem 53:7486–7493

Zabalza A, Gaston S, Ribas-Carbó M, Orcaray L, Igal M, Royuela M (2006) Nitrogen assimilation studies using ^{15}N in soybean plants treated with imazethapyr, an inhibitor of branched-chain amino acid biosynthesis. J Agric Food Chem 54:8818–8823

Zabalza A, van Dongen JT, Froehlich A, Oliver S, Faix B, Kapuganti JG, Schmalzlin E, Igal M, Orcaray L, Royuela M, Geigenberger P (2009) Regulation of respiration and fermentation to control the plant internal oxygen concentration. Plant Physiol 149:1087–1098

Zabalza A, Orcaray L, Igal M, Schauer N, Fernie AR, Geigenberger P, van Dongen JT, Royuela M (2011) Unraveling the role of fermentation in the mode of action of acetolactate synthase inhibitors by metabolic profiling. J Plant Physiol 168:1568–1575

Zhu J, Patzoldt WL, Shealy RT, Vodkin LO, Clough SJ, Tranel PJ (2008) Transcriptome response to glyphosate in sensitive and resistant soybean. J Agric Food Chem 56:6355–6363

Part VII
Technical Advances

Methods and Techniques to Measure Molecular Oxygen in Plants

Cindy Ast and Arie Draaijer

Abstract Designing and developing sensors for molecular oxygen (O_2) has turned into a large, interdisciplinary field of research, with significant progress seen in the past decades. Until the early 1980s, the field of O_2 sensing was dominated by polarographic electrode sensors, among which the most popular Clark-type electrode found wide application in plant science. Nevertheless, the great demand for more sophisticated, intracellularly applicable O_2 sensors for real-time measurements in plants cannot be satisfied by the predominant techniques. Thus, optical sensors applying an O_2-specific reduction of luminescent probes or dyes provide novel, promising tools and open new perspectives on the cellular or even subcellular level of O_2 measurements. This chapter aims to give a comprehensive overview on the variety of methods and systems available in the field of O_2 sensing with respect to application in plant tissue. Different types of the earlier polarographic electrode technique as well as emerging alternatives will be discussed, including fluorescent proteins as potential, genetically encoded intracellular O_2 sensors. Due to the tremendous variety of materials and formats, the young field of optical O_2 sensing will receive particular attention directing the focus towards the progress that has been made in developing new probes and dyes. Moreover, the current state of fluorescence measurements will be explored, particularly novel, plant-specific measurement modalities that mask plant autofluorescence. For the potential user, important practical aspects are also presented, revealing the limitations of the existing methods and further encouraging more interdisciplinary research in O_2 sensing.

C. Ast (✉)
NanoPolyPhotonik, Fraunhofer Institute for Applied Polymer Research, Geiselbergstr. 69, 14476 Potsdam-Golm, Germany

Energy Metabolism Research Group, Max Planck Institute of Molecular Plant Physiology, Am Mühlenberg 1, 14476 Potsdam-Golm, Germany
e-mail: cindy.ast@iap.fraunhofer.de

A. Draaijer
Gompenstraat, 25a 5145 RM Waalwijk, The Netherlands

J.T. van Dongen and F. Licausi (eds.), *Low-Oxygen Stress in Plants*, Plant Cell Monographs 21, DOI 10.1007/978-3-7091-1254-0_21, © Springer-Verlag Wien 2014

Abbreviations

AD	Analog-to-digital
CPP	Cell-penetrating peptides
FbFP	Flavin-binding fluorescent protein
FLIM	Fluorescence lifetime imaging
FluBO	Fluorescent protein-based biosensor for O_2
FP	Fluorescent protein
FRET	Förster resonance energy transfer
GFP	Green fluorescent protein
O_2	Molecular oxygen
TCSPC	Time-correlated single photon counting
TPE	Two-photon excitation
YFP	Yellow fluorescent protein

1 Methods in O_2 Sensing

Two major methods dominate the field of O_2 measurements in plant science, namely electrochemical sensors in the shape of electrodes and optical sensors in various formats. However, the third promising tool of genetically encoded biosensors is evolving and will be introduced here briefly.

1.1 Polarographic Electrode Sensors

The basic principle of polarography is the detection of a current flow caused by the electrolytical reduction of an analyte in solution, such as dissolved O_2, on the sensing cathode. The cathode is in conjunction with a reference electrode, the anode, and when suitable potentials are applied, O_2 will be reduced to hydroxide ions resulting in a detectable current. Due to the linear relationship between the O_2 concentration and the electrical current, the unknown O_2 concentration in the measured media can be determined (Kolthoff and Jordan 1952). The use of polarographic electrodes to measure tissue O_2 tension was first described by Davies and Brink in 1942 and found broad application also in plant science since (Davies and Brink 1942). In 1964, William Armstrong published the first data on O_2 diffusion from the roots of bog plants using cylindrical platinum electrodes surrounding the root (Armstrong 1964) and 3 years later a detailed report on how to employ polarographic techniques to assay O_2 diffusion from roots into anaerobic media followed (Armstrong 1967).

In general, all polarographic O_2 sensors are variations of two basic designs. The most popular one is the Clark-type electrode (Clark et al. 1953), where cathode and reference electrode are submersed in an electrolytic solution. Both are combined behind an O_2-permeable membrane, rendering it more sensitive and less prone to local O_2 changes that are due to the reduction of O_2 on the cathode. Other electrodes contain the cathode separated from the anode and are sometimes referred to as cathode-type electrodes. Both types exist as macro- and microelectrodes whereas macroelectrodes are rather suitable for external measurements, having a diameter of 5–10 mm and microelectrodes being 1–10 μm in tip diameter can be used for intracellular investigations. The main types of those electrodes are summarized in Table 1. Most of the electrodes applied are custom-made with detailed descriptions of fabrication methods in the literature, but there are commercially available electrodes, too. These delicate microelectrodes have already been in use half a century ago and can even be used for single cell measurements. A coated Platinum-microelectrode of a diameter of 1 μm was used to measure the O_2 partial pressure in the vacuole sap of root cells for O_2 concentration profiles across intact sunflower roots (Bowling 1973).

Much of the present knowledge on root oxygenation (Ober and Sharp 1996; Verslues et al. 1998) as well as O_2 flux and transport (Armstrong 1964, 1979; Mancuso et al. 2000) during hypoxia and anoxia we owe to the use of in vivo O_2 microelectrodes (Armstrong et al. 1993; Adamec 2005). Aeration of wetland plants by measuring radial O_2 loss has been very well studied (Armstrong et al. 2000, 2006; Garthwaite et al. 2008) as well as environmental effects such as oil pollution on wetland plant growth (Armstrong et al. 2009).

1.1.1 Advantages and Limitations of Polarographic Electrode Sensors

O_2 measurements with polarographic oxygen sensors enjoy great popularity because of their simple and cost-inexpensive equipment, little technical complexity, and the option of commercial purchase. The large amount of publications and manuals on polarographic electrode sensors makes it easy for the new user to enter the field of O_2 sensing. However, several disadvantages hamper further development (also see Table 1), a major one being the constant consumption of O_2 by the electrode itself, which can result in artifacts. Another issue is stability. For a stable current proportional to the investigated O_2 concentration, careful calibration is important to avoid effects caused by, e.g., temperature, drift, or voltage changes. They are prone to poisoning by proteins and heavy metals, reducing their lifetime and restricting reproducibility of data. Additionally, the invasive character due to tissue penetration, which can cause tissue trauma and quantitative errors, limits particularly intracellular measurements with high spatial resolution. In general, these electrode sensors are also tedious to fabricate and very delicate, and thus need careful handling. To overcome these limitations and to satisfy the demand for wireless, reliable and reproducible O_2 measurements, new nonpolarographic techniques are emerging among which optical O_2 sensing has improved greatly within the past years and already found application in plant science.

Table 1 Polarographic electrode sensors

	Electrode type	Material	Application	Advantages	Disadvantages	References
Macro-electrodes	*Sleeving cylindrical Pt-electrode* that surrounds the root	Thermo-pure Pt-cylinder insulated by epon-epoxy resin and enclosed by acrylic cylinder	– Quantification of O_2 flux into and out of the root system	– Sampling is non-destructive and continuous	– Prone to significant O_2 shifts, since naked	Armstrong (1964, 1979, 1994), Armstrong and Wright (1975)
		Remote anode: Ag:AgCl	– Investigation of O_2 transport from the shoot to the root – Quantification of O_2 changes in the root due to growth and temperature	– Maintains sensitivity over long periods of time	– Require stirring of media	
	Naked wire electrode	Thermo-pure Pt-wire Pt-Rhodium and Pt-Iridium alloys for strength	– Soil O_2 fluxes in plant-soil aeration studies	– Measures O_2 flux, since it integrates the influence of concentration and diffusivity in media	– Most reliable in nearly or saturated soil conditions – Require stirring	Armstrong (1979, 1994)
	Membrane-covered *Clark-type electrode*	*Cathode*: Pt-wire	– Used as respirometry cell sensor and probe-type sensor	– Reduces current fluctuations due to local O_2 changes	– Slower response times	Clark et al. (1953), Davies (1962), Armstrong and Gaynard (1976), Laan et al. (1990)
		Anode: Ag:AgCl	– Measure O_2 uptake by root	– Membrane protects electrode from aging	– Roots are excised and suspended	
		Membrane: Teflon, silicone rubber, polyethylene	– O_2 transport differences between wetland and non-wetland plants	– Operates in gas-phase	– Electrode can only sample from bulk solution	

Micro-electrodes	*Naked wire electrode*	Pt-, gold wire or wood metal alloy	– Sample of O_2 partial pressure gradients across soy bean nodules	– Small size increases the spatial resolution	– More susceptible to poisoning by proteins than macro-electrodes	Tjepkema and Yocum (1974), Armstrong (1993, 1994)
		Pt-wire with gold tip Tip can be exposed, beveled or recessed	– Study of radial O_2 distribution in maize roots		– Possibility of heavy metal contamination	
	Membrane-coated electrode	Pt-wire with gold tip coated with conducting membrane	– O_2 partial pressure in the vacuole sap of root cells was observed across intact sunflower roots			Bowling (1973), Armstrong (1994)
	Clark-type micro-electrode	Core is a naked wire electrode, outer body filled with electrolyte such as 1M KCl and plugged with silicone rubber	– O_2 micro-profiles in natural sediments and algal biofilms	– Guard electrode can be included to minimize residual current in the electrode	– Construction is more difficult than macro-electrodes	Revsbech and Ward (1983), Revsbech and Jørgensen (1986), Witty et al. (1987), Glud et al. (1992), Armstrong (1994)
		Anode: AgCl-wire	– O_2 distribution in the root nodule of pea and French bean		– Slight damage due to penetration of tissue causes artifacts	

1.2 Optical O_2 Sensors

Optical O_2 measurements rely on the principle of photoluminescence quenching by O_2, a process leading to the decrease of luminescence intensity and lifetime of an O_2-sensitive probe (Demas et al. 1999). Both phenomena can be detected in an O_2 concentration-dependent manner. In general, O_2 is an efficient quencher of molecular luminescence processes, because it exhibits exceptional electronic properties. Due to its biradical electronic character, it exists in a triplet ground state which makes it prone to interaction with excited molecules in the triplet state. This phenomenon, first described as "amortization" of phosphorescence by Kautsky and Hirsch in 1935 (Kautsky and Hirsch 1935), is a dynamic process of the collisional interaction of molecular O_2 with an indicator molecule, also termed dynamic quenching. This molecule can be either in the triplet state or in a long-lived singlet state (Vaughan and Weber 1970). Upon collision, energy transfer takes place, resulting in the elevation of the O_2 molecule to its excited state, the singlet state. At the same time, the indicator dye decays freely of radiation to the ground state. This energy transfer is concentration dependent and highly efficient and can thus be employed for the determination of the O_2 concentration. Because of the unique properties of O_2, only minor cross-reactions might interfere, rendering this detection method highly specific.

The optical measurement setup exists of the luminescent (phosphorescent or a fluorescent) dye, mostly embedded in a polymer matrix, an excitation source, usually a blue laser or LED, and a photodetector to collect the phosphorescent signal. A glass fiber can be used to combine transmission of both excitation light and the phosphorescence signal to the detector, constraining the setup to one fiber only. Additionally to glass fiber-optic microsensors, which are also termed micro-optodes, it has been demonstrated that those sensor materials are also suitable for microscopic fluorescence imaging when used as soluble sensors or miniaturized sensory beads (Sect. 1.2.1). The basic techniques to measure luminescent lifetimes in a nanosecond range are phase modulation or time-correlated single photon counting (TCSPC). Different detection methods to determine O_2 concentrations via phosphorescent sensor properties will be discussed in Sect. 2, including a dual-frequency phase modulation technique, particularly developed to mask the strong autofluorescence of the green plant tissue.

1.2.1 Materials and Sensor Formats of Optical O_2 Sensors

A large selection of O_2-sensitive probes is available (Dmitriev and Papkovsky 2012; Ast et al. 2012) and constantly new probes are being discovered with improved signal intensity and photostability. The most prominent probes are the phosphorescent Platinum (II)-, Palladium (II)-porphyrin dyes, Ruthenium (II)-complexes as well as the fluorescent pyrene family and their specific properties are summarized in Table 2.

Table 2 Selection of some O_2-sensitive probes and their phosphorescence characteristics

Indicator probe	Excitation (nm)	Emission (nm)	Lifetime (μs)	References
Pt-porphyrins (e.g., PtPFPP, PtCP)	380–400, 500–550	630–700	40–100	Papkovsky and O'Riordan (2005), Dmitriev and Papkovsky (2012)
Pd-porphyrins (e.g., PdPFPP, PdCP)	390–400, 500–550	630–700	400–1,000	Papkovsky and O'Riordan (2005), Dmitriev and Papkovsky (2012)
Ru(II)-bipyridyl-complexes	450–460	610	<1	Zhang et al. (2002), Neugebauer et al. (2008)
Ru(II)-dpp-complexes (e.g., (Ru[dpp]$_3$)X)	400–480	600	<5	Castellano and Lakowicz (1998)
Ir(III)-complexes (e.g., Ir(III) $(C_x)_2$(acac))	450–480, 420–450	480–570	10–13	Borisov and Klimant (2007)
Pyrenebutyric acid	350	390–440	<200	Vaughan and Weber (1970)

PtPFPP Pt(II)tetrapentafluorophenylporphyrin, *PtCP* Pt(II)coproporphyrin, *PdPFPP* Pd(II)tetrapentafluorophenylporphyrin, *PdCP* Pd(II)coproporphyrin, *(Ru[dpp]$_3$)X* Tris(4,7-diphenyl-1,10-phenanthroline)-ruthenium (II) derivatives, *Ir(III)$(Cx)_2$(acac)* cyclometalated Ir(III)μ-chloro-bridged dimer coumarin complexes

Also, the encapsulating matrix that is used to adhere the sensor dye to a carrier or to entrap it in beads has to fulfill certain properties, most important being O_2 permeability. The other properties are chemical and photochemical stability, rigidity, and transparency. An advantage of the matrix is that the dye intensity increases after polymerization which makes them additionally favorable to employ. Established matrices are silicones, polystyrene, polymethylacrylate, fluoropolymers, and glass-like materials formed by sol–gel processes. The embedding procedure enables fabrication of a variety of sensor formats, thus opening new application dimensions for monitoring of O_2 in real time (Grist et al. 2010; Dmitriev and Papkovsky 2012; Ast et al. 2012).

For visualization of O_2 profiles of intact, living cells and tissues, or whole plants over a period of days or weeks, thin optical sensor films are well suited. They can be created in the shape of small dots to apply them, e.g., on glass slides, microtiter plates and jars, up to the size of foils that can be casted on walls of whole containers. The sensor foil thickness can be of a few micrometers only, preventing measurement artifacts that are due to long sensor response times (see Sect. 3.1). The great advantage here is that the biological sample can be grown directly on the sensor film or in close proximity, like it was successfully demonstrated for plant rhizosphere investigations (Mills and Lepre 1997; Frederiksen and Glud 2006; Blossfeld et al. 2011; Rudolph et al. 2012; Tschiersch et al. 2012) (see Sect. 4). Application in solid and liquid media is possible, allowing simultaneous cultivation of cells or tissue and monitoring of O_2. Marine research has particularly benefited from this sensor design (Precht et al. 2004). Since a time series of images with changing O_2

patterns can be obtained, even movies as new visualization medium may be realized as a tool to investigate dynamic behavior of multidimensional O_2 availability (Tschiersch et al. 2012).

Extra- and intracellular measurements of O_2 can be performed by micro-optodes. They consist of a tapered glass fiber with a tip-diameter of 10–50 μm the end of the tip being coated with the sensor material (Rosenzweig and Kopelman 1995). The latter may consist of small foil spots or sensor solution attached to the tip. Experiments on living root nodules using very delicate fibers demonstrate the applicability for intracellular measurements (Löhmannsröben et al. 2005).

Another design is the fabrication of optical sensor beads, which can be miniaturized to a few tens of a nanometer (Clark et al. 1999). For this purpose, the phosphorescent dyes are entrapped in a polymer matrix, which protects and stabilizes the dye, but at the same time renders it water soluble. Successful O_2 concentration studies applying microbeads were performed on cells of the green algae *Chara corallina L.* and on *Arabidopsis thaliana* leaf epidermal cells, revealing that increased photosynthetic activity increases the O_2 concentration well above ambient conditions (Schmälzlin et al. 2005). Intracellular delivery techniques are micro- or picoinjection and gene gun that not only allow translocation of the beads into living cells but can also cause tissue damage and thus artifacts. Strategies to overcome the invasive character are liposomal transfer, endocytosis and targeting via cell-penetrating peptides (CPPs) (Lee and Kopelman 2012). CPPs are short, positively charged peptides which are capable to translocate across cell membranes, carrying a cargo of a much higher mass and size (Rothbard et al. 2002). CPP-translocation has been described for plant cells, in particular for plant protoplasts, which lack the cell wall (Vivès et al. 2003; Mäe et al. 2005; Chang et al. 2005). An illustrative overview on a selection of the current sensor formats is presented in Fig. 1.

Optical sensors additionally enable multiple sensing by taking advantage of probes whose signals may be differentiated by spectral and/or temporal resolution. Multisensor systems can be established for films or nanosensor. For sensor films, single- or multilayer approaches are described. Here, two different sensor spots can be aligned in close proximity or as "sandwich" with two sensor layers immobilized on a solid support. For single layer sensors, beads containing different dyes are incorporated into a single polymer matrix. Immobilizing different dye molecules only in the same matrix can lead to FRET (Förster resonance energy transfer). This energy transfer between luminescent probes is extremely efficient for small distances, leading to loss of luminescence but can be avoided when beads are employed, since they increase the proximity of the dyes (Stich et al. 2010). Fig. 1 Dual sensor films were established for O_2 and pH (Borchert et al. 2011; Kocincová et al. 2008), temperature (Köse et al. 2005; Baleizão et al. 2008), as well fiber-optic microsensors (Kocincova et al. 2007). Also, triple and quadruple sensor layers for the detection of O_2, pH, and temperature or additionally carbon dioxide were described (Stich et al. 2010; Borisov et al. 2011). A dual nanosensor for O_2 and pH for intracellular measurements was reported as well (Wang et al. 2012).

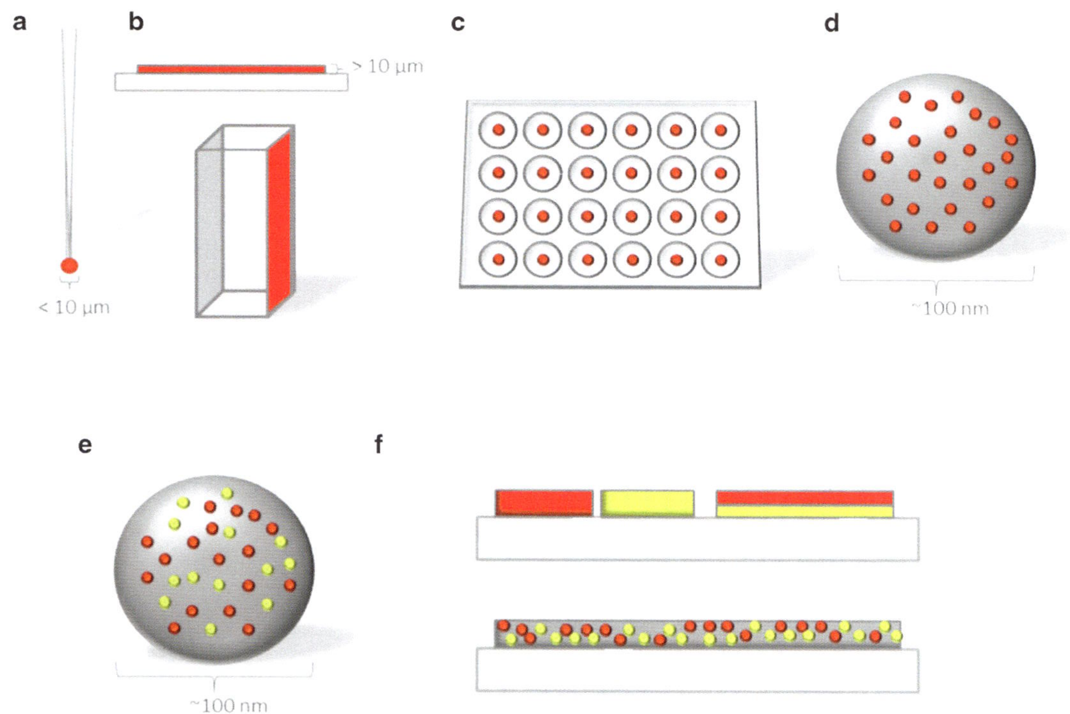

Fig. 1 Selection of sensor formats. The *upper row* showing single sensor applications, the *bottom row* showing multiple sensor systems: (**a**) micro-optode; (**b**) thin sensor foil on planar support (*top*) and on container wall (*bottom*); (**c**) sensor spots in wells of microtiter plates; (**d**) beads with sensor dye entrapped; (**e**) dual sensor probes entrapped in polymer bead; (**f**) dual sensor system as single layer (*top*, *left*), sandwich multilayer (*top*, *right*), or single layer with different beads (*bottom*) on support

1.2.2 Advantages and Limitations of Optical O_2 Sensors

The method of optical O_2 sensing bears several advantages over polarographic measurements, due to the huge variety of sensor materials and formats, which allow intracellular measurements and miniaturization down to the nanoscale. The high surface-to-volume ratio of micro- and nanoparticle sensors results in only minimal physical interference but offers high accessibility for the analyte to the sensing components. Also multiplexing of additional parameters like pH and temperature controls, but also references to eliminate concentration- or instrument-related signal interferences are easier realized, as they can be combined in one matrix.

Photobleaching, caused by a large number of absorption and emission cycles eventually leading to diminished luminescence of the probe, is a common concern in luminescence measurements. It can be reduced by the choice of light condition and the time period of the measurement. A major advantage to be stressed is that most materials presented here are compatible with two-photon excitation (TPE), a new technique decreasing photobleaching and -damage (Denk et al. 1990). TPE is a process of excitation by simultaneous absorption of two photons by the luminophore, thus reducing background fluorescence from the luminophore outside the focal volume. So far, it is mainly applied in laser scanning fluorescence microscopy but was also tested for optical fibers (Helmchen et al. 2002).

A limitation of the optical O_2-sensing technique is the occurrence of singlet O_2 as a by-product during the quenching process, which can be harmful for biological samples, as demonstrated for studies on piglet eyes (Stepinac et al. 2005). The severity of tissue damage is dose dependent, but can nowadays be reduced since the sensitivity of luminescence detection techniques is improving. More importantly is the employment of encapsulation matrices, which cause quenching of the reactive O_2 species before they can interact with components of the sample, thus reducing toxic effects.

1.3 Genetically Encoded O_2 Sensors

Especially in biology-related disciplines of science like molecular biology and biochemistry, genetically encoded sensors are popular tools to measure intracellular compounds in vivo and in real time. The number of successfully applied sensors of this kind is tremendous, reaching from the detection of ions, pH-value, metabolites, and hormones (Frommer et al. 2009; Tantama et al. 2011). A popular sensor design is the FRET-based sensor. It usually consists of a binding protein with a different conformation in the bound and unbound state, fused to two fluorescent proteins (FPs), one FP being the FRET-donor and the other being the FRET-acceptor. The FRET-process depends on three properties, the spectral overlap of the donor emission and acceptor excitation, the distance between both FPs, and the relative orientation of the FP's transition dipole moments (Miyawaki 2003). In absence or presence of the investigated compound, the FRET-ratio varies, giving information on, e.g., concentration and localization. First attempts on developing FRET-based O_2 sensors were carried out on arthropod hemocyanins coupled to different FRET-donors. Hemocyanins bind O_2 with very high cooperativity and were already employed as binding protein for an in vitro FRET-sensor (Erker et al. 2004).

The first in vivo FRET-based sensor for O_2 applies a different principle. FluBO (fluorescent protein-based biosensor for O_2) is genetically encoded and takes advantage of the sensitivity of FPs towards O_2. The FRET-donor is a cyan flavin-binding fluorescent protein (FbFP), which is combined with its FRET-acceptor, a yellow fluorescent protein (YFP), by a short linker. While maturation of FbFP is O_2-independent, YFP requires O_2 for the maturation of the fluorescent chromophore. Thus, efficient FRET from the FbFP-donor to the YFP-acceptor can only occur in the presence of O_2, when the fusion protein is fully expressed, but not in its absence. FluBO allows monitoring of low O_2 conditions in living bacteria but involves rapid turnover of new protein or even new bacteria (Potzkei et al. 2012).

The O_2 sensitivity is a feature of most of the FPs, a class of chemiluminescent proteins that originally derived from the jellyfish *Aequorea victoria*. The tripeptide sequence, which is crucial for the fluorescence of the FP, is formed autonomously in an autocatalytic reaction, the only requirement being O_2 for the oxygenation step to complete the π-conjugated, fluorescent chromophore. The green fluorescent protein (GFP) is the most popular of that class and has been employed to access low O_2 levels in a different way. For a variety of GFPs, an O_2-dependent red shift in fluorescence upon irradiation with blue light was reported. A low O_2 environment and a short exposure to blue light are sufficient to generate a stable red fluorescent form of the GFP protein. This phenomenon described with the term anaerobic redding is only observed at low O_2 concentrations of 0–2 % and is reversible after reoxygenation, leading to a fast disappearance of the red fluorescent form. The underlying mechanism is not fully understood, especially for the reversible photoconversion. For the irreversible activation of the red derivative, it was suggested that the irradiation with blue light cleaves the backbone of the chromophore, introducing an additional double bond that extends the π-conjugated system and red shifts the emission of the chromophore (Ha and Tinnefeld 2012). So far, in vivo photoconversion was demonstrated using various GFP mutants, localized in different cellular compartments and organisms like *Escherichia coli* (Elowitz et al. 1997), fission yeast (Sawin and Nurse 1997), and mammalian cell lines (Bogdanov et al. 2009).

1.3.1 Advantages and Disadvantages of Genetically Encoded O_2 Sensors

The field of genetically encoded O_2 sensors is a very young one with few applications and little practical experience. Thus, the advantages listed here refer to the general benefits of FPs as biological sensors. First, since genetically encoded, these sensors can be regarded as least invasive, because they do not have to be introduced externally and are thus better tolerated by the cells. Second, they exist in all hues of the visible spectrum, thus allowing multiplexing, FRET or single FP measurements in any spectral range. This has the advantage that disturbing background fluorescence, e.g., red chlorophyll fluorescence, can be circumvented by choosing an FP whose emission does not interfere with common background signals of the investigated sample. Third, as already described for optical O_2 probes, FPs greatly favor from the improved fluorescence measurement and imaging techniques, including TPE, which causes less photobleaching. The latter is a common obstacle in particular for long-term measurement (Sect. 1.2.2).

The major limitation to be listed is the need of O_2 for the chromophore formation of GFP and its derivatives, which limits growth assays and measurements of cells or tissues that are kept under hypoxic conditions. Those measurements however are crucial, e.g., for fluorescence imaging in anaerobic systems or understanding metabolic changes upon O_2 depletion. Therefore, for future genetically encoded O_2 sensors, O_2-insensitive FPs are highly desirable and the novel class of FbFPs is a promising alternative. FbFPs bind flavin mononucleotide as a chromophore, have comparable fluorescence intensities to enhanced GFP, and already have been employed as reporters under anaerobic and aerobic conditions in microbes (Drepper et al. 2007, 2010) and mammalian cells (Walter et al. 2012). Additionally, FbFP has been exploited for the construction of the FRET-based O_2 sensor FluBO, where it serves as a suitable FRET-donor for YFP (Potzkei et al. 2012).

2 Measurement Modalities for Luminescent Probes

Incorporating luminescent dyes into the field of O_2 sensing greatly benefits from the tremendous progress that has been made in the field of luminescence detection and imaging. Phosphorescent and fluorescent dyes exhibit a variety of measureable properties, e.g., luminescence intensity and lifetime, excitation, and emission wavelengths. While intensity-based measurements are accompanied with problems arising from photobleaching, light scattering, autofluorescence of the sample, and brightness of the light source, recording the luminescence lifetime is an attractive alternative. The latter is unaffected by the probe concentration, static quenching, or instrumental instabilities (Schäferling 2012). Therefore, this section will focus on lifetime-based sensing of O_2, facilitated either by phase modulation or by TCSPC under pulsed excitation.

2.1 Dual-Frequency Phase Modulation Technique for the Detection of Optical Sensors

Phase modulation techniques in general apply the excitation of the sensor with sinusoidally modulated light at a frequency approximately reciprocal to the decay time. The emission of a luminescent signal does not occur simultaneously with the excitation, but with a certain delay, which depends on the lifetime of the excited state of the molecule. This is measured as shift of the phase of the emission light as compared to the original modulation phase of the excitation light. Accordingly, the lifetime can be calculated from the phase angle or the modulation ratio (Schmälzlin et al. 2005; Schäferling 2012).

In plants, the strong autofluorescence of the chlorophyll generally interferes with the detection of sensor phosphorescence and thus severely hampers optical sensing techniques. The red chlorophyll fluorescence superimposes the optical sensor signal and thus reduces the apparent phase shift. To overcome this limitation for optical sensor applications in plants, a special dual-frequency phase modulation technique was developed to discriminate between the O_2-dependent luminescence signal of a sensor probe and the autofluorescence of the plant tissue. It is based on the respective lifetime differences: The lifetime of chlorophyll ranges around nanoseconds and can be regarded as negligible in comparison to the long-lived phosphorescence lifetimes of the dyes in the range of microseconds. If the respective phase shifts are measured simultaneously at two different, modulated frequencies, quantification and subsequent mathematical removal of the background signals can be performed (Schmälzlin et al. 2005). This method works with two sinus generators that each modulate the amplitude of the excitation light and thus provide superimposed sinus signals with two different frequencies. After excitation of the sensor, a lock-in amplifier unit in the detector separates the two frequencies to measure the respective phase shifts in reference to the excitation signal. Since all fluorescent background signals are in phase with the excitation light, this fraction can be calculated and eliminated. Thus, not only the autofluorescence signals are being removed, but also residual excitation light, which might reach the detector unintentionally and would otherwise be detected as signal noise. This is particularly advantageous for the application of the optical micro- and nanoprobes with very low phosphorescent intensities, but the technique is well compatible with fiber optical sensors as well (Steinbrück et al. 2011).

2.2 Pulse Ratio Technique

A different approach to determine the decay time is the time-gated technology, one example being TCSPC. This method is based on recording the decay curve after the light pulse coming from an excitation source. A very precise timer is started with the excitation pulse and stopped as soon as one single photon of the excited fluorophore is detected. Repeating this measurement several tens of thousands times delivers very precise decay curves in the picoseconds to millisecond range (Draaijer et al. 1995). For the long-lived dyes used in O_2 measurements, the luminescence lifetime can be measured using a pulsed LED and a photodiode with appropriate amplifier circuitry and AD (analog-to-digital) converters. If the AD converter is suitable to record, e.g., the intensity of the decay curve in two subsequent points of time within that curve, the ratio of these two measured intensities is proportional to the fluorescence lifetime. For multiplexing, multiexponential decays can be analyzed too. If the measured signal is solely coming from the O_2-sensitive probe and not from stray light from the LED or other fluorescent substances, the conversion from raw signals to O_2 is easily realized. TCSPC applicability for recording fluorescence decay curves of chlorophyll was demonstrated for various plant samples (Becker 2008). The versatility of this method makes it possible to use it in many fields amongst which confocal laser scanning microscopy, making fluorescence lifetime imaging (FLIM) for optical dyes possible. TCSPC FLIM measurements were, e.g., performed on GFP-labeled microtubules on tobacco tissue and protoplasts (Brandner et al. 2008).

3 Practical Aspects

O_2 is a gaseous compound that also solubilizes in aqueous solutions. Since O_2 is in the air all around us, a particular experimental setup is needed to control the O_2 environment for experiments in the gas phase. A closed box, e.g., glove box with O_2 control, makes it possible to study O_2-dependent phenomena. If different O_2 concentrations in liquids are desired, these can be prepared by mixing nitrogen with O_2 or air and “flushing” this through the water phase of the investigated system.

3.1 O_2 Diffusion in Different Media (Water and Air)

In general, it should be taken into account that diffusion rates of O_2 in water are roughly 10,000 times slower compared to the diffusion rates in gases. This fact is of great importance for the response time of O_2 sensors and affects polarographic and optical sensors differently. In the Clark cell for instance O_2 is consumed, so ultimately the O_2 concentration at the electrode interface is very close to zero, maximizing the driving force for continuous O_2 diffusion towards the electrode. For optical sensors that usually employ the dye entrapped in a polymer matrix, O_2 has to enter the matrix first to interact with the dye. The graph in Fig. 2 shows the influence of the coating thickness on the response time of a sensor. While a thin sensor layer of 10 μm leads to a full response in less than 8 s, the response is delayed several tens of seconds with increasing coating thickness. Thus, thin sensor coating is highly desirable. Additionally, a common obstacle to be named for O_2 measurements in the gas phase is the formation of a condensation layer on the measuring tip. This dramatically slows down the response time and may cause artifacts if fast O_2 production or consumption processes are studied.

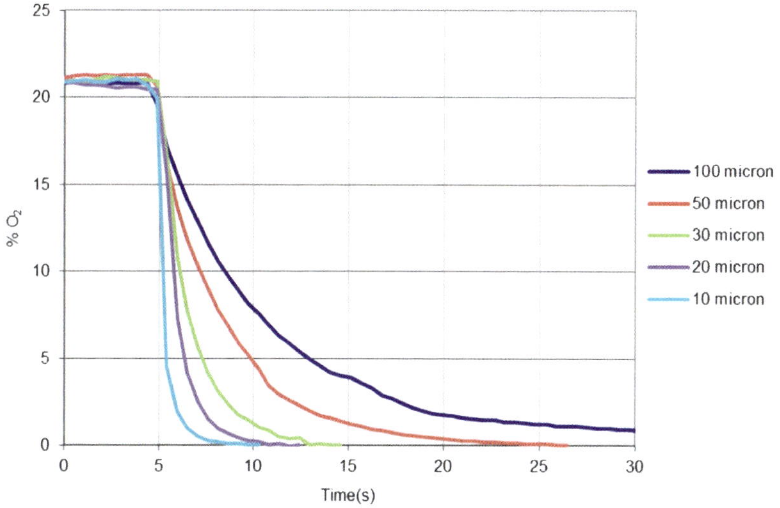

Fig. 2 Effect of coating thickness on the sensor response time in gas phase at room temperature. Graphs were obtained from coatings of Tris-(4,7-diphenyl-1,10-phenantrolin)-Ruthenium (II) dissolved in polydimethylsiloxane (PDMS)

3.2 *Temperature Effects/Thermal Quenching*

Fluctuations of the temperature can have a severe effect on the result of O_2 measurements. First, temperature has a significant impact on the concentration of O_2 in water since it dissolves less well with increasing temperatures. Second, dynamic quenching and luminescence are temperature dependent as well. The general consequence of thermal quenching is that increasing temperatures decrease the luminescence and lifetime of the luminophore, because nonradiative relaxation of the excited state is facilitated (Schäferling 2012). Therefore, the temperature has to be monitored either by supplied external temperature sensors that come with the optical sensing setup or, as described in Sect. 1.2.1, dual sensors for O_2 and temperature for simultaneous assessment of both parameters are employed. In all cases, it is necessary that the response time of the optical sensor is the same as the temperature sensor to prevent over and under compensation problems and thus artifacts.

3.3 *Drift of the Sensor Output*

A change in the signal output over time, also referred to as drift, is commonly observed for O_2 measurements. Drift can occur for instance if the dye exists in more than one chemical state or if the physical properties of the dye change due to the immobilization matrix or excessive laser excitation. Therefore, the sensor material utilized should be pure and should maintain its chemical form throughout sensor fabrication. The host or immobilization matrix should protect the probe and should be inert to not influence the sensitivity of the sensor probe towards O_2 over time. Also, differing humidity can have an impact on the sensor signal as well as, e.g., oils and alcohols. Thus, the chemical robustness is a major challenge for the coating design and drift should be taken into account during data analysis.

4 Current Plant Application in Optical O_2 Sensing

A major gain of optical O_2 sensing is the complexity of data that can be obtained. This section will focus on two novel, highly sophisticated strategies for O_2 mapping. By using micro-optodes and planar sensor foils, respectively, detailed O_2 maps were obtained, demonstrating the great potential of optical O_2 sensors to access complex O_2 dynamics in the field of plant science. Therefore, the number of future application will increase without doubt.

4.1 O_2 Maps of Seeds

The applicability of optical microsensors in plant seed O_2 studies was well demonstrated by Rolletschek et al. Several crop plant species, ranging from maize to wheat, were investigated by a reported standard procedure (review Rolletschek et al. 2009). The micro-optode was stepwise inserted into the seed and local O_2 measurements were carried out. To prevent diffusion of O_2 into the seed, the entry point had to be sealed. Following the measurement, seeds were dissected to identify the particular zones to set up the corresponding O_2 profile. With the obtained data, O_2 maps of seeds were established to show tissue O_2 levels and changes in response to environmental factors. The benefits from these maps are the identification of regions within the seed that are susceptible to diffusional impedance of gas exchange, gas diffusion pathways, as well as O_2 release due to photosynthesis (Rolletschek et al. 2009). A novel approach that greatly simplifies the delicate measurements of microsensors was recently reported. Here, sensor foils were directly placed on the sample surface and the O_2 pattern was visualized by a miniature microscope. Detailed maps of the O_2 pattern across developing seeds, but also of the rhizosphere, stem and leaves were obtained that were consistent with previous microsensor data (Tschiersch et al. 2012). This information is of particular interest for crop growth and development, since O_2 deficiency inside seeds limits their germination and development.

4.2 O_2 Maps of the Plant Root Zone

Planar sensor foils are particularly versatile for experiments on larger surfaces, e.g., on the plant root system in its natural sediment. Reported results are detailed, two-dimensional maps of the O_2 distribution around buried intact root systems as function of various environmental controls during active root growth for many days or even weeks. Frederiksen and Glud were the first to apply this system by attaching planar Ruthenium luminophore foils along one side of an aquarium and monitoring the O_2 distribution around living roots under various light regimes and O_2 concentrations in the overlying water column over a period of three weeks. They were able to thoroughly document the O_2 dynamics during rhizosphere development and to obtain data even from the microniches around the root tip by a series of O_2 images (Frederiksen and Glud 2006). The advantage of multiple optical sensing was implemented by Blossfeld and coworkers, who simultaneously assessed O_2 and pH dynamics in the root zone over a long-term period of eight weeks. The aim was to investigate the dynamic interplay between O_2 concentration, pH, and organic acids as main factors of the natural rhizosphere, particularly for wetland plants growing in O_2-deficient waterlogged soils. Their elegant approach combines optical measurements of the two parameters with sterile sampling of rhizosphere solution across and along the growing root by employing a modified rhizobox for plant growth (Blossfeld et al. 2011). Rudolf et al. reported a strategy for simultaneous measurements employing neutron radiography in combination with optical O_2 measurements. The purpose of this work was to gain detailed insights into root respiration by investigating the O_2 and water distribution of the rhizosphere of lupine plants over two weeks. This approach utilizes the fabrication of planar, thin sensor foils casted on the inner wall of glass containers which then can be readily used for the cultivation of intact plants. A time series of fluorescence images was obtained to visualize the O_2 depletion due to root respiration in relation to the water content. The latter was monitored in parallel by neutron radiography, a noninvasive method that visualizes water distribution with high resolution (Rudolph et al. 2012).

5 Concluding Remarks

Biological sensors in general are versatile tools, providing not only information on concentration and localization, but also monitoring metabolic changes and influences of external stimuli on a cellular basis and in real time. However, the conventional polarographic electrodes for O_2 sensing are facing some limitations, which cannot be fully overcome yet, and the rapid development of alternative tools is expected to offer more freedom and to open up new experimental possibilities. Optical O_2 techniques allow miniaturization, reduced invasiveness and overall more convenient measurements, including simultaneous monitoring of O_2 and other parameters of interest with high spatial and temporal resolution in living biological samples. The development of novel imaging techniques and measurement modalities, especially for plant application favored from the interaction between different science disciplines and further intensification of this interdisciplinary interaction of chemistry, physics, and biology would greatly contribute to the advancement of O_2 sensing in plant science.

References

Adamec L (2005) Ecophysiological characterization of carnivorous plant roots: oxygen fluxes, respiration, and water exudation. Biol Plant 49:247–255

Armstrong W (1964) Oxygen diffusion from the roots of some British bog plants. Nature 204:801–802

Armstrong W (1967) The use of polarography in the assay of oxygen diffusing from roots in anaerobic media. Physiol Plant 20:540–553

Armstrong W (1979) Aeration in higher plants. In: Woolhouse HWW (ed) Advances in botanical research, vol 7. Academic, London, pp 225–332

Armstrong W (1994) Polarographic oxygen electrodes and their use in plant aeration studies. Proc R Soc Edinb 102B:511–521

Armstrong W, Gaynard TJ (1976) The critical oxygen pressure for root respiration in intact plants. Physiol Plant 37:200–206

Armstrong W, Wright EJ (1975) The theoretical basis for the manipulation of flux data obtained by the cylindrical platinum electrode technique. Physiol Plant 35:21–26

Armstrong W, Cringle S, Brown M, Greenway H (1993) A micro-electrode study of oxygen distribution in the roots of intact maize seedlings. In: Jackson MB, Black CR (eds) Interacting stresses on plants in a changing climate, vol 16, NATO ASI Series 1. Springer, Berlin, pp 287–304

Armstrong W, Cousins D, Armstrong J, Turner DW, Beckett PM (2000) Oxygen distribution in wetland plant roots and permeability barriers to gas-exchange with the rhizosphere: a microelectrode and modelling study with *Phragmites australis*. Ann Bot 86:687–703

Armstrong J, Jones RE, Armstrong W (2006) Rhizome phyllosphere oxygenation in Phragmites and other species in relation to redox potential, convective gas flow, submergence and aeration pathways. New Phytol 172:719–731

Armstrong J, Keep R, Armstrong W (2009) Effects of oil on internal gas transport, radial oxygen loss, gas films and bud growth in *Phragmites australis*. Ann Bot 103:333

Ast C, Schmälzlin E, Löhmannsröben HG, van Dongen JT (2012) Optical oxygen micro- and nanosensors for plant applications. Sensors (Basel) 12:7015–7032

Baleizão C, Nagl S, Schäferling M, Berberan-Santos MN, Wolfbeis OS (2008) Dual fluorescence sensor for trace oxygen and temperature with unmatched range and sensitivity. Anal Chem 80:6449–6457

Becker W (2008) The bh TCSPC handbook, 3rd edn. Becker & Hickl GmbH, Berlin

Blossfeld S, Gansert D, Thiele B, Kuhn AJ, Lösch R (2011) The dynamics of oxygen concentration, pH value, and organic acids in the rhizosphere of Juncus spp. Soil Biol Biochem 43:1186–1196

Bogdanov AM, Mishin AS, Yampolsky IV, Belousov VV, Chudakov DM, Subach FV, Verkhusha VV, Lukyanov S, Lukyanov KA (2009) Green fluorescent proteins are light-induced electron donors. Nat Chem Biol 5:459–461

Borchert NB, Ponomarev GV, Kerry JP, Papkovsky DB (2011) O(2)/pH multisensor based on one phosphorescent dye. Anal Chem 83:18–22

Borisov SM, Klimant I (2007) Ultrabright oxygen optodes based on cyclometalated iridium(III) coumarin complexes. Anal Chem 19:7501–7509

Borisov SM, Seifner R, Klimant I (2011) A novel planar optical sensor for simultaneous monitoring of oxygen, carbon dioxide, pH and temperature. Anal Bioanal Chem 400:2463–2474
Bowling DJF (1973) Measurement of a gradient of oxygen partial pressure across the intact root. Planta 111:323–328
Brandner K, Sambade A, Boutant E, Didier P, Mély Y, Ritzenthaler C, Heinlein M (2008) Tobacco mosaic virus movement protein interacts with green fluorescent protein-tagged microtubule end-binding protein 1. Plant Physiol 147:611–623
Castellano FN, Lakowicz JR (1998) A water-soluble luminescence oxygen sensor. Photochem Photobiol 67:179–183
Chang M, Chou JC, Lee HJ (2005) Cellular internalization of fluorescent proteins via arginine-rich intracellular delivery peptide in plant cells. Plant Cell Physiol 46:482–488
Clark LC, Wolf R, Granger D, Taylor Z (1953) Continuous recording of blood oxygen tensions by polarography. J Appl Physiol 6:189–193
Clark HA, Hoyer M, Philbert MA, Kopelman R (1999) Optical nanosensors for chemical analysis inside single living cells. 1. Fabrication, characterization, and methods for intracellular delivery of PEBBLE sensors. Anal Chem 71:4831–4836
Davies PW (1962) The oxygen cathode. In: Nastuk WH (ed) Physical techniques in biological research, vol IV, Special methods. Academic, New York, pp 137–179
Davies PW, Brink F (1942) Microelectrodes for measuring local oxygen tension in animal tissues. Rev Sci Instrum 13:524
Demas JN, DeGraff BA, Coleman PB (1999) Oxygen sensors based on luminescence quenching. Anal Chem 71:793A–800A
Denk W, Strickler JH, Webb WW (1990) Two-photon laser scanning fluorescence microscopy. Science 248:73–76
Dmitriev RI, Papkovsky DB (2012) Optical probes and techniques for O2 measurement in live cells and tissue. Cell Mol Life Sci 69:2025–2039
Draaijer A, Sanders R, Gerritsen HC (1995) Fluorescence lifetime imaging, a new tool in confocal microscopy. In: Pawley JB (ed) Handbook of biological confocal microscopy, 2nd edn. Plenum Press, New York, pp 491–505
Drepper T, Eggert T, Circolone F, Heck A, Krauss U, Guterl JK, Wendorff M, Losi A, Gärtner W, Jaeger KE (2007) Reporter proteins for in vivo fluorescence without oxygen. Nat Biotechnol 25:443–445
Drepper T, Huber R, Heck A, Circolone F, Hillmer AK, Büchs J, Jaeger KE (2010) Flavin mononucleotide-based fluorescent reporter proteins outperform green fluorescent protein-like proteins as quantitative in vivo real-time reporters. Appl Environ Microbiol 76:5990–5994
Elowitz MB, Surette MG, Wolf PE, Stock J, Leibler S (1997) Photoactivation turns green fluorescent protein red. Curr Biol 7:809–812
Erker W, Schoen A, Basché T, Decker H (2004) Fluorescence labels as sensors for oxygen binding of arthropod hemocyanins. Biochem Biophys Res Commun 324:893–900
Frederiksen SM, Glud RN (2006) Oxygen dynamics in the rhizosphere of Zostera marina: a two-dimensional planar optode study. Limnol Oceanogr 51:1072–1083
Frommer WB, Davidson MW, Campbell RE (2009) Genetically encoded biosensors based on engineered fluorescent proteins. Chem Soc Rev 38:2833–2841
Garthwaite AJ, Armstrong W, Comer TD (2008) Assessment of O_2 diffusivity across the barrier to radial O_2 loss in adventitious roots of *Hordeum marinum*. New Phytol 179:405–416
Glud RN, Ramsing NB, Revsbech NP (1992) Photosynthesis and photosynthesis-coupled respiration in natural biofilms quantified with oxygen microsensors. J Phycol 28:51–60
Grist SM, Chrostowski L, Cheung KC (2010) Optical oxygen sensors for applications in microfluidic cell culture. Sensors (Basel) 10:9286–9316
Ha T, Tinnefeld P (2012) Photophysics of fluorescent probes for single-molecule biophysics and super-resolution imaging. Annu Rev Phys Chem 63:595–617
Helmchen F, Tank DW, Denk W (2002) Enhanced two-photon excitation through optical fiber by single-mode propagation in a large core. Appl Opt 41:2930–2934

Kautsky H, Hirsch A (1935) Nachweis geringster Sauerstoffmengen durch Phosphoreszenztilgung. Z Anorg Allg Chem 222:126–134

Kocincova AS, Borisov SM, Krause C, Wolfbeis OS (2007) Fiber-optic microsensors for simultaneous sensing of oxygen and pH, and of oxygen and temperature. Anal Chem 79:8486–8493

Kocincová AS, Nagl S, Arain S, Krause C, Borisov SM, Arnold M, Wolfbeis OS (2008) Multiplex bacterial growth monitoring in 24-well microplates using a dual optical sensor for dissolved oxygen and pH. Biotechnol Bioeng 100:430–438

Kolthoff IM, Jordan J (1952) Oxygen induced electroreduction of hydrogen peroxide and reduction of oxygen at the rotated gold wire electrode. J Am Chem Soc 74:4801–4805

Köse ME, Carroll BF, Schanze KS (2005) Preparation and spectroscopic properties of multiluminophore luminescent oxygen and temperature sensor films. Langmuir 21:9121–9129

Laan P, Tosserams M, Blom CWPM, Armstrong W (1990) Internal oxygen transport in *Rumex* species and its significance for respiration under hypoxic conditions. Plant Soil 122:39–46

Lee YE, Kopelman R (2012) Nanoparticle PEBBLE sensors in live cells. Methods Enzymol 504:419–470

Löhmannsröben H-G, Beck M, Hildebrandt N, Schmälzlin E, van Dongen JT (2005) New challenges in biophotonics: laser-based fluoroimmuno analysis and in vivo optical oxygen monitoring. Proc SPIE 6157:20–25

Mäe M, Myrberg H, Jiang Y, Paves H, Valkna A, Langel U (2005) Internalization of cell-penetrating peptides into tobacco protoplasts. Biochim Biophys Acta 1669:101–107

Mancuso S, Papeschi G, Marras AM (2000) A polarographic, oxygen-selective, vibrating-micro-electrode system for the spatial and temporal characterisation of transmembrane oxygen fluxes in plants. Planta 211:384–389

Mills A, Lepre A (1997) Controlling the response characteristics of luminescent porphyrin plastic film sensors for oxygen. Anal Chem 69:4653–4659

Miyawaki A (2003) Visualization of the spatial and temporal dynamics of intracellular signaling. Dev Cell 4:295–305

Neugebauer U, Pellegrin Y, Devocelle M, Forster RJ, Signac W, Moran N, Keyes TE (2008) Ruthenium polypyridyl peptide conjugates: membrane permeable probes for cellular imaging. Chem Commun 42:5307–5309

Ober ES, Sharp RE (1996) A microsensor for direct measurement of O_2 partial pressure within plant tissues. J Exp Bot 47:447–454

Papkovsky DB, O'Riordan TC (2005) Emerging applications of phosphorescent metalloporphyrins. J Fluoresc 4:569–584

Potzkei J, Kunze M, Drepper T, Gensch T, Jaeger KE, Büchs J (2012) Real-time determination of intracellular oxygen in bacteria using a genetically encoded FRET-based biosensor. BMC Biol 10:28

Precht E, Franke U, Polerecky L, Huettel M (2004) Oxygen dynamics in permeable sediments with wave-driven pore water exchange. Limnol Oceanogr 49:693–705

Revsbech NP, Jørgensen BB (1986) Micro-electrodes: their use in microbial ecology. Adv Microb Ecol 9:293–352

Revsbech NP, Ward DM (1983) Oxygen micro-electrode that is insensitive to medium chemical composition: use in an acid microbial mat dominated by Cyanidium caldarium. Appl Environ Microbiol 45:755–759

Rolletschek H, Stangelmeyer A, Borisjuk L (2009) Methodology and significance of microsensor-based oxygen mapping in plant seeds—an overview. Sensors (Basel) 9:3218–3227

Rosenzweig Z, Kopelman R (1995) Development of a submicrometer optical fiber oxygen sensor. Anal Chem 67:2650–2654

Rothbard JB, Kreider E, Pattabiraman K, Pelkey ET, VanDeusen CL, Wright L, Wylie BL, Wender PA (2002) Arginine-rich molecular transporters for drugs: the role of backbone and sidechain variations on cellular uptake. In: Langel Ü (ed) Cell-penetrating peptides: processes and applications. CRC Press, Boca Raton, FL, pp 141–160

Rudolph N, Esser HG, Carminati A, Moradi AB, Hilger A, Kardjilov N, Nagl S, Oswald SE (2012) Dynamic oxygen mapping in the root zone by fluorescence dye imaging combined with neutron radiography. J Soil Sediment 12:63–74

Sawin KE, Nurse P (1997) Photoactivation of green fluorescent protein. Curr Biol 7:606–607

Schäferling M (2012) The art of fluorescence imaging with chemical sensors. Angew Chem Int Ed Engl 51:3532–3554

Schmälzlin E, van Dongen JT, Klimant I, Marmodée B, Steup M, Fisahn J, Geigenberger P, Löhmannsröben HG (2005) An optical multifrequency phase-modulation method using microbeads for measuring intracellular oxygen concentrations in plants. Biophys J 89:1339–1345

Steinbrück D, Schmälzlin E, Peinemann F, Kumke MU (2011) An innovative laser-based sensing platform for realtime optical monitoring of oxygen. In: Proceedings of joint international IMEKO TC1+ TC7+ TC13 symposium, Jena, 31 Aug to 2 Sept 2011

Stepinac TK, Chamot SR, Rungger-Brändle E, Ferrez P, Munoz JL, van den Bergh H, Riva CE, Pournaras CJ, Wagnières GA (2005) Light-induced retinal vascular damage by Pd-porphyrin luminescent oxygen probes. Invest Ophthalmol Vis Sci 46:956–966

Stich MI, Fischer LH, Wolfbeis OS (2010) Multiple fluorescent chemical sensing and imaging. Chem Soc Rev 39:3102–3114

Tantama M, Hung YP, Yellen G (2011) Imaging intracellular pH in live cells with a genetically encoded red fluorescent protein sensor. J Am Chem Soc 133:10034–10037

Tjepkema JD, Yocum CS (1974) Measurement of oxygen partial pressure within soybean nodules by oxygen microelectrodes. Planta 119:351–360

Tschiersch H, Liebsch G, Borisjuk L, Stangelmayer A, Rolletschek H (2012) An imaging method for oxygen distribution, respiration and photosynthesis at a microscopic level of resolution. New Phytol 196:926–936

Vaughan WM, Weber G (1970) Oxygen quenching of pyrenebutyric acid fluorescence in water. A dynamic probe of the microenvironment. Biochemistry 9:464–473

Verslues PE, Ober ES, Sharp RE (1998) Root growth and oxygen relations at low water potentials. Impact of oxygen availability in polyethylene glycol solutions. Plant Physiol 116:1403–1412

Vivès E, Richard J-P, Rispal C, Lebleu B (2003) TAT peptide internalization: seeking the mechanism of entry. Curr Protein Pept Sci 4:125–132

Walter J, Hausmann S, Drepper T, Puls M, Eggert T, Dihné M (2012) Flavin mononucleotide-based fluorescent proteins function in mammalian cells without oxygen requirement. PLoS One 7:e43921

Wang XD, Stolwijk JA, Lang T, Sperber M, Meier RJ, Wegener J, Wolfbeis OS (2012) Ultra-small, highly stable, and sensitive dual nanosensors for imaging intracellular oxygen and pH in cytosol. J Am Chem Soc 134:17011–17014

Witty JF, Skot L, Revsbech NP (1987) Direct evidence for changes in the resistance of legume nodules to oxygen diffusion. J Exp Bot 38:1129–1140

Zhang P, Guo J, Wang Y, Pang W (2002) Incorporation of luminescent tris(bipyridine)ruthenium (II) complex in mesoporous silica spheres and their spectroscopic and oxygen sensing properties. Mater Lett 53:400–405

Index

J.T. van Dongen and F. Licausi (eds.), *Low-Oxygen Stress in Plants*, Plant Cell Monographs 21, DOI 10.1007/978-3-7091-1254-0, © Springer-Verlag Wien 2014

Printed by Printforce, the Netherlands